Horizons in World Physics
Volume 252

Chemical Physics Research Trends

Horizons in World Physics

Volume 252 **Chemical Physics Research Trends**
Samuel V. Arnold

Volume 251 **Quantum Dots: Research Developments**
Peter A. Ling

Volume 250 **Quantum Gravity Research Trends**
Albert Reimer

Volume 249 **General Relativity Research Trends**
Albert Reimer

Volume 248 **Spacetime Physics Research Trends**
Albert Reimer

Volume 247 **New Developments in Quantum Cosmology Research**
Albert Reimer

Volume 246 **Quantum Cosmology Research Trends**
Albert Reimer

Volume 245 **Horizons in World Physics**
Tori V. Lynch

Volume 244 **Horizons in World Physics**
Albert Reimer

Volume 243 **Theoretical Physics 2002, Part 2**
Henk F. Arnoldus and Thomas F. George

Volume 242 **Models and Methods of High-T_c Superconductivity: Some Frontal Aspects, Volume 2**
Edited by J. K. Srivastava and S. M. Rao

Volume 241 **Models and Methods of High-T_c Superconductivity: Some Frontal Aspects, Volume 1**
Edited by J. K. Srivastava and S. M. Rao

Volume 240 **Horizons in World Physics**
Albert Reimer

Volume 239 **Theoretical Physics 2002, Part 1**
Henk F. Arnoldus and Thomas F. George

Volume 238 **Theoretical Physics 2001**
Henk F. Arnoldus and Thomas F. George

Volume 237 **Generalized functions in mathematical physics: Main ideas and concepts**
A.S. Demidov

Volume 236 **Edge Excitations of Low-Dimensional Charged Systems**
Oleg Kirichek

Volume 235 **On the Structure of the Physical Vacuum and a New Interaction in Nature (Theory, Experiment, Applications)**
Yu. A. Baurov

Volume 234 **Electron Scattering on Complex Atoms (Ions)**
V. Lengyel, O. Zatsarinny and E. Remeta

Volume 233 **An Introduction to Hot Laser Plasma Physics**
A.A. Andreev, A.A. Mak and N.A. Solovyev

Volume 232 **Electron-Atom Scattering: An Introduction**
Maurizio Dapor

Volume 231 **Theory of Structure Transformations in Non-equilibrium Condensed Matter**
Alexander L. Olemskoi

Volume 230 **Topics in Cosmic-Ray Astrophysics**
M. A. DuVernois (Ed.)

Volume 229 **Interactions of Ions in Condensed Matter**
A. Galdikas and L. Pranevièius

Volume 228 **Optical Vortices**
Edited by M. Vasnetsov and K. Staliunas

Volume 227 **Frontiers in Field Theory, Quantum Gravity and Strings**
R.K. Kaul, J. Maharana, S. Mukhi and S. K. Rama

Volume 226 **Quantum Groups, Noncommutative Geometry and Fundamental Physical Interactions**
D. Kastler, M. Russo and T. Schücker

Volume 225 **Aspects of Gravitational Interactions**
S.K. Srivastava and K.P. Sinha

Volume 224 **Dynamics of Transition Metals and Alloys**
S. Prakash

Volume 223 **Fields and Transients in Superhigh Pulse Current Devices**
G.A. Shneerson

Volume 222 **Gaussian Beams and Optical Resonators**
A. N. Oraevsky

Volume 221 **Laser Cathode-Ray Tubes**
Edited by Yu. M. Popov

Volume 220 **Quantum Electrodynamics with Unstable Vacuum**
Edited by V.L. Ginzburg

Volume 219 **The Muon Method in Science**
V.P. Smilga and Yu.M. Belousov

Volume 218 **Radiative and Collisional Characteristics of Ions in Hot Plasmas**
Edited by I.I. Sobel'man

Volume 217 **Nonlinear and Quantum Optical Phenomena in Nonequilibrium Media**
Edited by V.A. Shcheglov

Volume 216 **Semiconductors and Insulators: Optical and Spectroscopic ...**
Edited by Yu.I. Koptev

Volume 215 **Relativistic Theory of Gravity**
A.A. Logunov

Volume 214 **Research in Quantum Field Theory**
Edited by V.I. Man'ko

Volume 213 **Nonlinear Theory of Strong Electromagnetic Wave-plasma...**
Edited by O.N. Krokhin

Volume 212 **Dynamics of Elementary Atomic-Molecular...**
Edited by O. S. Scheglov

Volume 211 **Optics and Lasers**
Edited by G.G. Petrash

Volume 210 **Theory and Technology of High Temperature Superconductors**
Edited by Yu.I. Koptev

Volume 209 **Theory of Interaction of Multilevel Systems...**
Edited by V.I. Man'ko

Volume 208 **Physical Properties of High Temperature Superconductors**
Edited by Yu.I. Koptev

Volume 207 **Generalized Mayer Series in Classical Statistical Mechanics**
Igor J. Ivanchik

Volume 206 **Mathematical Physics, Applied Mathematics and Informatics**
Edited by Yu.I. Koptev

Volume 205 **Squeezed and Correlated States of Quantum Systems**
Edited by M.A. Markov

Volume 204 **Sakharov Memorial Lectures in Physics**
Edited by L.V. Keldysh and V.Ya. Fainberg

Volume 203 **Fiber Optics: Research and Development**
Edited by Ye.M. Dianov

Volume 202 **Gas Dynamics**
Edited by Yu.I. Koptev

Volume 201 **Generation of Nonlinear Waves and Quasistationary ...**
Edited by L.M. Kovrizhnykh

Volume 200 **Laser Thermonuclear Targets and Superdurable Microballoons**
Edited by A.I. Isakov

Volume 199 **Spatiotemporal Characteristics of Laser Emission**
M.V. Pyatakhin and A.F. Suchkov

Volume 198 **Theory of Nonstationary Quantum Oscillators**
Edited by M.A. Markov

Volume 197 **Reliability Problems of Semiconductor Lasers**
P.G. Eliseev

Volume 196 **Pulsars**
Edited by A.D. Kuzmin

Volume 195 **Physical Characteristics and Critical Temperature of High Temperature Superconductors**
Edited by M.M. Sushchinskiy

Volume 194 **Biophysical Approach to Complex Biological Phenomena**
Edited by E. Volkov

Volume 193 **Research on Chemical Lasers**
Edited by A.N. Orayevskiy

Volume 192 **Atomic and Ionic Spectra and Elementary Processes in Plasma**
Edited by I.I. Sobelman

Volume 191 **Polarization of the Vacuum and a Quantum Relativistic Gas ...**
A.Ye. Shabad

Volume 190 (Supp. Vol.) **Superconductors with A15 Lattice and Bridge Contacts Based ...**
Edited by M. M. Sushchinskiy

Volume 190 **Metal-Optics and Superconductivity**
Edited by A.I. Golovashkin

Volume 189 **Clusters of Galaxies and Extragalactic Radio Sources**
Edited by A.D. Kuz'min

Volume 188 **Integral Systems, Solid State Physics and Theory of Phase Transitions. Pt. 2. Symm. and Alg. Structures in Physics**
Edited by V.V. Dodonov and V.I. Man'ko

Volume 187 **Quantum Field Theory, Quantum Mechanics and Quantum Optics: Pt. 1. Symm. and Alg. Structures in Physics**
Edited by V.V. Dodonov and V.I. Man'ko

Volume 186 **Photoproduction of Pions on Nucleons and Nuclei**
Edited by A.A. Komar

Volume 185 **Injection Lasers in Optical Communication and Information Processing Systems**
Edited by Yu. M. Popov

Volume 184 **Electron Processes in MIS-Structure Memories**
Edited by A.F. Plotnikov

Volume 183 **Invariants and the Evolution of Nonstationary Quantum Systems**
Edited by M.A. Markov

Volume 182 **Luminescence of Wideband Semiconductors**
Edited by M.D. Galanin

Volume 181 **Pulsed Metal and Metal Halogenide Vapor Lasers**
Edited by G.G. Petrash

Volume 180 **Inelastic Light Scattering in Crystals**
Edited by M.M. Sushchinskiy

Volume 179 (Supp. Vol. 2) **Electron-Excited Molecules in Nonequilibrium Plasma**
Edited by N.N. Sobolev

Volume 179 (Supp. Vol. 1) **Interaction of Ultrashort Pulses with Matter**
Edited by M.D. Galanin

Volume 179 **X-Ray Plasma Spectroscopy and the Properties of Multiply-Charged Ions**
Edited by I.I. Sobel'man

Volume 178 **The Delfin Laser Thermonuclear Installation: Operational Complex ...**
Edited by G.V. Sklizkov

Volume 177 **Stoichiometry in Crystal Compounds and Its Influence on Their Physical Properties**
Edited by Yu. M. Popov

Volume 176 (Supp. Vol) **Quantization, Gravitation and Group Methods in Physics**
Edited by A.A. Komar

Volume 176 **Classical and Quantum Effects in Electrodynamics**
Edited by A.A. Komar

Volume 175 (Supp. Vol.) **Luminescence and Anisotropy of Zinc Sulfide Crystals**
Edited by M.D. Galanin

Volume 175 **Luminescence Centers of Rare Earth Ions in Crystal Phosphors**
Edited by M.D. Galanin

Volume 174 (Supp. Vol. 2) **Electron Liquid Theory of Normal Metals**
Edited by V.P. Silin

Volume 174 (Supp. Vol. 1) **Nonequilibrium Superconductivity**
Edited by V.L. Ginzburg

Volume 174 **Thermodynamics and Electrodynamics of Superconductors**
Edited by V.L. Ginzburg

Volume 173 **Quantum Mechanics and Statistical Methods**
Edited by M.M. Sushchinskiy

Volume 172 **Phase Conjugation of Laser Emission**
Edited by N.G. Basov

Volume 171 **Research on Laser Theory**
Edited by A.N. Orayevskiy

Volume 170 **The Theory of Target Compression by Longwave Laser Emission**
Edited by G.V. Sklizkov

Volume 169 **The Physical Effects in the Gravitational Field Black Holes**
Edited by M.A. Markov

Volume 168 **Issues in Intense-Field Quantum Electrodynamics**
Edited by V.L. Ginzburg

Volume 167 **Group Theory, Gravitation and Elementary Particle Physics**
Edited by A.A. Komar

Volume 166 **The Nonlinear Optics of Semiconductor Lasers**
Edited by N.G. Basov

Volume 165 **Solitons and Instantons, Operator Quantization**
Edited by V.L. Ginzburg

Horizons in World Physics
Volume 252

Chemical Physics Research Trends

Samuel V. Arnold
Editor

Nova Science Publishers, Inc.
New York

For permission to use material from this book please contact us:
Telephone 631-231-7269; Fax 631-231-8175
Web Site: http://www.novapublishers.com

Library of Congress Cataloging-in-Publication Data

Chemical physics research trends/Samuel V. Arnold (editor)
p. cm. – (Horizons in world physics series : V. 252)
Includes index.
ISBN-13: 978-1-60021-554-4 (hardcover)
ISBN-10: 1-60021-554-4 (hardcover)
1. Chemistry, Physical and theoretical—Research. I.Arnold, Samuel V.
CD455.5.C545 2006
541—dc22

2006103012

Published by Nova Science Publishers, Inc. ✢ New York

CONTENTS

PREFACE

Chemical physics and physical chemistry are closely related fields of study. Together they are distinguished from other disciplines by the incredible range of problems addressed by their practitioners. An effective physical chemist or chemical physicist is a "jack-of-all-trades," able to apply the principles and techniques of the field to everything from high-tech materials to biology. Just as the fields of chemistry and physics have expanded, so have chemical physics subject areas, which include polymers, materials, surfaces/interfaces, and biological macromolecules, along with the traditional small molecule and condensed phase systems. This book gathers new and important research from around the world.

Chapter 1 - In this review article, the electron–phonon interactions in the charged molecular systems such as polyacenes, polyfluoroacenes, B, N-substituted polyacenes, and polycyanodienes are discussed. We estimated the electron–phonon coupling constants and the frequencies of the vibronic active modes playing an essential role in the electron–phonon interactions in order to discuss how CH–CF, CC–BN, and CC–CN substitutions are closely related to the essential characteristics of the electron–phonon interactions in these molecules by comparing the calculated results for charged polyacenes with those for charged B, N-substituted polyacenes and polycyanodienes, respectively. The C–C stretching modes around 1500 cm^{-1}strongly couple to the highest occupied molecular orbitals (HOMO), and the lowest frequency modes and the C–C stretching modes around 1500 cm^{-1} strongly couple to the lowest unoccupied molecular orbitals (LUMO) in polyacenes. The C–C stretching modes around 1500 cm^{-1}strongly couple to the HOMO and LUMO in polyfluoroacenes. The B–N stretching modes around 1500 cm^{-1} strongly couple to the HOMO and LUMO in B, N-substituted polyacenes. The C–C and C–N stretching modes around 1500 cm^{-1} strongly couple to the HOMO and LUMO in polycyanodienes. The total electron–phonon coupling constants for the monocations (l_{HOMO}) and monoanions (l_{LUMO}) decrease with an increase in molecular size in polyacenes, polyfluoroacenes, B, N-substituted polyacenes, and polycyanodienes. In general, we can expect that monocations and monoanions, in which number of carriers per atom is larger, affords larger value. The CH–CF, CC–BN, and CC–CN atomic substitutions are effective way to seek for larger l_{HOMO} values, and the CH–CF and CC–CN atomic substitutions are effective way to seek for larger l_{LUMO} values in polyacenes. The logarithmically averaged phonon frequencies (ω_{ln}) which measure the frequencies of the vibronic active modes playing an essential role in the electron–phonon interactions for the monocations ($\omega_{ln,HOMO}$) and monoanions ($\omega_{ln,LUMO}$) in polyacenes, polyfluoroacenes, B, N-substituted polyacenes, and polycyanodienes are investigated. The $\omega_{\text{ln,HOMO}}$ values decrease with an increase in molecular size in polyacenes, polyfluoroacenes, and polycyanodienes, and the $\omega_{\text{ln,LUMO}}$ values decrease with an increase in molecular size in polyacenes, polyfluoroacenes, B, N-substituted polyacenes, and polycyanodienes. We can expect that in the hydrocarbon molecular systems, the ω_{ln} values would basically decrease by substituting hydrogen atoms by heavier atoms. This can be understood from the fact that the frequencies of all vibronic active modes in polyacenes downshift by H–F substitution. However, considering that the ω_{ln} value for the LUMO rather localized on carbon atoms in large sized polyfluoroacenes becomes larger by H–F substitution, we can expect that the ω_{ln} value for a molecular orbital localized on carbon atoms has a possibility to increase by substituting hydrogen atoms by heavier atoms if the phase patterns of the molecular orbital do not significantly change by such atomic substitution. Therefore, the detailed properties of the vibrational modes and the electronic structures as well as the molecular weights are closely related to the frequencies of the vibronic active modes playing an important role in the electron–phonon interactions in the monoanions of polyfluoroacenes.

Chapter 2 - Block Copolymers are the focus of intense research due to their ability to self-assemble into nanostructures with well-defined morphology and size. Comparing with bulk states, block copolymers can self-assemble into vast variety of copolymer micelles in selective solvents. The interest in block copolymer micelles stems from the numerous applications of such systems.

This chapter deals with recent progress including authors' contributions in the field of block copolymer self-assembly in solution and also tries to highlight some new directions in this field. Generalities about the critical micelle concentration, micellar structures, micelles preparation techniques, characterization techniques, dynamics of block copolymer micelles in solution and cross-linked copolymer micelles are firstly presented. Selected examples of micelles formation in aqueous and organic medium are given for block copolymers assemblies with cornucopian architectures. Strategies about how to tune the block copolymer micelles morphology in solution are addressed. New trends in the self-assembly of block copolymers in selective solvent, such as multicompartmennt micelles, co-micellization of block copolymers in selective solvent, and non-covalent complexes in micellar system are shown in the following part. Current and potential applications of block copolymer colloidal assemblies as templates for nanostructures fabrication, controlled drug delivery of bioactive agents, among others are finally discussed.

The aim of this chapter is to provide an overview of these fields of self-assembled block copolymers systems, illustrated by some selected examples.

Chapter 3 - This paper presents the results of the investigation of the luminescence of ladder-type methyl-, phenyl-poly(para-phenylene) (Me-, PhLPPP), and poly(9,9-di(ethylhexyl)fluorine) (PF2/6) activated by phosphorescent and fluorescent dyes obtained under their optical excitation. It is shown that the use of molecules of the type of diketones (benzil and its derivatives) makes it possible to realize conversion of singlet excitations of the polymer to its triplet state and thus upgrade the efficiency of its population. This process results from the singlet-singlet energy transfer from the polymer to the dopant and the effective intersystem crossing in it followed by the triplet-triplet energy transfer back to the polymer. The use of these converters has made it possible to investigate in detail the spectral-kinetic properties of the phosphorescence of MeLPPP. It is shown that the main channel of polymer fluorescence quenching is the dipole-dipole controlled single step transfer from an excited donor to an acceptor. In the case of a planar structure of the polymer and the activator, along with the energy transfer, the charge transfer makes a great contribution to the polymer fluorescence quenching. The latter process is accompanied by the formation of an intermolecular complex (exciplex). The introduction of volume substituents into the molecule of both the polymer and the activator decreases the efficiency of this process. The use, as a singlet-triplet converter, of organometallic complexes containing heavy atoms leads to an increase in the phosphorescence intensity and in its decay rate, which is due to the remote resonantly enhanced external heavy atom effect. The investigation of the system MeLPPP/Pt(II)octaethylporphyrin (PtOEP) has shown that the main channel of population of the activator triplet state is the singlet-singlet energy transfer, while both the singlet and triplet levels of the dopant are located below the respective levels of the polymer. The reason for this effect is the low population efficiency of the polymer triplet state. The addition of benzil leads to its appreciable increase and, as a consequence, to an increase in the contribution of the triplet channel to the excitation of the PtOEP phosphorescence. When the activator is excited to the absorption band proper, delayed fluorescence of the polymer is observed, in other words, up-conversion takes place. On the basis of the data on the quenching of delayed fluorescence by an electric field it has been shown that its source is the exciplex between PtOEP and MeLPPP generated by excitation of the S_1 state of PtOEP. On the basis of PF2/6 activated by fluorescent dyes two systems with white color luminescence have been proposed.

Chapter 4 - The development of the Quantum Theory of Atoms in Molecules (QTAIM) by Bader's research group has provided chemists with a rigorous physical basis for getting insight about many problems. This paper reviews results obtained (mostly by our research group) in the topological analysis of the electron density of a large set of organic molecules with QTAIM. The results are connected and discussed in relation to several basic concepts of Chemistry. We concentrate on those concepts where QTAIM results are not keeping in line with widespread traditional explanations that are still common in textbooks and research papers. In particular, we deal with some recent findings about approximate transferability (one of the first subjects attacked by the author of this theory), strain energy, anomeric effect, hydrogen bonding, and resonance structures. Our results on approximate transferability indicate the need for defining at least four specific methylene groups in the vicinity of a functional group when the chain is in antiperiplanar arrangement. Nevertheless, we can only distinguish two specific methylene groups when the chain is in gauche arrangement. Also atomic energies cannot be used as a transferable property unless they are computed from an electron density that verifies the virial theorem very strictly. QTAIM study of heterocycles reveals that homodesmotic strain energies comprise energy terms that are not related to ring opening. These terms are not included in a QTAIM based definition of strain energy. The partitioning of molecular energy into one- and two-center atomic terms, with atoms defined in terms of the QTAIM, gives rise to a new interpretation of the anomeric effect, with results that are not in line with either the electrostatic or the stereoelectronic models developed for it. Also, classical ideas about hydrogen bonding have to be re-examined according to QTAIM results. Finally, the study of several protonation and hydride addition process has revealed several trends that are not in line with the resonance model.

Chapter 5 - Dendrimers became of rising interest in the past decade in different fields of biophysical research due to their highly branched and well-defined architecture, which consists of a high number of functional end groups. On the other hand the unique photophysical and photochemical properties of fullerenes have led to the synthesis of a large number of compounds in which fullerenes are covalently linked to different kinds of photoactive groups. Some of these compounds could serve as artificial light harvesting or photosynthesis model systems due to the high capability of the fullerene moiety to act as an electron acceptor or even an electron accumulator. Additionally, C_{60} can be used as a versatile building block for the construction of globular dendrimers, which opens up ways for biomedical applications of fullerene based high branched molecular systems. One possibility is the use of the fullerene as functional core of an efficient multiplier in modular drug delivery systems for the photodynamic therapy (PDT) of tumors.

In this work we report on the photophysical investigations of dendritic fullerene-dye complexes with up to 24 pyropheophorbide *a* units attached to a fullerene core. Their electronic properties are discussed in detail. Depending on the addition pattern on the fullerene, after light absorption in these molecular systems either internal electron or/and energy transfer processes occur. Due to the length and high flexibility of the carbon chains different energy transfer processes are possible. Beside Förster resonance transfer excitonic interactions occur due to intramolecular stacking effects of the dye molecules. The formation of different types of energy traps was postulated and proved experimentally. Their strong influence on the photophysical properties of fullerene-dye compounds in dependence of the number of appended dye molecules is discussed. A phenomenological model describing the different energy transfer processes in dendritic dye systems is given. Furthermore the potential of the fullerene-dye-complexes as a combined artificial light-harvesting and charge separation unit is discussed. Finally, their possible use for PDT applications is shown.

Chapter 6 - 'Search for potential materials' for third-order nonlinear optics has been of continuing interest in recent years. In this context, organic molecules are increasingly being recognized as the materials of the future because their molecular nature combined with the versatility of synthetic chemistry can be used to alter and optimize molecular structure to maximize third order nonlinear response. Chalcones have received considerable interest as materials for second-order nonlinear optical applications due to their ability to crystallize in noncentrosymmetric structure and their blue light transmittance. Being charge transfer compounds, chalcones can also possess large third-order nonlinearities due to their π-conjugated structure. In this article, we discuss the structure-nonlinear response relationship among a few chalcones and the possibility of using them for third-order applications. A meager or no work, to our knowledge, has been done so far on these molecules in this regard. Z-scan and degenerate four-wave mixing techniques were employed to investigate third-order optical nonlinearities of chalcone derivatives. Some of these molecules possess large $\chi^{(3)}$ of magnitude as high as 10^{-12} esu and exhibit strong optical limiting properties. Possible mechanisms that are responsible for optical limiting property of these molecules have been discussed.

Chapter 7 - Modeling of molecular behavior can be generally categorized into two broad groups – the quantum mechanics approach and the molecular mechanics approach. Although the latter does not consider the electronic distribution, it can provide solutions that are reasonably accurate in a fraction of computation time as compared to the former. Indeed, "simple potential functions such as Morse, Rydberg, Born-Mayer, etc., give a quantitatively correct description of the potential and only modest extensions are needed to obtain functions which stand up to the most stringent experimental tests" (Murrell et al, 1984). Arising from the ease to which the molecular energy can be calculated, numerous potential energy functions have been proposed. It follows that comparative advantage exists among these potential functions. As such it is important to understand the characteristics and extent of discrepancies among normalized potential functions. An understanding of such discrepancies aid the selection of appropriate potential energy functions for computation under specific environmental conditions and for conservative computation in molecular design modeling. Although the discrepancies observation is straightforward, the methods for normalizing different potential energy functions must firstly be addressed. To normalize these potential energy functions, it is imperative that they be equated under a certain logical criteria. This is unlike bridging parameters that allow a generic potential function to be reduced into specific ones by mere substitution of numerical values into the bridging parameters [5-7]. Rather, the process of normalizing refers to the mathematical formalism that enables parameters from different potential functions to be related. In this way, a set of parameters from one potential function can be converted into parameters of another potential function. Theoretical plots of these normalized potential functions reveal their characteristics discrepancies and shed light into their relative performance in modeling under specified conditions and applications. This chapter covers the mathematical procedure for normalizing molecular potential among 2-body, 3-body, 4-body and non-bonded potential energy functions.

Chapter 8 - Electric polarizability is a fundamental characteristic of atomic and molecular systems. The theory of electric polarizability is a key element of the rational interpretation of a wide range of phenomena, from nonlinear optics and electron scattering to phenomena induced by intermolecular interactions. Recently, segmented contracted basis sets of double, triple, and quadruple zeta valence quality plus polarization functions for the atoms from H to Ar were presented by Jorge *et al.* These sets were augmented with diffuse functions to more accurately describe the long range behavior of the wave functions. We report a systematic study of the basis sets required to obtain accurate values for the electric dipole moment and static dipole polarizability and first dipole hyperpolarizability for the five molecules H_2, CO, CH_4, NH_3, and HCP at their equilibrium geometries. Several methods were examined, including Hartree-Fock, Mϕller-Plesset second-order perturbation theory, and density functional theory. Comparison with theoretical and experimental data reported in the literature was done. One verifies that the B3LYP/AQZP mean dipole polarizabilities, anisotropies, and first dipole hyperpolarizabilities are in good agreement with the experimental values, and in very reasonable agreement with results obtained through higher theory levels.

Chapter 9 - The Clapeyron equation cannot describe a number of phase transitions in ice: it gives huge mistakes. Even the sign of dT/dP is often wrong. It is shown that the modified Clapeyron equation earlier obtained by the author gives much better results. The following transitions are considered: Low-Density Amorph (LDA) → I_c, LDA → High-Density Amorph (HDA) at 0.1 and 600 MPa, ice III → ice V, ice V→ ice VI, ice VI → ice VII, ice VI → ice VIII.

In: Chemical Physics Research Trends, Vol. 252
Editor: Samuel V. Arnold, pp. 1-3

ISBN: 978-1-60021-554-4

Expert Commentary

A NEW FORMULA FOR AVERAGING PHYSICAL QUANTITIES

*I. A. Stepanov**
School of Chemistry, University of Nottingham, University Park, Nottingham, NG7 2RD, UK

ABSTRACT

A new method for averaging physical quantities is discovered. It is shown that the traditional method of finding the average value of a physical quantity often gives the wrong results.

Consider the flow of liquid in a tube which has the form of a truncated cone. Its left broad side has radius R_1 and its right narrow side has radius R_2 and its length is L. It is necessary to find the average cross section of the tube. By definition, the mean value of the function f(x) at the interval $[x_1, x_2]$ is given by (1):

$$\overline{f(x)} = 1 / (x_2 - x_1) \int_{x_1}^{x_2} f(x)dx \tag{1}$$

[1]. It is the only method of averaging functions. By this equation, the average cross section $\pi \overline{R^2}$ is equal to $\pi(R_1^2 + R_1R_2 + R_2^2) / 3$. If $R_2 = 0$, then the flow through the tube is impossible, but (1) gives $\pi \overline{R^2} = \pi R_1^2 / 3$. The flow through the tube is proportional to the average cross section, therefore, according to (1), flow in the tube must exist. It is a contradiction, and one sees that (1) is not applicable in this case.

In [2], the equation of the new type of averaging was found:

* E-mail: pcxis@nottingham.ac.uk, igstepanov@yahoo.com

$$\langle f(x)\rangle = (x_2 - x_1) / \int_{x_1}^{x_2} dx / f(x). \qquad (2)$$

Let us use (2) for calculation of the average cross section of a tapering tube. The dependence of the radius on the distance along the axis of the tube is

$R(x) = (R_2 - R_1) / L \cdot x + R_1$, $x_2 - x_1 = L$ hence

$$\pi\langle R^2\rangle = \pi R_1 R_2. \qquad (3)$$

Consider the energy of interaction between two charges q at distance r_1 at time t_1: $E_1 = q^2 / r_1$, at distance r_2 at time t_2: $E_2 = q^2 / r_2$, ..., at distance r_n at time t_n: $E_n = q^2 / r_n$, $n \to \infty$, $r_i \in [R_1, R_2]$. It is necessary to find the average radius $\langle r\rangle$ such that

$$\langle E\rangle = \lim_{n\to\infty} \sum_{i=1}^{n} E_i / n = 1 / (R_2 - R_1) \int_{R_1}^{R_2} q^2 / r dr = q^2 \ln (R_2 / R_1) / (R_2 - R_1)$$
$$= q^2 / \langle r\rangle. \qquad (4)$$

According to the traditional method,

$$\langle r\rangle = \lim_{n\to\infty} \sum_{i=1}^{n} r_i / n = (R_2 + R_1) / 2. \qquad (5)$$

The result is wrong because if $R_2 \to 0$ then $\sum_{i=1}^{n} E_i / n \to \infty$ but from (4) and (5)

$$\lim_{n\to\infty} \sum_{i=1}^{n} E_i / n = q^2 / (R_2 + R_1) / 2 \neq \infty.$$

According to (2), $\langle r\rangle = (R_2 - R_1) / \ln (R_2 / R_1)$ and introducing it to (4) we find that $\langle E\rangle = q^2 / \langle r\rangle$.

In discrete case, equation (1) is

$$\bar{f} = \sum_{i=1}^{n} f_i / n \qquad (6)$$

and equation (2) is

$$\langle f\rangle = n / \sum_{i=1}^{n} 1/f_i \qquad (7)$$

REFERENCES

[1] G. A. Korn and T. M. Korn, *Mathematical Handbook for Scientists and Engineers*. 2nd ed. McGraw Hill Book Company. NY. 1968.

[2] I. A. Stepanov, http://arxiv.org/abs/physics/0603187. Accepted in Transactions of the Canad. Soc. Mech. Engineers.

In: Chemical Physics Research Trends, Vol. 252
Editor: Samuel V. Arnold, pp. 5-64
ISBN: 978-1-60021-554-4

Chapter 1

ELECTRON–PHONON INTERACTIONS AND THEIR APPLICATIONS TO NOVEL NANOELECTRONICS I: THE EFFECT OF ATOMIC SUBSTITUTIONS

Takashi Kato*
Institute for Innovative Science and Technology,
Graduate School of Engineering, Nagasaki Institute of Applied Science,
3-1, Shuku-machi, Nagasaki 851-0121, Japan

ABSTRACT

In this review article, the electron–phonon interactions in the charged molecular systems such as polyacenes, polyfluoroacenes, B, N-substituted polyacenes, and polycyanodienes are discussed. We estimated the electron–phonon coupling constants and the frequencies of the vibronic active modes playing an essential role in the electron–phonon interactions in order to discuss how CH–CF, CC–BN, and CC–CN substitutions are closely related to the essential characteristics of the electron–phonon interactions in these molecules by comparing the calculated results for charged polyacenes with those for charged B, N-substituted polyacenes and polycyanodienes, respectively. The C–C stretching modes around 1500 cm^{-1}strongly couple to the highest occupied molecular orbitals (HOMO), and the lowest frequency modes and the C–C stretching modes around 1500 cm^{-1} strongly couple to the lowest unoccupied molecular orbitals (LUMO) in polyacenes. The C–C stretching modes around 1500 cm^{-1}strongly couple to the HOMO and LUMO in polyfluoroacenes. The B–N stretching modes around 1500 cm^{-1} strongly couple to the HOMO and LUMO in B, N-substituted polyacenes. The C–C and C–N stretching modes around 1500 cm^{-1} strongly couple to the HOMO and LUMO in polycyanodienes. The total electron–phonon coupling constants for the monocations (l_{HOMO}) and monoanions (l_{LUMO}) decrease with an increase in molecular size in polyacenes, polyfluoroacenes, B, N-substituted polyacenes, and polycyanodienes. In general, we can expect that monocations and monoanions, in which number of carriers per atom is larger, affords larger value. The CH–CF, CC–BN, and CC–CN atomic substitutions are effective way to seek for larger l_{HOMO} values, and the CH–CF and

* E-mail: kato@cc.nias.ac.jp

CC–CN atomic substitutions are effective way to seek for larger l_{LUMO} values in polyacenes. The logarithmically averaged phonon frequencies (ω_{ln}) which measure the frequencies of the vibronic active modes playing an essential role in the electron–phonon interactions for the monocations ($\omega_{ln,HOMO}$) and monoanions ($\omega_{ln,LUMO}$) in polyacenes, polyfluoroacenes, B, N-substituted polyacenes, and polycyanodienes are investigated. The $\omega_{ln,HOMO}$ values decrease with an increase in molecular size in polyacenes, polyfluoroacenes, and polycyanodienes, and the $\omega_{ln,LUMO}$ values decrease with an increase in molecular size in polyacenes, polyfluoroacenes, B, N-substituted polyacenes, and polycyanodienes. We can expect that in the hydrocarbon molecular systems, the ω_{ln} values would basically decrease by substituting hydrogen atoms by heavier atoms. This can be understood from the fact that the frequencies of all vibronic active modes in polyacenes downshift by H–F substitution. However, considering that the ω_{ln} value for the LUMO rather localized on carbon atoms in large sized polyfluoroacenes becomes larger by H–F substitution, we can expect that the ω_{ln} value for a molecular orbital localized on carbon atoms has a possibility to increase by substituting hydrogen atoms by heavier atoms if the phase patterns of the molecular orbital do not significantly change by such atomic substitution. Therefore, the detailed properties of the vibrational modes and the electronic structures as well as the molecular weights are closely related to the frequencies of the vibronic active modes playing an important role in the electron–phonon interactions in the monoanions of polyfluoroacenes.

I. Introduction

In modern physics and chemistry, the effects of vibronic interactions [1] and electron–phonon interactions [1–3] in molecules and crystals have been an important topic. Analysis of vibronic interaction [1–3] is important for the prediction of electronic control of nuclear motions in degenerate electronic systems. Application of vibronic interaction theory covers a large variety of research fields such as spectroscopy,[4] instability of molecular structure, electrical conductivity,[5] and superconductivity.[5, 6] Vibronic interactions in discrete molecules can be viewed as the coupling between frontier orbitals and molecular vibrations, while those in solids are the coupling between free electrons near the Fermi level and acoustic phonons. There is a close analogy between them.

Electron–phonon coupling [1–3] is the consensus mechanism for attractive electron–electron interactions in the Bardeen–Cooper–Schrieffer (BCS) theory of superconductivity.[5,6] Since Little's proposal for a possible molecular superconductor based on exciton mechanism,[7] the superconductivity of molecular systems has been extensively investigated. Although such a unique mechanism has not yet been established, advances in design and synthesis of molecular systems have yielded a lot of BEDT-TTF-type organic superconductors, [8, 9] where BEDT-TTF is *bis*(ethylenedithio)tetrathiafulvalene. An inverse isotope effect due to substituting hydrogen by deuterium in organic superconductivity was observed by Saito et al.[10] Goddard et al. proposed that the mechanism for superconductivity of BEDT–TTF type organic molecules involves the coupling of charge transfer to the boat deformation mode.[11] It was found that the alkali-doped A_3C_{60} complexes [12] exhibit superconducting transition temperatures (T_cs) of more than 30 K (Ref.

[13]) and 40 K under pressure.[14] In superconductivity in alkali-doped fullerenes,[15] pure intramolecular Raman-active modes have been suggested to be important in a BCS-type [6] strong coupling scenario.

It was proposed that the electron–phonon interactions dominate the charge transport in the crystals of naphthalene ($C_{10}H_8$) (2a), anthracene (3a), tetracene (4a), and pentacene (5a).[16–18] Interestingly, from a theoretical viewpoint, possible superconductivity of polyacene has been proposed.[19,20] Saito *et al.*[10] observed an inverse isotope effect due to substituting hydrogen by deuterium in organic superconductivity. It is important to consider how intramolecular or intermolecular vibrations play a role in the occurrence of superconductivity. If the intermolecular vibrations are important, the phonon-frequency dependence of the transition temperature appears in the prefactor through Debye frequency in the formula for the superconducting transition temperatures (T_cs) in the Bardeen–Cooper–Schrieffer (BCS) theory so that the normal isotope effects are expected. If the intramolecular vibrations are important, the phonon-frequency dependence appears in the denominator of the expression for the electron–phonon coupling constant so that the inverse isotope effect can be expected in some cases. From the inverse isotope effect on deuterium substitution observed by Saito et al.,[10] we expect that such inverse isotope effects can be widely observed in molecular organic superconductors. The origin of such inverse isotope effects in organic superconductors has not yet been fully elucidated. Shortly after the discovery of superconductivity in palladium hydrides,[21] an isotope effect in T_c was found by Stritzker and Buckel.[22] The T_c of Pd-D was higher than that of Pd-H,[23] contrary to the expectations from a simple BCS theory After that, an even larger inverse isotope effect for Pd-T was measured by Schirber et al.[24]

In previous work, we have analyzed the vibronic interactions and estimated possible T_cs in the monocations of polyacenes based on the hypothesis that the vibronic interactions between the intramolecular vibrations and the highest occupied molecular orbitals (HOMO) play an essential role in the occurrence of superconductivity in positively charged nanosized molecular systems.[25] On the basis of an experimental study of ionization spectra using the high-resolution gas-phase photoelectron spectroscopy, the electron–phonon interactions in the positively charged polyacenes were well studied recently.[26] Our predicted frequencies for the vibrational modes which play an essential role in the electron–phonon interactions [25] as well as the predicted total electron–phonon coupling constants [25] are in excellent agreement with those obtained from the experimental research.[26]

In this review article, we discuss the electron–phonon interactions in the charged molecular systems such as polyacenes such as C_6H_6 (1a), $C_{10}H_8$ (2a), $C_{14}H_{10}$ (3a), $C_{18}H_{12}$ (4a), and $C_{22}H_{14}$ (5a),[25] polyfluoroacenes such as C_6F_6 (1fa), $C_{10}F_8$ (2fa), $C_{14}F_{10}$ (3fa), $C_{18}F_{12}$ (4fa), and $C_{22}F_{14}$ (5fa),[27] B, N-substituted polyacenes such as $B_3N_3H_6$ (1bn), $B_5N_5H_8$ (2bn), $B_7N_7H_{10}$ (3bn),[28] and polycyanodienes such as $C_4N_2H_6$ (1cn), $C_6N_4H_8$ (2cn), $C_8N_6H_{10}$ (3cn), and $C_{10}N_8H_{12}$ (4cn)[29] (Scheme 1). We will estimate the electron–phonon coupling constants and the frequencies of the vibronic active modes playing an essential role in the electron–phonon interactions. These physical values are essential to discuss the several physical phenomena such as intramolecular electrical conductivity, intermolecular charge transfer, attractive electron–electron interactions and Bose–Einstein condensation, and superconductivity, which will be discussed in detail in the next review article.

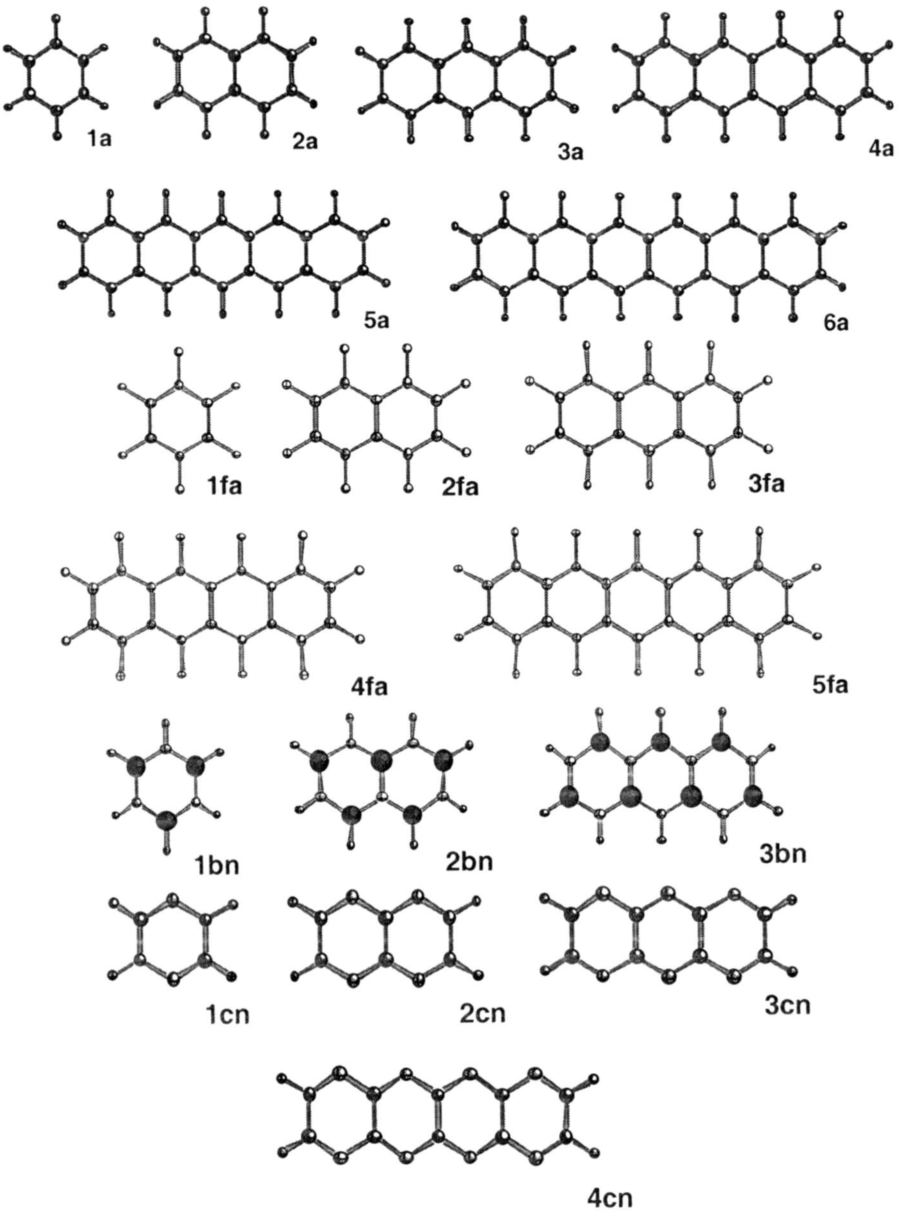

Scheme 1.

Motivated by the possible inverse isotope effects in Pd-H, Pd-D, and Pd-T superconductivity [21–24] and organic superconductivity observed by Saito et al.,[10] we discuss how the H–F substitution are closely related to the essential characteristics of the electron–phonon interactions in these molecules by comparing the calculated results for charged polyacenes with those for charged polyfluoroacenes, since fluorine atoms are much heavier than D and T atoms, and the phase patterns of the frontier orbitals such as the HOMO and LUMO are not expected to be significantly changed. Furthermore, we discuss how C–BN and C–N substitutions are closely related to the essential characteristics of the electron–phonon interactions in these molecules by comparing the calculated results for charged polyacenes with those for charged B, N-substituted polyacenes and polycyanodienes, respectively. We can expect that the characteristics of electron–phonon interactions are significantly changed by such atomic substitutions because of electronegativity perturbation[30] in polyacenes. These physical values are essential to discuss the several physical phenomena such as intramolecular electrical conductivity, intermolecular charge transfer, attractive electron–electron interactions and Bose–Einstein condensation, and superconductivity, which will be discussed in detail in the next review article.

II. THEORETICAL BACKGROUND

We describe the theoretical background for the vibronic coupling in polyacenes, polyfluoroacenes, B, N-substituted polyacenes, and polycyanodienes. We will use small letters for "one-electron orbital symmetries" and capital letters for symmetries of both "electronic" and "vibrational" states, as usual. The vibronic matrix element, $E_{xy}(r,Q)$,[1–3] is given by

$$E_{xy}(r,Q) = \varepsilon_{xy}(r,Q) - \varepsilon_{xy}(r,0)$$
$$= \sum_{\alpha}\left(\frac{\partial \varepsilon_{xy}}{\partial Q_\alpha}\right)_0 Q_\alpha + \frac{1}{2}\sum_{\alpha,\beta}\left(\frac{\partial^2 \varepsilon_{xy}}{\partial Q_\alpha \partial Q_\beta}\right)_0 Q_\alpha Q_\beta \tag{1}$$

where $\varepsilon_{xy}(r,Q)$ is defined as

$$\varepsilon_{xy}(r,Q) = \langle \phi_x | h(r,Q) | \phi_y \rangle \tag{2}$$

Here, $h(r,Q)$ is the Hamiltonian of one-electron orbital energy, and ϕ_x and ϕ_y are one-electron wave functions. r and Q signify the whole set of coordinates of the electrons and nuclei, respectively. What we see in the first term on the right-hand side of Eq. (1) is the linear orbital vibronic coupling constant.

We discuss a theoretical background for the orbital vibronic interactions in polyacenes, polyfluoroacenes, B, N-substituted polyacenes, and polycyanodienes. The potential energy for the neutral ground state, negatively and positively charged electronic states in polyacenes, polyfluoroacenes, B, N-substituted polyacenes, and polycyanodienes are shown in Figure 1. Here, we take a one-electron approximation into account; the vibronic coupling constants of the vibrational modes to the electronic states in the monoanions and cations of polyacenes, polyfluoroacenes, B, N-substituted polyacenes, and polycyanodienes are defined as a sum of orbital vibronic coupling constants from all the occupied orbitals, [1(a)]

$$g_{\text{electronic state}} = \sum_{i}^{\text{occupied}} g_i \, . \tag{3}$$

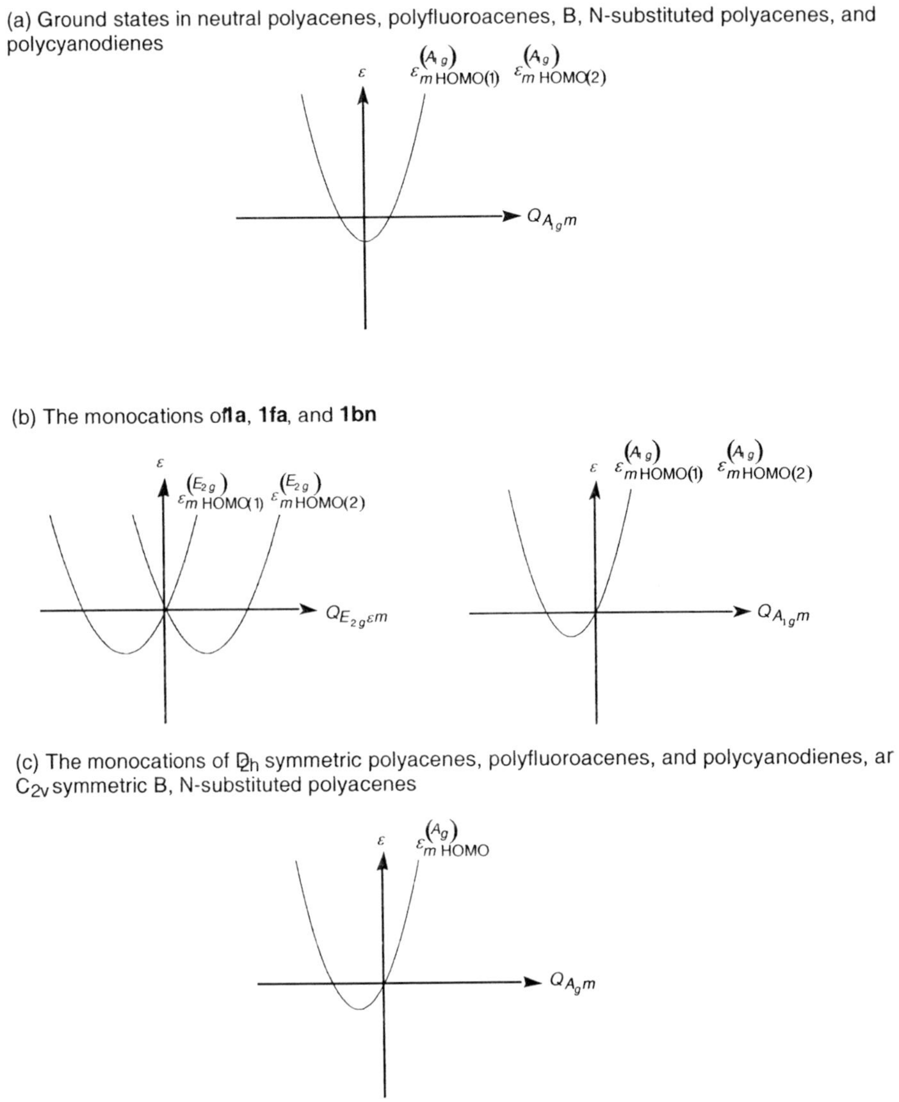

Figure 1. Potential energy for the neutral ground state and positively charged electronic states in polyacenes, polyfluoroacenes, B, N-substituted polyacenes, and polycyanodienes.

Considering the one-electron approximation and that the first derivatives of the total energy vanish in the ground state at the equilibrium structure in neutral polyacenes, polyfluoroacenes, B, N-substituted polyacenes, and polycyanodienes (i.e., $g_{\text{neutral}} = \sum_{i}^{\text{HOMO}} g_i = 0$) (Figure 1 (a)), and one electron must be injected into (removed from) the LUMO (HOMO) to generate the monoanions (monocations), the vibronic coupling constants of the totally symmetric vibrational modes to the electronic states of the monoanions and cations of polyacenes, polyfluoroacenes, B, N-substituted polyacenes, and polycyanodienes can be defined by Eqs. (4) and (5), respectively,

$$g_{\text{monoanion}}(\omega_m) = g_{\text{LUMO}}(\omega_m), \quad (4)$$

$$g_{\text{monocation}}(\omega_m) = g_{\text{HOMO}}(\omega_m). \quad (5)$$

A. Vibronic Interactions between the Twofold Degenerate Frontier Orbitals and the E_{2g} Vibrational Modes In Benzene

Benzene (**1a**) has the twofold degenerate HOMO and LUMO due to its high D_{6h} symmetry. The symmetry labels of the HOMO and LUMO are e_{1g} and e_{2u}, respectively. Now we take two approximations into account in Eq. (1). First, we ignore the nondiagonal matrix elements containing the quadratic vibronic constants; only the terms of type Q_α^2 are taken into account. Second, we focus upon the diagonal processes; we consider the direct product of the orbital symmetries, which can be reduced as

$$e_{1g} \times e_{1g} = e_{2u} \times e_{2u} = A_{1g} + A_{2g} + E_{2g}. \quad (6)$$

Thus, the A_{1g} and E_{2g} modes can linearly couple to the e_{1g} HOMO and e_{2u} LUMO. The numbers of the A_{1g} and E_{2g} modes are 2 and 4, respectively. We must consider multimode problems, but in the limit of linear vibronic coupling we can treat each twofold degenerate set of modes (mode index m as shown below) independently.[1]

Let us first consider the orbital vibronic coupling of the twofold degenerate e_{1g} HOMO to the E_{2g} vibrational mode. Taking two approximations mentioned above and Eq. (1) into account, the two energy sheets of the e_{1g} HOMO that can couple to the mth E_{2g} vibrational mode are given in the form of Eq. (7), where $Q_{E_{2g}\gamma m}$ is the vibrational normal coordinate belonging to row γ of the vibrational mode of the irreducible representation E_{2g}:

$$\varepsilon_{mxy}^{(E_{2g})}\left(Q_{E_{2g}\theta m}, Q_{E_{2g}\varepsilon m}\right)$$

$$= \frac{1}{2}\sum_{\gamma} K_{E_{2g}m} Q_{E_{2g}\gamma m}^2 \delta_{xy} + \sum_{\gamma}\left\langle e_{1g\,x}\ \text{HOMO}\left|\left(\frac{\partial h}{\partial Q_{E_{2g}\gamma m}}\right)_0\right| e_{1g\,y}\ \text{HOMO}\right\rangle Q_{E_{2g}\gamma m} \quad (7)$$

$$(m = 1, 2, \ldots, 4)$$

Here, γ takes two vibrational states θ and ε. The first term on the right-hand side of Eq. (7) is the elastic term, and $\left\langle e_{1g\,x}\ \text{HOMO}\left|\left(\frac{\partial h}{\partial Q_{E_{2g}\gamma m}}\right)_0\right| e_{1g\,y}\ \text{HOMO}\right\rangle$ in the second term corresponds to the linear orbital vibronic coupling constant, which is a good measure of the interaction between the vibration labeled by γ in the mth mode of the irreducible representation E_{2g} and the molecular orbitals labeled by x and y in the irreducible representation e_{1g}. According to the Wigner–Eckart theorem, [1] we can rewrite the second term on the right-hand side of Eq. (7) using the Clebsch–Gordan coefficients and we obtain the vibronic coupling matrix[1] as

$$\varepsilon_{E_{2g}m} = A_m \begin{pmatrix} -Q_{E_{2g}\varepsilon m} & Q_{E_{2g}\theta m} \\ Q_{E_{2g}\theta m} & Q_{E_{2g}\varepsilon m} \end{pmatrix}, \tag{8}$$

where A_m corresponds to the reduced matrix element for the mth mode of vibration:

$$A_m = \left\langle e_{1g}\ \mathrm{HOMO} \left\| \left(\frac{\partial h}{\partial Q_{E_{2g}m}} \right)_0 \right\| e_{1g}\ \mathrm{HOMO} \right\rangle \tag{9}$$

A_m depends on the irreducible representation and not on rows. Using Eq. (8), Eq. (7) can be transformed into Eq. (10):

$$\varepsilon_m^{(E_{2g})}\left(Q_{E_{2g}\theta m}, Q_{E_{2g}\varepsilon m}\right) = \frac{1}{2} K_{E_{2g}m}\left(Q_{E_{2g}\theta m}^2 + Q_{E_{2g}\varepsilon m}^2\right)\begin{pmatrix} 1 & 0 \\ 0 & 1 \end{pmatrix} + A_m \begin{pmatrix} -Q_{E_{2g}\varepsilon m} & Q_{E_{2g}\theta m} \\ Q_{E_{2g}\theta m} & Q_{E_{2g}\varepsilon m} \end{pmatrix} \tag{10}$$

Now we consider one member of the twofold degenerate vibrational mode. For example, when we consider only the $Q_{E_{2g}\varepsilon m}$ mode ($Q_{E_{2g}\theta m} = 0$), the energy sheets of the molecular orbital in row ε of the mth mode become

$$\varepsilon_m^{(E_{2g})}\left(Q_{E_{2g}\varepsilon m}\right) = \frac{1}{2} K_{E_{2g}m} Q_{E_{2g}\varepsilon m}^2 \begin{pmatrix} 1 & 0 \\ 0 & 1 \end{pmatrix} + A_m \begin{pmatrix} -Q_{E_{2g}\varepsilon m} & 0 \\ 0 & Q_{E_{2g}\varepsilon m} \end{pmatrix} \tag{11}$$

This has already been diagonalized, and the energy sheets of the molecular orbital (HOMO (1) and HOMO (2)) in row ε cut by the plane $Q_{E_{2g}\theta m} = 0$ become

$$\varepsilon_{m\ \mathrm{HOMO\ (1)}}^{(E_{2g})}\left(Q_{E_{2g}\varepsilon m}\right) = \frac{1}{2} K_{E_{2g}m} Q_{E_{2g}\varepsilon m}^2 + A_m Q_{E_{2g}\varepsilon m} \tag{12}$$

and

$$\varepsilon_{m\ \mathrm{HOMO\ (2)}}^{(E_{2g})}\left(Q_{E_{2g}\varepsilon m}\right) = \frac{1}{2} K_{E_{2g}m} Q_{E_{2g}\varepsilon m}^2 - A_m Q_{E_{2g}\varepsilon m} \tag{13}$$

$$Q_{E_{2g}\theta m} = 0. \tag{14}$$

These are illustrated in Figure 1. Only the cases where the quadratic vibronic constants are positive are shown as examples here. We see from this illustration that the E_{2g} modes lift the degeneracy of the twofold degenerate orbitals. The dimensionless diagonal linear orbital vibronic coupling constants of the e_{1g} HOMO for its mth mode is defined by Eq. (15).

$$g_{e_{1g}\ \mathrm{HOMO}}(\omega_m) = \frac{1}{\hbar\omega_m}\left\langle e_{1g}\ \mathrm{HOMO}\left\|\left(\frac{\partial h}{\partial q_{E_{2g}m}}\right)_0\right\| e_{1g}\ \mathrm{HOMO}\right\rangle \tag{15}$$

In these equations, $q_{E_{2g}m}$ is the dimensionless normal coordinate [31] defined by

$$q_{E_{2g}m} = \sqrt{\omega_m / \hbar} Q_{E_{2g}m} \tag{16}$$

Similar discussions can be made in the vibronic interactions between the E_{2g} modes and the e_{2u} LUMO, and the dimensionless diagonal linear orbital vibronic coupling constants of the e_{2u} LUMO for the mth mode is defined by

$$g_{e_{2u}\ \mathrm{LUMO}}(\omega_m) = \frac{1}{\hbar\omega_m}\left\langle e_{2u}\ \mathrm{LUMO}\left\|\left(\frac{\partial h}{\partial q_{E_{2g}m}}\right)_0\right\| e_{2u}\ \mathrm{LUMO}\right\rangle \tag{17}$$

B. Vibronic Interactions between the Twofold Degenerate Frontier Orbitals and the A_{1g} Vibrational Modes in Benzene

In this case, the two energy sheets of the molecular orbitals are given in the form:

$$\varepsilon_m^{(A_{1g})}\left(Q_{A_{1g}m}\right) = \frac{1}{2} K_{A_{1g}m} Q_{A_{1g}m}^2 \begin{pmatrix} 1 & 0 \\ 0 & 1 \end{pmatrix} + B_m \begin{pmatrix} Q_{A_{1g}m} & 0 \\ 0 & Q_{A_{1g}m} \end{pmatrix} \tag{18}$$

where B_m corresponds to the reduced matrix element written as

$$B_m = \left\langle e_{1g}\ \mathrm{HOMO}\left\|\left(\frac{\partial h}{\partial Q_{A_{1g}m}}\right)_0\right\| e_{1g}\ \mathrm{HOMO}\right\rangle \tag{19}$$

This has already been diagonalized, and the energy sheets of the molecular orbital (HOMO (1) and HOMO (2)) become

$$\varepsilon_{m\ \mathrm{HOMO}\ (1)}^{(A_{1g})}\left(Q_{A_{1g}m}\right)=\varepsilon_{m\ \mathrm{HOMO}\ (2)}^{(A_{1g})}\left(Q_{A_{1g}m}\right)=\frac{1}{2}K_{A_{1g}m}Q_{A_{1g}m}^{2}+B_{m}Q_{A_{1g}m} \tag{20}$$

These are illustrated in Figure 1. We can see from this illustration that the A_{1g} modes do not lift the degeneracy of the twofold degenerate orbitals. The dimensionless diagonal linear orbital vibronic coupling constants of the e_{1g} HOMO for its mth mode is defined by Eq. (21),

$$g_{e_{1g}\ \mathrm{HOMO}}(\omega_m)=\frac{1}{\hbar\omega_m}\left\langle e_{1g}\ \mathrm{HOMO}\left\|\left(\frac{\partial h}{\partial q_{A_{1g}m}}\right)_0\right\|e_{1g}\ \mathrm{HOMO}\right\rangle \tag{21}$$

In this equation, $q_{A_{1g}m}$ is the dimensionless normal coordinate [31] of the mth mode defined by using the normal coordinate $Q_{A_{1g}m}$ as

$$q_{A_{1g}m}=\sqrt{\omega_m/\hbar}\,Q_{A_{1g}m} \tag{22}$$

In a similar way, the dimensionless diagonal linear orbital vibronic coupling constants of the e_{2u} LUMO for the mth mode is defined by Eq. (23),

$$g_{e_{2u}\ \mathrm{LUMO}}(\omega_m)=\frac{1}{\hbar\omega_m}\left\langle e_{2u}\ \mathrm{LUMO}\left\|\left(\frac{\partial h}{\partial q_{A_{1g}m}}\right)_0\right\|e_{2u}\ \mathrm{LUMO}\right\rangle \tag{23}$$

C. Vibronic Interactions between the Nondegenerate Frontier Orbitals and the A_g Vibrational Modes in Polyacenes

Let us look into orbital vibronic coupling in polyacenes that have D_{2h} geometries, and the nondegenerate HOMOs and LUMOs, the symmetry labels of the HOMOs (LUMOs) of 2a, 4a, and 6a being a_u (b_{1g}) and those of 3a and 5a being b_{2g} (b_{3u}). Thus, the direct product of the HOMOs (LUMOs) symmetries can be reduced as

$$a_u\times a_u=b_{2g}\times b_{2g}=b_{1g}\times b_{1g}=b_{3u}\times b_{3u}=A_g \tag{24}$$

Therefore, the totally symmetric A_g modes of vibration couple to the HOMOs (LUMOs). The numbers of the A_g vibrational modes are 9, 12, 15, 18, and 21 for **2a**, **3a**, **4a**, **5a**, and **6a**, respectively. The symmetry labels of nondegenerate frontier orbitals of these polyacenes are abbreviated as "a" in the following discussion.

The energy sheet of the HOMO is given in the form:

$$\varepsilon_{m\,\mathrm{HOMO}}^{(A_g)}\left(Q_{A_g m}\right) = \frac{1}{2}K_{A_g m}Q_{A_g m}^2 + C_m Q_{A_g m} \tag{25}$$

where C_m corresponds to the reduced matrix element written as

$$C_m = \left\langle a\ \mathrm{HOMO} \left\| \left(\frac{\partial h}{\partial Q_{A_g m}} \right)_0 \right\| a\ \mathrm{HOMO} \right\rangle \tag{26}$$

This is illustrated in Figure 1. The dimensionless diagonal linear orbital vibronic coupling constants of the HOMO for its *m*th mode is defined by Eq. (27),

$$g_{a\,\mathrm{HOMO}}\left(\omega_m\right) = \frac{1}{\hbar\omega_m}\left\langle a\ \mathrm{HOMO} \left\| \left(\frac{\partial h}{\partial q_{A_g m}} \right)_0 \right\| a\ \mathrm{HOMO} \right\rangle \tag{27}$$

In this equation, $q_{A_g m}$ is the dimensionless normal coordinate [31] of the *m*th mode defined by using the normal coordinate $Q_{A_g m}$ as

$$q_{A_g m} = \sqrt{\omega_m / \hbar}\, Q_{A_g m} \tag{28}$$

In a similar way, the dimensionless diagonal linear orbital vibronic coupling constant of the LUMO for the *m*th A_g mode is defined by Eq. (29),

$$g_{a\,\mathrm{LUMO}}\left(\omega_m\right) = \frac{1}{\hbar\omega_m}\left\langle a\ \mathrm{LUMO} \left\| \left(\frac{\partial h}{\partial q_{A_g m}} \right)_0 \right\| a\ \mathrm{LUMO} \right\rangle \tag{29}$$

D. Vibronic Interactions in Polyfluoroacenes, B, N-Substituted Polyacenes, and Polycyanodienes

Similar discussions can be made in the monocations and anions of polyfluoroacenes, B, N-substituted polyacenes, and polycyanodienes. The numbers of vibronic active modes are 6, 9, 12, 15, 18, and 21 for 1fa, 2fa, 3fa, 4fa, 5fa, and 6fa, respectively, those are 11, 17, and 23 for 1bn, 2bn, and 3bn, respectively, and those are 5, 7, 9, and 11 for 1cn, 2cn, 3cn, and 4cn, respectively.

III. Electron–Phonon Coupling Constants for the Charged Electronic States of Polyacenes, Polyfluoroacenes, B, N-Substituted Polyacenes, and Polycyanodienes

Let us next discuss the total electron–phonon coupling constants (l_{total}) in the monocation and monoanion crystals. Since the l_{total} is the sum of the electron–phonon coupling constants originating from both intramolecular vibrations (l_{intra}) and intermolecular vibrations (l_{inter}), the l_{total} is defined as

$$l_{\text{total}} = l_{\text{intra}} + l_{\text{inter}}. \tag{30}$$

However, it should be noted that the intramolecular orbital interactions are much stronger than the intermolecular orbital interactions. Therefore, it is rational that the l_{intra} values are much larger than the l_{inter} values in molecular systems. In fact, it is considered by several researchers that the contribution from the intramolecular modes in molecular systems is decisive in the pairing process in the superconductivity in doped C_{60}. [15,32] For example, it was reported that the l_{intra} values are much larger than the l_{inter} values in K_3C_{60} and Rb_3C_{60} ($l_{\text{intra}} \geq 10 l_{\text{inter}}$). [32] Furthermore, it has also been shown from a neutron-scattering investigation [33] that the electron–libration intramolecular-mode coupling is small in alkali-metal-doped C_{60}. Therefore, we consider only intramolecular electron–phonon coupling in this study. The l_{total} value for the charged and excited electronic states can be defined as

$$l_{\text{total}} \approx l_{\text{intra}} = l_{\text{LUMO}} \text{ (for the electronic state of the monoanion)}, \tag{31}$$

$$= l_{\text{HOMO}} \text{ (for the electronic state of the monocation)}, \tag{32}$$

In the previous section, the vibronic interactions in free polyacenes, polyfluoroacenes, B, N-substituted polyacenes, and polycyanodienes were discussed. We can derive the l_{total} by using the vibronic coupling constants defined in the previous section as follows. As described above, since polyacenes, polyfluoroacenes, B, N-substituted polyacenes, and polycyanodienes would consist of strongly bonded molecules arranged on a lattice with weak van der Waals intermolecular bonds, we can derive the dimensionless electron–phonon coupling constant λ in a similar way as in theory in previous research. [15,25,27–29] We use a standard expression for λ, [15,25,27–29]

$$\lambda = \frac{2}{N(0)} \sum_{m,q} \sum_{k,k'} \frac{1}{2\omega_{m,q}^2} \left| h_{k,k'}(m,q) \right|^2 \delta(E_k)\delta(E_{k'}), \tag{33}$$

where $\omega_{m,q}$ is the vibrational frequency for the mth phonon mode of wave vector $\boldsymbol{q}$; $h_{k,k'}$ is the corresponding electron–phonon matrix element between the electronic states of wave vectors $\boldsymbol{k}$ and $\boldsymbol{k}'$; $E_{\boldsymbol{k}}$ and $E_{\boldsymbol{k}'}$ are the corresponding energies measured from the Fermi level (original point in Figure 1 (b)) of polyacenes, polyfluoroacenes, B, N-substituted polyacenes, and polycyanodienes; $N(0)$ is the total density of states (DOS) at the Fermi level per spin. The charged electronic states in polyacenes, polyfluoroacenes, B, N-substituted polyacenes, and polycyanodienes are essentially composed of the charged electronic states at the equilibrium structures of the ground states in polyacenes, polyfluoroacenes, B, N-substituted polyacenes, and polycyanodienes and we can write in the form of Bloch sum:

$$\psi(\boldsymbol{k}) = \frac{1}{\sqrt{N}} \sum_{\boldsymbol{R}} c(\boldsymbol{k}) e^{i\boldsymbol{kR}} \phi_{\boldsymbol{R}} \tag{34}$$

where $\boldsymbol{R}$ denotes the cell position; N is the number of molecules in the crystal; $\phi_{\boldsymbol{R}}$ is wave function which denotes the charged electronic states at the equilibrium structures of the ground state of polyacenes, polyfluoroacenes, B, N-substituted polyacenes, and polycyanodienes at the cell position $\boldsymbol{R}$. If we neglect the intermolecular electron–lattice coupling, $h_{\boldsymbol{k},\boldsymbol{k}'}$ can be reduced to

$$h_{\boldsymbol{k},\boldsymbol{k}'}(m,\boldsymbol{q}) = \langle \psi(\boldsymbol{k}) | h | \psi(\boldsymbol{k}') \rangle$$

$$= \frac{1}{N} \sum_{\boldsymbol{R}} c^*(\boldsymbol{k}) c(\boldsymbol{k}') e^{i(\boldsymbol{k}-\boldsymbol{k}')\boldsymbol{R}} h_{\boldsymbol{R},\boldsymbol{R}}(m,\boldsymbol{q}) \tag{35}$$

where $h_{\boldsymbol{R},\boldsymbol{R}}$ is the intramolecular coupling matrix and

$$h_{\boldsymbol{R},\boldsymbol{R}}(m,\boldsymbol{q}) = \langle \phi_{\boldsymbol{R}} | h | \phi_{\boldsymbol{R}} \rangle \tag{36}$$

For the one-phonon mode with wave vector $\boldsymbol{q}$, this term takes the following form:

$$h_{\boldsymbol{R},\boldsymbol{R}}(m,\boldsymbol{q}) = \left(1/\sqrt{N}\right) e^{i\boldsymbol{qR}} h_{00m} \tag{37}$$

We insert Eq. (37) in Eq. (35) taking the condition $\boldsymbol{k}' = \boldsymbol{k} - \boldsymbol{q}$ into account to get

$$h_{\boldsymbol{kk}-\boldsymbol{q}} = \frac{1}{\sqrt{N}} h_{00m} c^*(\boldsymbol{k}) c(\boldsymbol{k}-\boldsymbol{q}) \tag{38}$$

We now proceed to calculate λ by inserting Eq. (38) in Eq. (33) considering that $\omega_{m,\boldsymbol{q}}$ is independent of $\boldsymbol{q}$. We then obtain

$$\lambda = \frac{2}{N(0)} \sum_{k,k-q} \sum_{m} \frac{1}{2\omega_{m,q}^2} \frac{h_{00m} h_{00m}^*}{N} c^*(k)c(k)c^*(k-q)c(k-q)\delta(E_k)\delta(E_{k-q}) \tag{39}$$

The partial DOS per molecule at the Fermi level of polyacenes, polyfluoroacenes, B, N-substituted polyacenes, and polycyanodienes, $n(0)$ can be rewritten as

$$n(0) = (1/N)\sum_{k} c^*(k)c(k)\delta(E_k) \tag{40}$$

We can derive Eq. (41) from Eqs. (39) and (40) using $n(0) = N(0)/N$,

$$\lambda = \frac{2}{n(0)} \sum_{m} \frac{1}{2\omega_m^2} h_{00m} h_{00m}^* n(0)^2 \tag{41}$$

where $n(0)$ is now the DOS per spin and per molecule.

A. Twofold Degenerate Electronic States of The Monocations of Monoanions of Benzene, Hexafluorobenzene, and Borazine

Here, for example, let us consider twofold degenerate electronic states such as the monocation of benzene, In such a case, we can assume that $n_{\nu\nu'}(0) = \frac{1}{2} n(0)\delta_{\nu\nu'}$, λ takes the form of Eq. (42):

$$\lambda = \frac{n(0)}{4} \sum_{m} \frac{h_{00m}^2}{\omega_m^2} \tag{42}$$

where

$$h_{00m} = h_{E_{2g}m}' \tag{43}$$

Here $h_{E_{2g}m}'$ is the derivative matrix of the vibronic coupling matrix, $h_{E_{2g}m}$, derived by Eq. (45) with respect to the mode amplitude, $Q_{E_{2g}\gamma m}$, as

$$h_{E_{2g}m}' = \frac{\partial}{\partial Q_{E_{2g}\gamma m}} \left[A_m \begin{pmatrix} -Q_{E_{2g}\varepsilon m} & Q_{E_{2g}\theta m} \\ Q_{E_{2g}\theta n} & Q_{E_{2g}\varepsilon m} \end{pmatrix} \right] = \frac{\partial}{\partial Q_{E_{2g}\gamma m}} h_{E_{2g}m} \tag{44}$$

where

$$h_{E_{2g}m} = A_m \begin{pmatrix} -Q_{E_{2g}\varepsilon m} & Q_{E_{2g}\theta m} \\ Q_{E_{2g}\theta m} & Q_{E_{2g}\varepsilon m} \end{pmatrix} \tag{45}$$

and A_m is the reduced matrix element and is the slope in the original point (i.e., equilibrium structures of the ground states of **1a** ($Q_{E_{2g}\varepsilon m} = Q_{E_{2g}\theta m} = 0$)) on the potential energy surface of the charged electronic state along each vibrational mode (Figure 1 (b)), and is defined as

$$A_m = \left\langle e_{1g}\ \mathrm{HOMO} \left\| \left(\frac{\partial h}{\partial Q_{E_{2g}m}} \right)_0 \right\| e_{1g}\ \mathrm{HOMO} \right\rangle \tag{46}$$

Since $h_{00m}^2 = 4A_m^2$, one can rewrite Eq. (42) as

$$\lambda = n(0) \sum_m \frac{A_m^2}{\omega_m^2} \tag{47}$$

Using Eqs. (15) and (32), and after some simple transformations, we finally get the relation between the nondimensional electron–phonon coupling constant λ, and the intramolecular vibronic coupling constant, $g_{e_{1g}\ \mathrm{HOMO}}(\omega_m)$, as

$$\lambda = \sum_m \lambda_m,\ \lambda_m = n(0) l_{e_{1g}\ \mathrm{HOMO}}(\omega_m), \tag{48}$$

where $l_{e_{1g}\ \mathrm{HOMO}}(\omega_m)$ is the electron–phonon coupling constant defined as

$$l_{e_{1g}\ \mathrm{HOMO}}(\omega_m) = g_{e_{1g}\ \mathrm{HOMO}}^2(\omega_m)\hbar\omega_m,\ (m = 1,2,3,4). \tag{49}$$

In a similar way, the $l_{e_{1g}\ \mathrm{HOMO}}(\omega_m)$ for the A_{1g} vibronic active modes can be defined as

$$l_{e_{1g}\ \mathrm{HOMO}}(\omega_m) = \frac{1}{2} g_{e_{1g}\ \mathrm{HOMO}}^2(\omega_m)\hbar\omega_m,\ (m = 5,6). \tag{50}$$

In a similar way, $l_{e_{2u}\ \mathrm{LUMO}}(\omega_m)$ for the monoanion of benzene can be defined as

$$l_{e_{2u}\ \mathrm{LUMO}}(\omega_m) = g_{e_{2u}\ \mathrm{LUMO}}^2(\omega_m)\hbar\omega_m,\ (m = 1,2,3,4). \tag{51}$$

$$l_{e_{2u}\,\mathrm{LUMO}}(\omega_m)=\frac{1}{2}g^2_{e_{2u}\,\mathrm{LUMO}}(\omega_m)\hbar\omega_m,\ (m=5,6). \tag{52}$$

Similar discussions can be made in the monocations and monoanions of **1fa** and **1bn**.

B. Nondegenerate Electronic Systems of Polyacenes, Polyfluoroacenes, B, N-Substituted Polyacenes, and Polycyanodienes

In a similar way, by using Eq. (41), the $l_{\mathrm{HOMO}}(\omega_m)$ and $l_{\mathrm{LUMO}}(\omega_m)$ for the nondegenerate electronic states in the monocations and monoanions of polyacenes, polyfluoroacenes, B, N-substituted polyacenes, and polycyanodienes can be defined as

$$l_{\mathrm{HOMO}}(\omega_m)=g^2_{\mathrm{HOMO}}(\omega_m)\hbar\omega_m, \tag{53}$$

$$l_{\mathrm{LUMO}}(\omega_m)=g^2_{\mathrm{LUMO}}(\omega_m)\hbar\omega_m. \tag{54}$$

IV. OPTIMIZED STRUCTURES

A. Benzene and Polyacenes

The structures of neutral 1a and polyacenes were optimized under D_{6h} and D_{2h} symmetries, respectively, using the hybrid Hartree–Fock (HF)/density-functional-theory (DFT) method of Becke [34] and Lee, Yang, and Parr [35] (B3LYP) and the 6-31G* basis set. [36] GAUSSIAN 98 program package [37] was used for our theoretical analyses. This level of theory is, in our experience, sufficient for reasonable descriptions of the geometric, electronic, and vibrational structures of hydrocarbons. An optimized D_{6h} structure of *1a* and D_{2h} structures of polyacenes are shown in Figure 2. These structures were confirmed to be a minimum on each potential energy surface from vibrational analyses. We can see that in the D_{6h} structure of 1a there is no bond alternation and all C–C bond lengths are approximately 1.4 Å. In the D_{2h} structure of 2a, there is a distinct variation in the C–C distances. This result can be understood in view of the orbital patterns of the HOMO. Selected vibronic active modes and the phase patterns of the HOMOs and LUMOs of the neutral molecules under consideration are shown in Figure 3. We can see from this figure that the atomic orbitals between two neighboring C_{2a} and C_{3a} atoms are combined in phase and thus form a strong π-bonding in the HOMO of 2a. On the other hand, the HOMO contributes nonbonding interactions between C_{1a} and C_{1b} atoms and between C_{1a} and C_{2a} atoms, and an antibonding interaction between C_{3a} and C_{3b} atoms. This is the reason why the C_{2a}–C_{3a} bond is much shorter (1.377 Å) than any other C–C bond (1.417–1.434 Å). Similar discussions can be made in 3a, 4a, 5a, and 6a; the C–C bond distances between two neighboring carbon atoms whose atomic orbitals are combined in phase (out of phase) are short (long).

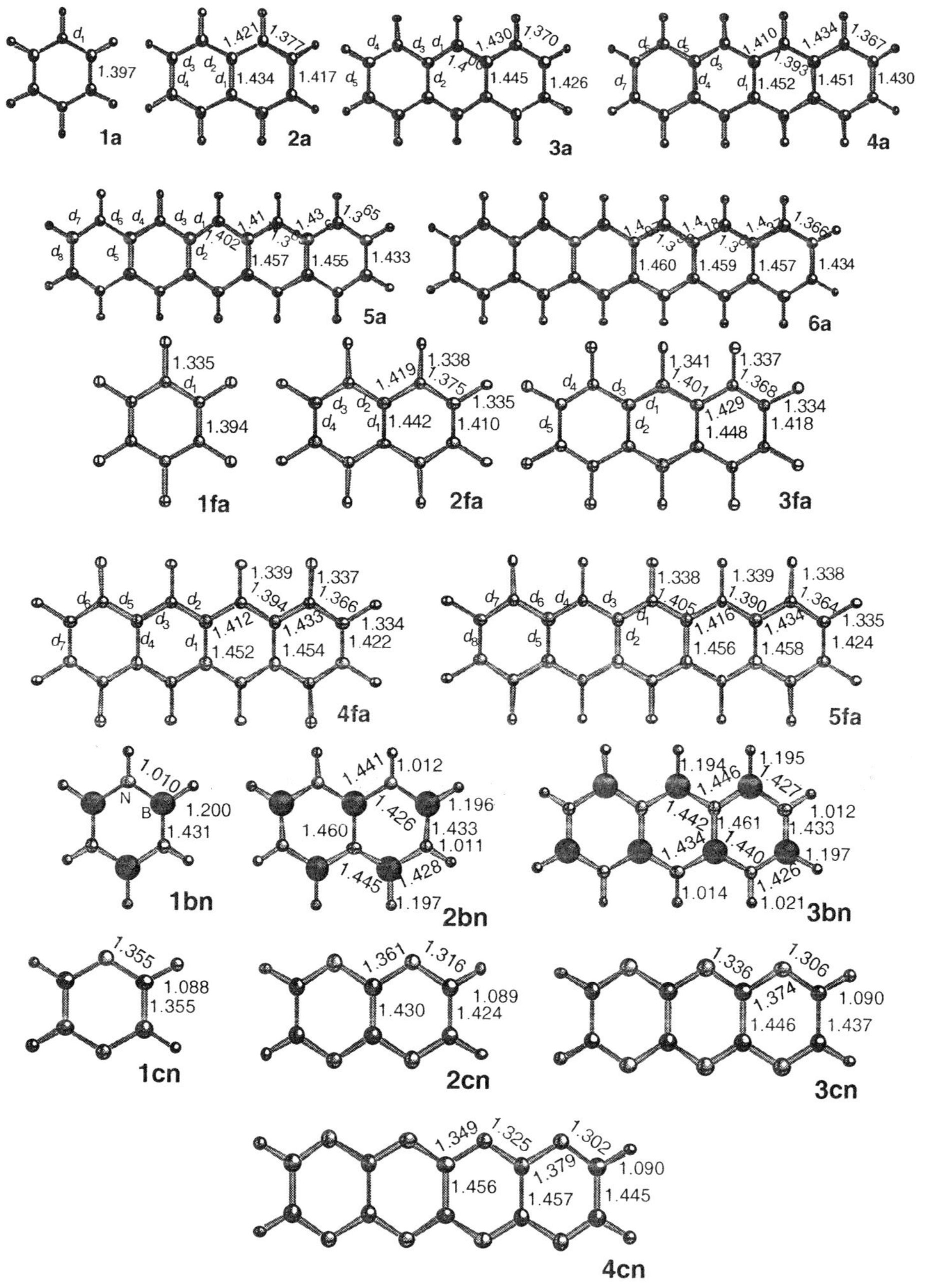

Figure 2. Optimized structures of the neutral polyacenes, polyfluoroacenes, B, N-substituted polyacenes, and polycyanodienes.

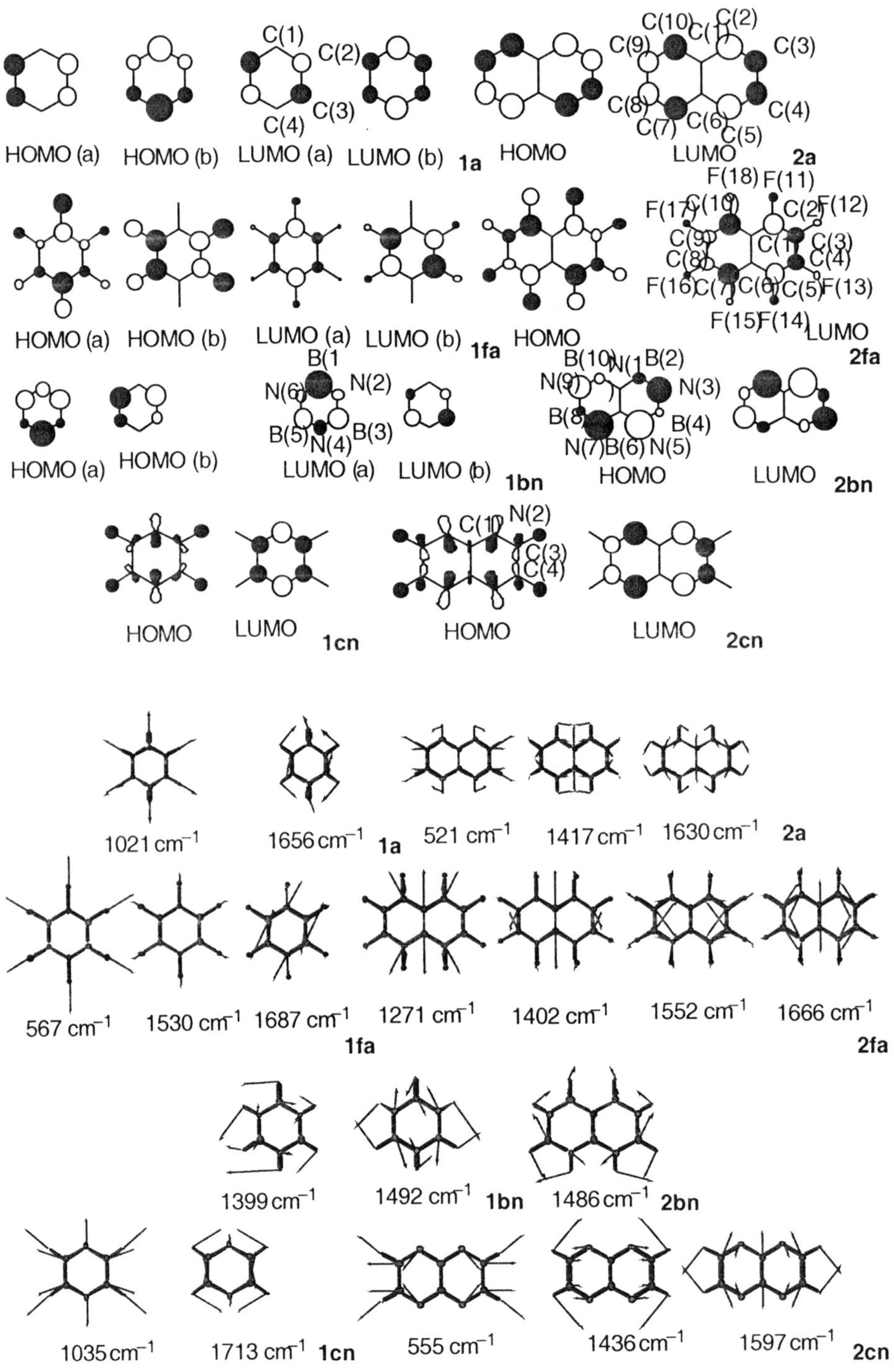

Figure 3. The phase patterns of the HOMO and LUMO and selected vibronic active modes in polyacenes, polyfluoroacenes, B, N-substituted polyacenes, and polycyanodienes.

B. Polyfluoroacenes

Hexafluorobenzene (C_6F_6) (1fa) has been well studied for a long time. [38–53] 1a and 1fa have nearly the same polarizations and polarizability anisotropies. [50] As well as being similar in molecular geometry and shape, 1a and 1fa have similar physical properties. [51] Their electric quadruple moments are roughly equal in magnitude but opposite in sign. [52,53] In recent review, [54] the study of Jahn–Teller and pseudo-Jahn–Teller effects in various molecular systems including the monoanion [55] and cation [56,57] of 1fa are well summarized.

The structures of neutral 1fa was optimized under D_{6h} symmetry, and 2fa, 3fa, 4fa, and 5fa were optimized under D_{2h} symmetry. According to our calculations, the energy differences between the HOMO and LUMO of 1fa, 2fa, 3fa, 4fa, and 5fa are 6.52, 4.56, 3.35, 2.57, and 2.03 eV, respectively, and those of 1a, 2a, 3a, 4a, and 5a are 6.80, 4.83, 3.59, 2.78, and 2.21 eV, respectively. Therefore, the HOMO–LUMO gaps in polyfluoroacenes are slightly smaller than those in polyacenes. This can be understood in view of the orbital patterns of the HOMO and LUMO in these molecules. We can see from Figure 3 that the phase patterns of the LUMO of polyfluoroacenes are not significantly different from those of polyacenes. The phase patterns of the LUMO in carbon framework of polyfluoroacenes are similar to those of polyacenes in that the LUMO of polyfluoroacenes is rather localized on carbon framework and the electron density on fluorine atoms is low, and the LUMO of polyacenes is completely localized on carbon atoms. The phase patterns of the HOMO in carbon framework of polyfluoroacenes are also similar to those of polyacenes. However, the HOMO of polyfluoroacenes is delocalized and the electron density on fluorine atoms is not small, while the HOMO of polyacenes is completely localized on carbon atoms. The atomic orbitals between two neighboring carbon and fluorine atoms are combined out of phase in the HOMO in polyfluoroacenes. Therefore, the HOMO is slightly destabilized in energy measured from the LUMO in polyfluoroacenes by H–F substitution. This is the reason why the energy difference between the HOMO and LUMO in polyfluoroacenes is slightly smaller than that in polyacenes.

Let us look into optimized structures of polyfluoroacenes. We can see from Figure 2 that the calculated C–C distances in 1fa are 1.394 Å. There is a distinct variation in the C–C distances in polyfluoroacenes. This result is reasonable in view of the HOMOs. The C–C distances between two neighboring carbon atoms whose atomic orbitals are combined in phase (out of phase) in the HOMOs of polyfluoroacenes are short (long). It should be noted that the C–C distances in polyfluoroacenes are similar to those in polyacenes; most of the C–C bond lengths differences between polyacenes and polyfluoroacenes are smaller than 0.003 Å. Only the d_4 value difference between 2a and 2fa, d_5 value difference between 3a and 3fa, d_7 value difference between 4a and 4fa, and d_8 value difference between 5a and 5fa are somewhat large (0.007–0.009 Å). Therefore, carbon framework structures in polyfluoroacenes are not significantly different from those in polyacenes as a consequence of H–F substitution.

C. B, N-Substituted Polyacenes

Substitution of carbon with boron and nitrogen have been carried out, producing totally substituted boron nitride nanotubes (BN-NTs). [58] Many researchers have studied and compared the structure of these new boron–nitrogen-containing fullerenes and nanotubes with that of carbon nanotubes (C-NTs). [59–73] A large number of molecular boron–nitrogen compounds have been prepared. [74–79] Because borazine ($B_3N_3H_6$) (1bn) is both isoelectronic and isosteric with 1a, it is a very interesting compound. Stock and Pohland [80] first reported formation of borazine 1bn from the thermolysis (200 °C) of the addition complex $[H_2B(NH_2)_2^+][BH_4^-]$, [81] formulated in 1926 as $B_2N_6{\cdot}2NH_3$. Laubengayer and co-workers [82] studied the gas-phase pyrolysis of 1bn in greater detail, and they proposed the intermediate formation of a naphthalene analogue, $B_5N_5H_8$ (2bn).

The structure of neutral "inorganic benzene", $B_3N_3H_6$ (1bn) was optimized under D_{3h} symmetry, and "inorganic-" naphthalene ($B_5N_5H_8$) (2bn) and anthracene ($B_7N_7H_{10}$) (3bn) were optimized under C_{2v} symmetry. According to our calculations, the energy difference between the HOMO and the LUMO of 1bn (8.27 eV) is larger than that of 1a (6.80 eV). Furthermore, that of 2bn (7.12 eV) is larger than that of 2a (4.83 eV), and that of 3bn (6.68 eV) is larger than that of 3a (3.59 eV). This can be understood as follows. The degenerate 1a levels have not been split apart in energy but the HOMO contains more nitrogen character than boron character while opposite is true for the LUMO.[30] This is a result clearly in keeping with an electronegativity perturbation on 1a. [30] Due to the electronegativity perturbation, the HOMO (LUMO) of 1bn is stabilized (destabilized) in energy with respect to the HOMO (LUMO) of 1a. This is the reason why the HOMO–LUMO gap is larger in 1bn than in 1a. Similar discussions can be made in 2bn and 3bn. Let us next look into optimized structure of 1bn. We can see from Figure 2 that the calculated B–N, B–H, and N–H distances in 1bn are 1.431, 1.200, and 1.010 Å, respectively, and those determined from experimental method [83] are 1.44, 1.20, and 1.02 Å, respectively. Therefore, the structure optimized at the B3LYP/6-31G* level of theory is in excellent agreement with that determined from the experimental method. There is a distinct variation in the B–N distances in 2bn and 3bn. This result is reasonable in view of the HOMOs shown in Figure 3. The B–N distances between two neighboring B and N atoms whose atomic orbitals are combined in phase (out of phase) in the HOMOs of 2bn and 3bn are short (long). It should be noted that the B–N distance (1.431 Å) in 1bn is larger than the C–C distance (1.397 Å) in 1a, but smaller than the B–N single bond distance (1.60 Å). Therefore, it can be said that each B–N bond in 1bn acquires double bond character. Similar discussion can be made in 2bn and 3bn; all B–N bonds in 2bn and 3bn are longer than all C–C bonds in 2a and 3a, but smaller than 1.6 Å.

D. Polycyanodienes

We can expect that in the monocations of polycyanodienes, the electron–phonon interactions become much stronger than in the monocations of polyacenes because the strong σ-orbital interactions as well as the π-orbital interactions between two neighboring atoms in the HOMO are significant, [84] and because the HOMO is delocalized and the electron density on hydrogen atoms as well as on carbon and nitrogen atoms is high in the HOMO of polycyanodienes. We call this system "σ-conjugated momocations".

The structures of polycyanodienes were optimized under D_{2h} symmetry. According to our calculations, the energy differences between the HOMO and LUMO of 1cn, 2cn, 3cn, and 4cn are 5.16, 3.97, 3.19, and 2.68 eV, respectively. Therefore, the HOMO–LUMO gaps in polycyanodienes are slightly smaller than those in polyacenes.

V. Electron–Phonon Coupling Constants

A. The Monocations and Monoanions of Benzene and Polyacenes

We next carried out vibrational analyses of 1a and polyacenes at the B3LYP/6-31G* level. Agreement with experiment is excellent for the vibrational frequencies, calculated wave numbers of 1a being accurate within a range of 2.5–4 %. Since the conventional Hartree–Fock method overestimates molecular vibrational frequencies by typically 10 %, this hybrid DFT method appears to be very useful for vibrational analyses. We next calculated first-order derivatives at this equilibrium structure on each orbital energy surface by distorting the molecule along the A_{1g} and E_{2g} modes of 1a and the A_g modes of polyacenes. In these calculations, the step size of the normal-mode displacements was taken to be in the order of 10^{-1}. What we obtained from the first-order derivatives are the dimensionless diagonal linear orbital vibronic coupling constants $g_{\mathrm{HOMO}}(\omega_m)$ and $g_{\mathrm{LUMO}}(\omega_m)$. We can estimate the electron–phonon coupling constants $l_{\mathrm{HOMO}}(\omega_m)$ and $l_{\mathrm{LUMO}}(\omega_m)$ from the dimensionless diagonal linear orbital vibronic coupling constants by using Eqs. (49)–(54). The calculated electron–phonon coupling constants in the monocatins and anions of 1a and polyacenes are shown in Figures 4 and 5, respectively.

Let us next take a look at the electron–phonon coupling of the A_{1g} and E_{2g} vibrational modes to the e_{1g} HOMO in 1a. E_{2g} modes of 1a are classified into a C–C–C in-plane bending (622 cm^{-1}), a C–H in-plane bending (1208 cm^{-1}), a C–C stretching (1656 cm^{-1}), and a C–H stretching (3184 cm^{-1}). Figure 4 demonstrates that the E_{2g} vibrational modes of 622 and 1656 cm^{-1} strongly couple to the e_{1g} HOMO in 1a. This can be understood in view of the orbital patterns of the e_{2u} LUMO and vibrational modes of 1a. When 1a is distorted along this E_{2g} mode toward the direction as shown in Figure 3, the antibonding (bonding) interactions between C_{1b} and C_{1c} in the HOMO (a) (HOMO (b)) become strong, and the bonding interactions between C_{1a} and C_{1b} and between C_{1c} and C_{1d} in the HOMO (a) become weak. Therefore, the HOMO (a) (HOMO (b)) is significantly destabilized (stabilized) in energy. On the other hand, when 1a is distorted toward the opposite direction along the arrow of this mode, the HOMO (a) (HOMO (b)) is significantly stabilized (destabilized) in energy. This is the reason why the E_{2g} mode of 1656 cm^{-1} strongly couples to the e_{1g} HOMO in 1a. Figure 4 demonstrates that the C–C stretching A_g modes of 1417 and 1630 cm^{-1} strongly couple to the HOMO in 2a. When 2a is distorted along the A_g mode of 1417 cm^{-1} toward the direction as shown in Figure 3, the antibonding interactions between C_{3a} and C_{3b} in the HOMO become strong, and the bonding interactions between C_{2a} and C_{3a} and between C_{2b} and C_{3b} become weak.

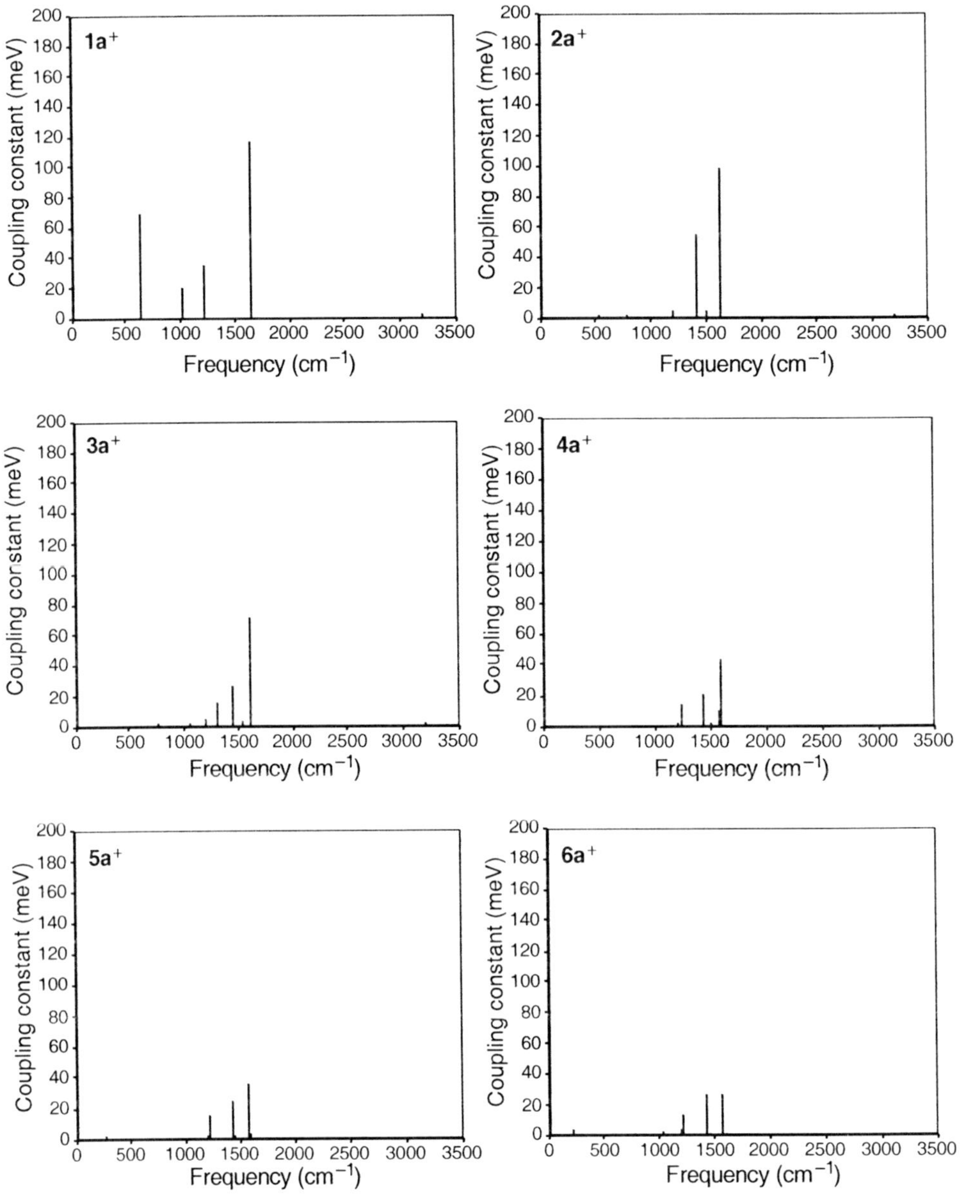

Figure 4. Electron–phonon coupling constants for the monocations of polyacenes.

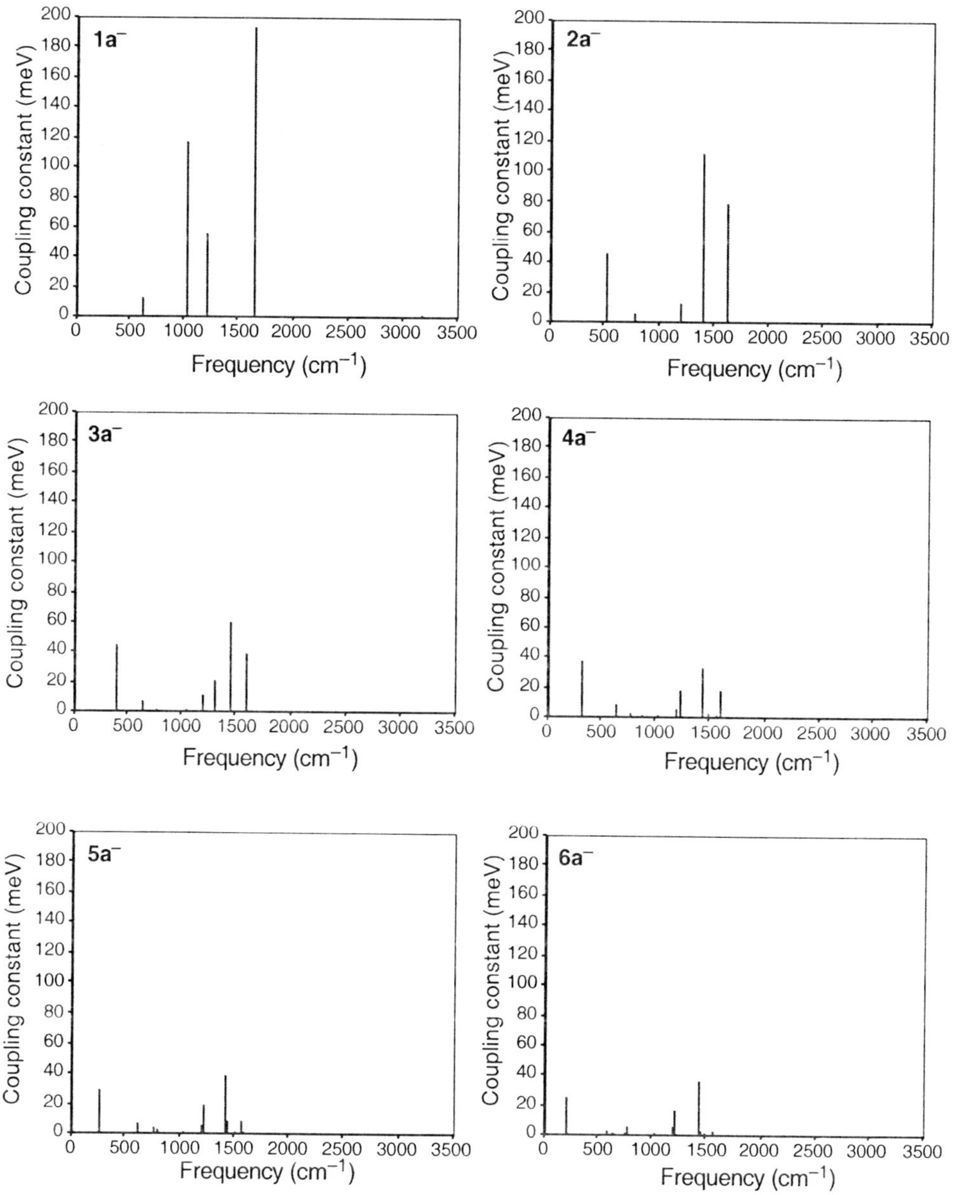

Figure 5. Electron–phonon coupling constants for the monoanions of polyacenes.

Therefore, the HOMO of 2a is significantly destabilized in energy. On the other hand, when 2a is distorted toward the opposite direction, the HOMO is significantly stabilized in energy. This is the reason why the C–C stretching A_g mode of 1417 cm^{-1} strongly couples to the HOMO. In a similar way, the C–C stretching A_g mode of 1630 cm^{-1} strongly couples to the HOMO. In the two frequency modes, the displacements of carbon atoms are very large compared with any other A_g mode of 2a, and the HOMO is completely localized on carbon atoms. It is rational that a frequency mode in which normal displacements are large for atoms where there exists considerable orbital amplitude can strongly couple to the molecular orbital. Identical discussions can be made in 3a, 4a, 5a, and 6a; the C–C stretching A_g modes around 1500 cm^{-1} and the lowest frequency A_g mode strongly couple to the HOMOs.

Let us take a look at the electron–phonon coupling of the A_{1g} and E_{2g} vibrational modes to the e_{2u} LUMO in 1a. We can see from Figure 5 that the E_{2g} mode of 1656 cm^{-1} strongly couples to the e_{2u} LUMO in 1a. This can be also understood in view of the phase patterns of the e_{2u} LUMO and vibrational modes of 1a. Figure 5 demonstrates that the C–C stretching A_g modes of 1417 and 1630 cm^{-1} strongly couple to the b_{1g} LUMO. In the two frequency modes, the displacements of carbon atoms are very large compared with any other A_g mode of 2a, and the b_{1g} LUMO is completely localized on carbon atoms. It is rational that a frequency mode in which normal displacements are large for atoms where there exists considerable orbital amplitude can strongly couple to the molecular orbital. Identical discussions can be made in 3a, 4a, 5a, and 6a; the C–C stretching A_g modes around 1500 cm^{-1} and the lowest frequency A_g mode strongly couple to the LUMOs. The C–C stretching A_g modes around 1500 cm^{-1} strongly couple to the HOMOs as well as to the LUMOs in polyacenes. But it should be noted that the lowest frequency modes as well as the C–C stretching modes strongly couple to the LUMO in polyacenes, while the only C–C stretching modes strongly couple to the HOMO in polyacenes.

B. Polyfluoroacenes

1. Monocations

We carried out vibrational analyses of polyfluoroacenes at the B3LYP/6-31G* level of theory. There are four E_{2g} modes in 1fa (C–C–F in-plane bending mode of 259 cm^{-1}; C–C–C in-plane bending mode of 444 cm^{-1}; C–F stretching mode of 1192 cm^{-1}; C–C stretching mode of 1687 cm^{-1}). There are four E_{2g} vibrational modes in 1a (C–C–C in-plane bending mode of 622 cm^{-1}; C–C–H in-plane bending mode of 1208 cm^{-1}; C–C stretching mode of 1656 cm^{-1}; C–H stretching mode of 3184 cm^{-1}). Therefore, the frequencies for the C–C–H in-plane bending mode (1208 cm^{-1}) and for the C–H stretching mode (3184 cm^{-1}) in 1a are much larger than those for the C–C–F in-plane bending mode (259 cm^{-1}) and for the C–F stretching mode (1192 cm^{-1}) in 1fa, respectively, as expected. Furthermore, the frequency for the C–C–C in-plane bending mode (622 cm^{-1}) in 1a is larger than that (444 cm^{-1}) in 1fa. The frequency for the C–C stretching mode (1656 cm^{-1}) in 1a is similar to that (1687 cm^{-1}) in 1fa.

The calculated electron–phonon coupling constants in the monocations and monoanions of 1fa, 2fa, 3fa, 4fa, and 5fa are shown in Figures 6 and 7, respectively. We can see from Figure 6 that the C–C stretching E_{2g} mode of 1687 cm^{-1} can strongly couple to the e_{1g} HOMO in 1fa. This can be understood in view of the phase patterns of the HOMO in 1fa. The reduced mass for the E_{2g} mode of 1687 cm^{-1} is 12.05, and the displacements of carbon atoms are much larger than those of fluorine atoms in the mode.

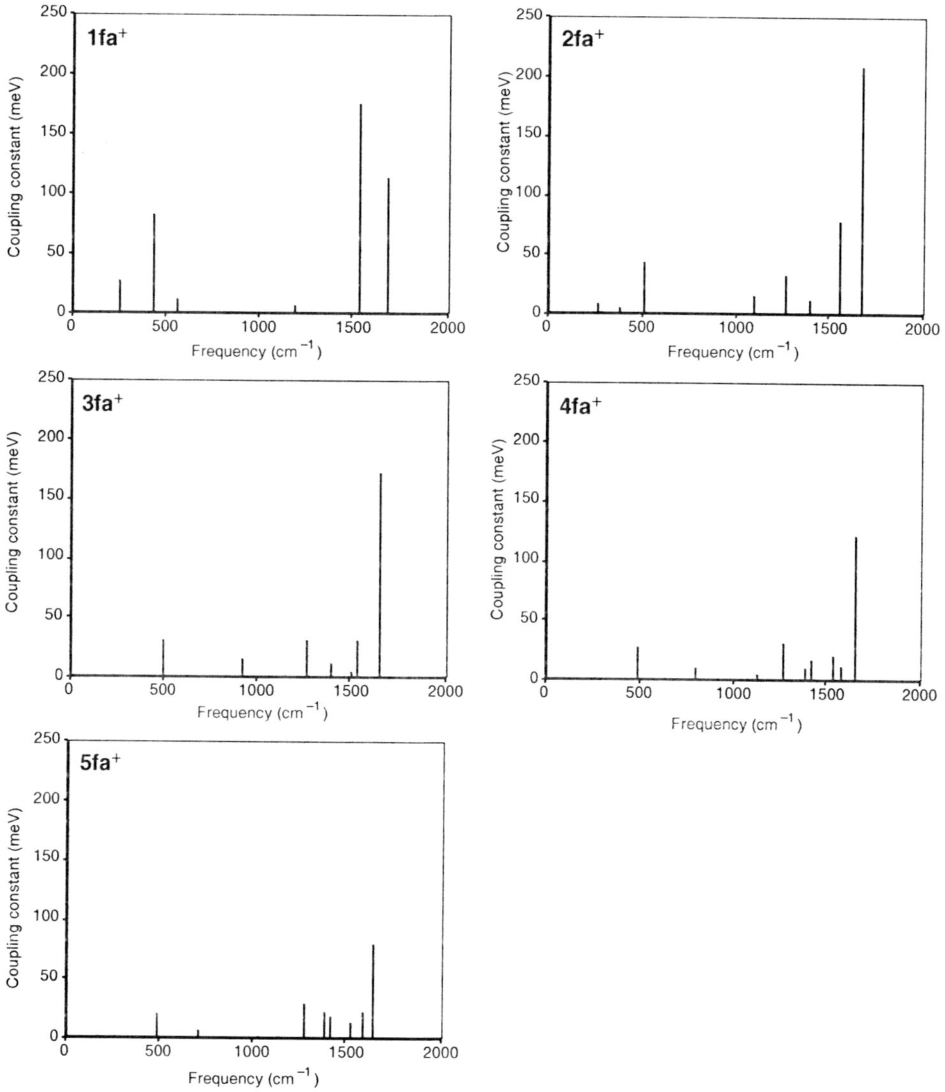

Figure 6. Electron–phonon coupling constants for the monocations of polyfluoroacenes.

When 1fa is distorted along the E_{2g} mode of 1687 cm^{-1} toward the same direction as shown in Figure 3, the antibonding (bonding) interactions between two neighboring carbon atoms in the HOMO (a) (HOMO (b)) become significantly stronger, and the bonding interactions between two neighboring carbon atoms in the HOMO (a) become significantly weaker, and thus the HOMO (a) (HOMO (b)) is significantly destabilized (stabilized) in energy by such a distortion. This is the reason why the E_{2g} mode of 1687 cm^{-1} the most strongly couples to the e_{1g} HOMO in 1fa.

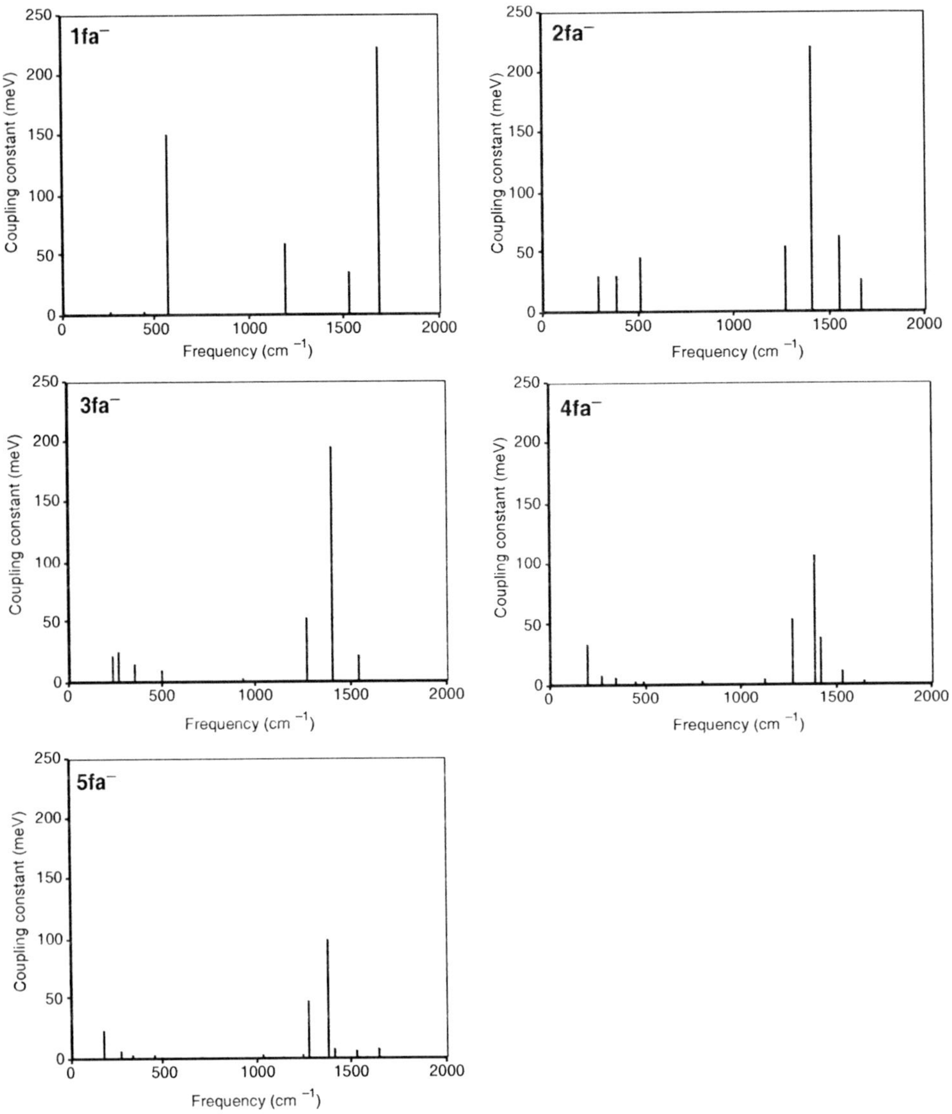

Figure 7. Electron–phonon coupling constants for the monoanions of polyfluoroacenes.

We can see from Figure 6 that the A_{1g} mode of 1530 cm^{-1} can very strongly couple to the e_{1g} HOMO in 1fa. This can be understood as follows. When 1fa is distorted along the A_{1g} mode of 1530 cm^{-1}, the bonding interactions between two neighboring carbon atoms become significantly weaker, and the antibonding interactions between two neighboring carbon and fluorine atoms in the e_{1g} HOMO become significantly stronger, and thus the e_{1g} HOMO is significantly destabilized in energy by such a distortion in 1fa. This is the reason why the A_{1g} mode of 1530 cm^{-1} can strongly couple to the e_{1g} HOMO in 1fa. The A_{1g} mode of 567 cm^{-1} much less strongly couples to the e_{1g} HOMO than the A_{1g} mode of 1530 cm^{-1} in 1fa. This can be understood as follows. When 1fa is distorted along the A_{1g} mode of 567 cm^{-1}, the bonding interactions between two neighboring carbon atoms and the antibonding interactions between two neighboring carbon and fluorine atoms in the e_{1g} HOMO become weaker. Since such weakened effects of bonding and antibonding interactions are compensated by each other, the

e_{1g} HOMO is slightly stabilized in energy by such a distortion in 1fa. This is the reason why the A_{1g} mode of 567 cm^{-1} much less strongly couples to the e_{1g} HOMO than the A_{1g} mode of 1530 cm^{-1} in 1fa.

We can see from Figure 6 that the C–C stretching A_g mode of 1666 cm^{-1} the most strongly couples to the HOMO in 2fa. This can be understood as follows. When 2fa is distorted along the A_g mode of 1666 cm^{-1}, the bonding and antibonding interactions between two neighboring carbon atoms become significantly stronger and weaker, respectively, and the antibonding interactions between two neighboring carbon and fluorine atoms in the HOMO in 2fa become weaker. Therefore, the HOMO is significantly stabilized in energy by such a distortion in 2fa. This is the reason why the A_g mode of 1666 cm^{-1} can the most strongly couple to the HOMO in 2fa. In a similar way, the A_g modes of 514 and 1552 cm^{-1} also strongly couple to the HOMO in 2fa. Similar discussions can be made in 3fa, 4fa, and 5fa; the low frequency mode around 500 cm^{-1}, the high frequency modes around 1200 cm^{-1}, and the frequency modes around 1600 cm^{-1} afford large electron–phonon coupling constants in the monocations of polyfluoroacenes.

2. Monoanions

Let us first look into the electron–phonon interactions between the E_{2g} modes and the e_{2u} LUMO in 1fa. We can see from this figure that the C–C stretching E_{2g} mode of 1687 cm^{-1} can strongly couple to the e_{2u} LUMO in 1fa. This can be understood in view of the phase patterns of the LUMO in 1fa. When 1fa is distorted along the E_{2g} mode of 1687 cm^{-1} toward the same direction as shown in Figure 3, the bonding (antibonding) interactions between two neighboring carbon atoms in the LUMO (a) become stronger (weaker), therefore, the LUMO (a) is significantly stabilized in energy. In a similar way, the antibonding interactions between two neighboring carbon atoms in the LUMO (b) become stronger by such a distortion, therefore, the LUMO (b) is significantly destabilized in energy. This is the reason why the E_{2g} mode of 1687 cm^{-1} can strongly couple to the e_{2u} LUMO in 1fa. In addition to this mode, the C–F stretching E_{2g} mode of 1192 cm^{-1} can strongly couple to the e_{2u} LUMO, but the E_{2g} modes of 259 and 444 cm^{-1} hardly couple to it. This can be understood as follows. The reduced masses for the E_{2g} modes of 1192 and 1687 cm^{-1} are 13.69 and 12.05, respectively, therefore, the displacements of carbon atoms are much larger than those of fluorine atoms in these modes. On the other hand, the reduced masses for the E_{2g} modes of 259 and 444 cm^{-1} are 18.62 and 16.08, respectively, and therefore, the displacements of fluorine atoms are more dominant than those of carbon atoms in these modes. Since the e_{2u} LUMO is rather localized on carbon atoms, it is reasonable that a frequency mode in which normal displacements of carbon atoms are large (small), can strongly (weakly) couple to the e_{2u} LUMO. This is the reason why the E_{2g} modes of 1192 and 1687 cm^{-1} can strongly couple to the e_{2u} LUMO, while the E_{2g} modes of 259 and 444 cm^{-1} hardly couple to it.

Let us next look into the electron–phonon interactions between the A_{1g} modes and the e_{2u} LUMO in 1fa. We can see from Figure 7 that the A_{1g} mode of 567 cm^{-1} much more strongly couples to the e_{2u} LUMO than the A_{1g} mode of 1530 cm^{-1} in 1fa. This can be understood as follows. When 1fa is distorted along the A_{1g} mode of 567 cm^{-1}, the antibonding interactions between two neighboring carbon atoms, and between two neighboring carbon and fluorine atoms, both in the LUMO (a) and LUMO (b), become weaker. Therefore, both the LUMO (a) and LUMO (b) are significantly stabilized in energy by such a distortion. When 1fa is distorted along the A_{1g} mode of 1530 cm^{-1}, the antibonding interactions between two neighboring carbon atoms, and between two neighboring carbon and fluorine atoms, become weaker and stronger, respectively, both in the LUMO (a) and LUMO (b). Since such strengthened and weakened effects of antibonding interactions are compensated by each other, the A_{1g} mode of 1530 cm^{-1} less strongly couples to the e_{2u} LUMO than the A_{1g} mode of 567 cm^{-1}.

Let us next look into the electron–phonon interactions between the A_g modes and the b_{1g} LUMO in 2fa. We can see from Figure 7 that the C–C stretching A_g mode of 1402 cm^{-1} much more strongly couples to the LUMO than the other modes in 2fa. This can be understood as follows. The reduced mass for the A_g mode of 1402 cm^{-1} is 12.11, therefore, the displacements of carbon atoms are much larger than those of fluorine atoms in the mode. When 2fa is distorted along the A_g mode of 1402 cm^{-1}, the bonding (antibonding) interactions between two neighboring carbon atoms become stronger (weaker), and the antibonding interactions between two neighboring carbon and fluorine atoms in the LUMO become weaker. Therefore, the LUMO is significantly stabilized in energy by such a distortion. This is the reason why the A_g mode of 1402 cm^{-1} strongly couples to the LUMO in 2fa. We can see from Figure 7 that the A_g modes of 1271 and 1552 cm^{-1} also somewhat strongly couple to the LUMO in 2fa. This can be understood as follows. When 2fa is distorted along the A_g mode of 1552 cm^{-1}, the bonding interactions between C(3) and C(4) atoms, and between C(8) and C(9) atoms, become stronger, and the antibonding interactions between C(3) and F(12) atoms, between C(4) and F(13) atoms, between C(8) and F(16) atoms, and between C(9) and F(17) atoms, in the LUMO, become weaker. Therefore, the LUMO is stabilized in energy by such a distortion. This is the reason why the A_g mode of 1552 cm^{-1} also somewhat strongly couples to the LUMO in 2fa. But it should be noted that the antibonding interactions between C(2) and C(3) atoms, between C(4) and C(5) atoms, between C(7) and C(8) atoms, and between C(9) and C(10) atoms in the LUMO, become slightly stronger, and thus the LUMO would be slightly destabilized in energy by such an effect. The destabilization effect and the stabilization effect described above are compensated by each other, the LUMO is slightly stabilized in energy by such a distortion. This is the reason why the A_g mode of 1552 cm^{-1} much less strongly couples to the LUMO than the A_g mode of 1402 cm^{-1} even though both A_g modes are C–C stretching modes and the reduced mass for the A_g mode of 1552 cm^{-1} is similar to that for the A_g mode of 1402 cm^{-1}. When 2fa is distorted along the C–F stretching A_g mode of 1271 cm^{-1}, the antibonding interactions between C(2) and F(11) atoms, between C(5) and F(14) atoms, between C(7) and F(15) atoms, and between C(10) and F(18) atoms in the LUMO, become stronger, therefore, the LUMO is destabilized in energy by such a distortion. But there are not significant changes in the strengths of the orbital interactions between two neighboring carbon atoms in the mode, and thus the C–F stretching A_g mode of 1271 cm^{-1} much less strongly couples to the LUMO than the C–C stretching A_g mode of 1402 cm^{-1} in 2fa. Similar discussions can be made in 3fa, 4fa, and 5fa; the C–C stretching A_g modes of 1401, 1389, and 1379 cm^{-1} the most strongly couple to the LUMO in 3fa, 4fa, and 5fa, respectively.

Let us next compare the calculated results in polyfluoroacenes with those in polyacenes. As in case of the monoanion of 1fa, the C–C stretching E_{2g} mode of 1656 cm^{-1} affords the largest electron–phonon coupling constant (193 meV) in the monoanion of 1a. But it should be noted that the C–C stretching E_{2g} mode of 1687 cm^{-1} in the monoanion of 1fa affords a larger electron–phonon coupling constant (223 meV) than the C–C stretching E_{2g} mode of 1656 cm^{-1} in the monoanion of 1a. This can be understood as follows. The reduced mass for the E_{2g} mode of 1656 cm^{-1} in 1a is 5.32, and therefore, the displacements of hydrogen atoms as well as those of carbon atoms are large in the mode. On the other hand, the reduced mass for the E_{2g} mode of 1687 cm^{-1} in 1fa is 12.05, and thus, the displacements of carbon atoms are very large and those of fluorine atoms are very small in the mode. The LUMO is completely localized on carbon atoms in 1a, and is rather localized on carbon atoms in 1fa. Therefore, it is rational that the E_{2g} mode of 1687 cm^{-1} in 1fa in which the displacements of carbon atoms are larger, more strongly couples to the LUMO localized on carbon atoms than the E_{2g} mode of 1656 cm^{-1} in 1a in which the displacements of carbon atoms are smaller.

Similar discussions can be made in 2fa, 3fa, 4fa, and 5fa; the C–C stretching modes afford larger electron–phonon coupling constants in the monoanions of 2fa, 3fa, 4fa, and 5fa than in the monoanions of 2a, 3a, 4a, and 5a, respectively. This is because the displacements of carbon atoms in the modes are larger in 2fa, 3fa, 4fa, and 5fa than those in 2a, 3a, 4a, and 5a, respectively, and the orbital patterns of the LUMO in polyacenes hardly change by H–F substitution.

It should be noted that the C–C stretching modes around 1500 cm^{-1} and the low frequency modes, less and more, respectively, strongly couple to the LUMO with an increase in molecular size in polyacenes, while the low frequency modes as well as the C–C stretching modes around 1500 cm^{-1} less strongly couple to the LUMO with an increase in molecular size in polyfluoroacenes.

It should be noted that the A_{1g} mode of 567 cm^{-1} (1530 cm^{-1}) much more (less) strongly couples to the e_{2u} LUMO than to the e_{1g} HOMO in 1fa. This can be understood as follows. When 1fa is distorted along the A_{1g} mode of 567 cm^{-1}, the antibonding interactions between two neighboring carbon atoms and between two neighboring carbon and fluorine atoms in the e_{2u} LUMO in 1fa become significantly weaker, and thus the e_{2u} LUMO is significantly stabilized in energy, while the energy level of the e_{1g} HOMO does not significantly change by such a distortion because the weakened effects of bonding and antibonding interactions are compensated by each other, as described above. This is the reason why the A_{1g} mode of 567 cm^{-1} much more strongly couples to the e_{2u} LUMO than to the e_{1g} HOMO in 1fa. When 1fa is distorted along the A_{1g} mode of 1530 cm^{-1}, the antibonding interactions between two neighboring carbon atoms and between two neighboring carbon and fluorine atoms in the e_{2u} LUMO in 1fa become weaker and stronger, respectively. Since such weakened and strengthened effects of antibonding interactions are compensated by each other, the e_{2u} LUMO in 1fa is slightly destabilized in energy by such a distortion. On the other hand, such a compensation does not occur in the electron–phonon interactions between the e_{1g} HOMO and the A_{1g} mode of 1530 cm^{-1}, as described above. This is the reason why the A_{1g} mode of 1530 cm^{-1} much more strongly couples to the e_{1g} HOMO than to the e_{2u} LUMO in 1fa. But it should be noted that the high frequency E_{2g} modes of 1192 and 1687 cm^{-1} (the low frequency E_{2g} modes of 259 and 444 cm^{-1}) much more (less) strongly couple to the e_{2u} LUMO than to the e_{1g} HOMO in 1fa. This can be understood as follows. The reduced masses for the high frequency E_{2g} modes of 1192 and 1687 cm^{-1} are 13.69 and 12.05, respectively, and thus the displacements of carbon atoms are much larger than those of fluorine atoms in these modes in 1fa, while those for the low frequency E_{2g} modes of 259 and 444 cm^{-1} are 18.62 and 16.08, respectively, and thus the displacements of fluorine atoms are larger than those of carbon atoms in these modes in 1fa. The e_{2u} LUMO is rather localized on carbon atoms and in which the electron density on fluorine atoms are low, on the other hand, the e_{1g} HOMO is delocalized and in which the electron density on fluorine atoms are as high as those on carbon atoms in 1fa. It is rational that the higher (lower) frequency E_{2g} modes, in which the displacements of carbon (fluorine) atoms are very large, more (less) strongly couple to the e_{2u} LUMO rather localized on carbon atoms than to the delocalized e_{1g} HOMO in 1fa. This is the reason why the high frequency E_{2g} modes of 1192 and 1687 cm^{-1} (the low frequency E_{2g} modes of 259 and 444 cm^{-1}) much more (less) strongly couple to the e_{2u} LUMO than to the e_{1g} HOMO in 1fa.

It should be noted that the A_g mode of 1666 cm^{-1} (1402 cm^{-1}) much more (less) strongly couples to the HOMO than to the LUMO in 2fa. This can be understood as follows. When 2fa is distorted along the A_g mode of 1666 cm^{-1}, the bonding and antibonding interactions between two neighboring carbon atoms in the LUMO in 2fa become weaker and stronger, respectively, and thus the LUMO would be destabilized in energy by such a distortion. On the other hand, the antibonding interactions between two neighboring carbon and fluorine atoms in the LUMO in 2fa become weaker, and thus the LUMO would be stabilized in energy by such a distortion. Since such destabilization and stabilization effects are compensated by each other, the LUMO is slightly destabilized in energy by such a distortion in 2fa. Such a compensation does not occur in the electron–phonon interactions between the A_g mode of 1666 cm^{-1} and the HOMO in 2fa, and thus the A_g mode of 1666 cm^{-1} much more strongly couples to the HOMO than to the LUMO in 2fa. In a similar way, such a compensation occurs (does not occur) in the electron–phonon interactions between the A_g mode of 1402 cm^{-1} and the HOMO (LUMO), and thus the energy level of the LUMO much more significantly changes than that of the HOMO by such a distortion in 2fa. This is the reason why the A_g mode of 1402 cm^{-1} can much more strongly couple to the LUMO than to the HOMO in 2fa. That is, the significant phase patterns difference between the HOMO, in which the atomic orbitals between two neighboring carbon atoms are mainly combined in phase and those between two neighboring carbon and fluorine atoms are combined out of phase, and the LUMO, in which the atomic orbitals between two neighboring carbon atoms and between two neighboring carbon and fluorine atoms are combined out of phase, is the main reason why the vibrational modes which play an essential role in the electron–phonon interactions in the monocations are significantly different from those in the monoanions in 2fa.

In a similar way, we can rationalize in view of the phase patterns of the HOMO and LUMO the calculated results that the frequency modes lower than 500 cm^{-1} and the high frequency modes around 1400 cm^{-1} much more strongly couple to the LUMO than to the HOMO in 2fa, 3fa, 4fa, and 5fa, on the other hand, the frequency modes around 500 cm^{-1} and the frequency modes around 1600 cm^{-1} much more strongly couple to the HOMO than to the LUMO in 2fa, 3fa, 4fa, and 5fa.

C. B, N-Substituted Polyacenes

1. Monocations

There are five kinds of E' vibrational modes in 1bn (B–N–B or N–B–N in-plane bending mode of 520 cm^{-1}; B–N–H or N–B–H in-plane bending modes of 951 and 1073 cm^{-1}; B–N stretching modes of 1399 and 1492 cm^{-1}; B–H stretching mode of 2625 cm^{-1}; N–H stretching mode of 3640 cm^{-1}). The frequencies of the vibrational modes in B, N-substituted polyacene-series are slightly lower than those in polyacene-series. For example, the frequencies of the B–N stretching modes of 1399 and 1492 cm^{-1} in 1bn are lower than that of the C–C stretching mode of 1656 cm^{-1} in 1a. This can be understood as follows. The B–N bonds in 1bn are longer than the C–C bonds in 1a. In general, the frequency of the stretching mode between two neighboring atoms whose bond lengths are larger, is lower. The C–H (1.087 Å) bond lengths in 1a lie between the N–H (1.010 Å) and B–H (1.200 Å) bond lengths in 1bn, and the frequency for the C–H (3184 cm^{-1}) stretching mode in 1a also lies between those for the N–H (3640 cm^{-1}) and B–H (2625 cm^{-1}) stretching modes in 1bn.

The calculated electron–phonon coupling constants in the monocations and monoanions of 1bn, 2bn, and 3bn are shown in Figure 8. We can see from Figure 3 that the HOMO and LUMO in B, N-substituted polyacene-series are rather localized on N and B atoms, respectively, due to the electronegativity perturbation on polyacene-series. We can see from Figure 8 that the B–N stretching E' mode of 1492 cm^{-1} can strongly couple to the e' HOMO in 1bn. This can be understood in view of phase patterns of the HOMO in 1bn. When 1bn is distorted along the E' mode of 1492 cm^{-1} toward the same direction as shown in Figure 3, the bonding (antibonding) interactions between two neighboring B and N atoms in the HOMO (a) become stronger (weaker), therefore, the HOMO (a) is significantly stabilized in energy. In a similar way, the bonding interactions between two neighboring B and N atoms in the HOMO (b) become weaker by such a distortion, therefore, the HOMO (b) is significantly destabilized in energy. This is the reason why the E' mode of 1492 cm^{-1} can strongly couple to the e' HOMO in 1bn. In addition to this mode, the E' modes of 520 and 1399 cm^{-1} and the A_1' mode of 949 cm^{-1} can also couple to the e' HOMO in 1bn. But it should be noted that the E' mode of 1399 cm^{-1} much less strongly couples to the e' HOMO than the E' mode of 1492 cm^{-1} even though both vibrational modes are the B–N stretching modes. This can be understood as follows. The reduced mass for the E' mode of 1399 cm^{-1} is 1.89, while that for the E' mode of 1492 cm^{-1} is 4.93, and thus the displacements of B and N atoms are larger in the E' mode of 1492 cm^{-1} than in the E' mode of 1399 cm^{-1}. The e' HOMO is, on the other hand, localized on B and N atoms. It is rational that a frequency mode in which the displacements of B and N atoms are important, more strongly couples to the e' HOMO localized on B and N atoms. This is the reason why the E' mode of 1399 cm^{-1} much less strongly couples to the e' HOMO than the E' mode of 1492 cm^{-1} in 1bn. It should be noted that the B–H stretching E' mode of 2636 cm^{-1} somewhat couples to the e' HOMO in 1bn. Considering that the electron density on boron and hydrogen atoms in the e' HOMO is not high, such electron–phonon coupling constant for the E' mode of 2636 cm^{-1} is rather large.

Let us next look into the electron–phonon coupling in the monocations of 2bn and 3bn. We can see from Figure 8 that the B–N stretching A_1 mode of 1486 cm^{-1} can strongly couple to the HOMO in 2bn. Furthermore, the B–N stretching A_1 modes of 1480 and 1528 cm^{-1} strongly couple to the HOMO in 3bn. It should be noted that the electron–phonon coupling constants for the low frequency modes decrease with an increase in molecular size from the monocations of 1bn to 3bn more rapidly than those for the B–N stretching modes around 1500 cm^{-1}.

Let us next compare the calculated results for the monocations of B, N-substituted polyacenes with those for the monocations of polyacenes. We can see from Figure 8 that as in the monocation of 1bn, both the low frequency modes and the C–C stretching modes around 1500 cm^{-1} afford large electron–phonon coupling constants in the monocation of 1a. But in the monocations of large size of polyacenes such as 2a and 3a, only the C–C stretching modes around 1500 cm^{-1} afford large electron–phonon coupling constants, and the electron–phonon coupling constants for the C–C stretching modes around 1500 cm^{-1} decrease with an increase in molecular size from the monocations of 2a to 6a.

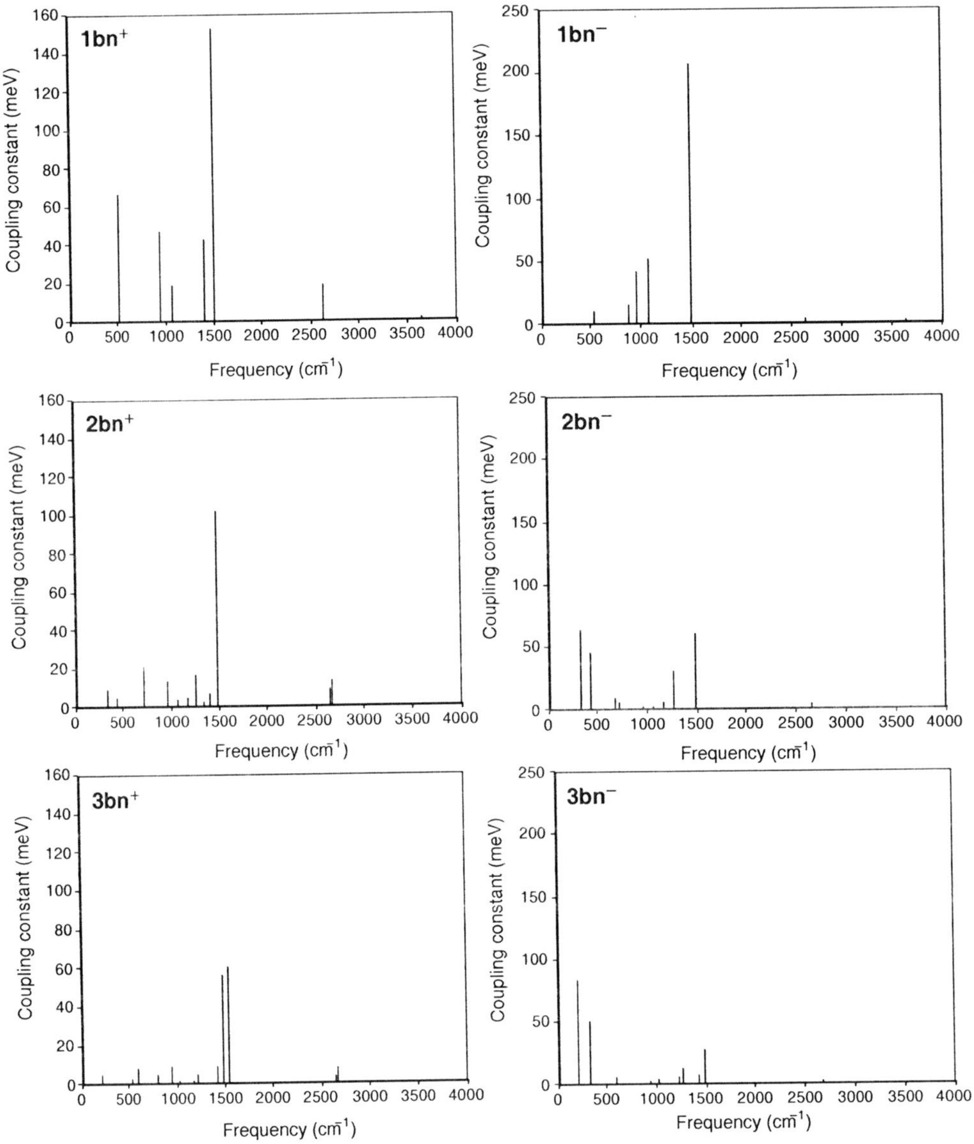

Figure 8. Electron–phonon coupling constants for the monocations and monoanions of B, N-substituted polyacenes.

2. Monoanions

We can see from Figure 8 that the B–N stretching E' mode of 1492 cm^{-1} can strongly couple to the e' LUMO in 1bn. This can be understood in view of the phase patterns of the LUMO in 1bn. When 1bn is distorted by the E' mode of 1492 cm^{-1} toward the same direction as shown in Figure 3, the bonding (antibonding) interactions between two neighboring B and N atoms in the LUMO (a) become weaker (stronger), therefore, the LUMO (a) is significantly destabilized in energy. In a similar way, the antibonding interactions between two neighboring B and N atoms in the LUMO (b) become weaker by such a distortion, therefore, the LUMO (b) is significantly stabilized in energy. That is the reason why the E' mode of 1492 cm^{-1} can strongly couple to the e' LUMO in 1bn. In addition to this mode, the E' mode of 1073 cm^{-1} and the A_1' mode of 949 cm^{-1} can also couple to the e' LUMO in 1bn. But it should be noted that the B–N stretching E' mode of 1399 cm^{-1} hardly couples to the e' LUMO in 1bn. This can be understood as follows. When 1bn is distorted by the E' mode of 1399 cm^{-1} toward the same direction as shown in Figure 3, the bonding (antibonding) interaction between N(1) and B(2) becomes stronger, while the bonding (antibonding) interaction between B(3) and N(3) becomes weaker in the LUMO (a) (LUMO (b)). Since such changes in the strengths of orbital interactions are compensated by each other, the energy levels of the LUMO (a) and LUMO (b) are not significantly changed by such a distortion. That is the reason why the B–N stretching E' mode of 1399 cm^{-1} hardly couples to the e' LUMO in 1bn. Let us next look into the electron–phonon coupling in the monoanions of 2bn and 3bn. We can see from Figure 8 that the lowest frequency A_1 modes of 327 and 447 cm^{-1} and the B–N stretching mode of 1486 cm^{-1} can strongly couple to the a_2 LUMO in 2bn. Furthermore, the lowest frequency A_1 modes of 212 and 345 cm^{-1} and the B–N stretching A_1 mode of 1480 cm^{-1} can strongly couple to the b_1 LUMO in 3bn. It should be noted that the B–N stretching modes around 1500 cm^{-1} and the low frequency modes, less and more, respectively, strongly couple to the LUMO with an increase in molecular size from 1bn to 3bn.

Let us next compare the calculated results for the monocations with those for the monoanions [74] in B, N-substituted polyacenes. As described above, the HOMO and LUMO in B, N-substituted polyacene-series are rather localized on N and B atoms, respectively, due to the electronegativity perturbation on polyacene-series. Therefore, significant vibronic interactions character differences between the monoanions and cations in B, N-substituted polyacene-series can be expected. As can be seen from Figure 8, both the low frequency modes and the B–N stretching modes around 1500 cm^{-1} afford large electron–phonon coupling constants in the monoanions and cations in B, N-substituted polyacenes. The low frequency modes play more important role in the electron–phonon interactions in the monocation than in the monoanion in 1bn. But the electron–phonon coupling constants for the low frequency modes decrease more rapidly with an increase in molecular size than those for the B–N stretching modes around 1500 cm^{-1} in the monocations of B, N-substituted polyacene-series. Therefore, the B–N stretching modes would play more important role in the electron–phonon interactions than the low frequency modes even in the large size of the monocations of B, N-substituted polyacenes. But the B–N stretching modes around 1500 cm^{-1} and the low frequency modes, less and more, respectively, strongly couple to the LUMO with an increase in molecular size from 1bn to 3bn. Therefore, the higher frequency modes would play an essential role in the electron–phonon interactions in the large size of the monocations, while the lower frequency modes play such a role in the large size of the monoanions in B, N-substituted polyacenes. This difference may come from the phase patterns difference between the HOMO and LUMO due to electronegativity perturbation on polyacene-series.

We can therefore conclude in this section that the lower frequency modes play an important role in the electron–phonon interactions in large size of the negatively charged polyacene-series and B, N-substituted polyacene-series, while the higher frequency modes such as B–N stretching modes around 1500 cm^{-1} play such a role in the positively charged polyacene-series and B, N-substituted polyacene-series.

D. Polycyanodienes

1. Monocations

The calculated electron–phonon coupling constants in the monocations of 2cn, 3cn, and 4cn are shown in Figure 9. Let us next look into the electron–phonon interactions between the a_g HOMO and the A_g modes in 2cn. We can see from Figure 9 that the A_g mode of 1597 cm^{-1} the most strongly couples to the a_g HOMO in 2cn. This can be understood in view of the phase patterns of the HOMO in 2cn. When 2cn is distorted along the A_g mode of 1597 cm^{-1} toward the same direction as shown in Figure 3, the bonding interactions between C(3) and C(4) atoms in the HOMO become weaker, and the antibonding interactions between N(2) and C(3) atoms become stronger. Furthermore, the characteristics of the bonding and antibonding interactions between two neighboring carbon and hydrogen atoms in the HOMO become less and more, respectively, significant.

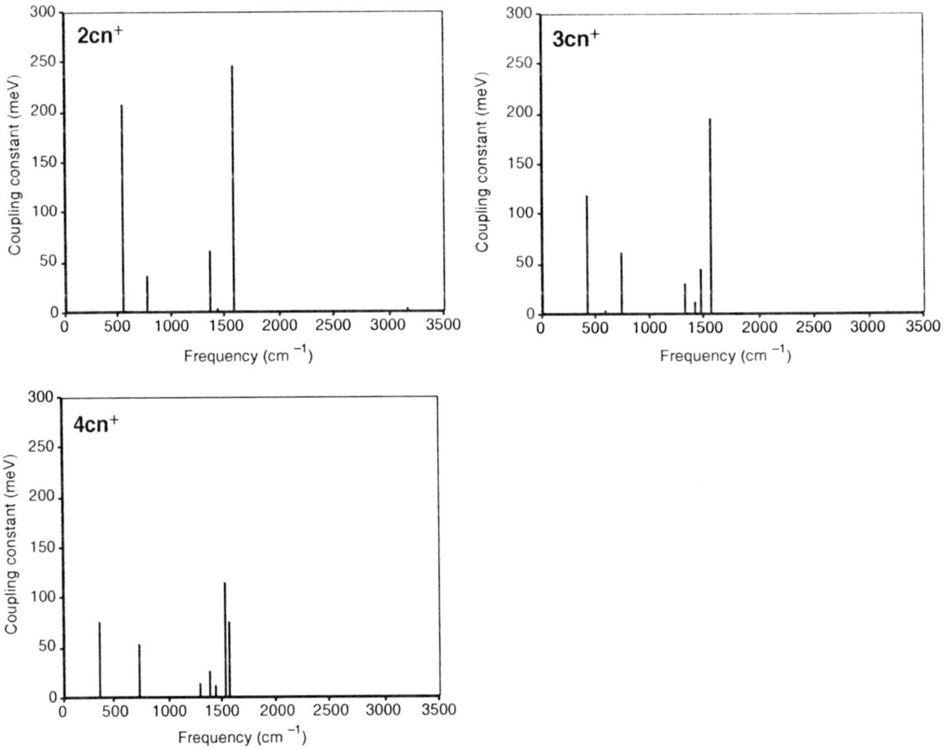

Figure 9. Electron–phonon coupling constants for the monocations of polycyanodienes.

Therefore, the a_g HOMO is significantly destabilized in energy by such a distortion. This is the reason why the A_g mode of 1597 cm^{-1} the most strongly couples to the a_g HOMO in 2cn. In addition to this, the lowest frequency A_g mode of 555 cm^{-1} affords large electron–phonon coupling constant in the monocation of 2cn. When 2cn is distorted along the A_g mode of 555 cm^{-1}, the characteristics of the antibonding interactions between N(2) and C(3) atoms become more significant, and the bonding interactions between two neighboring carbon and hydrogen atoms in the HOMO become weaker. Therefore, the a_g HOMO is significantly destabilized in energy by such a distortion. This is the reason why the A_g mode of 555 cm^{-1} affords large electron–phonon coupling constant in the monocation of 2cn. Similar discussions can be made in 3cn and 4cn; the C–C and C–N stretching modes around 1500 cm^{-1} and the lowest frequency modes strongly couple to the HOMO.

2. Monoanions

The calculated electron–phonon coupling constants in the monoanions of 1cn, 2cn, 3cn, and 4cn are shown in Figure 10. Let us first look into the electron–phonon interactions between the A_g modes and the b_{3u} LUMO in 1cn. We can see from Figure 10 that the A_g mode of 1713 cm^{-1} can the most strongly couple to the b_{3u} LUMO in 1cn. When 1cn is distorted along the A_g mode of 1713 cm^{-1} toward the same direction as shown in Figure 3, the bonding interactions between two neighboring carbon atoms in the b_{3u} LUMO become stronger, and the antibonding interactions between two neighboring carbon and nitrogen atoms become weaker, and thus the b_{3u} LUMO is significantly stabilized in energy. This is the reason why the A_g mode of 1713 cm^{-1} can the most strongly couple to the b_{3u} LUMO in 1cn. The A_g mode of 1035 cm^{-1} also strongly couples to the b_{3u} LUMO in 1cn. When 1cn is distorted along the A_g mode of 1035 cm^{-1}, the antibonding interactions between two neighboring carbon and nitrogen atoms in the b_{3u} LUMO become significantly weaker, and thus the b_{3u} LUMO is significantly stabilized in energy by such a distortion. On the other hand, the bonding interactions between two neighboring carbon atoms become also slightly weaker by such a distortion, the b_{3u} LUMO would be slightly destabilized in energy. Such stabilized and destabilized effects are compensated by each other. Therefore, the A_g mode of 1035 cm^{-1} affords slightly smaller electron–phonon coupling constant than the A_g mode of 1713 cm^{-1} in the monoanion of 1cn. Furthermore, the A_g mode of 623 cm^{-1} also somewhat strongly couples to the b_{3u} LUMO in 1cn. The reduced masses for the A_g modes of 623, 1035 and 1713 cm^{-1} are large, and are 8.80, 7.53 and 5.97, respectively, in 1cn. Therefore, the displacements of carbon and nitrogen are large in these modes. It is rational that the A_g modes of 623, 1035, and 1713 cm^{-1}, in which the displacements of carbon and nitrogen atoms are larger, can strongly couple to the b_{3u} LUMO localized on carbon and nitrogen atoms. On the other hand, the reduced masses for the A_g modes of 1252 and 3200 cm^{-1} are 1.11 and 1.10, respectively. It is rational that the A_g modes of 1252 and 3200 cm^{-1} in which the displacements of carbon and nitrogen atoms are small cannot strongly couple to the b_{3u} LUMO localized on carbon and nitrogen atoms. This is the reason why the A_g modes of 623, 1035, and 1713 cm^{-1} can more strongly couple to the b_{3u} LUMO than the A_g modes of 1252 and 3200 cm^{-1} in 1cn.

Let us look into the electron–phonon interactions between the A_g modes and the b_{1g} LUMO in 2cn. We can see from Figure 10 that the A_g mode of 1597 cm^{-1} strongly couples to the b_{1g} LUMO in 2cn. This can be understood as follows. When 2cn is distorted along the A_g mode of 1597 cm^{-1}, the bonding interactions between two neighboring carbon atoms in the b_{1g} LUMO become significantly weaker, and the antibonding interactions between two neighboring carbon and nitrogen atoms become significantly stronger, and thus the b_{1g} LUMO is significantly destabilized in energy by such a distortion in 2cn. This is the reason why the A_g mode of 1597 cm^{-1} can strongly couple to the b_{1g} LUMO in 2cn. The A_g mode of 1436 cm^{-1} can also strongly couple to the b_{1g} LUMO in 2cn. When 2cn is distorted along the

A_g mode of 1436 cm^{-1}, the antibonding interactions between two neighboring carbon and nitrogen atoms in the b_{1g} LUMO in 2cn become stronger, and thus the b_{1g} LUMO is significantly destabilized in energy by such a distortion in 2cn.

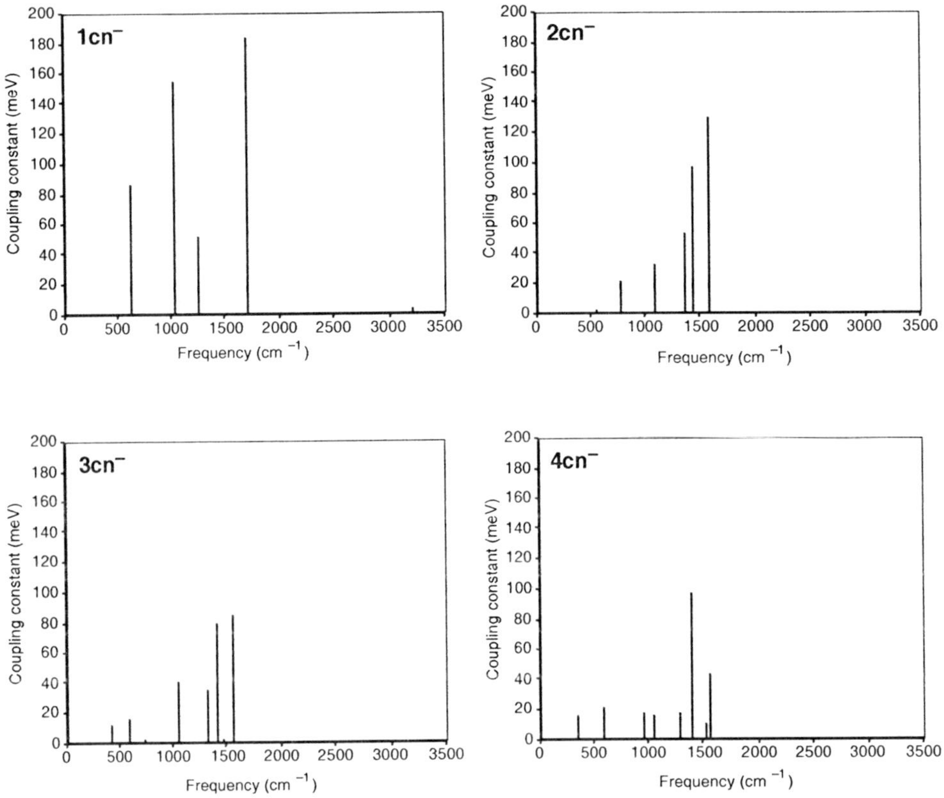

Figure 10. Electron–phonon coupling constants for the monoanions of polycyanodienes.

This is the reason why the A_g mode of 1436 cm^{-1} strongly couples to the b_{1g} LUMO in 2cn. Apart from the C–H stretching A_g mode of 3187 cm^{-1}, the electron–phonon coupling constant increases with an increase in frequency in 2cn. In the low frequency modes, which have similar characteristics to those of acoustic mode of phonon in solids, all atoms move toward the similar direction. Therefore, the orbital interactions between two neighboring atoms do not significantly change when 2cn is distorted along the low frequency A_g modes. This is the reason why the electron–phonon coupling decreases with a decrease in frequency in 2cn. The C–H stretching mode of 3187 cm^{-1} hardly couples to the b_{1g} LUMO in 2cn. The reduced mass for the A_g mode of 3187 cm^{-1} is 1.10, and in which the displacements of carbon and nitrogen atoms are very small. The b_{1g} LUMO is localized on carbon and nitrogen atoms in 2cn. It is rational that the A_g mode of 3187 cm^{-1}, in which the displacements of carbon and nitrogen atoms are very small, hardly couples to the b_{1g} LUMO localized on carbon and nitrogen atoms in 2cn. Similar discussions can be made in the monoanions of 3cn and 4cn; the C–C stretching modes around 1500 cm^{-1} can strongly couple to the LUMO.

E. Summary

In this section, we investigated the electron–phonon interactions in the monocations and monoanions of polyacenes, polyfluoroacenes, B, N-substituted polyacenes, and polycyanodienes. The C–C stretching modes around 1500 cm^{-1} strongly couple to the HOMO, and the lowest frequency modes and the C–C stretching modes around 1500 cm^{-1} strongly couple to the LUMO in polyacenes. The C–C stretching modes around 1500 cm^{-1} strongly couple to the HOMO and LUMO in polyfluoroacenes. The B–N stretching modes around 1500 cm^{-1} strongly couple to the HOMO and LUMO in B, N-substituted polyacenes. The C–C and C–N stretching modes around 1500 cm^{-1} strongly couple to the HOMO and LUMO in polycyanodienes.

VI. Total Electron–Phonon Coupling Constants

A. Polyacenes and Polyfluoroacenes

1. Monocations

Let us next discuss the total electron–phonon coupling constants in the monocations (l_{HOMO}) of polyfluoroacenes, and compare the calculated results for polyfluoroacenes with those for polyacenes. The l_{HOMO} values for *1a* and *1fa* are defined as

$$l_{\mathrm{HOMO}} = \sum_{m=1}^{6} l_{\mathrm{HOMO}}(\omega_m) = \sum_{m=1}^{4} g_{\mathrm{HOMO}}^2(\omega_m)\hbar\omega_m + \frac{1}{2}\sum_{m=5}^{6} g_{\mathrm{HOMO}}^2(\omega_m)\hbar\omega_m \tag{55}$$

and those for polyacenes and polyfluoroacenes (*2fa–5fa* and *2fa–5fa*) are defined as

$$l_{\mathrm{HOMO}} = \sum_{m} l_{\mathrm{HOMO}}(\omega_m) = \sum_{m} g_{\mathrm{HOMO}}^2(\omega_m)\hbar\omega_m \tag{56}$$

The calculated total electron–phonon coupling constants (l_{LUMO} and l_{HOMO}) in the monoanions and cations of polyacenes, polyfluoroacenes, B, N-substituted polyacenes, and polycyanodienes are shown in Figure 11. The l_{HOMO} values are estimated to be 0.418, 0.399, 0.301, 0.255, and 0.222 eV for 1fa, 2fa, 3fa, 4fa, and 5fa, respectively, and those are estimated to be 0.244, 0.173, 0.130, 0.107, and 0.094 eV for 1a, 2a, 3a, 4a, and 5a, respectively. Therefore, the l_{HOMO} values decrease with an increase in molecular size in both polyacenes and polyfluoroacenes. This can be understood as follows. The electron density per atom in the HOMO becomes lower with an increase in molecular size, and the orbital interactions between two adjacent atoms become weaker with an increase in number of atoms in polyacenes and polyfluoroacenes. Therefore, strengths of the orbital interactions between two adjacent atoms less significantly change with an increase in number of atoms when these monocations are distorted along the C–C stretching modes around 1500 cm^{-1}

playing an essential role in the electron–phonon interactions. This is the reason why the l_{HOMO} value decreases with an increase in number of atoms. Therefore, in general, we can expect that a monocation, in which number of carriers per atom is larger, affords larger l_{HOMO} value.

The l_{HOMO} values for polyfluoroacenes are much larger than those for polyacenes. This can be understood as follows. For example, let us compare the calculated results for 2fa with those for 2a. The reduced masses of the C–C stretching A_g modes of 1445 and 1610 cm^{-1} in 2a, which afford large electron–phonon coupling constants in the monocation of 2a, are 9.68 and 7.19, respectively, and thus the displacements of carbon atoms are not very large in these vibrational modes. On the other hand, the reduced masses of the C–C stretching modes of 1552 and 1666 cm^{-1} in 2fa, which afford large electron–phonon coupling constants in the monocation of 2fa, are 12.45 and 12.10, respectively, and thus the displacements of carbon atoms are very large in these vibrational modes. It is rational that the C–C stretching modes around 1500 cm^{-1}, in which the displacements of carbon atoms are larger in 2fa, can more strongly couple to the HOMO, in which the electron density on carbon atoms are very high, than the C–C stretching modes around 1500 cm^{-1}, in which the displacements of carbon atoms are smaller in 2a. Furthermore, the low frequency mode of 514 cm^{-1} affords much larger electron–phonon coupling constant in the monocation of 2fa than the low frequency modes around 500 cm^{-1} in the monocation of 2a. Similar discussion can be made in 3fa, 4fa, and 5fa.

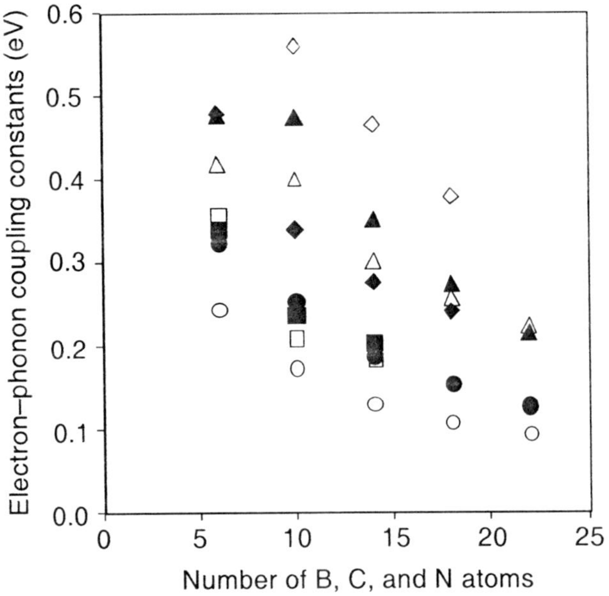

Figure 11. Total electron–phonon coupling constants for the monocations and monoanions of polyacenes, polyfluoroacenes, B, N-substituted polyacenes, and polycyanodienes. The opened circles, triangles, squares, and diamonds represent the l_{HOMO} values for polyacenes, polyfluoroacenes, B, N-substituted polyacenes, and polycyanodienes, respectively, and the closed circles, triangles, squares, and diamonds represent the l_{LUMO} values for polyacenes, polyfluoroacenes, B, N-substituted polyacenes, and polycyanodienes, respectively.

Therefore, we can conclude that the larger values of the electron–phonon coupling constants for the C–C stretching modes around 1500 cm^{-1} in the monocations of polyfluoroacenes than those in the monocations of polyacenes due to the larger displacements of carbon atoms in the C–C stretching modes in polyfluoroacenes than those in polyacenes, and the larger values of the electron–phonon coupling constants for the low frequency modes around 500 cm^{-1} in the monocations of polyfluoroacenes than those in the monocations of polyacenes, are the main reason why the l_{HOMO} values for polyfluoroacenes are much larger than those for polyacenes.

2. Monoanions

Let us next discuss the total electron–phonon coupling constants in the monoanions (l_{LUMO}) of polyfluoroacenes, and compare the calculated results for polyfluoroacenes with those for polyacenes. The l_{LUMO} value for **1fa** is defined as

$$l_{\text{LUMO}} = \sum_{m=1}^{6} l_{\text{LUMO}}(\omega_m) = \sum_{m=1}^{4} g_{\text{LUMO}}^2(\omega_m)\hbar\omega_m + \frac{1}{2}\sum_{m=5}^{6} g_{\text{LUMO}}^2(\omega_m)\hbar\omega_m \tag{57}$$

and that for polyfluoroacenes (*2fa–5fa*) is defined as

$$l_{\text{LUMO}} = \sum_{m} l_{\text{LUMO}}(\omega_m) = \sum_{m} g_{\text{LUMO}}^2(\omega_m)\hbar\omega_m \tag{58}$$

The l_{LUMO} values are estimated to be 0.475, 0.473, 0.350, 0.273, and 0.215 eV for 1fa, 2fa, 3fa, 4fa, and 5fa, respectively, while those are estimated to be 0.322, 0.255, 0.186, 0.154, and 0.127 eV for 1a, 2a, 3a, 4a, and 5a, respectively. Therefore, the l_{LUMO} values decrease with an increase in molecular size in both polyacenes and polyfluoroacenes.

The l_{LUMO} values for polyfluoroacenes are much larger than those for polyacenes. This can be understood as follows. As described above, the LUMOs of polyfluoroacenes are rather localized on carbon atoms, and those of polyacenes are completely localized on carbon atoms. And the displacements of carbon atoms in the C–C stretching modes around 1500 cm^{-1} in polyfluoroacenes become larger than those in polyacenes as a consequence of H–F substitution in polyacenes. That is, by substituting hydrogen atoms by heavier atoms which have the highest electronegativity in every element, fluorine atoms, the displacements of carbon atoms become significantly larger in the C–C stretching modes, while the electronic structure in the LUMO hardly changes. Therefore, the C–C stretching modes around 1500 cm^{-1} in polyfluoroacenes more strongly couple to the LUMO localized on carbon atoms than those in polyacenes. This is the main reason why the l_{LUMO} values for polyfluoroacenes are much larger than those for polyacenes. The l_{LUMO} value for polyfluoroacenes decreases with an increase in molecular size more rapidly than that for polyacenes, and the l_{LUMO} values difference between polyacenes and polyfluoroacenes decreases with an increase in molecular size. This is because the ratio of the number of fluorine atoms to that of carbon atoms becomes smaller from 1fa to 5fa, and the effects of H–F substitution become less important in this order.

Let us next compare the calculated results for the monocations with those for the monoanions in polyfluoroacenes. The l_{LUMO} values are larger than the l_{HOMO} values in 1fa, 2fa, 3fa, and 4fa. This can be understood as follows. The C–C stretching modes around 1500 cm^{-1}, in which the displacements of carbon atoms are very large, can more strongly couples to the LUMO rather localized on carbon atoms than to the delocalized HOMO in polyfluoroacenes. This is the reason why the electron–phonon coupling constants for the C–C stretching A_g modes around 1400 cm^{-1} in the monoanions are larger than those for the C–C stretching A_g modes around 1600 cm^{-1} in the monocations in polyfluoroacenes, and the reason why the l_{LUMO} values are larger than the l_{HOMO} values in 1fa, 2fa, 3fa, and 4fa. But it should be noted that the l_{LUMO} value decreases with an increase in molecular size more rapidly than the l_{HOMO} value in polyfluoroacenes, and the l_{LUMO} value is slightly smaller than the l_{HOMO} value in 5fa.

The l_{HOMO} and l_{LUMO} values for polyfluoroacene with D_{2h} geometry would converge and be estimated by extrapolation in Figure 11 to be 0.074 and 0.009 eV, respectively, assuming that the l_{HOMO} and l_{LUMO} values are approximately inversely proportional to the number of carbon atoms in each series.

The l_{HOMO} value is estimated to be much larger than the l_{LUMO} value in polyfluoroacene ($N \rightarrow \infty$), even though the l_{HOMO} value is smaller than the l_{LUMO} value in small size of polyfluoroacenes such as 2fa, 3fa, and 4fa. That is, the l_{HOMO} values decrease with an increase in molecular size less rapidly than the l_{LUMO} values in polyfluoroacenes. This is because the electron–phonon coupling constants originating from the C–C stretching modes around 1500 cm^{-1} decrease with an increase in molecular size more rapidly in the monoanions than in the monocations in polyfluoroacenes. The strengths of the orbital interactions between two neighboring carbon atoms in the HOMO and LUMO significantly change when polyfluoroacenes are distorted along the C–C stretching modes around 1500 cm^{-1}, and thus the C–C stretching modes around 1500 cm^{-1} can strongly couple to the HOMO and LUMO in polyfluoroacenes. Strengths of such orbital interactions in the HOMO become weaker with an increase in molecular size less rapidly than those in the LUMO in polyfluoroacenes because the electron density on carbon atoms in the LUMO rather localized on carbon atoms decrease with an increase in molecular size more rapidly than those in the delocalized HOMO in polyfluoroacenes. This is the reason why the electron–phonon coupling constants originating from the C–C stretching modes around 1500 cm^{-1} decrease with an increase in molecular size more rapidly in the monoanions than in the monocations in polyfluoroacenes, and the reason why the l_{HOMO} values decrease with an increase in molecular size less rapidly than the l_{LUMO} values in polyfluoroacenes.

The total electron–phonon coupling constants were estimated to be 0.322, 0.254, 0.186, 0.154, 0.127, and 0.106 eV (0.244, 0.173, 0.130, 0.107, 0.094, and 0.079 eV) in the monoanions (monocations) of 1a, 2a, 3a, 4a, 5a, and 6a, respectively. Therefore, the electron–phonon coupling constants decrease with an increase in molecular size. Figures 4 and 5 demonstrate that the electron–phonon coupling originating from the C–C stretching modes of 1400–1600 cm^{-1} becomes weaker as the molecular size becomes larger from 1a to 6a. Thus, the C–C stretching modes should play an important role in the electron–phonon coupling in the monoanions and cations of 1a and polyacenes. The total electron–phonon coupling constants in the monocations of polyacenes (l_{HOMO}) are smaller than those in the monoanions (l_{LUMO}). This is because the lowest frequency mode, which plays an important role in the electron–phonon coupling in the monoanions of polyacenes, does not play a role in the electron–phonon coupling in the monocations.

B. B, N-Substituted Polyacenes

The l_{HOMO} for *1bn* is defined as

$$l_{HOMO} = \sum_{m=1}^{11} l_{HOMO}(\omega_m) = \sum_{m=1}^{7} g_{HOMO}^2(\omega_m)\hbar\omega_m + \frac{1}{2}\sum_{m=8}^{11} g_{HOMO}^2(\omega_m)\hbar\omega_m \tag{59}$$

That for 2bn is defined as

$$l_{HOMO} = \sum_{m=1}^{17} l_{HOMO}(\omega_m) = \sum_{m=1}^{17} g_{HOMO}^2(\omega_m)\hbar\omega_m \tag{60}$$

and that for 3bn is defined as

$$l_{HOMO} = \sum_{m=1}^{23} l_{HOMO}(\omega_m) = \sum_{m=1}^{23} g_{HOMO}^2(\omega_m)\hbar\omega_m \tag{61}$$

The l_{LUMO} for *1bn* is defined as

$$l_{LUMO} = \sum_{m=1}^{11} l_{LUMO}(\omega_m) = \sum_{m=1}^{7} g_{LUMO}^2(\omega_m)\hbar\omega_m + \frac{1}{2}\sum_{m=8}^{11} g_{LUMO}^2(\omega_m)\hbar\omega_m \tag{62}$$

That for 2bn is defined as

$$l_{LUMO} = \sum_{m=1}^{17} l_{LUMO}(\omega_m) = \sum_{m=1}^{17} g_{LUMO}^2(\omega_m)\hbar\omega_m \tag{63}$$

and that for 3bn is defined as

$$l_{LUMO} = \sum_{m=1}^{23} l_{LUMO}(\omega_m) = \sum_{m=1}^{23} g_{LUMO}^2(\omega_m)\hbar\omega_m \tag{64}$$

The l_{HOMO} (l_{LUMO}) values are estimated to be 0.357, 0.209, and 0.182 eV (0.340, 0.237, and 0.203 eV) for 1bn, 2bn, and 3bn, respectively, while those are estimated to be 0.244, 0.173, and 0.130 eV (0.322, 0.254, and 0.186 eV) for 1a, 2a, and 3a, respectively. Therefore, both the l_{HOMO} and l_{LUMO} values decrease with an increase in molecular size in B, N-substituted polyacene-series as well as in polyacene-series.

The l_{LUMO} values are much larger than the l_{HOMO} values in polyacene-series, while the l_{HOMO} values are similar to the l_{LUMO} values in B, N-substituted polyacene-series. This can be understood as follows. The low frequency modes as well as the C–C stretching modes around 1500 cm^{-1} afford electron–phonon coupling constants in the monoanions of polyacenes, while only the C–C stretching modes around 1500 cm^{-1} afford large electron–phonon coupling constants in the monocations of polyacenes except for 1a. Furthermore, the C–C stretching modes around 1500 cm^{-1} and the low frequency modes, less and more, respectively, strongly couple to the LUMO, while the C–C stretching modes around 1500 cm^{-1} less strongly couple to the HOMO with an increase in molecular size in polyacene-series. The electron–phonon coupling originating from the low frequency modes in the monoanions is the main reason why the l_{LUMO} values are much larger than the l_{HOMO} values in polyacene-series. Let us next look into B, N-substituted polyacene-series. The B–N stretching modes around 1500 cm^{-1} as well as the low frequency modes afford large electron–phonon coupling constants both in the monoanions and cations of B, N-substituted polyacene-series. The B–N stretching modes around 1500 cm^{-1} and the low frequency modes, less and more, respectively, strongly couple to the LUMO with an increase in molecular size, while the low frequency modes much less strongly couple to the HOMO than the B–N stretching modes around 1500 cm^{-1} with an increase in molecular size in B, N-substituted polyacene-series. Even though the low and high frequency modes afford large electron–phonon coupling constants in the monoanions and cations in B, N-substituted polyacene-series, respectively, there are not significant differences between the l_{LUMO} and l_{HOMO} values. This is the reason why the l_{LUMO} values are similar to the l_{HOMO} values in B, N-substituted polyacene-series.

We can see from Figure 11 that the l_{HOMO} values for B, N-substituted polyacenes are larger than those for polyacenes, and can expect that those for the large size of B, N-substituted polyacene-series are larger than those for the large size of polyacene-series. The l_{HOMO} values for B, N-substituted polyacene ($N \to \infty$) and polyacene ($N \to \infty$) are estimated to be 0.096 and 0.036 eV, respectively, assuming that the l_{HOMO} is approximately inversely proportional to the number of C, B, and N atoms in each series.[65] This can be understood as follows. For example, let us compare the HOMO of 3bn with that of 3a. We can see from Figure 3 that the HOMO of 3a is rather localized on C(1), C(3), C(5), C(8), C(10) and C(12) atoms which are located at the edge of the carbon framework, and the electron density on other carbon atoms is very low. Therefore, the HOMO of 3a has a non-bonding character and the orbital interactions between two neighboring carbon atoms are not so large. Such a non-bonding character in the HOMO becomes more significant with an increase in molecular size in polyacene-series. This is the reason why the l_{HOMO} values decrease with an increase in molecular size in polyacene-series. Let us next look into 3bn. Because of the electronegativity perturbation on 3a, the HOMO of 3bn is stabilized in energy with respect to the HOMO of 3a, and the electron density on N(7), N(9), N(11), and N(13) atoms in 3bn is higher than that on C(7), C(9), C(11), and C(13) atoms in 3a, respectively. Therefore, orbital interactions between N(7) and B(8), between B(8) and N(9), between N(9) and B(10), between B(10) and N(11), between N(11) and B(12), and between B(12) and N(13) atoms in the HOMO of 3bn, are stronger than those between two neighboring carbon atoms in the HOMO of 3a. This is the reason why the l_{HOMO} value for 3bn is larger than that for 3a. In a similar way, the large size of B, N-substituted polyacene-series would have larger l_{HOMO} values than the large size of polyacene-series. We can therefore conclude that because of the strong orbital interactions between B and N atoms in the HOMO in B, N-substituted polyacene-series as a consequence of the electronegativity perturbation on polyacene-series, the l_{HOMO} values for B, N-substituted polyacene-series would become larger than those for polyacene-series.

C. Polycyanodienes

1. Monocations

Let us next discuss the total electron–phonon coupling constants in the monocations (l_{HOMO}) of polycyanodienes, and compare the calculated results for polycyanodienes with those for polyacenes. The l_{HOMO} values for polycyanodienes are defined as

$$l_{\text{HOMO}} = \sum_{m} l_{\text{HOMO}}(\omega_m) = \sum_{m} g_{\text{HOMO}}^{2}(\omega_m)\hbar\omega_m \tag{65}$$

The l_{HOMO} values are estimated to be 0.558, 0.464, and 0.378 eV for 2cn, 3cn, and 4cn, respectively. Those were estimated to be 0.173, 0.130, and 0.107 eV for 2a, 3a, and 4a, respectively. Therefore, the l_{HOMO} values decrease with an increase in molecular size in the monocations of polycyanodienes as well as in the monocations of polyacenes. The l_{HOMO} values for polycyanodienes are much larger than those for polyacenes. This can be understood as follows. The HOMO of polyacenes is localized on edge part of carbon atoms, and thus the nonbonding characteristics are significant in the HOMO of polyacenes. Therefore, the energy levels of the HOMO in polyacenes do not significantly change even if polyacenes are distorted along the C–C stretching modes around 1500 cm^{-1} playing an essential role in the electron–phonon interactions. As in case of the HOMO of polyacenes, the HOMO of polycyanodienes are also somewhat localized on edge part of nitrogen atoms. On the other hand, the σ orbital interactions as well as the π orbital interactions are significant in the HOMO of polycyanodienes. Therefore, orbital interactions between two neighboring atoms are much stronger in the HOMO of polycyanodienes than in the HOMO of polyacenes. Furthermore, while the HOMO of polyacenes are completely localized on carbon atoms, the HOMO of polycyanodienes are delocalized, and the electron density on hydrogen atoms as well as that on carbon and nitrogen atoms in the HOMO of polycyanodienes is high. Therefore, the strengths of the orbital interactions between two neighboring carbon and hydrogen atoms are significantly changed by distortions by vibronic active modes. This is the reason why the l_{HOMO} values for σ-conjugated polycyanodienes are much larger than those for π-conjugated polyacenes. In summary, the strong σ orbital interactions between two neighboring carbon atoms and two neighboring carbon and nitrogen atoms and the orbital interactions between two neighboring carbon and hydrogen atoms in the HOMO of polycyanodienes are the main reason why the l_{HOMO} values for σ-conjugated polycyanodienes are much larger than those for π-conjugated polyacenes.

Let us next estimate the l_{HOMO} values for polycyanodienes ($N \to \infty$) with D_{2h} geometry. Assuming that the l_{HOMO} values for polycyanodienes with D_{2h} geometry are approximately inversely proportional to the number of carbon and nitrogen atoms in each series, as suggested in previous research, [21] the l_{HOMO} value for polycyanodiene ($N \to \infty$) is estimated to be 0.167 eV. The l_{HOMO} value for polyacene ($N \to \infty$) with D_{2h} geometry is estimated to be 0.027 eV. Therefore, the l_{HOMO} values for polycyanodienes ($N \to \infty$) are estimated to be much larger than those for polyacenes ($N \to \infty$). Strong σ orbital interactions between two neighboring atoms and the orbital interactions between two neighboring carbon and hydrogen atoms in the HOMO of polycyanodienes are the main reason why the l_{HOMO} values for σ-conjugated polycyanodienes are much larger than those for π-conjugated polyacenes.

2. Monoanions

Let us next discuss the total electron–phonon coupling constants in the monoanions (l_{LUMO}) of polycyanodienes, and compare the calculated results for polycyanodienes with those for polyacenes. The l_{LUMO} values for polycyanodienes are defined as

$$l_{\mathrm{LUMO}} = \sum_m l_{\mathrm{LUMO}}(\omega_m) = \sum_m g_{\mathrm{LUMO}}^2(\omega_m)\hbar\omega_m, \tag{66}$$

The l_{LUMO} values are estimated to be 0.478, 0.338, 0.275, and 0.240 eV for 1cn, 2cn, 3cn, and 4cn, respectively, while those are estimated to be 0.322, 0.255, 0.186, and 0.154 eV for 1a, 2a, 3a, and 4a, respectively, and those are estimated to be 0.340, 0.237, and 0.203 eV for 1bn, 2bn, and 3bn, respectively. Therefore, the l_{LUMO} values decrease with an increase in molecular size in polyacenes, B, N-substituted polyacenes, and polycyanodienes.

The l_{LUMO} values for polycyanodienes are much larger than those for polyacenes. This can be understood in view of the phase patterns of the LUMO in polyacenes and polycyanodienes. The LUMOs of polyacenes are rather localized on carbon atoms located at the edge part of carbon framework. Therefore, non-bonding characteristic is significant in the LUMOs in polyacenes. On the other hand, the LUMO of polycyanodienes is delocalized, and the electron density on carbon atoms as well as on nitrogen atoms located at the edge part of CN framework is high in the LUMO in polycyanodienes. Therefore, the orbital interactions between two neighboring atoms in the LUMO in polycyanodienes are stronger than those in the LUMO in polyacenes. This is the reason why the l_{LUMO} values for polycyanodienes are much larger than those for polyacenes. Let us next discuss why the LUMOs of polyacenes are rather localized on carbon atoms located at the edge part of carbon framework, while those of polycyanodienes are delocalized, and the electron density on carbon atoms as well as on nitrogen atoms is high in the LUMO in polycyanodienes. In general, due to electronegativity perturbation, [30] the π bonding orbitals are weighted more heavily on the atoms which are more electronegative, and the π antibonding orbitals weighted more heavily on the atoms which is less electronegative. As described above, the LUMO are rather localized on carbon atoms which are located at the edge part of carbon framework in polyacenes. In polycyanodienes, carbon atoms with high electron density in the LUMO, located at the edge part of carbon framework of polyacenes, are substituted by nitrogen atoms with the higher electronegativity. Therefore, the electron density on nitrogen atoms in the LUMO in polycyanodienes is lower than that on carbon atoms located at the edge part of carbon framework in the LUMO of polyacenes, and the electron density on carbon atoms which are not located at the edge part of CN framework in the LUMO in polycyanodienes is higher than that on carbon atoms which are not located at the edge part of carbon framework in the LUMO in polyacenes. This is the reason why the LUMOs of polyacenes are rather localized on carbon atoms located at the edge part of carbon framework, while those of polycyanodienes are delocalized, and the electron density on carbon atoms as well as on nitrogen atoms is high in polycyanodienes.

Let us next estimate the l_{LUMO} values for polycyanodienes ($N \rightarrow \infty$) with D_{2h} geometry. Assuming that the l_{LUMO} values for polycyanodienes with D_{2h} geometry are approximately inversely proportional to the number of carbon and nitrogen atoms in each series, as suggested before, the l_{LUMO} value for polycyanodiene ($N \rightarrow \infty$) is estimated to be 0.122 eV. The l_{LUMO} value for polyacene ($N \rightarrow \infty$) with D_{2h} geometry was estimated to be 0.019 eV.

In summary, the l_{LUMO} values for polycyanodienes are estimated to be much larger than those for polyacenes. Therefore, the orbital patterns difference between the LUMO localized on carbon atoms located at the edge part of carbon framework in polyacenes and the delocalized LUMO in polycyanodienes due to electronegativity perturbation, is the main reason why the l_{LUMO} values for polycyanodienes are much larger than those for polyacenes with D_{2h} geometry.

D. Summary

In this section, we estimated the total electron–phonon coupling constants in the monocations and monoanions of polyacenes, polyfluoroacenes, B, N-substituted polyacenes, and polycyanodienes. The l_{HOMO} and l_{LUMO} values decrease with an increase in molecular size in polyacenes, polyfluoroacenes, B, N-substituted polyacenes, and polycyanodienes. Electron density per atom in the HOMO becomes lower with an increase in molecular size, and the orbital interactions between two adjacent atoms become weaker with an increase in number of atoms in polyacenes, polyfluoroacenes, B, N-substituted polyacenes, and polycyanodienes. Therefore, strengths of the orbital interactions between two adjacent atoms less significantly change with an increase in number of atoms when the monocations and monoanions are distorted along the C–C stretching modes around 1500 cm^{-1} playing an essential role in the electron–phonon interactions. This is the reason why the l_{HOMO} value decreases with an increase in number of atoms. Therefore, in general, we can expect that monocations and monoanions, in which number of carriers per atom is larger, affords larger l_{HOMO} value. The l_{HOMO} and l_{LUMO} values for polyfluoroacenes are larger than those for polyacenes. Larger values of the electron–phonon coupling constants for the C–C stretching modes around 1500 cm^{-1} in the monocations and monoanions of polyfluoroacenes than those in the monocations and monoanions of polyacenes due to the larger displacements of carbon atoms in the C–C stretching modes in polyfluoroacenes than those in polyacenes, are main reason why the l_{HOMO} and l_{LUMO} values for polyfluoroacenes are much larger than those for polyacenes. The l_{HOMO} values for B, N-substituted polyacenes are larger than those for polyacenes. The l_{LUMO} values for B, N-substituted polyacenes are similar to those for polyacenes. The l_{LUMO} values are larger than the l_{HOMO} values in polyacenes, polyfluoroacenes, and B, N-substituted polyacenes. Therefore, electron doping rather than hole doping is more effective way to seek for large electron–phonon coupling constants in polyacenes, polyfluoroacenes, and B, N-substituted polyacenes. Both the l_{HOMO} and l_{LUMO} values for polycyanodienes are larger than those for polyacenes. Therefore, CH–CF, CC–BN, and CC–CN atomic substitutions are effective way to seek for large l_{HOMO} and l_{LUMO} values in polyacenes. The orbital patterns difference between the LUMO localized on carbon atoms located at the edge part of carbon framework in polyacenes and the delocalized LUMO in polycyanodienes due to electronegativity perturbation, is the main reason why the l_{LUMO} values for polycyanodienes are much larger than those for polyacenes. Strong σ orbital interactions between two neighboring carbon atoms and two neighboring carbon and nitrogen atoms and the orbital interactions between two neighboring carbon and hydrogen atoms in the HOMO of polycyanodienes are the main reason why the l_{HOMO} values for σ-conjugated polycyanodienes are much larger than those for π-conjugated polyacenes. The l_{HOMO} values are larger than the l_{LUMO} values in polycyanodienes. Strong σ orbital interactions between two neighboring carbon atoms and two neighboring carbon and nitrogen atoms and the orbital interactions between two neighboring carbon and hydrogen atoms in the HOMO and the weak π orbital interactions between two neighboring carbon atoms and two neighboring carbon and nitrogen atoms in the LUMO of polycyanodienes, are the main reason why the l_{HOMO} values are larger than the l_{LUMO} values in polycyanodienes.

VII. The Logarithmically Averaged Phonon Frequencies

Let us next look into the logarithmically averaged phonon frequencies ω_{ln} for the monocations ($\omega_{ln,HOMO}$) and monoanions ($\omega_{ln,LUMO}$) of polyacenes, polyfluoroacenes, B, N-substituted polyacenes, and polycyanodienes which measure the frequency of the vibrational modes which play an important role in the electron–phonon interactions. The $\omega_{ln,HOMO}$ values are defined by

$$\omega_{\ln,\mathrm{HOMO}} = \exp\left\{\sum_m \frac{l_{\mathrm{HOMO}}(\omega_m)\ln\omega_m}{l_{\mathrm{HOMO}}}\right\} \tag{67}$$

and the $\omega_{\ln,\mathrm{LUMO}}$ values are defined by

$$\omega_{\ln,\mathrm{LUMO}} = \exp\left\{\sum_m \frac{l_{\mathrm{LUMO}}(\omega_m)\ln\omega_m}{l_{\mathrm{LUMO}}}\right\} \tag{68}$$

The logarithmically averaged phonon frequencies for the monoanions ($\omega_{\ln,\mathrm{LUMO}}$) and cations of polyacenes, polyfluoroacenes, B, N-substituted polyacenes, and polycyanodienes as a function of molecular weight M_{w} are shown in Figure 12.

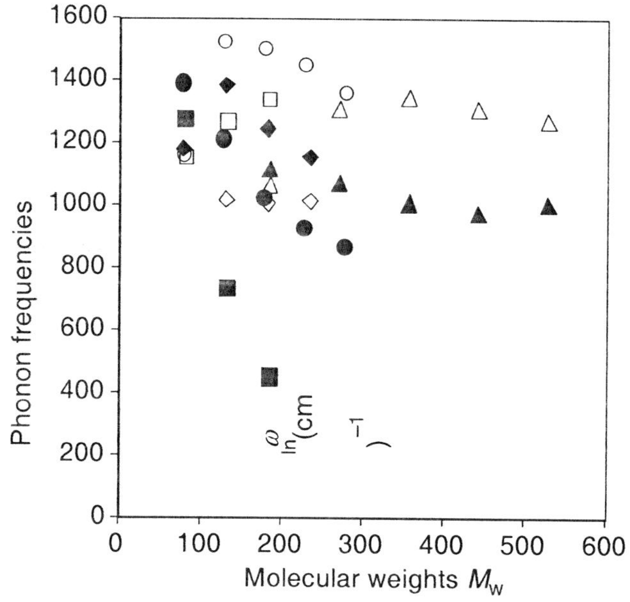

Figure 12. Logarithmically averaged phonon frequencies versus molecular weights in the monocations and monoanions of polyacenes, polyfluoroacenes, B, N-substituted polyacenes, and polycyanodienes. The opened circles, triangles, squares, and diamonds represent the $\omega_{\ln,\mathrm{HOMO}}$ values for polyacenes, polyfluoroacenes, B, N-substituted polyacenes, and polycyanodienes, respectively, and the closed circles, triangles, squares, and diamonds represent the $\omega_{\ln,\mathrm{LUMO}}$ values for polyacenes, polyfluoroacenes, B, N-substituted polyacenes, and polycyanodienes, respectively.

A. Polyacenes and Polyfluoroacenes

1. Monocations

The $\omega_{\ln,\mathrm{HOMO}}$ values are estimated to be 1060, 1303, 1341, 1305, and 1270 cm^{-1} for 1fa, 2fa, 3fa, 4fa, and 5fa, respectively, and those are estimated to be 1164, 1521, 1501, 1450, and 1359 cm^{-1} for 1a, 2a, 3a, 4a, and 5a, respectively. Therefore, the $\omega_{ln,HOMO}$ values decrease with an increase in molecular size in polyacenes (from 2a to 5a) and polyfluoroacenes (from 2fa to 5fa). This is in qualitatively agreement with a tendency in conventional superconductors; light masses will lead to higher values of ω_{ln}.

The $\omega_{ln,HOMO}$ values for polyfluoroacenes become smaller than those for polyacenes by H–F substitution in polyacenes. This can be understood as follows. The frequencies of the vibronic active modes in polyfluoroacenes are much smaller than those in polyacenes, as expected. Furthermore, the low frequency mode around 500 cm^{-1} as well as the C–C stretching modes around 1500 cm^{-1} can strongly couple to the HOMO in polyfluoroacenes, while only the C–C stretching modes around 1500 cm^{-1} can strongly couple to the HOMO in polyacenes. This is the reason why the $\omega_{ln,HOMO}$ values for polyacenes are larger than those for polyfluoroacenes.

2. Monoanions

Let us next look into the logarithmically averaged phonon frequencies ω_{ln} for the monoanions of polyfluoroacenes. The $\omega_{ln,LUMO}$ values are estimated to be 1112, 1070, 1005, 972, and 1003 cm^{-1} for 1fa, 2fa, 3fa, 4fa, and 5fa, respectively, and those are estimated to be 1390, 1212, 1023, 926, and 869 cm^{-1} for 1a, 2a, 3a, 4a, and 5a, respectively. Therefore, the $\omega_{ln,LUMO}$ values for polyacenes and polyfluoroacenes decrease with an increase in molecular size. This is in qualitatively agreement with a tendency in conventional superconductors; light mass will lead to higher values of ω_{ln}. But it should be noted that the $\omega_{ln,LUMO}$ values for polyfluoroacenes decrease with an increase in molecular size less rapidly than those for polyacenes. This can be understood as follows. In the monoanions of polyacenes, the high frequency modes and the low frequency modes, less and more, respectively, strongly couple to the LUMO with an increase in molecular size, and thus, the $\omega_{ln,LUMO}$ values significantly decrease with an increase in molecular size. In the monoanions of polyfluoroacenes, on the other hand, the low frequency modes as well as the high frequency modes less strongly couple to the LUMO with an increase in molecular size, and thus, the $\omega_{ln,LUMO}$ values slightly decrease with an increase in molecular size. The $\omega_{ln,LUMO}$ values for 1fa, 2fa, and 3fa become smaller than those for 1a, 2a, and 3a, respectively, by H–F substitution in polyacenes. However, considering the large M_w values difference between 1fa ($M_w = 186$) and 1a ($M_w = 78$), between 2fa ($M_w = 272$) and 2a ($M_w = 128$), and between 3fa ($M_w = 358$) and 3a ($M_w = 178$), the $\omega_{ln,LUMO}$ values difference between polyacenes and polyfluoroacenes is not large. This can be understood as follows. Even though the frequencies of all vibronic active modes in polyfluoroacenes are much smaller than those in polyacenes, the displacements of hydrogen and fluorine atoms do not play an essential role in the electron–phonon interactions in the monoanions of polyacenes and polyfluoroacenes, respectively, because the LUMOs are localized on carbon atoms and the electron density on hydrogen and fluorine atoms are very small in polyacenes and polyfluoroacenes, respectively. This is the reason why the $\omega_{ln,LUMO}$ values difference between polyacenes and polyfluoroacenes is not large even though the M_w values difference between them is large. Furthermore, the $\omega_{\mathrm{ln,LUMO}}$ values for 4fa and 5fa are larger than those for 4a and 5a even though the M_{w} values for 4fa ($M_{\mathrm{w}} = 444$) and 5fa ($M_{\mathrm{w}} = 530$) are much larger than those for 4a ($M_{\mathrm{w}} = 228$) and 5a ($M_{\mathrm{w}} = 278$), respectively. This is because the low frequency modes as well as the C–C stretching modes around 1500 cm^{-1} strongly couple to the LUMO in 4a and 5a, while the C–C stretching modes around 1500 cm^{-1} much more strongly couple to the LUMO than the low frequency modes in 4fa and 5fa.

We can expect that in the hydrocarbon molecular systems, the ω_{ln} values would basically decrease by substituting hydrogen atoms by heavier atoms. This can be understood from the fact that the frequencies of all vibronic active modes in polyacenes downshift by H–F substitution. However, considering that the ω_{ln} value for the LUMO rather localized on carbon atoms in 4fa and 5fa becomes larger by H–F substitution, we can expect that the ω_{ln} value for a molecular orbital localized on carbon atoms has a possibility to increase by substituting hydrogen atoms by heavier atoms if the orbital patterns of the molecular orbital do not significantly change by such substitution. Therefore, the detailed properties of the vibrational modes and the electronic structures as well as the molecular weights are closely related to the frequencies of the vibrational modes which play an important role in the electron–phonon interactions in the monoanions of polyfluoroacenes.

The $\omega_{\mathrm{ln,HOMO}}$ values are larger than the $\omega_{\mathrm{ln,LUMO}}$ values in polyfluoroacenes. This can be understood as follows. The frequency modes lower than 500 cm^{-1} and the high frequency modes around 1400 cm^{-1} much more strongly couple to the LUMO than to the HOMO in 2fa, 3fa, 4fa, and 5fa, while the frequency modes around 500 cm^{-1} and the frequency modes around 1600 cm^{-1} much more strongly couple to the HOMO than to the LUMO in 2fa, 3fa, 4fa, and 5fa. In particular, the C–C stretching modes around 1600 cm^{-1} the most strongly couple to the HOMO, while the C–C stretching modes around 1400 cm^{-1} the most strongly couple to the LUMO in 2fa, 3fa, 4fa, and 5fa. This is the reason why the $\omega_{\mathrm{ln,HOMO}}$ values are larger than the $\omega_{\mathrm{ln,LUMO}}$ values in polyfluoroacenes. As described above, the significant phase patterns difference between the HOMO and the LUMO is the main reason why the vibrational modes which play an essential role in the electron–phonon interactions in the monocations are significantly different from those in the monoanions in 2fa, 3fa, 4fa, and 5fa. Therefore, the detailed properties of the vibrational modes and the electronic structures as well as the molecular weights are closely related to the frequencies of the vibrational modes which play an important role in the electron–phonon interactions in charged polyfluoroacenes.

B. B, N-Substituted Polyacenes

The $\omega_{\mathrm{ln,HOMO}}$ ($\omega_{\mathrm{ln,LUMO}}$) values are estimated to be 1154, 1268, and 1337 cm^{-1} (1273, 737, and 449 cm^{-1}) for 1bn, 2bn, and 3bn, respectively. Therefore, the $\omega_{\mathrm{ln,HOMO}}$ and $\omega_{\mathrm{ln,LUMO}}$ values, increase and significantly decrease, respectively, with an increase in molecular size from 1bn to 3bn. As in polyacene-series, the $\omega_{\mathrm{ln,HOMO}}$ values are larger than the $\omega_{\ln,\mathrm{LUMO}}$ values in large size of B, N-substituted polyacene-series. And as in 1a, the $\omega_{\ln,\mathrm{LUMO}}$ value is slightly larger than the $\omega_{\ln,\mathrm{HOMO}}$ value in 1bn.

Let us next compare the calculated results for B, N-substituted polyacene-series with those for polyacene-series. The differences between the $\omega_{\ln,\mathrm{HOMO}}$ and $\omega_{\ln,\mathrm{LUMO}}$ values become larger with an increase in molecular size in B, N-substituted polyacene-series, while those in polyacene-series hardly change. This can be understood as follows. Let us first look into polyacene-series. [64] The C–C stretching modes around 1500 cm^{-1} and the low frequency modes strongly couple to the LUMO, and less and more, respectively, strongly couple to the LUMO with an increase in molecular size in polyacene-series. On the other hand, the C–C stretching modes around 1500 cm^{-1} strongly couple to the HOMO, and less strongly couple to the HOMO with an increase in molecular size in polyacene-series. Therefore, both the $\omega_{\ln,\mathrm{LUMO}}$ and $\omega_{\ln,\mathrm{HOMO}}$ values decrease with an increase in molecular size, and the difference between the $\omega_{\ln,\mathrm{HOMO}}$ and $\omega_{\ln,\mathrm{LUMO}}$ values hardly changes in polyacene-series. Let us next look into B, N-substituted polyacene-series. The B–N stretching modes around 1500 cm^{-1} and the low frequency modes strongly couple to the LUMO, and less and more, respectively, strongly couple to the LUMO with an increase in molecular size in B, N-substituted polyacene-series. On the other hand, the B–N stretching modes around 1500 cm^{-1} and the low frequency modes strongly couple to the HOMO, and the low frequency modes much less strongly couple to the HOMO than the B–N stretching modes around 1500 cm^{-1} with an increase in molecular size in B, N-substituted polyacene-series. This is the reason why the $\omega_{\ln,\mathrm{HOMO}}$ and $\omega_{\ln,\mathrm{LUMO}}$ values, increase and significantly decrease, respectively, with an increase in molecular size from 1bn to 3bn, and the reason why the differences between the $\omega_{\ln,\mathrm{HOMO}}$ and $\omega_{\ln,\mathrm{LUMO}}$ values significantly increase with an increase in molecular size in B, N-substituted polyacene-series. Such different properties in the $\omega_{\ln,\mathrm{HOMO}}$ and $\omega_{\ln,\mathrm{LUMO}}$ values between polyacene-series and B, N-substituted polyacene-series may come from the electronegativity perturbation on polyacene-series.

The $\omega_{\ln,\mathrm{HOMO}}$ values for polyacene-series are larger than those for B, N-substituted polyacene-series. This is because only the C–C stretching modes around 1500 cm^{-1} play an essential role in the electron–phonon interactions in polyacene-series, while the low frequency modes as well as the B–N stretching modes around 1500 cm^{-1} play an important role in the electron–phonon interactions in B, N-substituted polyacene-series. The results for the $\omega_{\ln,\mathrm{HOMO}}$ values in B, N-substituted polyacene-series are not in agreement with a tendency in solids, as described above. That is, the electronic structures rather than the molecular weights are closely related to the frequencies of the vibrational modes which play an important role in the electron–phonon interactions in the monocations of B, N-substituted polyacene-series. The $\omega_{\ln,\mathrm{HOMO}}$ value increases with an increase in molecular size from 1a to 2a, but the $\omega_{\ln,\mathrm{HOMO}}$ value slightly decreases with an increase in molecular size from 2a to 6a. On the other hand, the $\omega_{\ln,\mathrm{HOMO}}$ values for B, N-substituted polyacene-series increase with an increase in molecular size from 1bn to 3bn. But from analogy with the calculated results for polyacene-series, the $\omega_{\ln,\mathrm{HOMO}}$ values may begin to decrease at the finite size of B, N-substituted polyacene-series, according to the general tendency in solids; light mass will lead to higher values of $\omega_{\ln}$.

C. Polycyanodienes

The $\omega_{\ln,HOMO}$ values are estimated to be 1015, 1002, and 1009 cm^{-1} in 2cn, 3cn, and 4cn, respectively. Those are estimated to be 1521, 1501, and 1450 cm^{-1} in 2a, 3a, and 4a, respectively. Therefore, the $\omega_{\ln,HOMO}$ values do not significantly change with an increase in molecular size in polycyanodienes. Furthermore, the $\omega_{\ln,HOMO}$ values for polycyanodienes are smaller than those for polyacenes.

The $\omega_{\ln,LUMO}$ values are estimated to be 1179, 1381, 1242, and 1150 cm^{-1} for 1cn, 2cn, 3cn, and 4cn, respectively, and those are estimated to be 1390, 1212, 1023, and 926 cm^{-1} for 1a, 2a, 3a, and 4a, respectively. Therefore, apart from 1cn, the $\omega_{\ln,LUMO}$ values for polyacenes and polycyanodienes decrease with an increase in molecular weights. This is in qualitative agreement with a tendency; light mass will lead to higher values of $\omega_{\ln}$. But it should be noted that the $\omega_{\ln,LUMO}$ values for polycyanodienes with D_{2h} geometry are larger than those for polyacenes with D_{2h} geometry. As described above, the LUMO of polyacenes is rather localized on carbon atoms located at the edge part of carbon framework in polyacenes, and thus the non-bonding characteristics are significant. Therefore, the orbital interactions between two neighboring carbon atoms are weak. On the other hand, due to electronegativity perturbation, the LUMO of polycyanodienes is delocalized, and the electron density on carbon atoms as well as on nitrogen atoms located at the edge part of CN framework is high, and thus the orbital interactions between two neighboring carbon and nitrogen atoms and between two neighboring carbon atoms in the LUMO are strong. This is the reason why the C–N and C–C stretching modes around 1500 cm^{-1} in the monoanions of polycyanodienes afford larger electron–phonon coupling constants than the C–C stretching modes around 1500 cm^{-1} in the monoanions of polyacenes, and the reason why the $\omega_{\ln,LUMO}$ values for polycyanodienes are larger than those for polyacenes. Therefore, the orbital patterns difference between the delocalized LUMO in polycyanodienes and the LUMO localized on carbon atoms located at the edge part of carbon framework in polyacenes is the main reason why the l_{LUMO} and $\omega_{\ln,LUMO}$ for polycyanodienes are larger than those for polyacenes.

D. Summary

In this section, we investigated the logarithmically averaged phonon frequencies which measure the frequencies of the vibronic active modes playing an essential role in the electron–phonon interactions in polyacenes, polyfluoroacenes, B, N-substituted polyacenes, and polycyanodienes. The $\omega_{\ln,HOMO}$ and $\omega_{\ln,LUMO}$ values decrease with an increase in molecular size in polyacenes and polyfluoroacenes. This is in qualitative agreement with a tendency in conventional superconductors; light mass will lead to higher values of $\omega_{\ln}$. The $\omega_{\ln,HOMO}$ values for polyfluoroacenes become smaller than those for polyacenes by H–F substitutions. The $\omega_{\ln,LUMO}$ values for polyfluoroacenes decrease with an increase in molecular size less rapidly than those for polyacenes. The $\omega_{\ln,LUMO}$ values for 1fa, 2fa, and 3fa become smaller than those for 1a, 2a, and 3a, by H–F substitutions. However, considering the large M_w values difference between 1fa and 1a, 2a and 2fa, and 3a and 3fa, the $\omega_{\ln,LUMO}$ values difference between polyacenes and polyfluoroacenes is not large. Furthermore, the $\omega_{\ln,LUMO}$ values for 4fa and 5fa are larger than those for 4a and 5a even though the M_w values for 4fa and 5fa are larger than those for 4a and 5a, respectively. We can expect that in the hydrocarbon molecular systems, the $\omega_{\ln}$ values would basically decrease by substituting hydrogen atoms by heavier atoms. This can be understood from the fact that the frequencies of all vibronic active modes in polyacenes downshift by H–F substitution. However, considering that the $\omega_{\ln}$ value for the LUMO rather localized on carbon atoms in 4fa and 5fa becomes larger by H–F substitution, we can expect that the $\omega_{\ln}$ value for a molecular orbital localized on carbon atoms has a possibility to increase by substituting hydrogen atoms by heavier atoms if the phase patterns of the molecular orbital do not significantly change by such atomic substitution. Therefore, the detailed properties of the vibrational modes and the electronic structures as well as the molecular weights are closely related to the frequencies of the vibronic active modes playing an important role in the electron–phonon interactions in the monoanions of polyfluoroacenes. The $\omega_{\ln,HOMO}$ values are larger than the $\omega_{\ln,LUMO}$ values in polyfluoroacenes. The significant phase patterns difference between the HOMO and LUMO is the main reason why the frequencies of vibronic active modes playing an essential role in the electron–phonon interactions in the monocations are significantly different from those in the monoanions in 2fa, 3fa, 4fa, and 5fa. Therefore, the detailed properties of the vibrational modes and the electronic states as well as the molecular weights are closely related to the frequencies of the vibronic active modes playing an important role in the electron–phonon interactions in charged polyfluoroacenes. The $\omega_{\ln,HOMO}$ and $\omega_{\ln,LUMO}$ values, increase and significantly decrease, respectively, with an increase in molecular size in B, N-substituted polyacenes. As in polyacenes, the $\omega_{\ln,HOMO}$ values are larger than the $\omega_{\ln,LUMO}$ values in large size of B, N-substituted polyacenes. As in 1a, the $\omega_{\ln,HOMO}$ value is slightly larger than the $\omega_{\ln,LUMO}$ value in 1bn. The differences between the $\omega_{\ln,HOMO}$ and $\omega_{\ln,LUMO}$ values become larger with an increase in molecular size in B, N-substituted polyacenes, while those in polyacenes hardly change. Such different properties in the $\omega_{\ln,HOMO}$ and $\omega_{\ln,LUMO}$ values between polyacenes and B, N-substituted polyacenes may come from the electronegativity perturbation on polyacenes. The $\omega_{\ln,HOMO}$ values for polyacenes are larger than those for B, N-substituted polyacenes. The results for the $\omega_{\ln,HOMO}$ values in B, N-substituted polyacenes are not in agreement with a tendency in solids, as described above. That is, the electronic structures rather than the molecular weights are closely related to the frequencies of the vibrational modes playing an important role in the electron–phonon interactions in the monocations of B, N-substituted polyacenes. The $\omega_{\text{ln,HOMO}}$ values do not significantly change with an increase in molecular size in polycyanodienes. Furthermore, the $\omega_{\text{ln,HOMO}}$ values for polycyanodienes are smaller than those for polyacenes. Apart from 1cn, the $\omega_{\text{ln,LUMO}}$ values for polyacenes and polycyanodienes decrease with an increase in molecular weights. The $\omega_{\text{ln,LUMO}}$ values for D_{2h} symmetric polycyanodienes are larger than those for D_{2h} symmetric polyacenes. The phase patterns difference between the delocalized LUMO in polycyanodienes and LUMO localized on carbon atoms located at the edge part of carbon framework in polyacenes is the main reason why the $\omega_{\text{ln,LUMO}}$ values for polycyanodienes are larger than those for polyacenes.

VIII. Concluding Remarks

In this review article, we discussed the electron–phonon interactions in the charged molecular systems such as polyacenes, polyfluoroacenes, B, N-substituted polyacenes, and polycyanodienes. We estimated the electron–phonon coupling constants and the frequencies of the vibronic active modes playing an essential role in the electron–phonon interactions. These physical values are essential to discuss several physical phenomena such as intramolecular electrical conductivity, intermolecular charge transfer, attractive electron–electron interactions and Bose–Einstien condensation, and superconductivity, which will be discussed in detail in the next review article. Motivated by the possible inverse isotope effects in Pd-H, Pd-D, and Pd-T superconductivity, and organic superconductivity observed by Saito et al., we discussed how the H–F substitution are closely related to the essential characteristics of the electron–phonon interactions in these molecules by comparing the calculated results for charged polyacenes with those for charged polyfluoroacenes, since fluorine atoms are much heavier than D and T atoms, and the phase patterns of the frontier orbitals such as the HOMO and LUMO are not expected to be significantly changed. Furthermore, we discuss how CC–BN and CC–CN substitutions are closely related to the essential characteristics of the electron–phonon interactions in these molecules by comparing the calculated results for charged polyacenes with those for charged B, N-substituted polyacenes and polycyanodienes, respectively. These physical values are essential to discuss the several physical phenomena such as intramolecular electrical conductivity, intermolecular charge transfer, attractive electron–electron interactions and Bose–Einstien condensation, and superconductivity, which will be discussed in detail in the next review article.

In this review article, we investigated the electron–phonon interactions in the monocations and monoanions of polyacenes, polyfluoroacenes, B, N-substituted polyacenes, and polycyanodienes. The C–C stretching modes around 1500 cm^{-1} strongly couple to the HOMO, and the lowest frequency modes and the C–C stretching modes around 1500 cm^{-1} strongly couple to the LUMO in polyacenes. The C–C stretching modes around 1500 cm^{-1} strongly couple to the HOMO and LUMO in polyfluoroacenes. The B–N stretching modes around 1500 cm^{-1} strongly couple to the HOMO and LUMO in B, N-substituted polyacenes. The C–C and C–N stretching modes around 1500 cm^{-1} strongly couple to the HOMO and LUMO in polycyanodienes.

We estimated the total electron–phonon coupling constants in the monocations and monoanions of polyacenes, polyfluoroacenes, B, N-substituted polyacenes, and polycyanodienes. The l_{HOMO} and l_{LUMO} values decrease with an increase in molecular size in polyacenes, polyfluoroacenes, B, N-substituted polyacenes, and polycyanodienes. In general, we can expect that monocations and monoanions, in which number of carriers per atom is larger, affords larger l_{HOMO} value. The l_{HOMO} and l_{LUMO} values for polyfluoroacenes are larger than those for polyacenes. The l_{HOMO} values for B, N-substituted polyacenes are larger than those for polyacenes. The l_{LUMO} values for B, N-substituted polyacenes are similar to those for polyacenes. The l_{LUMO} values are larger than the l_{HOMO} values in polyacenes, polyfluoroacenes, and B, N-substituted polyacenes. Therefore, electron doping rather than hole doping is more effective way to seek for large electron–phonon coupling constants in polyacenes, polyfluoroacenes, and B, N-substituted polyacenes. Both the l_{HOMO} and l_{LUMO} values for polycyanodienes are larger than those for polyacenes. Therefore, CH–CF, CC–BN, and CC–CN atomic substitutions are effective way to seek for large l_{HOMO} and l_{LUMO} values in polyacenes. The orbital patterns difference between the LUMO localized on carbon atoms located at the edge part of carbon framework in polyacenes and the delocalized LUMO in polycyanodienes due to electronegativity perturbation, is the main reason why the l_{LUMO} values for polycyanodienes are much larger than those for polyacenes. Strong σ orbital interactions between two neighboring carbon atoms and two neighboring carbon and nitrogen atoms and the orbital interactions between two neighboring carbon and hydrogen atoms in the HOMO of polycyanodienes are the main reason why the l_{HOMO} values for σ-conjugated polycyanodienes are much larger than those for π-conjugated polyacenes. The l_{HOMO} values are larger than the l_{LUMO} values in polycyanodienes. Strong σ orbital interactions between two neighboring carbon atoms and two neighboring carbon and nitrogen atoms and the orbital interactions between two neighboring carbon and hydrogen atoms in the HOMO and the weak π orbital interactions between two neighboring carbon atoms and two neighboring carbon and nitrogen atoms in the LUMO of polycyanodienes, are the main reason why the l_{HOMO} values are larger than the l_{LUMO} values in polycyanodienes.

We also investigated the logarithmically averaged phonon frequencies which measure the frequencies of the vibronic active modes playing an essential role in the electron–phonon interactions in polyacenes, polyfluoroacenes, B, N-substituted polyacenes, and polycyanodienes. The $\omega_{\text{ln,HOMO}}$ and $\omega_{\text{ln,LUMO}}$ values decrease with an increase in molecular size in polyacenes and polyfluoroacenes. This is in qualitative agreement with a tendency in conventional superconductors; light mass will lead to higher values of ω_{ln}. The $\omega_{\text{ln,HOMO}}$ values for polyfluoroacenes become smaller than those for polyacenes by H–F substitutions. The $\omega_{\text{ln,LUMO}}$ values for polyfluoroacenes decrease with an increase in molecular size less rapidly than those for polyacenes. The $\omega_{\text{ln,LUMO}}$ values for 1fa, 2fa, and 3fa become smaller than those for 1a, 2a, and 3a, by H–F substitutions. However, considering the large M_{W} values difference between 1fa and 1a, 2a and 2fa, and 3a and 3fa, the $\omega_{\text{ln,LUMO}}$ values difference between polyacenes and polyfluoroacenes is not large.

Furthermore, the $\omega_{\text{ln,LUMO}}$ values for 4fa and 5fa are larger than those for 4a and 5a even though the M_{w} values for 4fa and 5fa are larger than those for 4a and 5a, respectively. We can expect that in the hydrocarbon molecular systems, the ω_{ln} values would basically decrease by substituting hydrogen atoms by heavier atoms. This can be understood from the fact that the frequencies of all vibronic active modes in polyacenes downshift by H–F substitution. However, considering that the ω_{ln} value for the LUMO rather localized on carbon atoms in 4fa and 5fa becomes larger by H–F substitution, we can expect that the ω_{ln} value for a molecular orbital localized on carbon atoms has a possibility to increase by substituting hydrogen atoms by heavier atoms if the phase patterns of the molecular orbital do not significantly change by such atomic substitution. Therefore, the detailed properties of the vibrational modes and the electronic structures as well as the molecular weights are closely related to the frequencies of the vibronic active modes playing an important role in the electron–phonon interactions in the monoanions of polyfluoroacenes. The $\omega_{\text{ln,HOMO}}$ values are larger than the $\omega_{\text{ln,LUMO}}$ values in polyfluoroacenes. The significant phase patterns difference between the HOMO and LUMO is the main reason why the frequencies of vibronic active modes playing an essential role in the electron–phonon interactions in the monocations are significantly different from those in the monoanions in 2fa, 3fa, 4fa, and 5fa. Therefore, the detailed properties of the vibrational modes and the electronic states as well as the molecular weights are closely related to the frequencies of the vibronic active modes playing an important role in the electron–phonon interactions in charged polyfluoroacenes. The $\omega_{\text{ln,HOMO}}$ and $\omega_{\text{ln,LUMO}}$ values, increase and significantly decrease, respectively, with an increase in molecular size in B, N-substituted polyacenes. As in polyacenes, the $\omega_{\text{ln,HOMO}}$ values are larger than the $\omega_{\text{ln,LUMO}}$ values in large size of B, N-substituted polyacenes. As in 1a, the $\omega_{\text{ln,HOMO}}$ value is slightly larger than the $\omega_{\text{ln,LUMO}}$ value in 1bn. The differences between the $\omega_{\text{ln,HOMO}}$ and $\omega_{\text{ln,LUMO}}$ values become larger with an increase in molecular size in B, N-substituted polyacenes, while those in polyacenes hardly change. Such different properties in the $\omega_{\text{ln,HOMO}}$ and $\omega_{\text{ln,LUMO}}$ values between polyacenes and B, N-substituted polyacenes may come from the electronegativity perturbation on polyacenes. The $\omega_{\text{ln,HOMO}}$ values for polyacenes are larger than those for B, N-substituted polyacenes. The results for the $\omega_{\text{ln,HOMO}}$ values in B, N-substituted polyacenes are not in agreement with a tendency in solids, as described above. That is, the electronic structures rather than the molecular weights are closely related to the frequencies of the vibrational modes playing an important role in the electron–phonon interactions in the monocations of B, N-substituted polyacenes. The $\omega_{\text{ln,HOMO}}$ values do not significantly change with an increase in molecular size in polycyanodienes. Furthermore, the $\omega_{\text{ln,HOMO}}$ values for polycyanodienes are smaller than those for polyacenes. Apart from 1cn, the $\omega_{\text{ln,LUMO}}$ values for polyacenes and polycyanodienes decrease with an increase in molecular weights. The $\omega_{\text{ln,LUMO}}$ values for D_{2h} symmetric polycyanodienes are larger than those for D_{2h} symmetric polyacenes. The phase patterns difference between the delocalized LUMO in polycyanodienes and LUMO localized on carbon atoms located at the edge part of carbon framework in polyacenes is the main reason why the $\omega_{\text{ln,LUMO}}$ values for polycyanodienes are larger than those for polyacenes.

In summary, CH–CF, CC–BN, and CC–CN atomic substitutions are effective way to seek for larger l_{HOMO} values, and CH–CF and CC–CN atomic substitutions are effective way to seek for larger l_{LUMO} values in polyacenes.

Acknowledgment

This study was performed under the Project of Academic Frontier Center at Nagasaki Institute of Applied Science. This work is partly supported by a Grant-in-Aid for Scientific Research from the Japan Society for the Promotion of Science (JSPS-16560618, JSPS-17350094).

References

[1] (a) I. B. Bersuker, *The Jahn–Teller Effect and Vibronic Interactions in Modern Chemistry* (Plenum, New York, 1984); (b) I. B. Bersuker and V. Z. Polinger, *Vibronic Interactions in Molecules and Crystals* (Springer, Berlin, 1989); (c) I. B. Bersuker, Chem. Rev. 101, 1067 (2001); (d) I. B. Bersuker, *The Jahn–Teller Effect* (Cambridge University Press., Cambridge, 2006).

[2] G. Grimvall, *The Electron-Phonon Interaction in Metals* (North-Holland, Amsterdam, 1981).

[3] G. Fischer, *Vibronic Coupling: The Interaction between the Electronic and Nuclear Motions* (Academic, London, 1984).

[4] (a) N. O. Lipari, C. B. Duke, and L. Pietronero, J. Chem. Phys. 65, 1165 (1976); (b) P. Pulay, G. Fogarasi, and J. E. Boggs, *J. Chem. Phys.* 74, 3999 (1981).

[5] Text books, (a) C. Kittel, *Quantum Theory of Solids* (Wiley, New York, 1963); (b) J. M. Ziman, *Principles of the Theory of Solids* (Cambridge University, Cambridge, 1972); (c) H. Ibach and H. Lüth, *Solid-State Physics* (Springer, Berlin, 1995).

[6] (a) J. R. Schrieffer, *Theory of Superconductivity* (Addison-Wesley, Massachusetts, 1964); (b) P. G. de Gennes, *Superconductivity of Metals and Alloys* (Benjamin, New York, 1966).

[7] W. A. Little, *Phys. Rev.* 134, A1416 (1964).

[8] (a) D. Jérome, A. Mazaud, M. Ribault, and K. Bechgaad, *J. Phys.* (France) Lett. 41, L95 (1980); (b) M. Ribault, G. Benedek, D. Jérome, and K. Bechgaad, ibid. 41, L397 (1980).

[9] Reviews and books: (a) D. Jérome and H. J. Schulz, *Adv. Phys.* 31, 299 (1982); (b) T. Ishiguro and K. Yamaji, *Organic Superconductors* (Springer, Berlin, 1990); (c) J. M. Williams, J. R. Ferraro, R. J. Thorn, K. D. Karlson, U. Geiser, H. H. Wang, A. M. Kini, and M. -H. Whangbo, *Organic Superconductors* (Prentice Hall, New Jersey, 1992).

[10] K. Oshima, H. Urayama, H. Yamochi, and G. Saito, *J. Phys. Soc. Jpn.* 57, 730 (1988).

[11] E. Demiralp, S. Dasgupta, and W. A. Goddard III, *J. Am. Chem. Soc.* 117, 8154 (1995).

[12] (a) A. F. Hebard, M. J. Rosseinsky, R. C. Haddon, D. W. Murphy, S. H. Glarum, T. T. M. Palstra, A. P. Ramirez, and A. R. Kortan, *Nature* 350, 600 (1991); (b) M. J. Rosseinsky, A. P. Ramirez, S. H. Glarum, D. W. Murphy, R. C. Haddon, A. F. Hebard, T. T. M. Palstra, A. R. Kortan, S. M. Zahurak, and A. V. Makhija, *Phys. Rev. Lett.* 66, 2830 (1991).

[13] K. Tanigaki, T. W. Ebbesen, S. Saito, J. Mizuki, J. S. Tsai, Y. Kubo, and S. Kuroshima, *Nature* 352, 222 (1991).

[14] T. T. M. Palstra, O. Zhou, Y. Iwasa, P. E. Sulewski, R. M. Fleming, and B. R. Zegarski, *Solid State Commun.* 93, 327 (1995).

[15] (a) C. M. Varma, J. Zaanen, and K. Raghavachari, *Science* 254, 989 (1991); (b) M. Lannoo, G. A. Baraff, M. Schlüter, and D. Tomanek, *Phys. Rev.* B 44, 12106 (1991); (c) Y. Asai and Y. Kawaguchi, *Phys. Rev.* B 46, 1265 (1992); (d) J. C. R. Faulhaber, D. Y. K. Ko, and P. R. Briddon, *Phys. Rev.* B 48, 661 (1993); (e) V. P. Antropov, O. Gunnarsson, and A. I. Lichtenstein, *Phys. Rev.* B 48, 7651 (1993); (f) A. Auerbach, N. Manini, and E. Tosatti, *Phys. Rev.* B 49, 12998 (1994); (g) N. Manini, E. Tosatti, and A. Auerbach, *Phys. Rev.* B 49, 13008 (1994); (h) O. Gunnarsson, *Phys. Rev.* B 51, 3493 (1995); (i) O. Gunnarsson, H. Handschuh, P. S. Bechthold, B. Kessler, G. Ganteför, and W. Eberhardt, *Phys. Rev. Lett.* 74, 1875 (1995); (j) J. L. Dunn and C. A. Bates, Phys. Rev. B 52, 5996 (1995); (k) O. Gunnarsson, *Rev. Mod. Phys.* 69, 575 (1997); (l) A. Devos and M. Lannoo, *Phys. Rev.* B 58, 8236 (1998); (m) O. Gunnarsson, *Nature* 408, 528 (2000).

[16] E. A. Silinch and V. Capek, *Organic Molecular Crystals* (AIP, New York, 1994).

[17] W. Warta and N. Karl, *Phys. Rev.* B 32, 1172 (1985).

[18] W. Warta, R. Stehle, and N. Karl, *Appl. Phys.* A 36, 163 (1985).

[19] S. Kivelson and O. L. Chapman, *Phys. Rev.* B 28, 7236 (1983).

[20] A. Mishima and M. Kimura, *Synth. Met.* 11, 75 (1985).

[21] T. Skoskiewicz, *Phys. Status Solidi A11*, K123 (1972).

[22] B. Stritzker and W. Buckel, Z. *Physik* 257, 1 (1972).

[23] W -H. Li, J. W. Lynn, H. B. Stanley, T. J. Udovic, R. N. Shelton, and P. Klavins, *Phys. Rev.* B 257, 4119 (1989).

[24] J. E. Schirber, J. M. Mintz, and W. Wall, *Solid State Commun.* 52, 837 (1984).

[25] (a) T. Kato and T. Yamabe, *J. Chem. Phys.* 115, 8592 (2001); (b) T. Kato and T. Yamabe, *Recent Research Developments in Quantum Chemistry* (Transworld Research Network, Kerala, 2004); (c) *Recent Research Developments in Physical Chemistry* (Transworld Research Network, Kerala, 2004).

[26] V. Coropceanu, M. Malagoli, D. A. da Silva Filho, N. E. Gruhn, T. G. Bill, and J. L. Brédas, *Phys. Rev. Lett.* 89, 275503 (2002).

[27] (a) T. Kato and T. Yamabe, J. Chem. Phys. 120, 7659 (2004); (b) T. Kato and T. Yamabe, *J. Chem. Phys.* 119, 11318 (2003).

[28] (a) T. Kato and T. Yamabe, Chem. Phys. 315, 97 (2005); (b) T. Kato and T. Yamabe, *J. Chem. Phys.* 118, 3300 (2003).

[29] (a) T. Kato and T. Yamabe, J. Chem. Phys. 123, 094701 (2005); (b) T. Kato and T. Yamabe, *J. Phys. Chem.* A 108, 11223 (2004).

[30] (a) T. A. Albright, J. K. Burdett, and M. -H. Whangbo, *Orbital Interactions in Chemistry* (Wiley, New York, 1985); (b) J. K. Burdett, *Chemical Bonding in Solids* (Oxford University, Oxford, 1995).

[31] E. M. Conwell, *Phys. Rev.* B 22, 1761 (1980).
[32] K. Tanaka, Y. Huang, and T. Yamabe, *Phys. Rev.* B 51, 12715 (1995).
[33] C. Christides, D. A. Neumann, K. Prassides, J. R. D. Copley, J. J. Rush, M. J. Rosseinsky, D. W. Murphy, and R. C. Haddon, *Phys. Rev.* B 46, 12088 (1992).
[34] (a) A. D. Becke, Phys. Rev. A 38, 3098 (1988); (b) J. Chem. Phys. 98, 5648 (1993).
[35] C. Lee, W. Yang, and R. G. Parr, *Phys. Rev.* B 37, 785 (1988).
[36] (a) R. Ditchfield, W. J. Hehre, and J. A. Pople, J. Chem. Phys. 54, 724 (1971); (b) P. C. Hariharan and J. A. Pople, *Theor. Chim. Acta* 28, 213 (1973).
[37] M. J. Frisch, G. W. Trucks, H. B. Schlegel, P. M. W. Gill, B. G. Johnson, M. A. Robb, J. R. Cheeseman, T. A. Keith, G. A. Petersson, J. A. Montgomery, K. Raghavachari, M. A. Al-Laham, V. G. Zakrzewski, J. V. Ortiz, J. B. Foresman, J. Cioslowski, B. B. Stefanov, A. Nanayakkara, M. Challacombe, C. Y. Peng, P. Y. Ayala, W. Chen, M. W. Wong, J. L. Andres, E. S. Replogle, R. Gomperts, R. L. Martin, D. J. Fox, J. S. Binkley, D. J. Defrees, J. Baker, J. J. P. Stewart, M. Head-Gordon, C. Gonzalez, and J. A. Pople, Gaussian 98, (Gaussian Inc., Pittsburgh, Pennsylvania, 1998).
[38] C. R. Patrick and G. S. Prosser, *Nature* 187, 1021 (1960).
[39] W. A. Duncan, J. P. Sheridan and F. L. Swinton, Trans. Faraday Soc. 62, 1090 (1966).
[40] J. R. Goates, J. B. Ott and J. Reeder, *J. Chem. Thermodyn.* 5, 135 (1973).
[41] J. S. Brennen, N. M. D. Brown and F. L. Swinton, *J. Chem. Soc. Faraday Trans.* 1 70, 1965 (1974).
[42] J. B. Ott, J. R. Goates and D. L. Carbon, *J. Chem. Thermodyn.* 8, 505 (1976).
[43] T. Dahl, *Acta Chem. Scand.* 25, 1031 (1971).
[44] T. Dahl, *Acta Chem. Scand.* 26, 1569 (1972).
[45] J. M. Steed, T. A. Dixon, and W. Klemperer, *J. Chem. Phys.* 70, 4940 (1979).
[46] J. H. Williams, *Mol. Phys.* 73, 99 (1991).
[47] J. H. Williams, *Chem. Phys.* 172, 171 (1993).
[48] M. Neelakandan, D. Pant, E. L. Quitevis, *Chem. Phys. Lett.* 265, 283 (1997).
[49] M. Neelakandan, D. Pant, E. L. Quitevis, *J. Phys. Chem.* A 101, 2936 (1997).
[50] C. G. Gray and K. E. Gubbins, *Theory of Molecular Fluids* (Clarendon Press, New York, 1984; Vol. 1, p 587).
[51] J. R. Riddick and W. B. Bunger, *Organic Solvents* (Wiley-Inter-science, New York, 1970).
[52] M. R. Battaglia, A. D. Buckingham, and J. H. Williams, *Chem. Phys. Lett.* 78, 421 (1981).
[53] (a) N. Del Campo, M. Besnard, and J. Yarwood, *Chem. Phys.* 184, 225 (1994); (b) M. Besnard, Y. Danten, and T. Tassaing, In *Collision- and Interaction-Induced Spectroscopy* G. C. Tabisz and M. N. Neuman Eds. (Kluwer, Dordrecht, 1995; p 201-213).
[54] (a) F. Williams, M. B. Yim, and D. E. Wood, *J. Am. Chem. Soc.* 95, 6475 (1973); (b) M. B. Yim and D. E. Wood, *J. Am. Chem. Soc.* 98, 2053 (1976); (c) M. Shiotani and F. Williams, J. Am. Chem. Soc. 98, 4006 (1976); (d) M. C. R. Symons, R. C. Selby, J.G. Smith, and S. W. Bratt, Chem. *Phys. Lett.* 48, 100 (1977); (e) J. T. Wang and F. Williams, *Chem. Phys. Lett.* 71, 471 (1980); (f) O. A. Anisimov, V. M. Grigoriants, and Yu. N. Molin, *Chem. Phys. Lett.* 74, 15 (1980); (g) L. N. Shchegoleva, I. I. Bilkis, and P. V. Schastnev, *Chem. Phys.* 82, 343 (1983); (h) W. E. Wentworth, T. Limero, and E. C. M. Chen, *J. Phys. Chem.* 91, 241 (1987); (i) P. V. Schastnev and L. N. Shchegoleva,

Molecular Distortions in Ionic and Excited States (CRC Press, Boca Raton, FL, 1995, p. 79); (j) K. Hiraoka, S. Mizume, and S. Yamabe, *J. Phys. Chem.* 94, 3689 (1990); (k) L. C. T. Shoute and J. P. Mittal, *J. Phys. Chem.* 97, 379 (1993); (l) A. Hasegawa, M. Shiotani, and Y. Hama, *J. Phys. Chem.* 98, 1834 (1994); (m) L. N. Shchegoeva, I. I. Beregovaya, and P. V. Schastnev, *Chem. Phys. Lett.* 312, 325 (1999).

[55] (a) W. D. Hobey and A. D. Mclachlan, *J. Chem. Phys.* 33, 1695 (1960); (b) L. N. Shchegoeva and P. V. Schastnev, *Chem. Phys. Lett.* 130, 115 (1989); (c) J. J. Nash and R. R. Squires, *J. Am. Chem. Soc.* 118, 11872 (1996); (d) V. V. Konovalov, S. S. Laev, I. V. Beregovaya, L. N. Shchegoleva, V. D. Shteingarts, Yu, D. Tsvetkov, and I. I. Bilkis, *J. Phys. Chem.* A 104, 152 (2000).

[56] I. B. Bersuker, *Chem. Rev.* 101, 1067 (2001).

[57] (a) T. A. Miller, Annu. Rev. Phys. Chem. 33, 257 (1982); (b) T. J. Sears, T. A. Miller, and V. E. Bondybey, *J. Chem. Phys.* 74, 3240 (1981).

[58] E. J. M. Hamilton, S. E. Dolan, C. M. Mann, H. O. Colijn, C. A. McDonald, and S. G. Shore, *Science* 260, 659 (1993).

[59] J. -Y. Yi and J. Bernholc, *Phys. Rev.* B 47, 1708 (1993).

[60] A. Rubio, J. L. Corkill, and M. L. Cohen, *Phys. Rev.* B 49, 5081 (1994).

[61] Y. Miyamoto, A. Rubio, M. L. Cohen, and S. G. Louie, *Phys. Rev.* B 50, 4976 (1994).

[62] Y. Miyamoto, A. Rubio, S. G. Louie, and M. L. Cohen, *Phys. Rev.* B 50, 18360 (1994).

[63] X. Blase, A. Rubio, S. G. Louie, and M. L. Cohen, *Europhys. Lett.* 28, 335 (1994).

[64] X. Blase, J. -C. Charlier, A. De Vita, and R. Car, *Appl. Phys. Lett.* 70, 197 (1997).

[65] B. -C. Wang, M. -H. Tsai, and Y.-M. Chou, *Synth. Met.* 86, 2379 (1997).

[66] M. Menon and D. Srivastava, *Chem. Phys. Lett.* 307, 407 (1999).

[67] D. L. Carroll, P. Redlich, P. M. Ajayan, S. Curran, S. Roth, and M. Rühle, Carbon 36, 753 (1998).

[68] X. Blase, J. -C. Charlier, A. De Vita, and R. Car, *Appl. Phys.* A 68, 293 (1999).

[69] P. W. Fowler, K. M. Rogers, G. Seifert, M. Terrones, and H. Terrones, *Chem. Phys. Lett.* 299, 359 (1999).

[70] V. V. Pokropivny, V. V. Skorokhod, G. S. Oleinik, A. V. Kurdyumov, T. S. Bartnitskaya, A. V. Pokropivny, A. G. Sisonyuk, and D. M. J. Scheichenko, *Solid State Chem.* 154, 214 (2000).

[71] T. Hirano, T. Oku, and K. Suganuma, *Diamond and Related Materials* 9, 625 (2000).

[72] S. Erkoc, *J. Mol. Struc.* (THEOCHEM) 542, 89 (2001).

[73] J. Kongsted, A. Osted, L. Jensen, P. Åstrand, and K. V. Mikkelsen, *J. Phys. Chem.* B 105, 10243 (2001).

[74] H. Steinberg and R. J. Brotherton, *Organoboron Chemistry* (Wiley, New York, 1966, Vol. 2).

[75] A. Meller, Top. *Cum. Chem.* 15, 146 (1970).

[76] W. Maringgele, *The Chemistry of Inorganic Homo- and Heterocycles* (I. Haiduc, Ed.; Academic Press, London, 1987).

[77] D. F. Gaïnes and J. Borlin, *Boron Hydride Chemistry* (E. L. Muet- teries, Ed.; Academic Press, New York, 1975, p 241).

[78] K. Niedenzu and J. W. Dawson, *Boron-Nitrogen Compounds* (Academic Press, New York, 1965).

[79] R. T. Paine and C. K. Narula, *Chem. Rev.* 90, 73 (1990).

[80] A. Stock and E. Pohland, *Chem. Rer.* 59, 2215 (1926).

[81] (a) S. G. Shore and R. W. Parry, *J. Am. Chem. Soc. 77, 6084 (1955); (b) S. G. Shore and R.W. Parry, J*. Am. Chem. Soc. 80, 8 (1958); (c) S. G. Shore and K. W. Bddeker, Inorg. Chem. 3, 914 (1964).
[82] A. W. Laubengayer, P. C. Jr. Moews, and R. F. Porter, *J. Am. Chem. Soc.* 83, 1337 (1961).
[83] Y. Matsunaga, *Bussei Kagaku* (In Japanese) (Syokabo, Tokyo, 1981).
[84] J. L. Brédas, B. Thémans, and J. M. André, *J. Chem. Phys.* 78, 6137 (1983).

In: Chemical Physics Research Trends, Vol. 252
Editor: Samuel V. Arnold, pp. 65-115

ISBN: 978-1-60021-554-4

Chapter 2

SELF-ASSEMBLY OF BLOCK COPOLYMERS IN SELECTIVE SOLVENTS

Jintao Zhu and Wei Jiang*
State Key Laboratory of Polymer Physics and Chemistry,
Changchun Institute of Applied Chemistry,
Chinese Academy of Sciences, Changchun, 130022,
Graduate School of the Chinese Academy of Sciences,
People's Republic of China

ABSTRACT

Block Copolymers are the focus of intense research due to their ability to self-assemble into nanostructures with well-defined morphology and size. Comparing with bulk states, block copolymers can self-assemble into vast variety of copolymer micelles in selective solvents. The interest in block copolymer micelles stems from the numerous applications of such systems.

This chapter deals with recent progress including authors' contributions in the field of block copolymer self-assembly in solution and also tries to highlight some new directions in this field. Generalities about the critical micelle concentration, micellar structures, micelles preparation techniques, characterization techniques, dynamics of block copolymer micelles in solution and cross-linked copolymer micelles are firstly presented. Selected examples of micelles formation in aqueous and organic medium are given for block copolymers assemblies with cornucopian architectures. Strategies about how to tune the block copolymer micelles morphology in solution are addressed. New trends in the self-assembly of block copolymers in selective solvent, such as multicompartmennt micelles, co-micellization of block copolymers in selective solvent, and non-covalent complexes in micellar system are shown in the following part. Current and potential applications of block copolymer colloidal assemblies as templates for nanostructures fabrication, controlled drug delivery of bioactive agents, among others are finally discussed.

---

* Author to whom all correspondences should be addressed; E-mail: Wjiang@ciac.jl.cn; Tel: 86-431-5262151; Fax: 86-431-5262126

The aim of this chapter is to provide an overview of these fields of self-assembled block copolymers systems, illustrated by some selected examples.

Keywords: Block Copolymers; Crew-cut; Cross-link; Drug Delivery; Janus Micelles; Micellization; Micelles; Multicompartment Micelles; Nanostructures; Selective Solvent; Self-Assembly.

ABBREVIATIONS

AFM	atomic force microscopy
b	grafting distance
CMC	critical micelle concentration
CMT	critical micelle temperature
CWC	critical water concentration
DLS	dynamic light scattering
DMF	N, N-dimethylformamide
f	melt volume fraction
IPECs	interpolyelectrolytes
LCST	low critical solubility temperature
MW	molecular weight
NMR	nuclear magnetic resonance
PAA	poly(acrylic acid)
PACs	poly (cesium acrylate)
PAMPS	poly(sodium-2-acrylamido-2-methyl-1-propanesulfonate)
PB	polybutadiene
PBO	poly(butylene oxide)
PCEMA	poly(cinnamoylethyl methacrylate)
PCL	poly(ε-caprolactone)
PDEAEMA	poly(N, N-diethylaminoethyl methacrylate)
PDMAEMA	poly(N, N-dimethylaminoethyl methacrylate)
PDMS	poly(dimethylsiloxane)
PDP	pentadecyl phenol
PDPAEMA	poly(N, N-diisopropylaminoethyl methacrylate)
PEE	poly(ethyl ethylene)
PEG	poly(ethylene glycol)
PEO	poly(ethylene oxide)
PEVP	poly(N-ethyl-4-vinyl pyridine bromide)
PFP	poly (penocenylphosphine)
PFPO	poly(perfluoropropylene oxide)
PFS	poly(penocenylphosphine)
PI	polyisoprene
PMAA	poly(methacrylic acid)
PMANa	poly(sodium methacrylate)
PMEMA	poly(2-(N-morpholinoethyl) methacrylate)
PMMA	poly(methyl methacrylate)

PMVE	poly(methyl vinly ether)
PNIPAM	poly(N-isopropyl acrylamide)
PPO	poly(propylene oxide)
PQVP	poly (vinyl pyridine) quaternized
PS	polystyrene
PSEMA	poly(2-succinyloxyethyl methacrylate)
PSGMA	poly(sulfonated glycidyl methacrylate)
PSS	poly(styrene sulfonate)
PtBA	poly(*tert*-buthylacrylate)
PtBMA	poly(*tert*-buthyl methacrylate)
PVA	poly(vinyl alchol)
PVP	poly(vinyl pyridine)
P2VP	poly(2-vinyl pyridine)
P4VP	poly(4-vinyl pyridine)
R_h	hydrodynamic radius
R_g	radius of gyration
R_c	core radium of micelle
R_m	overall radium of micelle
SANS	small-angle neutron scattering
SAXS	small-angle X-ray scattering
$scCO_2$	supercritical CO_2
SEC	size-exclusion chromatography
SEM	scanning electron microscopy
SLS	static light scattering
TEM	transmission electron microscopy
THF	tetrahydrofuran
T_g	glass transition temperature
UV-vis	Ultraviolet–Visible spectrometer
UCST	upper critical solubility temperature
χ	Flory-Huggins dimensionless interaction parameters
Z	aggregate number of block copolymer in the micelles

1. INTRODUCTION

Self-assembly is an essential process in biological systems; it provides the diverse range of highly ordered structures observed in living organisms [1]. Researchers have tried to mimic to fabricate a bottom-up approach to materials design and fabrication. In particular, the self-assembly of organic molecules and synthetic polymers, such as block copolymers, have been extensively investigated in recent years [2-3]. Block copolymers are generally defined as macromolecules with linear and/or radical arrangement of two or more different blocks of varying monomer composition connected at their termini. The covalent connection of thermodynamically incompatible segments leads to self-assembly of the components into ordered structures with periodicity or compositional heterogeneity on a nanometer (or even micrometer) length scale [4-5]. Many synthetic techniques are available for generating A-B, diblock, A-B-A and A-B-C triblock copolymers or even more complicated block architectures [6-7].

It has been known for many years that when diblock copolymers are dissolved in a solvent which is selective for one of the blocks, micelles or collodial size aggregates can form as a result of the aggregation of the insoluble blocks [8]. Self-assembly of block copolymers in selective solvent has been reviewed from experimental and theoretical points of view by Price [9], Piirma [10], Tuzar and Kratochvil [11], Riess [12-13], Webber [14], Hamley [15], Alexandridis and Hatton [16], Nace [17], Xie [18], Liu [19-20], Alexandridis and Lindman [21], Gohy [22], Förster and coworkers [23]. In this chapter, we will give a brief overview of recent results, including authors' contributions, about the self-assembly of block copolymers in selective solvents.

This chapter is divided into six parts. The first part deals with the generalities about critical micelle concentration, micellar structures, micelles preparation techniques, characterization techniques, dynamics of block copolymer micelles in solution, and cross-linked copolymer micelles in Section 2. Selected examples of block copolymer micelles with cornucopian architectures in aqueous and organic medium are given in the next two parts (Section 3 and 4). Strategies about how to tune the block copolymer micelles architectures in solution are addressed in section 5. New trends in the self-assembly of block copolymers in selective solvents, such as multicompartment micelles, co-micellization of block copolymers in selective solvents, and non-covalent complexes in micellar system are shown in the following section. Current and potential applications of block copolymer colloidal assemblies as templates for nanostructures fabrication, controlled drug delivery of bioactive agents, among others are finally discussed in section 7.

In the following discussion, block copolymers will be simply designed by the acronym A-B for a diblock copolymer, A-B-A for a triblock copolymer with two identical outer blocks, A-B-C for an ABC triblock copolymer, etc. Full names of block copolymers discussed in this chapter are provided in the abbreviations and symbols section.

2. Block Copolymer Micelles: Generalities

2.1. Critical Micelle Concentration

It is well know that block copolymers can self-assemble into micelles in selective solvents, as illustrated in Figure 1. The critical concentration which the first micelle forms under fixed temperature is called the critical micelle concentration (CMC) [24]. As the concentration of block copolymer chains increase in the solution, more micelles are formed while the concentration of nonassociated chains, the so-called unimers, remains constant and is equal to the value of the CMC. A critical micelle temperature or CMT is another useful value for block copolymer micelles system [25]. For low molecular weight surfactant, the CMC, as well as the CMT, can be determined by various scattering techniques (SLS, SAXS, etc), fluorescence or dye solubilization techniques, surface tension measurements, among others [12, 26]. These techniques can also be applicable to block copolymer micellar systems.

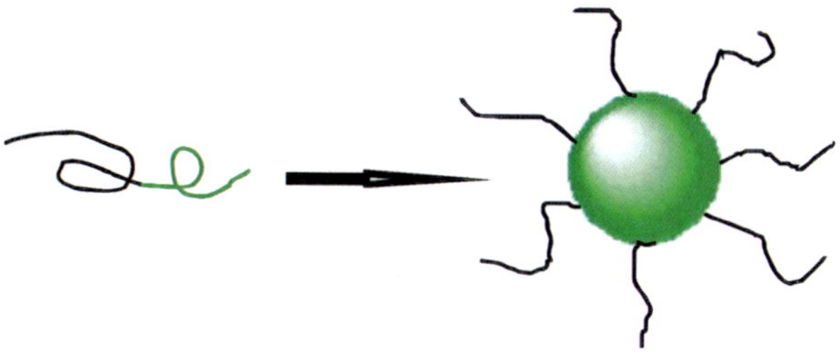

Figure 1. Schematic representation of self-assembly of block copolymer into micelles in selective solvent.

However, block copolymers have very low CMC values and furthermore that, due to their low diffusion coefficient, unimer-micelle equilibrium situations are only reached after a very long time period. In practical terms, the CMC is often determined from plots of the surface tension as a function of the logarithm of the copolymer concentration. In this case, the CMC is defined as the concentration at which the surface tension stops decreasing and reaches a plateau value.

Tsunashima and coworkers investigated the micellization of PS-PB diblock copolymer in various solvents and measured R_h by DLS [27]. It was confirmed that the PS-PB chains were molecularly dissolved and that both blocks adopted a stretched conformation due to intersegmented repulsive interactions in nonselective solvent (such as benzene). In a selective solvent (such as *n*-decane) for the PB block, however, PS-PB unimers with collapsed PS segments were observed at low concentration while aggregated PS-PB chains forming spherical micelles were observed at higher concentrations, thus defining the CMC. As outlined by Zana [28] and by Jada and coworkers [29], fluorescence technique, either with "free" probes like pyrene solubilization or covalently fixed fluoroprobes, has shown to be an another preferred technique for CMC and CMT measurements.

For highly asymmetric block copolymers with a large volume fraction of insoluble block, the copolymer chains can't be directly solubilized in a selective solvent. However, micelles can be obtained from these copolymers by the temporary use of a non-selective solvent, i.e., common solvent, which is finally removed by dialysis process [30]. In principle, all the copolymer chains are aggregated from these systems and the CMC is therefore estimated to be extremely low, as discussed by Eisenberg and coworkers [31]. In other case, when the crew-cut micelles are formed from a homogeneous copolymer solution by the addition of a precipitant (such as water) for one block, a micellization process is involved. Another way to present the micelle formation in this case is the critical water content (CWC) [32]. The CWC values are usually obtained by observation of the onset of turbidity of the block copolymer solutions or by light-scattering measurements. Uv-vis and DLS have been used to determine the CWC by measuring the turbidity increase rapidly of the copolymer solution [33]. Moreover, this method can also be used to study the dynamic of block copolymer micelles in solution.

Figure 2 shows a plot of the critical water content versus the logarithm of the copolymer concentration for a solution of ABC triblock copolymer PS_{755}-$P2VP_{130}$-PEO_{77} (the number in the subscripts indicate the number of repeat units of the blocks) in THF [34].

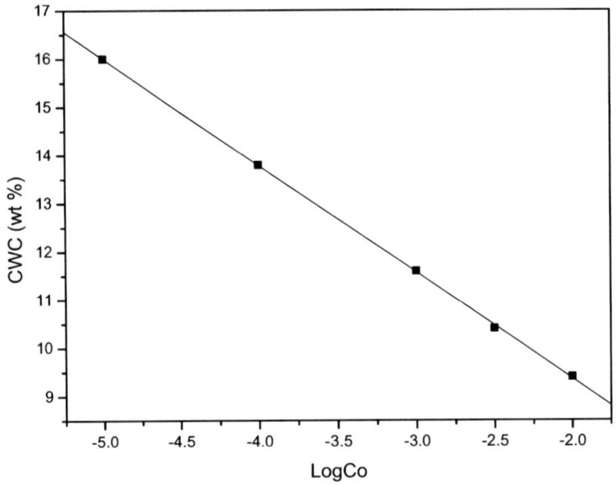

Figure 2. Critical water content vs logarithm of block copolymer concentration for a solution of PS_{755}-$P2VP_{130}$-PEO_{77} in THF. Adapted from Reference [34].

It is found from Figure 2 that the CWC depends on the copolymer concentration. The relationship between the CWC and the initial copolymer concentration, C_0, can thus be written as:

$$CWC= -A\log C_0+B \qquad \text{(eq.1)}$$

Where A and B are constants for a specific copolymers, at critical water content, the concentration of the copolymer represents its CMC. As the water content in the solution increases, the CMC decreases, and more and more polymer chains form micelles. The micelle fraction can be defined as a ratio of the concentration of associated polymer to the total polymer concentration, $(C_0\text{-}CMC)/C_0$. Therefore, the change in the micelles fraction as a function of the added water content can be calculated from equation 2 [32, 35]:

$$(C_0\text{-}CMC)/C_0=1\text{-}\exp[2.303(CMC\text{-}H_2O\%)/A] \qquad \text{(eq.2)}$$

When water content (H_2O%) is lower than the CMC, the ratio of $(C_0\text{-}CMC)/C_0$ should be zero, because all the block copolymer chains are unassociated. Utilizing the above procedure, one can also extrapolate the single-chain population (relative to total copolymer content) at water content of 15 wt %, 10^{-3}. From this result, we can conclude that the single-chain fraction in the triblock copolymer micellar system is quite low.

2.2. Micellar Structures

According to the polarity of the solvent used, block copolymer micelles can be divided into two types: regular and reverse micelles. In a polar solvent, such as water, regular micelles are formed, which have a hydrophobic core and a hydrophilic corona [36-39]. In an apolar (or low polarity) solvent, the locations of the blocks are reversed, i.e., the hydrophilic blocks form the core which is surrounded by a hydrophobic corona or shell. In this case, the micelles are referred to as reverse [40-42].

Depending on the composition of the starting block copolymers, each of the micelles types can be further divided into "star" and "crew-cut" micelles. Star micelles are made from block copolymers in which the corona-forming blocks are much longer than the core-forming blocks, while crew-cut micelles are characterized by a bulky core and a relatively short corona [43-46]. Figure 3 depicts schematic illustration of crew-cut and star block copolymer micelles.

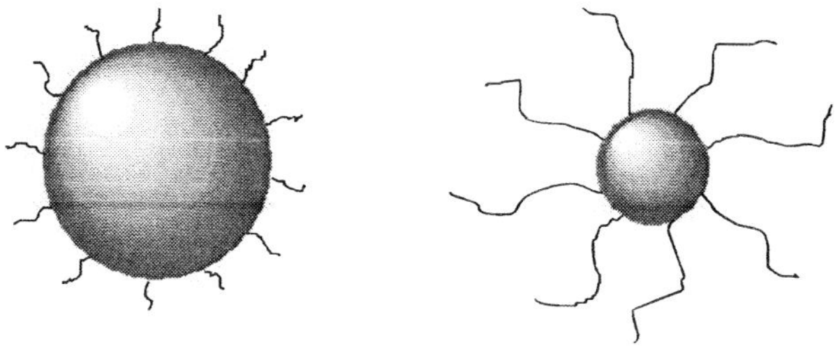

a) b)

Figure 3. Schematic illustration of (a) a star and (b) a crew-cut micelle. Note that the micellar structure displayed here is only valid for simple AB diblock copolymers. The situation can be much more complex for micelles prepared from block copolymers with complex architectures, as will be discussed in the followed parts.

Different systems of star micelles (both regular and reverse) have been studied extensively for decades [47-49]. Studies have mainly focused on the phenomenon of micellization of various block copolymers in selective solvents, the development of new experimental techniques and theories of micellization, and the application possibilities in industry and technology. By contrast, crew-cut micelles represent a new type of aggregates and have been receiving great attention recently after the research works of Eisenberg group [31, 32, 50, 51]. The concept of crew-cut micelles was proposed by Halperin [52], based on an earlier theoretical study of de Gennes [53].

There are two major differences between the crew-cut and star micelles, the process of the block copolymer micelles formation and their morphologies. The procedures of the copolymer micelles formation will be addressed in Section 2.3. Star micelles of block copolymers usually take on spherical geometry, consisting of a core of radium R_c, formed by the aggregation of insoluble blocks, surrounding by a spherical corona formed by the soluble blocks. In contrast, multiple morphologies of crew-cut aggregates, i.e., spheres, rods, vesicles, lamellae, large compound micelles, tubules, donuts and bowl-shaped micelles, among others have been observed in selective solvents [54].

The process of block copolymers micellization is usually characterized by the aggregation number of block copolymers (Z) in the micelles. The micellar structure is characterized by the core radius R_c, and the overall radius of the micelle R_m. There are other ways to define the overall dimension of a micelle core, the radium of gyration R_g and hydrodynamic radius R_h, as can be found elsewhere [55]. The distance between neighboring blocks at the core/corona interface is usually called the grafting distance b. b^2 is thus the area occupied by one chain at the core/corona interface. Important parameters that control Z, R_c, R_m, and b are the degree of polymerization of copolymer blocks, N_A and N_B, and the Flory-Huggins dimensionless interaction parameters χ. Aggregation number and core radii have been investigated in details for many block copolymer/solvent systems. The free energy of block copolymer micelles is mainly ascribed to three contributions: core chain stretching, corona chain repulsion, and the core/corona interfacial tension [32, 56]. Intrinsic properties, such as the χ, play an important role in balancing these contributions to free energy.

2.3. Micellar Preparation Techniques

Block copolymer micellar systems are generally produced by one of the following procedures. It has been pointed that the micellar characteristic features are strongly been influenced by the copolymer micelles preparation procedure [12].

In the first technique, direct method to prepare a block copolymer micellar solution refers to the directly dissolution of the bulk copolymer sample in a selective solvent for one of the blocks. However, this method usually only works if the total MW of the block copolymer is low and the length of the insoluble block is short enough. Generally, star-type micelles can be prepared directly by dissolving a highly asymmetric block copolymers in a solvent which is selective (good) for the long block. During the dissolution of the block copolymers, the insoluble (short) blocks form the core while the long blocks form the corona of the micelles. The micellar solution prepared in this method is always let to anneal by standing and/or the annealing process is made by thermal treatment, stirring or even under ultrasonic agitation [57].

Usually, the method of direct dissolution is not suitable for the preparation of crew-cut block copolymer micelles because of the very high fraction and the large relative size of the insoluble block. To prepare the stable solutions of crew-cut micelles, highly asymmetric block copolymers are firstly dissolved molecularly in a common solvent i.e., good solvent for both blocks. Subsequently, precipitant (such as deionized water) for the long block, but is good solvent for the short block, of the copolymers is added gradually to induce the aggregation of the long blocks, eventually followed by extracting the common solvent by dialysis process. This is the so-called indirect method for block copolymer micelles preparation [35].

The step-wise dialysis technique, pioneered by Tuzar and Kratochvil [11, 12], is another preparation technique for micellar systems, mainly in aqueous medium, as demonstrated by Munk [58] for PS-PMMA block copolymers micelles and as experienced by Laschewsky et al. [59] for multicompartment micelles formation through self-assembly of linear ABC triblock copolymer in aqueous solution.

Another way to preparing block copolymer micelles is the transfer of a cross-linked bulky assemblies into solution [60]. In this case, assembled structure is firstly formed in the bulk state and its core is stabilized by cross-linking. The resulting material is finally dissolved and separated in a good solvent, such as THF, for the coronal blocks. Liu et al. have prepared various block copolymer micelles by using this method for PCEMA-containing block copolymers [61-63]. Recently, a novel Janus-type [64] block copolymer micelle has also been prepared through similar procedure (see Section 6.2) [65, 66].

2.4. Characterization of Block Copolymer Micelles—Experimental Techniques

Experimental techniques which have been applied for given block copolymer micelles characterizations have been reviewed extensively by Tuzar [11], Munk [67], Chu [25], Webber [68], Mortensen [69], Zana [70], and Hamley [15]. According to the listing provided by Chu [25], these characterization techniques can be classified in scattering, spectroscopic and in a wide range of other physical techniques which are summarized in Table 1.

Table 1. Experimental techniques for block copolymer micelle characterization

Techniques	Micelles Characteristics
TEM, cryo-TEM, SEM, AFM	Shape, size
DLS	R_h
SANS and SAXS	Molecular weight, R_g, R_c, macrolattice structures
SLS	Molecular weight (weight-average), R_g
SEC	R_h, dynamics of micellar equilibrium
Ultracentrifugation	Micelle density, molecular weight (Z average)
Fluorescence	Chain dynamics, CMC, hybridization of micelles
NMR	Chain dynamics
Viscometry	R_h, intrinsic viscosity
Stop flow techniques	Kinetics of micelle formation and dissociation

It will not be possible to cover all these techniques in this chapter, and in the following part we will only outline the different methods that are often used for copolymer micellar morphologies characterization, illustrated by some selected examples.

A variety of experimental techniques can be used to characterize micellar morphologies of block copolymer formed in dilute solution. However, it is suitable to adopt TEM to directly observe the morphologies of complicated, irregular, and polydispersed copolymer aggregates [34]. Figure 4a show a typical TEM picture of vesicle structure from self-assembly of P4VP-PS-P4VP in dioxane/water mixtures [71]. From TEM techniques, one can obtain the internal structure information of block copolymer micelles, combining proper staining technique. There are different ways for sample preparation, some of them being stem from TEM observations on soft biological materials. Sometimes, weak electronic contrast can be obtained from these specimens because block copolymers micelles generally contain light elements such as C, H, N, and O. In addition, these soft materials are subjected to degradation during observation, a "ghost" picture of the structures being then recorded. Most of the preparation techniques have been devoted to enhancing both the electronic contrast and the stability of the micellar objects for TEM observations. Contrasting techniques with heavy metal have been widely used to reach these two goals. In these experiments the micelles are usually deposited on a TEM grid covered with a carbon film and further observed in the dried state. Heavy metals can interact with block copolymers micelles bearing suitable functional groups. Typical examples are the reactions of RuO_4 to aromatic double bonds and OsO_4 to aliphatic double bonds [72]. These staining techniques also allow the direct visualization of a specific compartment of the micelles, such as the core. Figure 5c illustrate the TEM image for aqueous micelles from a PS-P2VP-PEO triblock copolymer and contrasted with RuO_4 [72].

Negative contrasting techniques are based on the use of a contrasting agent that does not interact with the micelles and that therefore spreads onto the substrates. Micelles then appear as bright spots on a dark background. Phosphotungstic acid is a typical negative staining agent [72,73]. The dimension of the core and dried corona can be measured from this experiment, as shown in Figure 5a and 5b. Metal, for example palladium/platinum alloy, can be splattered on block copolymer micelles by exposing the TEM grid to metal vapor under a certain angle [31]. Each micelle will generate a shadow whose size and shape depend on the metal deposition angle and on the morphological characteristic features of the micelle. The size of the copolymer micelles can be measured directly from the prints of the microscope negatives.

TEM at liquid nitrogen temperature (cryo-TEM) technique is uniquely suited for identifying the local microstructures in complex fluids at high water content such as micellar and viscular dispersions [74].

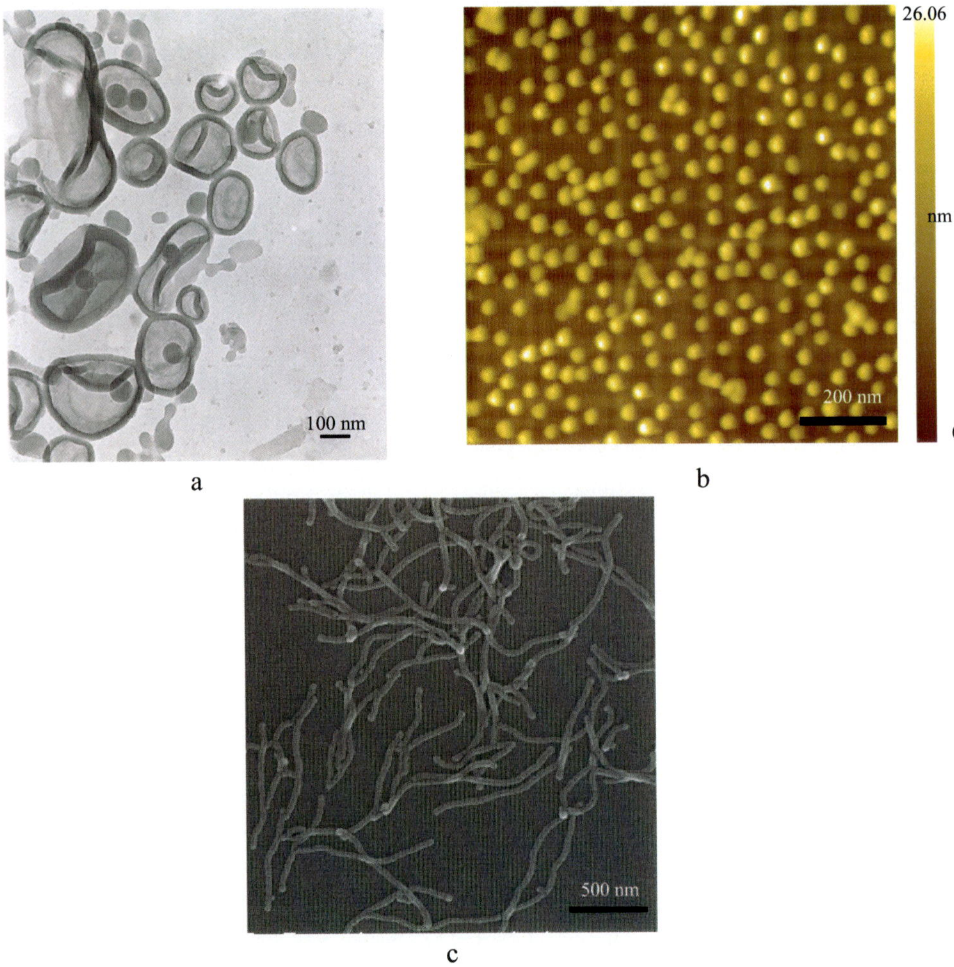

Figure 4. Representative block copolymer (P4VP-PS-P4VP) micellar morphologies of vesicles (a), obtained by TEM observation; sphere (b), obtained by AFM; and nanowires (c), obtained by SEM. Adapted from References [71, 84].

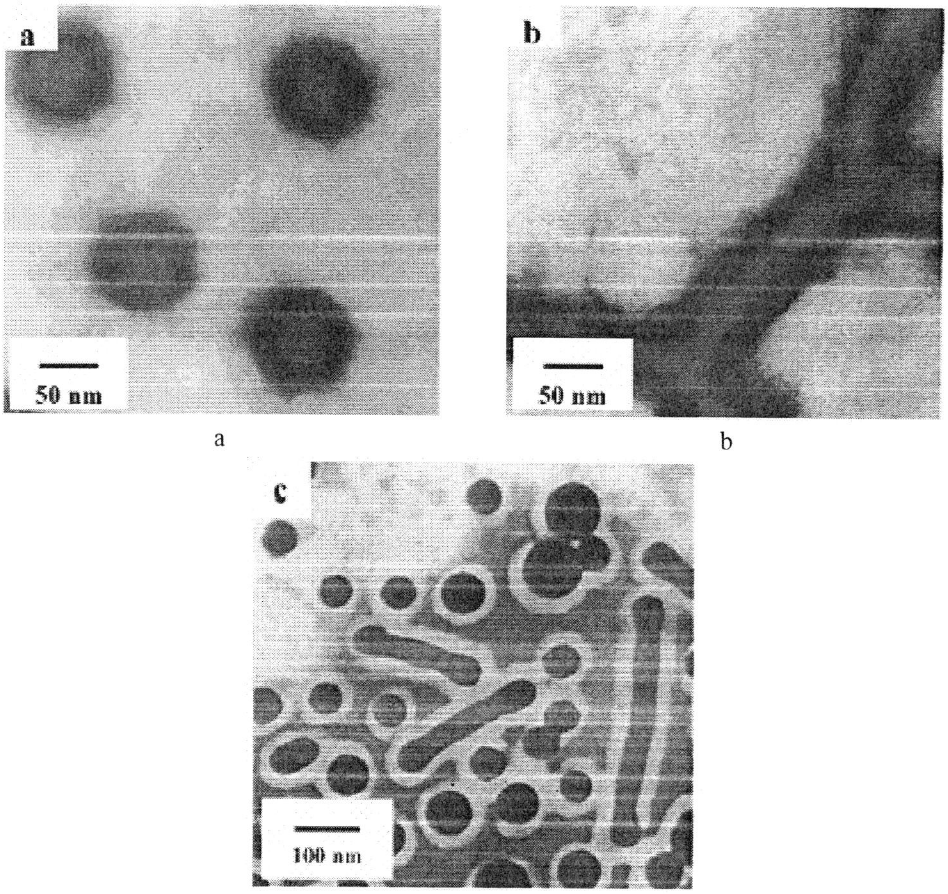

Figure 5. TEM images of PS_{200}-$P2VP_{140}$-PEO_{590} micelles contrasted by $H_3PO_4.12WO_2$ (a, b) and thoroughly contrasted by RuO_4 (c). Adapted from Reference [73].

Recently, Cryo-TEM technique has started to be used with increasing frequency for block copolymer micelle characterization in aqueous solution, as reported by Esselink and coworkers [75], Bates et al. [76], Talmon and coworkers [77], Lodge et al. [78], among other groups [79-80]. The advantage is that it allows for direct observation of micelles in a glassy water phase in situ and accordingly determines the characteristic dimensions of both the core and swollen corona provided that a sufficient electronic contrast is observed between these two domains. Recently, the core-corona structures (spheres and cylinders) of PB-PEO block copolymers micelles have been directly observed by cryo-TEM [81]. Images of both shapes of micelles show a dense core (PB segments) and a diffuse corona (PEO segments), as can be seen from Figure 6. In addition, when used for polymeric aggregates the size of which is nearly an order of magnitude greater, cryo-TEM is also a good quantitative probe that does not require any model-dependent analysis.

AFM has also become an efficient tool to directly visualize block copolymers micelles. It is possible to image the micellar strings and networks in more details by using AFM. AFM measurements can be performed either in the dried state [82] or directly in situ in solution (within a liquid cell) [83].

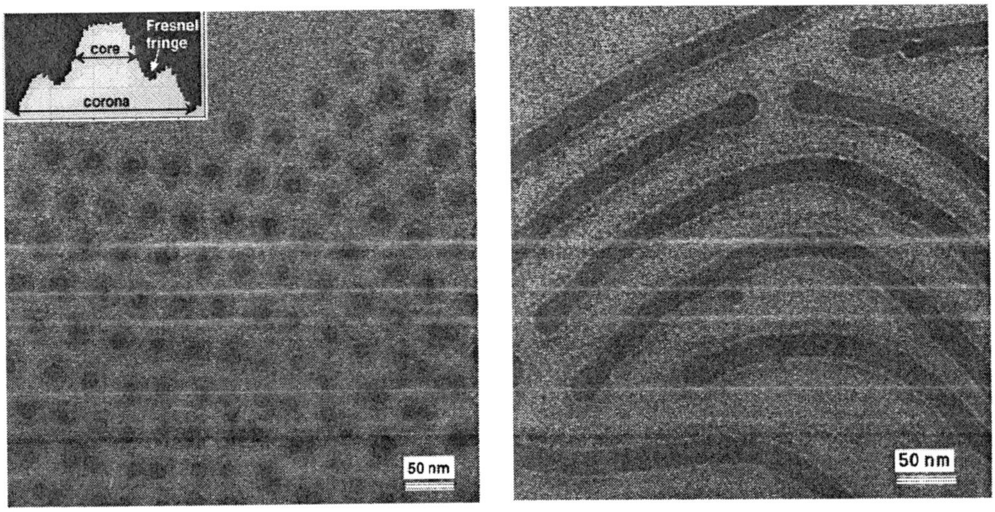

a b

Figure 6. Cryo-TEM micrograph of 1 wt % aqueous PB_{126}-PEO_{45} (sphere, a) and PB_{104}-PEO_{166} (cylinders, b) diblock copolymer solution. A dark core consisting of PB blocks and a corona of PEO blocks is evident. The segment density profile of one micelle for (a) is shown in the inset. Adapted from Reference [81].

Dry measurements are the most often used and realized on micelles absorbed or deposited on flat solid substrate (mica, silicon wafer, graphite, etc) that have been further dried. Typical AFM image (in the tapping mode) from dried aqueous block copolymer P4VP-PS-P4VP micelles are shown in Figure 4b [84]. The shape and size of the micelles can, however, be affected by tip-convolution effects, by specific interactions between the substrate and some moieties of block copolymer micelles, or by relaxation of a low T_g micellar core resulting in the flattening of the micelles on the substrate, as has been discussed recently [85].

SEM technique has proven to be another suitable tool for the visualization of block copolymer micelles morphologies. Typical SEM picture of cylindrical micelles obtained from P4VP-PS-P4VP is displayed in Figure 4c. However, SEM, like AFM, can only provide the micellar structure of surface morphology (external structure). By combining TEM technique, which can provide internal information of the copolymer micelles, with SEM (or AFM) which can provide the surface morphology, one can obtain the thorough information about the block copolymer micellar structures.

Light scattering is a very common technique used for block copolymer micelles characterization. A recent review on SANS on block copolymer micelles was published by Mortensen [69]. Information on micellar morphology can also be obtained by scattering techniques with the advantage of giving a mean value calculated over a large number of copolymer micelles. SLS allows one to determine the absolute weight-average MW and thus Z of the micelles, as well as R_g. Information about the quality of the selective solvent for the coronal chains through the second virial coefficient can also be obtained [55]. Meanwhile, R_h can be measured by DLS. In addition, SANS has also been widely used to elucidate micellar structures. In order to generate a sufficient scattering contrast, SANS measurements are generally conducted in deuterated solvents and/or with partially deuterated block copolymers. The drawback of this technique is that it need calculation and is model dependent.

In addition, analytical ultracentrifugation [86], and size-exclusion chromatography [87] have been used to characterize block copolymer micelles. NMR, stop-flow, and fluorescence technique have also been used to characterize different aspects related to the dynamics of block copolymer micelles in solution [12].

2.5. Dynamics of Block Copolymer Micelles in Solution

Reasonably fast dynamics of micelles formation and block copolymer molecules (unimers) exchange between micelles are important in reaching and establishing equilibrium. If the microstructure is to be maintained beyond equilibrium, then the exchange dynamics must be slow. Dynamics of copolymer micelle in solution include three important parts: micelle hybridization refer to the exchange rate of unimers between different micelles or so-called "mixed micelles"; unimer-micelle equilibrium; chain dynamics of copolymer blocks in either the core or the corona. These aspects are very important for the understanding of the mechanism of block copolymer micelles formation and particular micellar morphology. Kinetics studies of micelle formation and dissociation by direct methods are scare, as have been mentioned by Tuzar and Kratochvil [11], Hamley [15], Gohy [22], and Riess [12].

The micellization behavior of dispersions of block copolymers in solution is more complicated than that of low molecular weight surfactants because of the slow chain dynamics and large chain lengths. Many attempts have been made to understand the kinetics and mechanism of chain exchange between block copolymer micelles using a wide range of experimental techniques [88, 89], such as fluorescence nonradiative energy transfer, temperature-jump measurements, time-resolved light scattering, pulsed-field gradient NMR, and SANS. Briefly illustration will be given in the following discussion.

NMR has been proven useful in the measurement of coronal chain dynamics at the segmental level [90]. In systems of PS-PACs in carbon tetrachloride (CCl_4), the mobility of ^{2}H-labled polystyrene segments was probed at various distances from the ionic core [91]. ^{2}H relaxation time revealed that close to the ionic-nonionic junction there is a definite decrease in segmental mobility with respect to the "free" chain. Further from the ionic block, the segmental mobility is less restricted, and at a distance of ca. 40 units the mobility is close to that of nonionic homopolymer. It was also found that segmental mobility decreased with increasing length of the ionic block. Proton NMR has also been used to study the distribution coefficient of water between the copolymer micellar core and the solvent [92]. It has been found that reverse block copolymers micelles in nonpolar solvents are also able to solubilize small quantities of water. A single chemical shift for the water protons is observed, which is a weighted average of the chemical shift of water in the reverse micelles and in nonpolar solvent. The appearance of a single chemical shift indicates that the exchange of a water molecule between the ionic core and the solvent phase is very fast within the time scale of the NMR experiment.

SANS has been recently used to study copolymer micellar kinetics, as reported by Won and coworkers [93]. They have used time-resolved SANS to study block copolymer molecular exchange and micelle equilibrium for mixtures of normal and partially deuterated PEO-PB. The authors have shown that the micellar structures initially formed upon dissolution were completely locked in up to 8 days after preparation.

Stop-flow experiment, reported by Tuzar and Kratoch [11] and mostly recently by Kositza and coworkers [94], has been performed to study the kinetics of copolymer micellization. In analogy to low molar weight surfactants, it could be shown that two relaxation processes have to be considered for block copolymer micellar system: (1) a faster process attribute to unimer–micelle equilibrium at constant micelle concentrations. (2) a slower process assigned to an association-dissociation equilibrium, accompanied by changes in micellar concentrations, and accompanied by changes in micellar equilibrium. The differences observed between AB diblock and ABA triblock copolymers could be explained because two A bocks must escape from the micellar core in the case of ABA triblock copolymer chains.

Fluorescence quenching techniques have been proven useful in studying the exchange of block copolymers between micelles formed by PS-PCEMA diblock copolymers in THF/cyclopentane mixtures [95, 96]. DLS, SAXS, and others have also been used to monitor block copolymer micelle equilibrium in solution [69, 97-100].

2.6. Cross-Linked Copolymer Micelles

Although block copolymer micelles are stable in time at fixed conditions, their characteristic features depend for a given system on the thermodynamic quality of the solvent and on temperature [32, 93]. Block copolymer micelles will be diassociated under different conditions, such as in a different solvent, at different temperature, or different ionic conditions. In addition, optimization of block copolymer micelles in structural stability for practical applications, for example drug delivery, is challenging. In nature, stabilization of self-assembled structures is provided by a few covalent cross-links, for example the disulfide linkage that forms between cysteine residues in proteins. By analogy, the formation of covalent cross-links between the polymer chains that compose the multi-molecular polymer micelles generates a single, nanostuctured macromolecule from the original, self-assembled aggregates [101]. The obvious way to stabilize block copolymer micelles is cross-linked the micellar core or corona. Several procedures have been established to reach this goal. It has been shown that block copolymer micelles, for example, vesicles and wormlike micelles can be chemically cross-linked without structural disruption, leading to dramatic modifications of static and dynamic mechanical properties [102].

Liu and coworkers systematically used photo-cross-linking to stabilize copolymer micelles with a PCEMA block. They have investigated the micellization of PI-PCEMA-PtBA triblock copolymers in mixtures of THF and methanol and have used these micelles to create various nanoobjects. The shell-forming PCEMA block was then cross-linked by UV irradiation to lock the structure [103]. In a further step, hollow nanospheres were obtained by ozonolysis of the PI micellar core. If pure methanol is used as solvent, cylindrical micelles are formed which can also be transformed into nanotubes by the same procedure [104]. The micellization of the same copolymers in mixtures of THF and hexane was also studied. Under these conditions the corona is formed by the PI block, the shell by PCEMA, and the core by PtBA. After cross-linking the shell, the PI corona was hydroxylated to make the micelles dispersible in water, and the PtBA core was made compatible with inorganic species by cleaving the *tert*-butyl groups. These authors demonstrated that photo-cross-linking of PCEMA locked in the initial structure of the micelles without any significant change in their aggregation number and size distribution, proved by SLS, DLS, TEM, and SEC characterization [105]. The potential of these nanospheres as templates for the preparation of inorganic nanoparticles was demonstrated by incorporating iron oxide particles into the core. Prochaska and Baloch [106], Tuzar and coworkers [107] have tried to stabilize a particular micellar structure such as block copolymers with a PB blocks by cross-linking of the micellar core, either by UV or fast electron irradiation.

The other possibility, examined by Wooley and coworkers is to chemically cross-link the corona of the micelles [108]. These kinds of nanoparticles are named shell cross-linked (SCK) micelles by these authors. A large variety of block copolymers, mainly amphiphilic block copolymers with PAA or quaternized PVP as the water-soluble block, which can be chemically cross-linked in their micellar form have been applied this concept [109]. Armes and coworkers have used a similar strategy for the fabrication of shell-cross-linked micelles with hydrophilic core and shell [110, 111].

3. Self-Assembly of Block Copolymers in Aqueous Solutions

3.1. Self-Assembly of Amphiphilic Block Copolymers in Aqueous Solution

Amphiphilic block copolymers can be divided into two types according to the charge of blocks: nonionic amphiphilic block copolymers and ionic amphiphilic block copolymers.

Nonionic amphiphlic block copolymers in aqueous solution are typically formed of water soluble hydrophilic block, i.e., PEO, PMVE, PNIPAM, linked to a hydrophobic block, e.g., PPO, PBO, PS, PB, PMMA.

In the past decades, self-assembly of typical PEO-PPO and PEO-PBO amphiphilic diblock and triblock copolymers have been widely investigated in aqueous solution. Their micellization behavior has been studied quite extensively and reviewed by Chu [25], Hamley [15], Almgren et al. [112], Booth et al. [113], Alexandridis and coworkers [114], Wanka et al. [115], and Riess [12]. Depending on temperature and concentration, different structures have been reported for the corresponding micellar solutions, including micelles with various morphologies and physical gels. In marked contrast, heat-induced micellization has been observed in PEO-PPO diblock copolymers in water [116, 117]. Such entropy-driven micellization is related to the decreased solubility of PEO (and PPO) in water with increasing temperature, but its molecular origin remains in dispute [118]. The structure of PEO-PPO-PEO micelles is a weak function of concentration, but changes rapidly with temperature [119], with the association number increasing, and micelles becoming drier, as the temperature is increased and water is expelled from the micelle core [120]. Gradual replacement of water by block copolymer in the micelles core explains the broad nature of the observed transition from block copolymer unimers to micelles. While PEO-PPO-PEO micelles are typically spherical, a sphere-cylinder transition has also been documented at high temperature [121]. Mortensen and Pedersen have studied the self-assembly of PEO-PPO in aqueous solution and obtained detailed picture of the morphology of such micelles [122]. It has been found by them that in the case of spherical micelles, the PPO core is surrounded by a dense layer of PEO segments and an outer corona of flexible PEO chains. The role of various additives on the micellization behavior of PEO-PPO, PEO-PBO diblock copolymers has also been investigated recently [22].

Bates and coworkers have systemically studied the self-assembly of PB-PEO in aqueous solution [81, 123-126]. The core-corona structures (spheres and cylinders) of PB-PEO block copolymers micelles have been directly observed by cryo-TEM. Images of both shapes of micelles show a dense core (PB segments) and a diffuse corona (PEO segments) [81]. A low molecular weight PEO-PB diblock copolymer containing 50 weight percent PEO forms gigantic worm-like micelles at low concentrations in water [125]. Subsequent generation of free radicals with a conventional water-based redox reaction leads to chemical cross-linking of the PB cores without disruption of the cylindrical morphology, as evidenced by cryo-TEM and SANS experiments. It is observed that the bilayer-to-cylinder and cylinder-to-sphere boundaries will shift toward f_{EO} values when the overall molecular weights increase at a fixed PEO composition. Increasing the PEO composition causes a sequence of transitions from bilayers to cylinders and then to spheres [123, 124]. That is, interfacial curvature increases with increasing PEO composition to accommodate the increased asymmetry between hydrophobic and hydrophilic volumes per chain while respecting the constraints imposed by the connectively between hydrophilic and hydrophobic blocks and the incompressibility of the core region. More interestingly, Y-junctions and three-dimensional networks of PB-PEO micelles in water were formed at weight fractions of PEO intermediate to those associated with vesicles and cylindrical morphologies [126].

PS-PEO diblock and triblock copolymers have been investigated in details, as reviewed by Riess [12]. Eisenberg and coworkers have reported the formation of crew-cut micelles from the self-assembly of PS-PEO with a major PS blocks in aqueous solution [127-129]. Multiple micellar morphologies have been obtained by them through tuning the relative block length of PEO with PS and solvent properties. Mortensen et al. have studied a PS-PEO diblock copolymer by using SANS, DLS, and SLS and have evidenced the formation of anisotropic clusters in water at concentrations up to 10 wt % [130]. The clusters were considered to be the result of the merging of the initial spherical micelles into bigger aggregates [131], which has been confirmed by cryo-TEM investigation. It was suggested that forces leading to clustering of PEO chains could involve hydrogen bonding, the structure of water, and the hydrophobic effect. The aggregated PEO chains were then thought to form either spherulites or non-crystalline microgels particles [132].

Besides the block copolymers described above, nonionic hydrophilic PEO blocks have been combined with a variety of other hydrophobic blocks, including PI [133], PDMS [79], PCL [134]. In particular, PEO-PCL block copolymers have attracted much interest because they are biocompatible and partly biodegradable (for the PCL block) and they are thus ideal candidates for application in drug delivery [134]. Recently, Savic and coworkers reported on the study of PCL-PEO block copolymer micelles as delivery vehicles for drugs [134]. They also reported on the cellular distribution of PCL-PEO micelles whose core had been labled by a fluorescent dye. Confocal microscopy in living cells revealed the localization of the fluorescent-labled block copolymer micelles in several cytoplasmic organelles, including mitochondria, but not in the nucleus. [134] Moreover, micelles changed the cellular distribution and increased the amount of a drug delivered to the cell. This study therefore emphasized the possibility for PCL-PEO micelles to selectively deliver drugs to specific subcellular targets [135]. Another type of nonionic water-soluble PNIPAM and PMVE blocks have also been combined with hydrophobic blocks, such as PS and PtBMA [136,137].

Ionic block copolymers in solution can be divided into two main categories: block polyelectrolytes and block ionomers [138]. The classification of ionic block copolymers as either block polyelectrolytes or block ionomers is determined by the nature of the solvent. Ionic block copolymers in water are block polyelectrolytes, forming micelles with nonionic cores and coronas made of the soluble ionic blocks. In organic solvents, ionic block copolymers can be described as block ionomers, with ionic micelles cores surrounded by the nonionic soluble blocks. Block ionomers will be discussed in Section 4. Block copolymers micelles with a polyelectrolyte (ionic block) corona are a very important class of collodial particles in aqueous medium [23]. The interest in these block copolymer micelles arises from the polyelectrolyte coronal block whose intrinsic properties are strongly influenced by many parameters including pH, salt concentration, and polar interactions. The micellization behaviors of these charged micelles have been very recently reviewed by Eisenberg [138], Riess [12], and Förster et al. [23]. In this section, we will give a brief overview of the topic for self-assembly of amphiphilic block copolymers consisting of one hydrophobic block linked to one ionic hydrophilic block in aqueous solution.

According to the charge of the ionic block, the polyelectrolyte micelles can be further divided into two parts: "anionic" form, such as PS-PAA and PS-PMAA systems in their acidic or neutralized form reported by Tuzar [139] and Eisenberg groups [128, 140-142], and "cationic" micelles, for example PS-PQ4VP and other quaternized P2VP containing block copolymers reported by Selb and Gallot [143].

Tuzar and coworkers have shown for PS-PMAA the step increase of the hydrodynamic radius R_h and of the electrophoretic mobility at a pH around 7, corresponding to the increases in dissociation of the carboxy groups when the pH was changed from 5 to 10 in various buffer solutions [144, 145]. As reported by these authors, the shift of the inflexion point, from about 5 for PMAA homopolymer to a pH 7 for the PS-PMAA copolymer, may indicate that PMAA chains of the micellar systems, and more likely those close to the micellar core, experience a low "effective pH" than that of a given buffer. They concluded that the degree of dissociation decreases from the shell outer layer to the core-shell interface.

Eisenberg and coworkers investigated highly asymmetric PS-PAA copolymers with long PS block and a very short PAA segments [31, 32, 51, 54]. The block copolymers can self-assemble into the crew-cut micelles in mixture of a common solvent and a precipitant for core-forming blocks, as introduced in section 2.2. The morphology of crew-cut micelles could be changed from spheres to rods to vesicles and other complex micellar architectures by decreasing the PAA/PS ratio in the starting copolymer, by adding additives (such as salt, acid, base, homopolymers, and small molecular surfactants), or by changing the initial common solvent properties [31, 32]. They have thoroughly discussed the underlaying thermodynamic and kinetic principles explaining the formation of such morphologies. Micellar morphology transitions, such as spheres-to-cylinders and cylinders-to-vesicles transition, have been investigated through experimental method, and the micellar morphological transition mechanisms have been discussed [146-148].

Rager et al. reported studies on fluorescent labled PMMA-PAA copolymers in mixtures of water with organic solvents, such as dioxane or methanol [149]. From the fluorescence and NMR measurements, these authors came to conclusion that the block copolymers micellization is proceeded by the collapse of the hydrophobic block and that the micelle-unimer exchange equilibrium might not be kinetically hindered under given conditions, such as by the glassy state of the micellar core, but could be controlled by a strong thermodynamic preference for the aggregated state [12].

Another type of anionic copolymers derived from strong acids like PSS or PSGMA. PSS-containing copolymer micelles have been studied by Guenoun et al. [150], and Förster and coworkers [151, 152]. Jérôme and coworkers have investigated the self-assembly of PMMA-PSGMA and PtBMA-PSGMA copolymers in water [153, 154]. Formation of superaggregates of micelles was evidenced in these samples by a combination of DLS and size-exclusion chromatography experiments on the aqueous micellar solutions.

Recently, Förster et al. [23] and Kötz et al. [155] have reviewed the typical polyelectrolyte behavior of cationic block copolymers micelles in aqueous solutions. Particularly, the aggregation behavior of PVP (including P2VP and P4VP) based copolymers has received much attention recently [156]. The polymer PVP is generally a base. However, in a basic solvent such as DMF, α-H of VP may show some acidity through its special resonance structures [157]. Thus, PVP may show either acidic or basic properties in a basic solvent. When PVP is the corona-forming block, with changing pH, the nature of the corona will change from cationic to nonionic or to anionic. Thus, the aggregation behavior of copolymers with a PVP corona will change significantly. Because of the properties described above, PVP can be neutralized by an acid to form cationic blocks and by a base to form anionic blocks. However, in the systems observed past, amphiprotic properties were not observed frequently since PVP was either quaternized and in the corona or nonquaternized and in the core (reverse micelles). Although PVP block is not soluble in pure water, it is soluble in DMF, THF, dioxane containing up to 50 wt % water and is also soluble in water below a pH of 4.8 [34]. Thus, the PVP-containing block copolymer micelles can be prepared by indirect method, and the soluble micelles can be prepared in an acidic aqueous condition (addition of desired amount of HCl in the water). These types of block copolymer micelles can be ascribed as cationic micelles. Our group have systematically studied the self-assembly of PVP-based triblock copolymers (such as P4VP-PS-P4VP, PS-P2VP-PEO, PS-PB-P2VP) in common solvent/water solutions. Cornucopian micellar morphology, such as spheres, rods, Y-junctions, novel ring-shaped micelles, complex rings, vesicles, networks, giant segmented wormlike micelles, among others have been obtained through self-assembly of these triblock copolymers in solutions [34, 71, 82, 158-160]. Examples of the copolymer micellar morphologies are displayed in Figure 4, 7, 8. Micellar morphology transition from cylinders to ring-shaped structures has been observed when the annealing time is increased [82]. These micellar morphology formation mechanisms have been explored from an experimental and theoretical point of view [71, 82, 158, 159].

Recently, PB-P2VP micelles in which the P2VP block was reacted with HCl were investigated by cryo-TEM technique [23]. It was observed that the micellar PB core was surrounded by a thin dark layer containing a high density of condensed counterions, since the electronic contrast directly originated from the concentration of chlorine ions in the experiment. Furthermore this filaments consisting of protonated P2VP chain bundles were observed that extended from the interior shell into the dilute outer part of the corona. Micelles were connected by these filaments forming a random filament network. The formation of such structures was attributed to short-range attractive forces arising from transient fluctuations in the periphery of the corona, leading at times to attractive interactions between micelles. These aggregated states, i.e., strings and networks, were observed whenever the added salt concentration was increased.

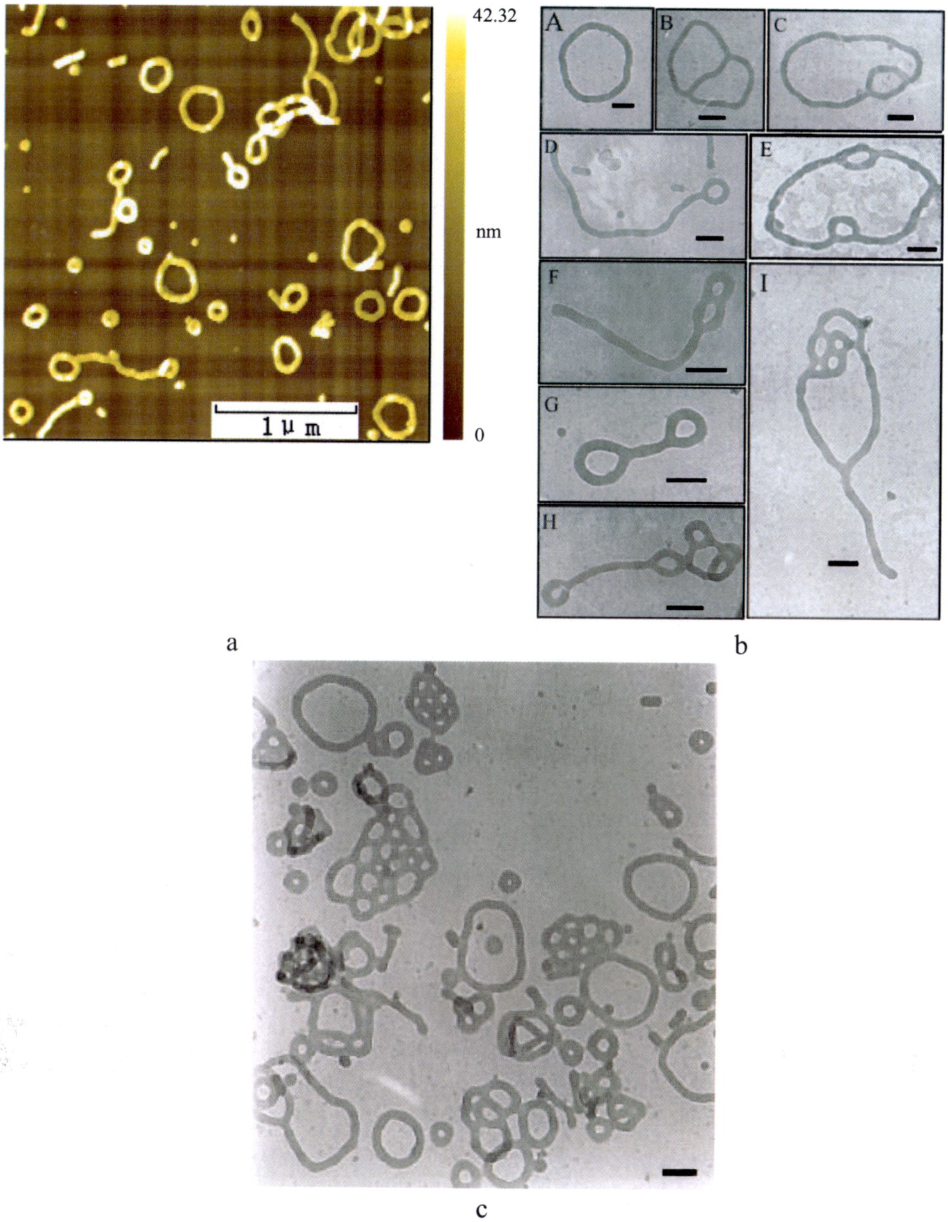

Figure 7. Cornucopia ring-shaped (a, AFM image) and compound rings (b and c, TEM images) morphologies through self-assembly of $P4VP_{43}$-PS_{260}-$P4VP_{43}$ in dioxane/water solution. The bar for TEM images (b and c) is 100 nm. Adapted from References [82, 159].

Besides the selected examples described above, there are another reports about the cationic block copolymers micelles. Selb and Callot have reviewed the micellar behavior of quaternized PS-P4VP block copolymers in water/methanol systems, among others [161]. Armes et al. investigated the micellar behavior of various amino-containing poly(methacrylates) with stimulus-responsive properties that were covalently linked with some “classical” hydrophobic blocks [162, 163].

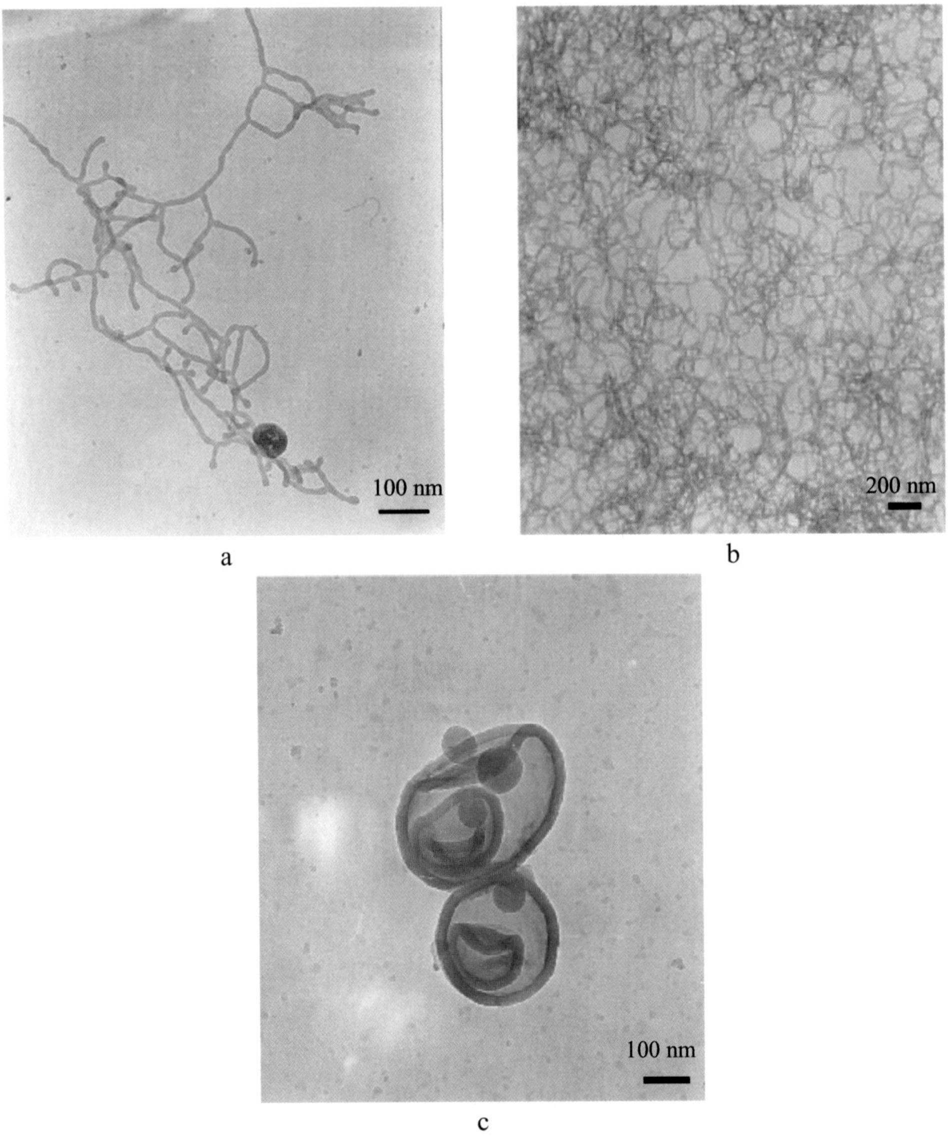

Figure 8. TEM images of branched cylinders (a), networks (b), and "pregnant vesicles" (c) obtained through self-assembly of P4VP-PS-P4VP in dilute aqueous solution. Adapted from Reference [159] for (a).

3.2. Self-Assembly of Double-Hydrophilic Block Copolymers in Aqueous Solution

Double hydrophilic block copolymers, also called bihydrophilic copolymers, refer to block copolymers (such as AB block copolymers) consists of two water-soluble blocks. Great efforts have been given to the study of double-hydrophilic block copolymers because one of the hydrophilic water-soluble blocks can be transformed into a hydrophobic block (that is amphiphilic block copolymers) whenever a given property of the aqueous medium is changed. Self-assembly of double-hydrophilic block copolymers in aqueous solvents has been reviewed recently by Cölfen [164]. In recent paper, Tauer et al. have explored the preparation, properties, and applications of double hydrophilic block copolymers [165]. Double-hydrophilic block copolymers have been successfully used as surface modifier, provided that one of the soluble blocks can specifically interact with the surface through hydrogen bonding interaction, and templates to control the crystallization of inorganic salts [165].

Stimulus-responsive block copolymers are based on copolymer blocks whose water solubility can be tuned by a variation in pH, temperature, or ionic strength of the surrounding aqueous medium. An early illustration of pH-responsive micellization can be found in P2VP-PEO copolymers that exist as unimers at pH<5 and form micelles at higher pH values, as demonstrated by Munk, Webber and coworkers [166]. Gohy et al. observed the self-assembly of P2VP-PEO in aqueous and found that a transition from spherical to cylindrical micelles and finally vesicles was observed for this block copolymers as the deprotonation degree increased [167]. Förster et al. recently reported giant vesicles (diameter: 5-10 μm, which can be imaged by optical microscopy) from the self-assembly of PEO-P2VP block copolymers in aqueous medium [168]. Moreover, these vesicles were stable down to pH 4.5, below which the P2VP block was protonated and the vesicles dissolved. Self-assembly of PEO-P2VP and PEO-P4VP copolymers at low pH in the presence of noble-metal salts have been investigated by Bronstein and coworkers [169-171]. The driving force of such micellization is the coordination of vinyl pyridine (VP) units with metal ions which proceeds by different basic scenarios, depending on the type of metal compound and the pH of the medium. The transformation of the PEO-PVP block copolymer micelles in water were found to depend on a number of parameters including the thermal history of the as-prepared block copolymer, the type of the metal compound, the metal loading [169]. For the $HAuCl_4$-filled PEO_{45}-$P4VP_{28}$ micelles, suggesting that slow nucleation (conformed by the formation of the large gold nanoparticles) and facilitated migration of gold ions yields the ideal conditions for sphere-to-rod micellar transition.

Another example of pH-driven aggregation is PDEAEMA-based block copolymers [172-174]. Lee et al. observed the structures of various PDEAEMA-based copolymers micelles in dilute solution by DLS and SANS as a function of pH and ionic strength [172]. These block copolymers carry a common pH-responsive core-block DEAEMA, and one of three coronal blocks, DMAEMA, PEO, and DMAEMA, whose side chain amine groups are selectively quaternized with benzyl chloride (Q-AMAEMA). It was found that the micelle aggregate number depends on the pH and ionic strength in aqueous medium. In all three copolymers, increases in the ionic strength cause the micelle aggregation number to increase by screening the electrostatic repulsions between chains. Another pH-response behavior was discovered by Cai and Armes for PEO-PDEAEMA-PSEMA copolymers [174]. Three types of micellar aggregates are found, depending on the pH of the solutions, with no second component being required. At low pH the core is formed due to hydrogen bonding between PEO and PSEMA, at pH=5 interpolyelectrolyte association (complex coacervation) determines core formation, whereas at high pH micelles with a hydrophobic PDEAEMA core are formed. Correspondingly, the corona becomes positively charged, neutral and negatively charged at low, medium and high pH respectively.

Another stimuli-response system for block copolymer micelles is the thermosensitive system, as reported by Armes and coworkers [162, 175]. Thermosensitive double-hydrophilic block copolymers micelles have been reported essentially for systems containing blocks exhibiting a LCST in water such as PMVE or PNIPAM. These blocks have been generally associated to PEO blocks [176]. At temperature below the LCST, both blocks are soluble and no aggregation is observed. At temperature above the LCST, the thermosensitive blocks turn hydrophobic and micellization takes place. For example, Poly(2-ethoxyethyl vinyl ether)-poly(2-methoxyethyl vinyl ether)-poly(2-(2-ethoxy) ethoxyethyl vinyl ether), PEOVE-PMOVE-PEOEOVE, forms micelles in water when temperature is increased above the LCST of the PEOVE block [177]. The micelle then associate to form a physical gel above the LCST of the PEOEOVE block and finally precipitate when the temperature is higher than the LCST of the PMOVE block. Another typical example of such a thermosensitive system has been reported for micelles formed by PEO-PNIPAM copolymers in water [178]. Temperature-dependent micellization of PMVE-PMTEGVE copolymers in water was monitored by DLS, ^{1}HNMR, and dye solubilization and found that unimers were detected below the LCST of the PMVE block, while well-defined micelles and a few superaggregation were present above the LCST [179]. Moreover, the LCST values for the PMVE-based copolymers were strongly influenced by the relative block length of the two blocks. Recently, more complex thermoresponsive copolymers consisting two blocks with each representing a LCST were investigated. In addition, these systems were sensitive to pH and ionic strength, paving the way to multi-responsive block copolymer micelles. In this respect, micelles sensitive to three stimuli: temperature, UV light and ionic strength were prepared in water from PAMPS-PNIPAM-PCEMA triblock copolymers [180]. The temperature dependence originated from the LCST of the PNIPAM block, UV light sensitivity from the cinnamoyl chromophores of the PCEMA block and ionic strength dependence from the ionic PAMPS block. Another interesting example was reported by Laschewsky and coworkers for block copolymer combining blocks with either a LCST or a UCST. The inside and outside blocks of the micelles can thus be tuned only by a simple thermal stimulus [181].

Another type of bihydrophilic block copolymers which have been investigated was diblock copolymer consisting of two different charged blocks, i.e., one blocks is negative charged while other block is positive charge. Interpolyelectrolyte (IPECs) between the oppositely charged blocks are observed in these systems, which can lead to insoluble material in aqueous solution [182]. Asymmetric copolymers with an excess of negative or positive charges can, however, give rise to water-soluble micellar systems, in which the micellar cores are formed by the insoluble IPECs, surrounded by a corona formed by the uncomplexed segments. The first example of these type copolymers has been reported for P2VP-P(M)AA copolymers [183]. The authors showed that the isoelectric point of these copolymers in water was controlled by the block length ratio of the two oppositely charged blocks. PDMAEMA-PMAA copolymers were also shown to behave similar [184-186]. Vesicular morphologies were observed around pH=9 for these copolymers containing a major PMAA block. These vesicles are thought to consisting of an insoluble layer formed by PDMAEMA/PMAA IPECs stabilized by tangling negatively charged PMAA segments. In addition, this type of block copolymers can complex with surfactants and has been used as a tool to generate well-defined micellar structures. Mahltig and coworkers have studied the P2VP-PDMAEMA copolymers, which is double-hydrophilic copolymers containing two basic blocks [187]. Self-assembly of the diblock copolymers as a function of pH have been investigated: at low pH, both the P2VP and PDMAEMA blocks were protonated and water soluble; at intermediate pH, polyelectrolyte block copolymer micelles were formed with the P2VP core and a positively charged PDMAEMA corona; at high pH, micelles were formed with the same insoluble P2VP core and thermoresponsive uncharged PDMAEMA coronal chains.

4. Self-Assembly of Block Copolymers in Organic Solvent

Basically, micelles in organic solvents can be considered into two types, those with one hydrophilic and one hydrophobic block, and with double hydrophobic blocks. The former one is called reverse micelles, containing hydrophilic core surrounding by a soluble hydrophobic corona. In recent publication of Liu et al. [19, 20], Hamley [15], and Chu et al. [25], the self-assembly of block copolymers in organic solvents has been reviewed. Generally, micellization of block copolymers in organic solvents are considered as an entropy-driven process.

In the past decades, reverse micelles from amphiphilic block copolymers have been widely investigated [139, 188]. In general, the core of reverse micelles typically consists of PEO, PVP, PAA, PMAA etc hydrophilic blocks while the corona formed from hydrophobic chains, such as PS, PPO.

PVP (include P2VP and P4VP)-containing reverse micelles have been widely investigated [189]. Recently, Förster et al. reported a very detailed investigation on PS-PVP copolymers dissolved in a selective solvent for the PS block, such as toluene [188]. PVP-PS blocks micelles have been widely used as microreactors or templates for the formation of metallic nanoparticles [189], as will be discussed in details in Section 7.1. Various metallic nanoparticles have been prepared in the case of PS-PVP reverse micelles, as illustrated by the works of Spatz, and Antonietti and coworkers [190-195]. Besides metallic nanoparticles synthesis, interaction of metal salts with PVP blocks has a deep effect on the micellization behavior of the corresponding PS-PVP copolymers [196]. Indeed, the incompatibility between the core and the shell in the micelles can be enhanced by neutralizing the PVP units of the core-forming block by $HAuCl_4$.

Minko and coworkers have investigated the self-assembly of PS-P2VP copolymers with a heteroarm starlike architecture in controlled environment and deposited onto mica or silicon wafers by using AFM with molecular resolution [197, 198]. They found that unimolecular micelles were formed with the P2VP arms in the micellar core and the swollen PS arms at the micellar shell in toluene. The reverse situation was observed in acidic water. Moreover, the interactions between those two inverse types of micelles were strongly modified by interaction with the mica substrate.

PEO-based copolymers have been received much attention. In this respect, PEO-PPO and PEO-PPO-PEO pluronic copolymers were investigated in organic solvents such as formamides, as illustrated by the works of Lindmann et al. and Alexandridis et al. [199-201]. Due to the ability of crystallization of PEO in organic solvent, the formation of reverse micelles from PEO-based block copolymers has been shown to be more complex phenomenon. Gast et al. studied the crystallization of PEO in PS-PEO and PB-PEO micelles [202-204]. Wu et al. demonstrated that a key factor in the crystallization of PEO was the amount of residual water [205, 206]. In the absence of water, lamellar microcrystals structures are formed by chain-folding crystallization of PEO as shown by Kovacs and coworkers [207, 208].

Reverse micelles formed from block copolymers with one ionic block are called "block ionomers". Reverse micelles from PAA and PMAA-containing block copolymers have been extensively studied by Eisenberg and coworkers [209, 210]. This type of micelles is formed from a major PS block linked to short ionized PAA and PMAA segments. Stable spherical micelles were formed by these block ionomers in organic solvent, such as toluene. The micelles were shown to consisting of an ionic core and a PS corona, characterized by TEM, SAXS, DLS, and SLS.

The self-assembly of block copolymers containing two hydrophobic blocks in organic solvents has been reviewed by Chu et al. [25] and Hamley [15]. The micellization of these types copolymers has been shown to be mainly controlled by the solvent and its interaction with the block copolymer blocks. Thus, the micellization can be tuned by changing the organic solvent properties. Very recently, Bang et al. observed the self-assembly of PS-PI diblock copolymers in various organic selective solvent for PS (such as dibuthyl phthalate, diethyl phthalate, and dimethyl phthalate) and their mixtures [211]. It has been found that with increasing solvent selectivity, the predominant micellar shape changes from spheres to cylinders to vesicles, reflecting the changing interfacial curvature. Recently, Larue et al. reported [212, 213] the self-assembly of PS-PI in heptane and found the morphology changed from spheres to cylinders by reducing the length of the isoprene corona block, analogous to the work of Won et al. [123]. Canham et al. dispersed a SBS triblock copolymer in ethyl acetate [214]. The solutions showed a distinct cloud point upon heating, and they found the onset of cloudness was due to the formation of long cylinders from spheres. Liu and coworkers have systematically studied the self-assembly of PCEMA-containing copolymers in organic solvents [103, 215-219]. Various shaped objects have been obtained from these type copolymers micelles. Moreover, these copolymer micelles can be easily cross-linked in the core because PCEMA block is photo-cross-linkable.

Another interesting study about the self-assembly of block copolymers in organic solvent is the use of novel solvents, such as supercritical CO_2 ($scCO_2$) and nematic liquid crystals. A main advantage of using CO_2 as solvent is the final products can be recovered by easy elimination of the "nonharmful" solvent. In addition, block copolymers micelles in supercritical CO_2 recently became a field of interest due to their capability to stabilize emulsion-like polymerization systems. For example, polystyrene-fluoropolymer block copolymers, employed as emulsifying agents in supercritical CO_2, form small (association number 6-7) micelles with a CO_2-phobic PS core and a CO_2-philic fluoropolymer corona [220]. In this case, the pressure of the supercritical CO_2 serves to control micellization. Specific experimental setups need to allow the study of micelles in $scCO_2$, which has been illustrated by Londono et al. for the study of micellization of PVAC-PTAN copolymers in $scCO_2$. Three regions have been observed in the phase diagrams of fluorinated block copolymer in $scCO_2$ [221, 222]: (1) a two-phase region at low CO_2 density; (2) solutions of spherical micelles at intermediate CO_2 densities; (3) solutions of individual copolymers at high CO_2 densities. Increasing pressure and decreasing temperature increase the density of CO_2 and thus improve its solvent quality. Micellization of block copolymers in CO_2 has been discerned to be endothermic and entropy-driven, surprisingly similar to that of amphiphilic block copolymers in water [221, 222].

The solution behavior of block copolymers with a liquid crystalline side-chain (LC) block and an isotropic (I) block has been studied in dilute nematic solutions composed of low-molar-mass liquid crystals [223]. Block copolymers with a high fraction of the I-block are insoluble in the nematic solvent, while copolymers with a high fraction of the LC-block are completely soluble. If the weight fraction of the LC-block lies in the range 0.46-0.85, thermoreversible thread-like aggregates are formed, with their long axis oriented along the director field of the nematic matrix.

5. CONTROL OF BLOCK COPOLYMER MICELLAR MORPHOLOGY

Great control over micellar shape and size is very important to obtain the desired functions and properties in practical application possibilities. In the fabrication of nanostructurd materials based on the copolymer micelles template, it would be highly desirable to control the morphology of the micelles. It is thus not surprising that great efforts have been devoted to the preparation of block copolymer micelles with various morphologies. Different strategies have been explored in the past few years, as will be illustrated in the following.

In a recent review, Eisenberg et al. have discussed the factors that affect the block copolymer micelle morphology and the strategies about how to control of micellar morphology for the crew-cut micelles [224]. It has been found that the block copolymer micellar morphology is mainly controlled by a force balance involving three factors:

(1) the stretching of the core-forming blocks;
(2) the interfacial tension between the core and corona chains;
(3) the repulsion among the corona chains.

There are many factors that affect the above three terms, and by variation of one or more of these, the block copolymers micellar morphologies can then, in principle, be fine-tuned. It has been found that the most important factors influencing the micellar morphology are the block copolymer composition, the initial copolymer concentration in solution, the relative block lengths, the type and concentration of additives (such as acid, base, salt, homopolymers, and small molecular surfactants), the nature of solvents, and temperature [52, 225, 226].

It is well know that relative block length play an important role in copolymer micellar architectures. Eisenberg and coworkers have systematically studied block length effect on series of PS-PAA and PS-PEO copolymers micellar morphology [224, 227]. Meanwhile, Bates et al. have intensively investigated the self-assembly of PB-PEO with various PB/PEO weight ratio in aqueous solution [123]. It has been found that for block copolymers with a constant hydrophobic block (such as PS or PB block) length, as the length of hydrophilic block (such as PAA or PEO block) decreases, the micelles at some point transit from spheres to rod-like micelles. As the length of hydrophilic block drops further, vesicular aggregates are formed. When the PAA block lengths become very short, the copolymers form highly polydispersed spheres, the so-called large compound micelles consist of a large number of reverse micelles with polar PAA cores in a PS matrix. Such micellar morphology transitions are well-estabilished on the basis of molecular packing considerations. That is, decreasing the hydrophilic block length induces lower interfacial curvature. A balance between interfacial tension and chain stretching while respecting the incompressibility of the core domain leads to an equilibrium domain morphology and size.

However, this approach requires the synthesis of a library of block copolymer with various relative block copolymer lengths, which sometimes can be challenging, expensive and time consuming. As an alternative, people wish to describe morphological control mechanisms based on the manipulation of solution conditions rather than block copolymers synthesis. With such an approach, the same copolymer can be used to prepare micelles of different shapes, and to transverse a spectrum of morphologies in a controlled and reversible fashion, giving a new dimension to the process of macromolecular self-assembly and design. In the following part, we will outline the control methods about tuning the solution conditions, instead of block copolymer synthesis.

5.1. Polymer Concentration

The onset of micellization and the morphology of the self-assembled micelles (at a given water content) are both dependent on the block copolymer concentration in solution [227, 228]. Figure 9 shows an example of micelles from self-assembly of P4VP-PS-P4VP in DMF/water mixtures [158]. The micellar morphology changed from spheres to short rods to cylindrical micelles (rings and stretched cylinders) when the initial copolymer concentration in common solvent was 0.5 wt %, 1 wt %, and 4 wt %, respectively.

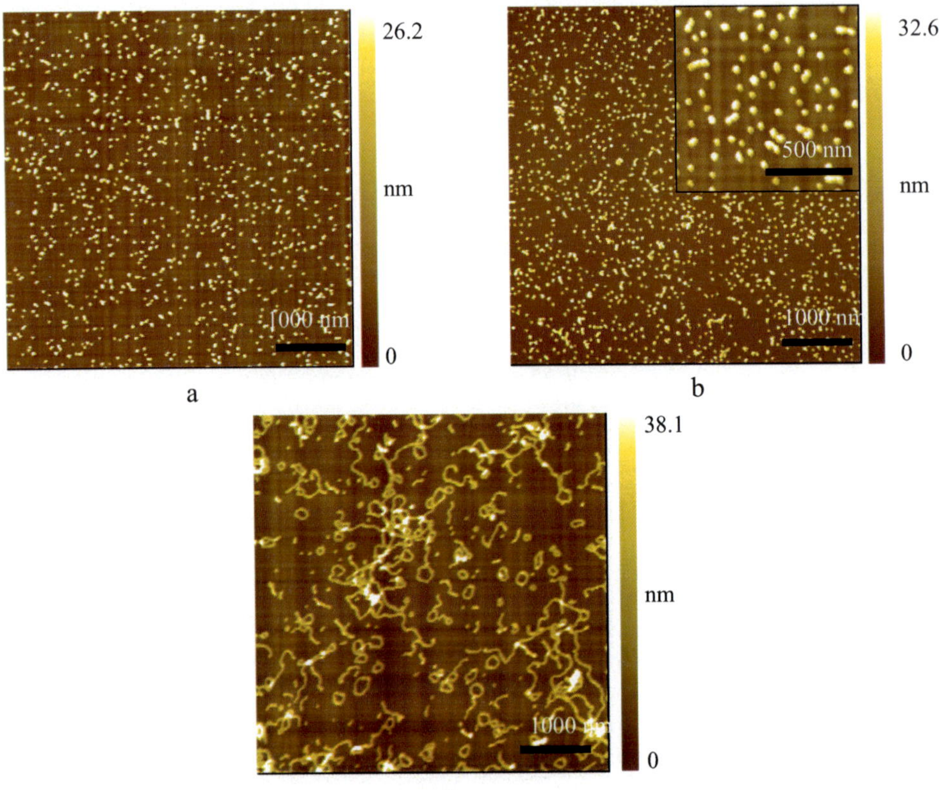

Figure 9. AFM height images of micelles formed from $P4VP_{43}$-PS_{366}-$P4VP_{43}$ with different initial block copolymer concentration: (a) 0.5 wt%; (b) 1wt %; (c) 4 wt %. Adapted from Reference [158].

The effect of the copolymer concentration on morphology is related to its effect on the aggregation number. For a given copolymer-solvent system (and therefore for a given value CMC), the aggregation number, N_{agg}, increases with the copolymer concentration, C. If one can extrapolate from the behavior of small molecule surfactants, then the relationship between the aggregation number and the copolymer concentration can be expressed in terms of the following equation [30]:

$$N_{agg}=2(C/CMC)^{1/2} \qquad \text{(eq.3)}$$

As the aggregation number increases, the degree of chain stretching increases and drives the observed transition from sphere to rods, and then to vesicles.

In addition, water content effect on the aggregate morphology of 1 % (w/w) PS_{310}-PAA_{52} in dioxane for the crew-cut micelles have been systemically studied [141]. It was found that when the water increased from 9.1 wt %, to 14.0 wt %, and then to 28.0 wt %, the aggregate was spheres, rods, and vesicles, respectively. More interesting, these aggregate morphology transition can be reversed. When the water decreased, similar aggregate morphology can be obtained. The ability to induce a series of morphological transitions by adding water can be explained that with increases the water content, reduced total free energy of the system drives spheres to transform into rods, and then to vesicles.

5.2. Solvent Properties

The micellar morphology can be easily tuned by changing the solvent properties [35, 160, 229-231]. For 2 wt % PS_{500}-PAA_{58} copolymers, when pure DMF is used as the common solvent, spherical micelles are formed; however, if the initial common solvent is THF or dioxane, vesicles are obtained [230]. The effect of the solvent effect on the morphology is also demonstrated by using a solvent mixture. As a typical example, PS_{200}-PAA_{18} yields spherical aggregates and large compound micelles when the initial common solvents are DMF and THF, respectively [35]. The intermediate morphologies, i.e., rods and vesicles, can be prepared by using DMF/THF mixtures of various ratios as the initial common solvent.

Recently, a convenient method of tuning the micellar morphology from triblock copolymer P4VP-PS-P4VP by adding selective solvent for core-forming blocks is introduced [160]. The micelles were obtained by dissolving the copolymer in a nonselective solvent (DMF), followed by the selective solvent (water) for P4VP blocks addition to induce the micellization of the copolymer. It was found that addition of selective solvent (such as toluene) for core-forming blocks (PS) has a significant effect on the micellar morphology. The micellar morphologies changed from initial spheres to rods, long tailed solid large compound spheres, and to long tailed hierarchical vesicles by adding 0.5, 10 and 30 wt % of toluene to the initial solvent, respectively. The toluene addition is expected to increase the stretching of the core-forming blocks (PS), and to modify the interfacial tension of core-corona interface, which are the main reasons for the micellar morphology transition.

5.3. Additives

Examples about the additives induced block copolymer (in particular polyelectrolytes) micellar morphology transition include the addition of ions (salt, base, or acid), homopolymers, or small molecular surfactants [42, 43, 52, 158, 232, 233]. The spherical micelles of PS_{410}-PAA_{25} increase in diameter from an initial value of 29 nm in the absence of any acid, to 38 nm when the solutions are presented at an HCl molar ratio R=0.035 (ratio of PAA unite to HCl molecules was named R). Preparing solutions at an HCl molar ratio R=0.04, 0.045 and 0.5 results in the formation of rods, vesicles and large compound vesicles, respectively. Base (such as NaOH), and salt (such as NaCl or $CaCl_2$) added to block copolymers can also have significant effect on the block copolymers micellar morphologies, as can be found elsewhere.

Recently, we have reported hydrogen-bonding-induced aggregate morphological transition of the triblock copolymers P4VP-PS-P4VP in dilute solution [158]. Small molecule surfactant PDP, were introduce to form H-bond with P4VP blocks. The interaction of both the hydrophobic chain and the hydrogen bonding of the PDP head-group with the block copolymer molecules affect the architecture of the copolymer aggregates. TEM observation shows that the micellar morphology transition occurred from spheres (initial one) to rods, network, vesicles, etc when increasing the PDP addition amount.

It has been found that homopolymer addition can also induce the micellar morphology transition for linear and even for cyclic block copolymer micelles [31, 52, 234-236]. Block copolymer micellar morphology was found to be influenced by the molecular weight and amount of the homopolymers added. We have observed that 20 wt % of homopolymer PS_{300} addition can induce the PS_{755}-$P2VP_{130}$-PEO_{77} micelles morphology from initial spheres to vesicles, as shown in Figure 10.

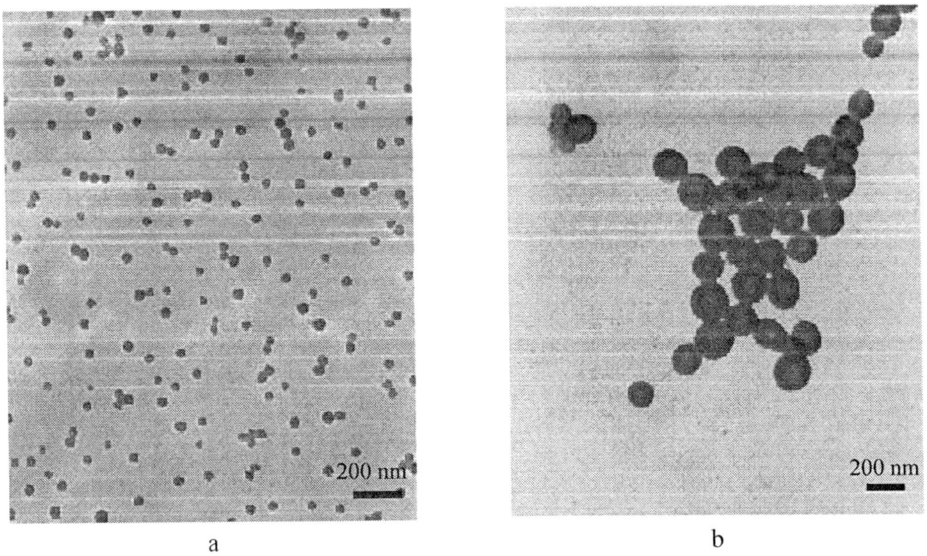

Figure 10. TEM images of initial spheres obtained from self-assembly of PS_{755}-$P2VP_{130}$-PEO_{77} in dioxane/water (a); and vesicles after 20 wt % homopolymer PS_{300} addition in the same micellar system.

Ouarti et al. recently studied the effect of PS homopolymer addition to the morphology of self-assembled block copolymer micelles made from linear or cyclic PS-PI in a selective solvent (heptane) for the PI block [236]. DLS and AFM results show that the addition of small amount PS homopolymer induces a dramatic change in the micellar organization. Indeed, a morphological transition, from spheres to cylinders for the linear copolymer, and from cylinders to vesicles for the cyclic copolymer, is observed.

5.4. Other Factors

In addition to the factors discussed above, the shape and size of block copolymer micelles are sensitive to other parameters, such as temperature [233, 237], molecular architecture, and polydispersity [238, 239]. These factors have been explored in the past reports and can also be used to tune the micellar morphology according to the practical applications.

The block copolymer morphogenic effects of the above-mentioned parameters can, in general, be ascribed to their influence on the force balance during the formation of the micelles. In particular, the length of corona-forming blocks variation or the addition of ions mainly affects the repulsions by the corona chains; the effect of the water content in solution is related to the change of the surface tension between the PS core and the solvent; and the effect of changing the solvent properties (such as changing the common solvent or addition of selective solvent for core-forming blocks) is due mainly to the increased dimensions of the core-forming blocks resulting from the changes of the solvent contents in the core and the aggregation numbers.

On the other hand, polymer, as a typical soft matter, its micellar structure and morphology strongly rely on the initial state, kinetic process of micellization and external field. These properties can also be used to control the micellar morphology. For examples, we have employed two methods to prepare micelles, i.e. indirect method and direct method. For indirect method, the copolymer ($P4VP_{43}$-*b*-PS_{260}-*b*-$P4VP_{43}$) was firstly dissolved in dioxane which is a good solvent for both PS and P4VP blocks and then deionized water was added gradually to the copolymer solution up to 25 wt % with stirring to induce the self-assembly of copolymer. For direct method, the same block copolymer was dissolved directly in dioxane/water mixture with the water content of 25 wt %. As a result, the copolymer is likely to form branched cylindrical micelles by indirect method, whereas the copolymer is likely to form complex ring-shaped micelles by direct method [159].

6. New Trends in Self-Assembly of Block Copolymers in Selective Solvent

6.1. Multicompartment Micelles

Multicompartment micelles, self-assembled aggregates with solvophobic cores that are further subdivided, drives from biological systems, where a single cell incorporates many different unites to perform distinct biological functions [240]. The concept of a multicompartment micelle was proposed by Ringsdorff in the midd-1990s to mimic an analogous biological behavior, and steps toward the realization of this vision have been reviewed by Laschewky [240]. Such systems are believed to be a perfect example for the use of bottom-up strategies in nanotechnology to tailor complex, functional materials and systems [241].

In general, synthetic ABC triblock copolymers with three mutually immiscible blocks can serve as a minimal model system to create several compartments in one micelle. Some initial progress toward this goal has recently been made [72, 242-250]. For example, Gohy et al. have reported the formation of core-shell-corona micelles (as illustrated in Figure 11a) from linear PS-P2VP-PEO. Lodge and coworkers have reported the formation of multicompartment micelles from ABC terpolymer comprising a hydrocarbon, a fluorocarbon, and a water-soluble block. For an ABC linear triblock copolymer, the fluorocarbon of the end A block induced a shape change from a core-corona spherical micelles to a disklike core-shell-corona structure, due to the superstrong internal segregation between the fluorocarbon (A) and hydrocarbon (B) segments. For an ABC mikto-arm starlike terpolymer, the mandatory convergence of the three domains at the molecular junction suppresses the formation of concentric core-shell-corona structures, leading to a new micellar morphology. They have observed disklike fluorocarbon domains sandwiched by two layers of hydrocarbon, while the hydrophilic blocks emanate from the disk/disk interfaces to stabilize the whole micellar core in aqueous solution. The overall structures that emerge depend on the relative lengths of the blocks and can be tuned from discrete multicompartment micelles to extended worm-like micelles with alternating fluorocabon and hydrocarbon cores.

We have reported a novel giant segmented wormlike micelles from the self-assembly of linear ABC triblock copolymers PS-P2VP-PEO in selective solution [34]. Giant segmented wormlike micelles (SWMs) have been observed, and the lengths and diameters of these giant SWMs were as high as 10 μm and 500 nm, respectively. Figure 12 shows representative TEM images of the segmented wormlike micelles formed from 1 % of PS_{755}-*b*-$P2VP_{130}$-*b*-PEO_{77} in THF/H_2O, the schematic drawing of an individual SWM micelle, and the proposed chain packing in the disk. TEM, AFM, SEM analysis revealed that the SWMs comprised sequences of repeated elemental parts, i.e., disks having a thickness of ca. 65 nm. In each disk, the hydrophobic PS units of polymer chains aggregate in the aqueous environment to form the core of the micelles, the hydrophilic P2VP units form a water-soluble shell that separate the core from its environment, and the hydrophilic PEO units form a corona.

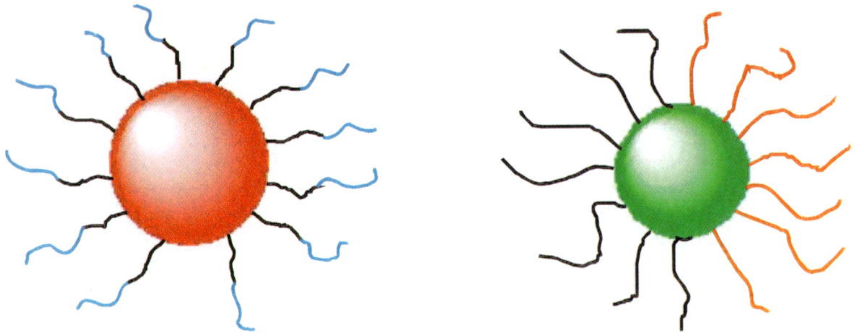

Figure 11. Schematic illustration of core-shell-corona copolymer micelles (a) and Janus copolymer micelles (b) obtained from self-assembly of triblock copolymers. Colors represent different components for the block copolymers.

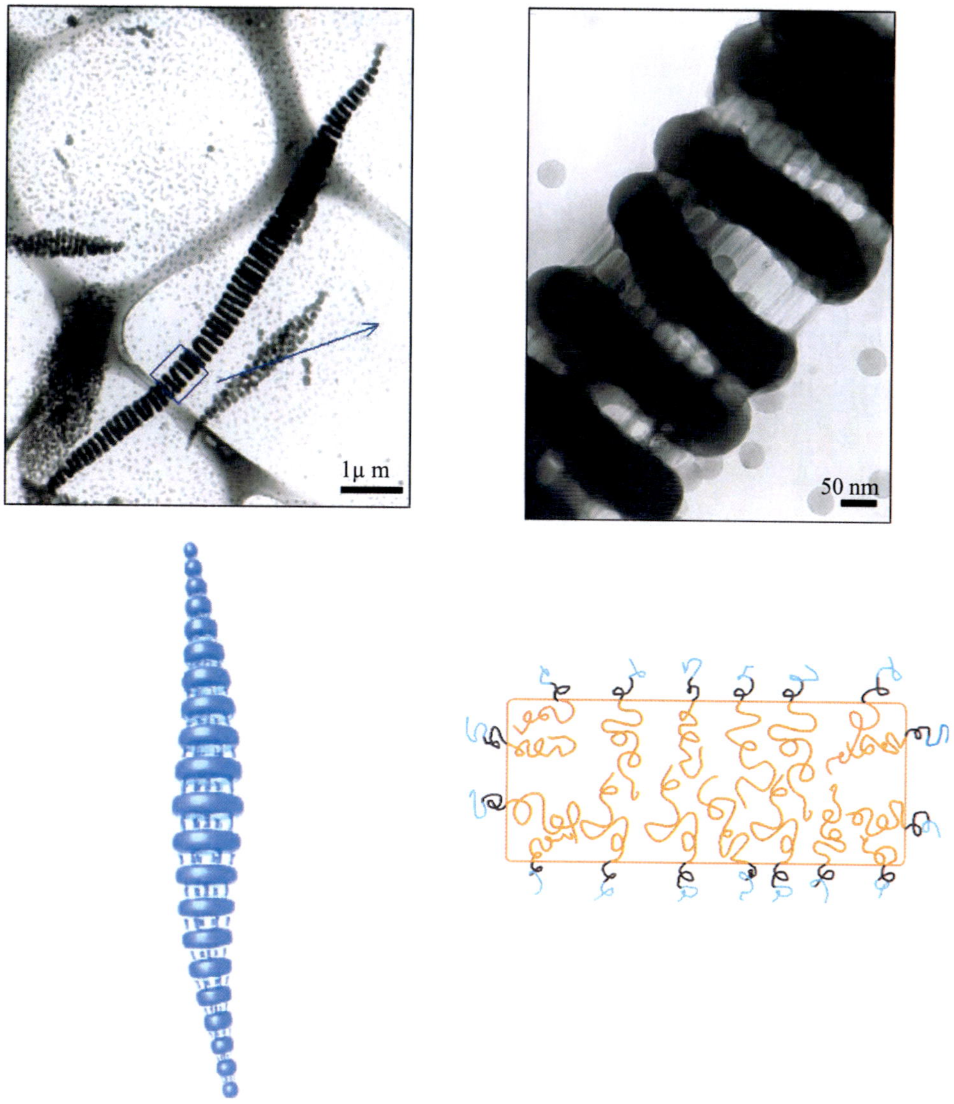

Figure 12. (a-b) Representative TEM images of the segmented wormlike micelles formed from 1 % of PS_{755}-*b*-$P2VP_{130}$-*b*-PEO_{77} in THF/H_2O mixtures (water content 15 wt %) annealing for 4 days; b is the enlarged part boxed in a, where the disks and threads can be seen most clearly. (c-d) Schematic drawing of an individual SWM micelles and proposed chain packing in the disks, the cross-section is parallel to the axes; the color of orange, black, ice blue represent PS, P2VP, PEO block in the cross-section, respectively. Adapted from Reference [34].

A most interesting feather is that disks having different diameters became connected through threads to form various segmented wormlike micelles. The micellar morphology has been known depended on the annealing time, solvent properties, pH value of the solvent.

Janus micelles are surface-compartmentalized nanoparticles, in which a cross-linked core is surrounded by two different corona hemispheres, as displayed in Figure 11b [60, 244, 247, 251, 252]. Recently, the synthesis of such structures was achieved by cross-linking of the center block of ABC triblock copolymers in the bulk state, using a morphology where the B block forms spheres between lamellae of the A and C blocks (see Figure 13a). Janus-type micelles with an interlocked PB core about 10 nm in diameter and a corona with two well-separated hemispheres PS and PMMA have been observed, as shown in Figure 13b.

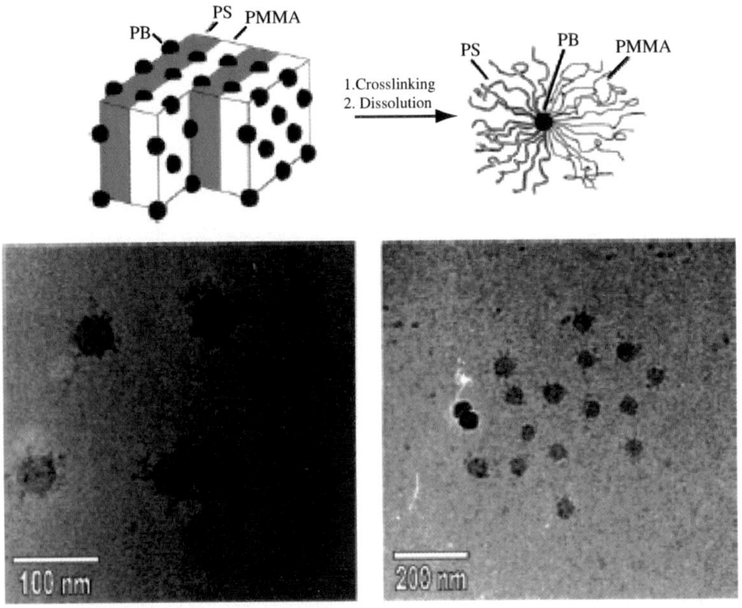

Figure 13. Scheme of the synthesis of a PS-PB-PMMA Janus micelle (a). Cyro-TEM images of the Janus micelles obtained in aqueous solution (b-c). Adapted from References [247, 251].

In solution, Janus micelles can further aggregate to large entities with a sharp size distribution, which can be considered as supermicelles. They coexist with single Janus micelles both in THF solution and on silicon and water surfaces. Cryo-TEM, DLS, and SLS observation indicate the presence of supermicelles and of significantly large structures in aqueous solution.

6.2. Co-Micellization of Block Copolymers in Solution

Most of studies have considered the micellization behavior of pure block copolymer chains. Some efforts have been given on the studies of co-aggregation behaviors of block copolymer mixtures in solution. This includes the co-micellization of copolymers with similar type, and co-aggregateion of copolymers with different type through non-covalent interactions.

Aqueous micellar dispersions of variety of binary blends of PEO-PB were prepared by premixing and postmixing diblock copolymers with varying composition and molecular weight, as reported by Bates and coworkers [253]. Using cryo-TEM, they observed there is no perceptible exchange of macromolecules between aggregates resulting in a nonergodic state, where overall equilibrium is never achieved. It has also been found that these nonergodic micelles relax to a state of local equilibrium through the redistribution of block copolymers within the topological framework established during dispersion. Binary PB-PEO mixtures under certain composition displayed intramicellar segregation leading to some novel aggregate morphology, such as cylindrical undulations and octopus-like aggregate with cylindrical micelles emanating from a single bilayer core.

Recently, we have investigated the self-assembly of binary mixtures of P4VP-PS-P4VP with various hydrophobic PS-block lengths in dioxane/water. It has been found that the aggregate morphologies can be tuned by only changing the composition of each copolymer in the binary mixtures. Some novel micellar morphologies, such as cage-like micelles, lamellae with protruding arms (rods) and unclosed vesicles with protruding arms (as displayed in Figure 14) have been observed for these binary block copolymer dispersions [254].

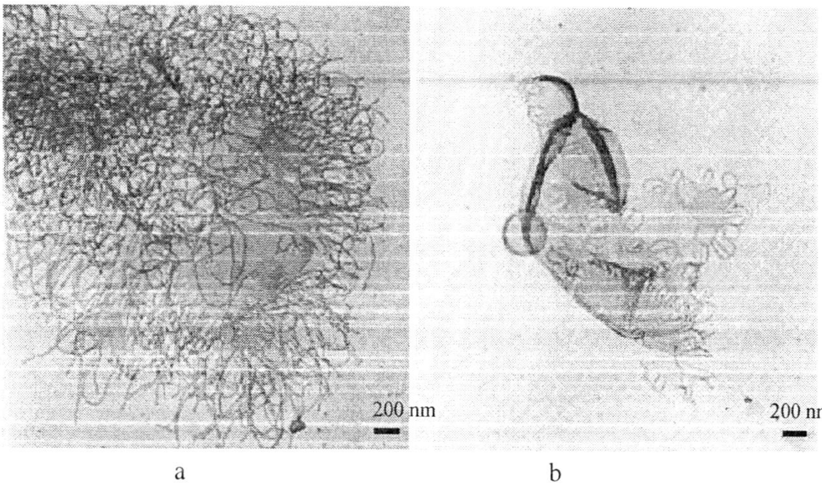

Figure 14. TEM images of block copolymer micelles with complex architectures obtained from self-assembly of $P4VP_{43}$-PS_{260}-$P4VP_{43}$ and $P4VP_{43}$-PS_{366}-$P4VP_{43}$ mixtures in dioxane/water solution.

Li and coworkers recently reported the self-assembly of mikto-arm terpolymer (form segmented cylindrical micelles in water) and AB diblock (form spheres in water) binary blends. The segmented cylindrical micelles shape changed from the initial one to prolatelike, to more oblatelike ellipsoids after the addition of AB block copolymers. The reasons were believed due to a collision/fusion/fission process of the two types of copolymers micelles [255]. Co-micellization of different type of block copolymers in dilute solution through complex formation will be illustrated in the following part.

6.3. Micelles with Non-Covalent Complexes in Solution

It have been known that micellization can be induced by complexation of one of the consistent blocks of the copolymer chains with other macromolecules or low MW molecules. In the following, we will briefly describe these contents.

Prochazka et al. have prepared onion-like micelles by mixing PtBA-P2VP micelles to P2VP-PEO chains in acidic water [256]. At pH below 4.8, the PtBA-P2VP micelles consisting of a PtBA core surrounding by a protonated P2VP corona, while PEO-P2VP chains behaved as double-Hydrophilic copolymers and thus existed as unimers. When the pH of the solution was raised above 4.8, insolubilization of the pH chains occurred. This process led to a co-precipitation of the P2VP chains from both the PtBA-P2VP micelle and the P2VP-PEO chains, which resulted in composite micelles with PtBA core, a collapsed P2VP shell, and a PEO corona. Similar onion type micelles were obtained by the combination of PS-P2VP heteroarm starlike copolymers with P2VP-PEO diblock copolymers, as reported recently by Tsitsilianis et al. [257].

Another possibility is that where the starting copolymers AB is already under its micellar form [258, 259]. By complexing the B chains in the corona with the homopolymer C the micellar structure could remain at low ratios of versus B. Lysenko et al. have reported the complexes formed between PS-PEVP and PMANa [259]. This lead to the formation of polyion complex micelles that are likely to have the multilayer structure: the PS core surrounded by the intermediate insoluble layer of mutually neutralized polyions and the lyophilizing shell of the polyion that is present in excess. The properties and stability of the complexes in aqueous dispersions depends on the ratio of charged units of the polycation and polyanion chains, Z=[PMANa]/[PEVP], confirmed by UV, TEM, SLS, DLS, and fluorescence spectroscopy. Zhang et al. investigated the co-micellization of P4VP and PEG-PAA in ethanol by SLS, DLS, and SEM [260]. It is found that co-micellization takes place due to strong hydrogen bonding between P4VP and the PAA block when mixing P4VP and PEG-PAA in ethanol. The resultant micellar complexes are core-corona spheres with the hydrogen-bonded PAA/P4VP as core and the PEG as corona. Furthermore, it is also found the hydrodynamic diameter R_h, the gyration radius R_g, and the mass weight MW of the micellar complexes can be easily tuned by changing the weight ratio of P4VP to PEG-PAA.

There are other possibilities for the micelle complexes. Jiang et al. reported the micellization of block copolymer and a random copolymer in a non-selective solvent caused by interpolymer hydrogen bonding complexation [261]. The same group recently reported on the preparation of "pseudo-graft" copolymers from a homopolymer pair with one bearing a terminal H-bond donor (acceptor) and the other containing H-bond acceptor (donors) along the chain. Such pseudo-graft copolymers formed nanoaggregates in block-selective solvents just like conventional graft copolymers or ever diblock copolymers [262, 263]. Mixtures of PS-P4VP copolymers with various low-MW molecules including perfluorooctanoic acid and formic acid have also been investigated [264, 265]. Antonieti et al. reported the interpolymer complex formation in organic medium of a system AB+CD [266], where A and D are long PS and PB segments, respectively, with B and C being PMAA and P4VP. "Janus" type micelles are claimed to be found under these conditions due to the incompatibility of the A and D blocks. Yan et al. quite recently reported the co-aggregation of BC and DC diblock copolymers through hydrogen bonding in selective solvent [267].

6.4. Others

In addition to contents described above, there are some other trends in self-assembly of block copolymers in selective solvents, such as rod-coil copolymer micelles [268-270], organometallic micelles [271-276], multiblock copolymers micelles [277, 278], star-like copolymer micelles [279, 280], H-shaped [281] and graft copolymer [282] micelles, among others.

The self-assembly of rod-coil block copolymers were extensively studied especially in the case with a helix type "rigid" block (such as polypeptide). Block copolymers containing a helical polypeptide block and a second coil-forming synthetic polymer block have been investigated for a long time, as reviewed by Lecommandoux and Klok [268]. Nolte and coworkers reported on the formation of micelles with a helical superstructure from AB amphiphilic diblock copolymers prepared from an amine-end-capped polystyrene that was as the initiator for the polymerization of various dipeptide-derived isocyanides [269].

An interesting development of polyferrocene-containing block copolymers have been extensively studied by Minners and Winnik groups [271-274]. These authors prepared a series of copolymers with an organometallic block either of poly(ferrocenylsilane)(PFS) or of poly(ferrocenylphosphine) (PFP). The solubilizing block of these copolymers could be either PDMS, PS, PI or PEO. Depend on the copolymer type and its composition, various micellar morphologies have been obtained ranging from spherical star-like micelles to worm-like micelles or hollow nanotubes, as well as flower-like micelles.

7. Applications of Block Copolymer Collodial Systems

Compared to small molecular weight molecules, micelles from block copolymer are more thermadynamically and kinetically stable, broadening their applications. There are broad range applications of block copolymer micelles in many industrial and pharmaceutical preparations field. A detailed description of all these application possibilities of block copolymers would exceed the scope of this chapter. We will briefly highlight some specific aspects of applications of block copolymer micelles.

7.1. Nanostructures Fabrication Based on Copolymer Micelles Templates

The assembly and fabrication of micro- and nanostructures have attracted a great deal of attention for the potential applications of creating novel materials and devices, such as optoelectronic devices, photonic bandgap materials, magnetic, and micromechanical devices [283, 284]. A number of methods have been successfully used to fabricate nanostructures. Among these methods, templates synthesis has been one of the most popular methods for nanostructures formation. In particular, block copolymer micelles have been widely used as soft template for nanostructures fabrication [40-42, 285].

The classical method to prepare such nanostructures is to trap metal salts in amphiphilic block copolymer micelles formed in selective solvents. The core is able to trap particles by complexation or association, and the shell provides stabilization. These micelles structures can then be considered as nanoreactors in which minute metal particles are formed by chemical transformation. In this case, the block copolymers must be selected in such a way that one block has affinity with the metal or metal salt, and other block with liquid medium. PVP (including P2VP and P4VP), for example, block (another often used block is PAA) is mostly used for its strong metal complexing ability. In this respect, the micelle is mostly reverse micelle and the medium is organic solvent. To achieve this goal, precursors of metals were firstly loaded into the micellar PVP core, which was achieved by simply stirring the precursor salt in the micellar solution. Transition metals were bound either directly to the PVP ligand in the micellar core or indirectly as counterion. In a further step, the metal precursor has to be reduced to form the metal nanoparticles. This step is rather easy because the redox potential of the coordinated metal is often reduced due to a partial charge transfer from the polarizable ligands to the positively charged metals. Common reducing agents are $NaBH_4$, $H_2N\text{-}NH_2$, H_2, or $LiAlH_4$. More interesting, semiconductor nanoparticles can also be prepared by addition of H_2S to the metal precusors (such as $Cd(ClO_2)_2$) to form the corresponding metal sulfides. The resulting nanoparticles further aggregate to form large clusters by nucleation and growth processes, which lead to one single colloid per micellar core or several small colloids within a micellar core. These have been reviewed by Förster and Antonietti [189]. Spatz and coworkers have reported the formation of Au-nanoclusters between 2 and 8 nm in diameter deposited on solid substrates in different pattern geometries [286]. The basis of this approach is the self-assembly of PS-P4VP ($HAuCl_4$) diblock copolymer micelles into uniform monomicellar films on solid supports such as Si-wafers or glass cover slips. $HAuCl_4$ as metallic precursor or a single solid Au-nanoparticles caused by reduction of the precursor are embedded in the centre of block copolymer micelles. Subsequent hydrogen, oxygen or argon gas plasma treatment of the dry film causes deposition of Au-nanoparticles onto the substrates by entire removel of the polymer. Separation distance between the dots is controlled by the molecular weight of the diblock copolymers.

Schmidt and coworkers reported the formation of gold clusters and nanowires based on the templates of core-shell cylindrical brushes from methacryl end-functionalized block copolymer consisting of PS and P2VP [287]. The block copolymers formed cylindrical brushes exhibiting a core of P2VP and a shell of PS in a selective solvent for PS (such as toluene). The core of the cylindrical brushes was loaded with $HAuCl_4$ in toluene, followed by the reduction of the noble metal salt by the electron beam, by UV light, or by chemical reducing agents. Depending on the amount of complexed noble metal ions and the reduction conditions, either a linear array of noble metals cluster or a continuous nanowires is formed.

Amphiphilic block copolymers micelles have also been used as negative template for the fabrication of nanostructures. Fahmi et al. fabricated metallized nanowires by coating metal nanolayers onto soft nanotemplates, which was formed through self-assembly of diblock copolymer PS-P4VP hydrogen bonded with small molecular amphiphile [288]. Recently, we have reported that assemblies (such as nanospheres, nanowires, nanorings) obtained from self-assembly of P4VP-PS-P4VP in selective solvent can be excellent templates for metal nanostructures and semiconductive nanomaterials fabrication [84, 289]. The micelles consist of PS cores, and P4VP shells which can be served as both stabilizing agent and a structure-directing agent. Figure 15 shows Ag@Au nanowires formed through this method and the diameter of the metal nanowires can be tuned by changing the concentration of the reduction agent solution. The ZnS nanocrystal could be aligned along the corona of the copolymer micelles in near-perfect structure through control of both the molar ratio of Zn^{2+} to P4VP and the reaction time, as shown in Figure 16.

Double-hydrophilic micelles like PEO-PEI (or PEO-P2VP, PEO-P(M)AA, PAA-P2VP) micelles are also suitable for metal complexation, as detailed by Tauer et al. [165]. Zhang and coworkers developed a wet chemical method for the synthesis of silver nanowires by using a double hydrophilic block copolymer PEO-PMAA as a directing and reducing agent [290]. TEM and UV-vis spectroscopy results confirmed the formation of silver nanowires in PEO-PMAA aqueous solution. It is revealed that the formation of PEO-PMAA complex micelles may play an important role in the directed aggregation growth of the silver nanowires. Moreover, double hydrophilic block copolymer can exert very effective control of the morphology of inorganic crystals such as $CaCO_3$, $BaSO_4$, and Calcium phosphate. Specifically, unusual bundles of $BaSO_4$ nanofibers can be easily prepared in aqueous solutions of a phosphonated double hydrophiphilic block copolymer [291-293].

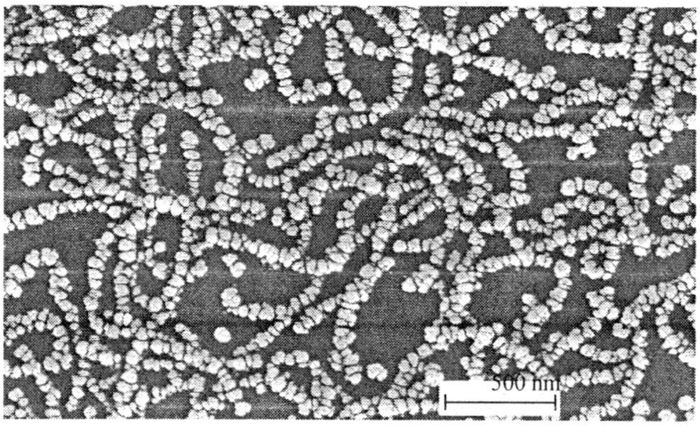

Figure 15. SEM image of Au@Ag bimetallic nanowires (average diameter: ca. 75 nm) based on the block copolymer $P4VP_{43}$-PS_{260}-$P4VP_{43}$ micelles templates. The nanowires were precoated with Au, and then coated with Ag. Adapted from Reference [84].

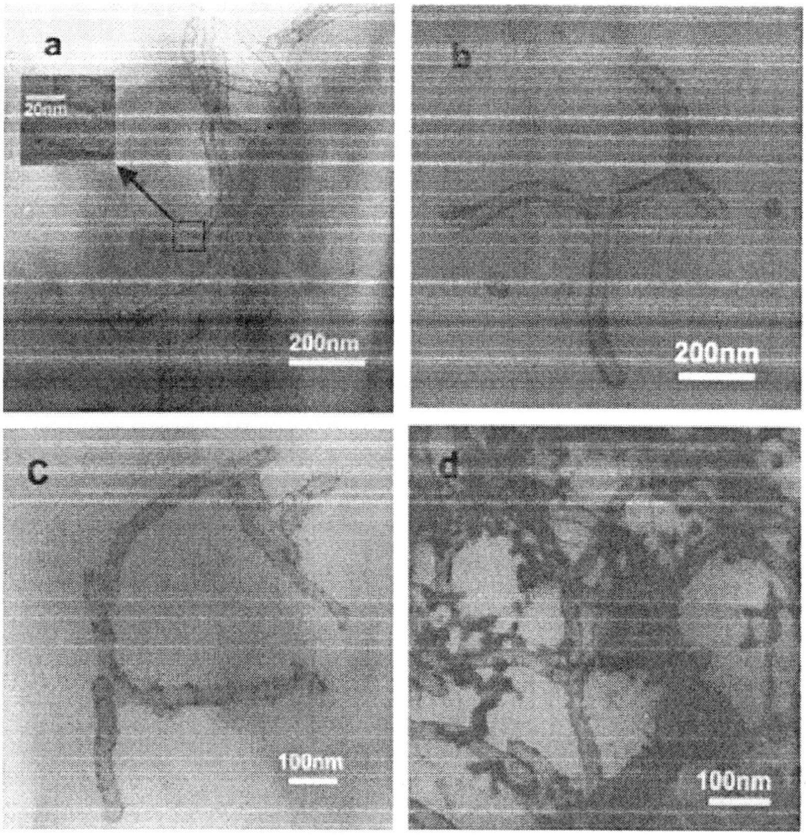

Figure 16. TEM images of ZnS nanocrystal assembled along block copolymer $P4VP_{43}$-PS_{260}-$P4VP_{43}$ cylindrical micelles at various Zn^{2+}/P4VP molar ratios: (a) 1:2 (the insert displays part of the original image at higher magnification); (b) 1:1; c (3:1); (d) 5:1. Adapted from Reference [289].

7.2. Drug Delivery System

In recent years, interest in micelles as potential delivery vehicles for pharmaceuticals, gene therapy agents and even food formulations has grown enormously [294]. The continuous development of new drug delivery systems is driven by the need to maximize therapeutic activity while minimizing negative side effects. One class of drug delivery vehicle that has received widespread attention over the past decades is micelles formed by self-assembly of amphiphilic block copolymers in aqueous solution. Various colloidal systems, such as liposomes (or vesicles), microspheres, emulsions have been described as drug targeting devices [295, 296]. Block copolymers micelles are of interest for drug delivery applications for a number of reasons. Firstly, hydrophobic drugs can be physically entrapped in the core of block copolymer micelles and transported at concentrations that can exceed their intrinsic water-solubility. Secondly, the hydrophilic blocks, such as PEO, can form hydrogen bonds with aqueous surroundings and form a tight shell around the micellar core. Block copolymer micelles with a PEO corona resist protein adsorption and cellular adhesion. As a result, the contents of the hydrophobic core are effectively protected against hydrolysis and enzymatic degradation. Thirdly, the PEO corona prevents recognition by the reticuloendothelial system and therefore preliminary elimination of the micelles from the bloodstream. Thus, these so-called "stealth" properties of the PEO corona result in increased blood circulation times and allow drugs to be administered over prolonged periods of time [294]. Last, block copolymer micelles' size and morphology can be controlled easily, as has been described in Section 5.

The first work on micelle-forming block copolymer, with the drug covalently linked to one of the blocks of the copolymer, was reported in early 80s by Bader et al. [297]. By using Poly(ethylene oxide)-poly(L-lysine) diblock copolymer, the authors fixed covalently the drug "cyclophosphamide" on the L-lysine block leading to micelles with a hydrophobic core of modified L-lysine and a hydrophilic fringe. Even more elaborate structures were developed for this type of copolymers in order to increase the activity of drug and to reach the level of the free drug, by introducing spacer groups and cleavable bonds, such as disulfide bonds, between the polymer backbone and the drug moiety [298]. If this approach is still of interest for water-soluble drugs, the main limitation for the controlled drug release of hydrophobic drug molecules by this strategy is that specific block copolymers, with suitable functional groups, spacers and cleavable bonds, have to be designed for a given kind of drug.

"Micellar microcontainers", where the drug is non-covalently fixed, in the hydrophobic core of block copolymers micelles, become therefore the preferred strategy during the last decade, especially with the possibility to provide these micellar systems with targeting function and stimuli sensitive properties.

In many case one might envision wishing to deliver two or more agents are in some sense incompatible, e.g., they might undergo chemical reaction before reaching the site of interest. Multicompartment micelles, however, offer this possibility: by dividing the core of a micelle into two or more distinct nanodomains, different agents might be transported simultaneously within core compartments. Lodge et al. very recently reported the simultaneous segregated storage of two agents in a multicompartment micelles, formed from self-assembly of mikto-arm starlike triblock copolymer consisting blocks of PEO, PEE, and PFPO in aqueous solution [299]. To demonstrate simultaneous selective uptake of two small molecules by the micelles, they chose two water-soluble chromophores: pyrene (Py), which is selectively dissolved in the PEE domains, and a fluoronaphthalene derivative, 1-naphthyl perfluoroheptanyl ketone (NFH), which is preferentially dissolved in the fluorinated PFPO domains. The UV-vis and DLS observations demonstrate that the multicompartment micelles can solubilize two distinct molecules within separate nanoscopic compartments.

7.3. Others

In addition to the above mentioned application possibilities, the variety of block copolymers which are now available, and the unique combination of properties they offer, make them further attractive for many chemical and industrial applications as emulsifiers, as stabilizer in latex technology, as compatibilizers in polymer blends, as surface adsorption and surface modification agents, and active component in separation processes [300, 301]. These contents have been recently reviewed by Riess [12] and Hamely [15].

8. Summary and Outlook

Block copolymers can self-assemble in selective solvents to form micelles with various morphologies, such as spheres, rods, vesicles, and others. Self-assembly of block copolymers in solution have been turned out to be a unique example to achieve self-assembled nanoparticles with defined morphologies, and be an area of increasing interest for the fundamental understanding and in view of practical applications. Because it was not possible to include every contribution to that field in the present chapter, only selected examples were discussed.

Although block copolymer micelles have been the subject of an enormous body of work during the past 30 years, some basic questions have not yet received a clear answer. Numerous examples of linear AB and ABA block copolymers were described due to the remarkable progress of synthesis of block copolymers. However, there is absolutely a need for libraries of a given block copolymer type, i.e., relative lengths of the different constitute blocks are systematically varied. Most of investigations on block copolymers micelles have indeed been realized on a limited number of samples. Up to now, only a limited number of examples have been reported in this respect.

On the other hand, there are nearly no limits in the design and optimization for novel types of block copolymers and of new architectures. Although many block copolymer architectures including star-shaped, H-shaped, graft, ABC triblock copolymer, multiblock copolymers, and dendrimers [302], can now be successfully synthesized, their micellization behavior remains poorly investigated. Micelles of complex architectures, such as ABC multicompartment micelles, Janus and onion micelles which have appeared in the past few years, are of interest and they need to be further investigated. The use of non-covalent interactions to trigger micellization from mutually interacting block copolymers is also a topic receiving increasing attention. In addition, some novel (so-called "hybrid") block copolymers comprising a "classical" synthetic polymer in one block, and synthetic or natural polypeptide (such as protein, enzyme, DNA, etc) [303-305], inorganic compound (such as metal, silicon) on the other block have been synthesized and their self-assembly behavior will be investigated in the next few years [306-308]. Another challenge for the synthesis chemist is to provide "superpure" block copolymers which could have a strong influence on the micellization process as point out by Eisenberg [238, 239] and Zana. [88].

There is substantial or theoretical work on micellization of block copolymers. Rather detailed mean-field theories have been developed and interesting studies are in progress by computer simulation. Insight into the origin of self-assembly, morphology transition, dynamics of micelles etc will be studied with the help of theories studies [71, 159, 309-313].

Block copolymers micelles and assemblies are of great technologies importance because of the broad applications and potential applications. Further efforts will be given on the studies of applications possibilities of block copolymer micelles in broader ranges. For all of the applications, the further challenge is to tune and to control the micellar architectures and to achieve the desired size and microstructure of the micellar system. Therefore, control over micellar morphology (spheres, rods, vesicles, tubules etc as well as size) and functionalization of blocks (such as stimulus-responsive materials, biological functions, reactive materials for cross-links) will be another driving forces for the design of novel micellar systems.

ACKNOWLEDGEMENT

We are grateful for the financial support provided by the Major Program (20490220) of National Natural Science Foundation of China (NSFC), the National Basic Research Program (2007CB808000) of China. Jiang W. thanks the Fund (50621302) for Creative Research Groups (NSFC).

REFERENCES

[1] Ilhan, F.; Galow, T. H.; Gray, M.; Clavier, G.; Rotello, V. M. *J. Am. Chem. Soc.* 2000, *122*, 5895-5896.

[2] Alivisatos, A. P.; Barbara, P. F.; Castleman, A. W.; Chang, J.; Dixon, D. A.; Klein, M. L.; McLendon, G. L.; Miller, J. S.; Ratner, M. A.; Rossky, P. J.; Stupp, S. I.; Thompson, M. E. *Adv. Mater.* 1998, *10*, 1297-1336.

[3] Lazzari, M.; López-Quintela, M. A. *Adv. Mater.* 2003, *15*, 1583-1594

[4] Ruzette, A.-V.; Leibler, A. L. *Nat. Mater.* 2005, *4*, 19-31.

[5] Hillmyer, M. A.; Lodge, T. P. *J. Polym. Sci. Part A: Polym. Chem.* 2002, *40*, 1-8.

[6] Riess, G.; Hurtrez, G.; Bahadur P.; *Block Copolymers*, 2nd ed. Encylopedia of Polymer Science and Engineering, Vlo. 2. New York: Wiley; 1985. p.324-434.

[7] Ruzette, A.-V.; Leibler, L. *Nat. Mater.* 2005, *4*, 19-31.

[8] Bates, F. S.; Fredrickson, G. H. *Annu. Rev. Phys. Chem.* 1990, *41*, 525-557.

[9] Price, C. *Colloidal Properties of Block Copolymers.* In: Goodman I (Editor). Developments in Block Copolymers. Applied Science, London. 1982.

[10] Piirma I (Ed) *Polymeric Surfactants. In: Surfactants Science Series*, Vol 42. Marcel Dekker, New York. 1992.

[11] Tuzar, Z.; Kratochvil, P. *Micelles of Block and Graft Copolymers in Solution.* In: Matijevic, E (Ed) Surface and Colloid Science, Vol 15. Plenum, New York. 1993.

[12] Riess, G. *Prog. Polym. Sci.* 2003, *28*, 1107-1170.

[13] Riess G.; Dumas, P. H.; Hurtrez, G. *Block Copolymer Micelles and Assemblies*, MML Series 5. Citus, London. 2002.

[14] Webber, S. E.; Munk, P.; Tuzar, Z. (Eds) *Solvents and Self-Organization of Polymers*. NATO ASI Series, Series E: Applied Sciences, Vol 327. Kluwer, Dordrecht. 1996.

[15] Hamley, I. W. (Ed) *The Physics of Block Copolymers*. Oxford Science Publications, Oxford. 1998.

[16] Alexandridis, P.; Hatton, T. A. *Block Copolymers*. In: Salamone, J. C. (Ed) Polymer Materials Encyclopedia. CRC Press, Boca Raton, FL. 1996.

[17] Nace, V. W. (Ed) NONIONIC Surfactants: *Polyoxyalkylene Block Copolymers*. In: Surfactant Science Series, Vol 60. Marcel Dekker, New York. 1996.

[18] Xie, H. Q.; Xie, D. *Prog. Polym. Sci.* 1999, *24*, 275-313.

[19] Liu, G. J. *"Polymeric Nanostructures" in Handbook of Nanostructured Materials and Nanotechnology*. pp.475-500, Vol. 5 Nalwa, H. S. (Ed), Academic Press: New York, 1999.

[20] Liu, G. J. *Curr. Opin. Colloid and Interface Sci* 1998, *3*, 200-208.

[21] Alexandridis, P.; Lindman, B. (Eds), *Amphiphilic Block Copolymers: Self-Assembly and Applications*. Elsevier, Amsterdam. 2000.

[22] Gohy, J.-F. *Adv. Polym. Sci.* 2005, *190*, 65-136.

[23] Förster, S.; Abetz, V.; Müller, A. H. E. *Adv. Polym. Sci.* 2004, *166*, 173-210.

[24] Elias, H. G. J *Macromol. Sci. Chem.* 1973, *7*, 601-622.

[25] Chu, B.; Zhou, Z. *Physical Chemistry of Polyoxyalkylene Block Copolymer Surfactants*. In: Nace V. N. (Ed), Nonionic Surfactants: Polyoxyalkylene Block Copolymers, Vol 60. Marcel Dekker, New York. 1996.

[26] Torchilin, V. P. *J. Controlled Release* 2001, *73*, 137-172.

[27] Tsunashima, Y.; Hirata, M.; Kawamata, Y. *Macromolecules* 1990, *23*, 1089-1096.

[28] Zana, R. *Fluorescence Studies of Amphiphilic Block Copolymers in Solution*. In: Alexandris, P.; Lindman, B. (Eds), Amphiphilic Block Copolymers: Self-Assembly and Applications. Amsterdam: Elsevier; 2000. p.221-252.

[29] Jada, A.; Siffert, B.; Riess, G. *Colloids Surf A: Physicochem Engng Aspects* 1993, *75*, 203-209.

[30] Zhang, L. F.; Shen, H. W.; Eisenberg, A. *Macromolecules* 1997, *30*, 1001-1011.

[31] Zhang, L. F.; Eisenberg, A. *J. Am. Chem. Soc.* 1996, *118*, 3168-3181.

[32] Zhang, L. F.; Eisenberg, A. *Polym. Adv. Technol.* 1998, *9*, 677-699.

[33] Liu, F.; Eisenberg, A. *J. Am. Chem. Soc.* 2003, *125*, 15059-15064.

[34] Zhu, J. T.; Jiang, W. *Macromolecules* 2005, *38*, 9315-9323.

[35] Yu, Y.; Zhang, L. F.; Eisenberg, A. *Macromolecules* 1998, *31*, 1144-1154.

[36] Xu, R.; Winnik, M. A.; Hallett, F. R.; Riess, G.; Croucher, M. D. Macromolecules 1991, 24, 87-93

[37] Prochazka, K.; Kiserow, D.; Ramireddy, C.; Tuzar, Z.; Munk, P.; Webber, S. E. *Macromolecules* 1992, *25*, 454-460.

[38] Khougaz, K.; Astafiecva, I.; Eisenberg, A. *Macromolecules* 1995, *28*, 7135-7147.

[39] Qin, A.; Tian, M.; Ramireddy, C.; Webber, S. E.; Munk, P.; Tuzar, A. *Macromolecules* 1994, *27*, 120-126.

[40] Spatz, J. P.; Mössmer, S.; Möller, M. *Angew. Chem., Int. Ed.* 1996, *35*, 1510-1512.

[41] Desjardins, A.; Eisenberg, A. *Macromolecules* 1991, *24*, 5779-5790.

[42] Förster, S.; Zisenis, M.; Wenz, E.; Antonietti, M. *J. Chem. Phys.* 1996, *104*, 9956-9970.

[43] Zhang, L. F.; Eisenberg, A. *Science* 1995, *268*, 1728-1731.
[44] Zhang, L. F.; Yu, K.; Eisenberg, A. *Science* 1996, *272*, 1777-1779.
[45] Lauw, Y.; Leermakers, F. A. M.; Cohen Stuart, M. A.; Borisov, O. V.; Zhulina, E. B. *Macromolecules* 2006, *39*, 3628-3641.
[46] Gao, A.; Liu, G. J.; Tao, J. *Macromolecules* 1996, *29*, 2487-2493.
[47] Roovers, J.; Toporowski, P.; Martin, J. *Macromolecules* 1989, *22*, 1897-1903.
[48] Bruns, W.; Carl, W. *Macromolecules* 1991, *24*, 209-212.
[49] Hadjichristidis, N.; Fetters, L. J. *Macromolecules* 1980, *13*, 191-193.
[50] Zhang, L. F.; Eisenberg, A. *Macromolecules* 1996, *29*, 8805-8815.
[51] Zhang, L. F.; Eisenberg, A. *Macromol. Symp.* 1997, *113*, 221-232.
[52] Halperin, A.; Tirrell, M.; Lodge, T. P. *Adv. Polym. Sci.* 1992, *100*, 31-71.
[53] de Gennes, P. G. *In Solid State Physics*, Liebert, L. (Ed) Academic Press, New York, Supplement 14, 1978, p. 1.
[54] Cameron, N. S.; Corbierre, M. K.; Eisenberg, A. *Can. J. Chem.* 1999, *77*, 1311-1326.
[55] Stepanek, P. In: Brown, W. (Ed) *Dynamic Light Scattering.* Oxford University Press, Oxford. 1972.
[56] Discher, D. E,; Eisenberg, A. *Science* 2002, *297*, 967-973.
[57] Tuzar, Z. *Overview of Polymer Micelles.* In: Webber, S. E.; Munk, P.; Tuzar, Z. (Eds), NATO ASI Series, Series E: Applied Sciences, Vol 327. Kluwer, Dordrecht, 1985.
[58] Munk, P. *Equilibrium and Nonequilibrium Polymer Micelles.* In: Webber, S. E.; Munk, P.; Tuzar, Z. (Eds) Applied Sciences, Vol. 327. Dordrecht: Kluwer Academic Publisher; 1996.p. 19-32.
[59] Kubowicz, S.; Baussard, J. -F.; Lutz, J. -F.; Thünemann, A. F.; von Berlepsch, H.; Laschewsky, A. *Angew. Chem., Int. Ed.* 2005, *44*, 5262-5265.
[60] Saito, R.; Fujita, A.; Ishizu, K. J. *J. Polym. Sci. Part A: Polym. Chem.* 2000, *38*, 2091-2097.
[61] Yan, X. H.; Liu, F.; Li, Z.; Liu, G. J. *Macromolecules* 2001, *34*, 9112-9116.
[62] Liu, G. J.; Ding, J.; Qiao, L.; Gao, A.; Dymov, B. P.; Gleeson, J. T.; Hashimoto, T.; Saijo, K. *Chem. Eur. J.* 1999, *5*, 2740-2749.
[63] Liu, G. J. *Adv. Mater.* 1997, *9*, 437-439.
[64] Roman Mythological Term "Janus" describes the properties of having two faces and was first used in chemistry by Casagrande, C.; Veyssié, C. R. Acad. Sci., Ser. 2: Mec., Phys., Chim., Sci. Terre Univers 1988, 306, 1423.
[65] Kim, Y.; Pyun, J.; Frechet, J. M. J.; Hawker, C. J.; Frank, C. W. *Langmuir* 2005, *21*, 10444-10458.
[66] Xu, H.; Erhardt, R.; Abetz, V.; Müller, A. H. E.; Goedel, E. A. *Langmuir* 2001, *17*, 6787-6793.
[67] Munk, P. *Classical methods for the study of block copolymer micelles.* In: Webber, S. E.; Munk, P.; Tuzar, Z. (Eds), NATO ASI Series E: Applied Sciences, Vol. 327. Kluwer, Dordrecht. 1996.
[68] Webber, S. E. *Use of fluorescence methods to characterize the interior of polymer micelles.* In: Webber, S. E.; Munk, P.; Tuzar, Z. (Eds), NATO ASI Series, Series E: Applied Micelles. Vol. 327. Kluwer, Dordrecht. 1996.
[69] Mortensen, K. *Small angle scattering studies of block copolymer micelles, micellar mesophases and networks.* In: Alexandridis, P.; Lindman, B. (Eds), Amphiphilic Block Copolymers: Self-assembly and Applications. Elsevier, Amesterdam, 2000.

[70] Zana, R. *Fluorescence studies of amphiphilic block copolymers in solution.* In: Alexandridis, P.; Lindman, B. (Eds), Amphiphilic Block Copolymers: Self-assembly and Applications. Elsevier, Amesterdam, 2000.
[71] Zhu, J. T.; Jiang, Y.; Liang, H. J.; Jiang, W. *J. Phys. Chem. B* 2005, *109*, 8619-8625.
[72] Gohy, J. -F.; Willet, N.; Varshney, S.; Zhang, J. -X.; Jérôme, R. *Angew. Chem., Int. Ed.* 2001, *40*, 3214-3216.
[73] Lei, L.; Gohy, J. -F.; Willet, N.; Zhang, J. -X.; Varshney, S.; Jérôme, R. *Macromolecules* 2004, *37*, 1089-1094
[74] Mortensen, K.; Talmon, Y. *Macromolecules* 1995, *28*, 8829-8834.
[75] Esselink, F. J.; Dormidontova, E.; Hadziioannou, G. *Macromolecules* 1998, *31*, 2925-2932.
[76] Zupancich, J. A.; Bates, F. S.; Hillmyer, M. A. *Macromolecules* 2006, *39*, 4286-4288.
[77] Goldraich, M.; Talmon, Y. *Direct-imaging cryo-transmission electron microscopy in the study of colloids and polymer solutions.* In: Alexandridis, P.; Lindman, B. (Eds), Amphiphilic Block Copolymers: Self-assembly and Applications. Elsevier, Amesterdam, 2000.
[78] Edmonds, W. F.; Li, Z. B.; Hillmyer, M. A. Lodge, T. P. *Macromolecules* 2006, *39*, 4526-4530.
[79] Kichelbick, G.; Bauer, J.; Huesing, N.; Andersson, M.; Holmberg, K. *Langmuir* 2003, *19*, 10073-10076.
[80] Eghbali, E.; Colombani, O.; Drechsler, M.; Muller, A. H. E.; Hoffmann, H. *Langmuir*, 2006, *22*, 4766-4776.
[81] Zheng, Y.; Won, Y. Y.; Bates, F. S.; Davis, H. T.; Scriven, L. E. Talmon, Y. *J. Phys. Chem. B* 1999, *103*, 10331-10334.
[82] Zhu, J. T.; Liao, Y. G.; Jiang, W. *Langmuir* 2004, *20,* 3809-3812.
[83] Connell, S. D.; Collins, S.; Fundin, J.; Yang, Z.; Hamley, I. W. Langmuir 2003, 19, 10449-10453.
[84] Zhu, J. T.; Jiang, W. *Mater. Chem. Phys.* 2007, *101*, 56-62..
[85] Hamley, I. W.; Connell, S. D.; Collins, S. *Macromolecules* 2004, *37*, 5337-5351.
[86] Vogel, V.; Gohy, J. -F.; Lohmeijer, B. G. G.; van den Broek, J. A.; Schubert, U. S.; Schubert, D. *J. Polym. Sci. Part A: Polym. Chem.* 2003, *41*, 3159-3168.
[87] Teo, H. H.; Styring, M. G.; Yeates, S. G.; Price, C.; Booth, C. *J. Colloid and Interface Sci.* 1986, *114*, 416-424.
[88] Zana, R. In: Alexandridis, P.; Lindman, B. (Eds), *Amphiphilic Block Copolymers: Self-assembly and Applications.* Elsevier, Amesterdam, 2000.
[89] Zhang, W.Q.; Shi, L. Q.; An, Y. L.; Gao, L. C.; He, B. L. *J. Phys. Chem. B* 2004, *108*, 200-204.
[90] Liu, S.; Billingham, N. C.; Armes, S. P. *Angew. Chem., Int. Ed.* 2001, *40*, 2328-2331.
[91] Gao, Z.; Zhong, X. F.; Eisenberg, A. *Macromolecules* 1994, *27*, 794-802.
[92] Gao, Z.; Desjardins, A.; Eisenberg, A. *Macromolecules* 1992, *25*, 1300-1303.
[93] Won, Y. Y.; Davis, H. T.; Bates, F. S. *Macromolecules* 2003, *36*, 953-955.
[94] Kositza, M. J.; Bohne, C.; Hatton, T. A.; Holzwarth, J. F. *Prog. Colloid. Polym. Sci.* 1999, *112*, 146-151.
[95] Underhill, R. S.; Ding, J. F.; Birss, V. I.; Liu, G. J. *Macromolecules* 1997, *30*, 8298-8303.

[96] Creutz, S.; Vanstam, J.; De Schryver, F. C.; Jéröme, R. *Macromolecules* 1998, *31*, 681-689.
[97] Stepanek, M.; Prochazka, K. *Langmuir* 2000, *16*, 2502-2507.
[98] Prochazka, K.; Limpouchova, Z.; Webber, S. E. *Block Copolymer Micelles: 2. Fluorimetric studies and cumpter modeling.* In: Salamone, J. C. (Ed), Polymer Materials encyalopedia, Vol. 1. CRC press, Boca Raton, FL. 1996.
[99] Holappa, S.; Kantonen, L.; Winnik, F. M.; Tenhu, H. *Macromolecules* 2004, *37*, 7008-7018.
[100] Calderara, F.; Hruska, Z.; Hurtrez, G.; Lerch, J. P.; Nugay, T.; Riess, G. *Macromolecules* 1994, *27*, 1210-1215.
[101] Thurmond, K. B.; Huang, H.; Clark Jr, C. G.; Kowalewski, T.; Wooley, K. L. *Colloids and Suf B: Biointer.* 1999, *16*, 45-54.
[102] Discher, B. M.; Bermudez, H.; Hammer, D. A.; Discher, D. E.; Won, Y. Y.; Bates, F. S. *J. Phys. Chem. B* 2002, *106*, 2848-2854.
[103] Stewart, S.; Liu, G. J. *Chem. Mater.* 1999, *11*, 1048-1054.
[104] Stewart, S.; Liu, G. J. *Angew. Chem., Int. Ed.* 2000, *39*, 340-344.
[105] Underhill, R. S.; Liu, G. J. *Chem. Mater.* 2000, *12*, 2082-2091.
[106] Prochazka, K.; Baloch, M. K. *Macromol. Chem.: Macromol. Chem. Phys.*1979, *180*, 2521-2523.
[107] Tuzar, Z.; Bednar, B.; Konak, C.; Kubin, M.; Svobodova, S.; Prochazka, K. *Macromol. Chem.: Macromol. Chem. Phys.* 1982, *183*, 399-408.
[108] ThurmondII, K. B.; Kowalewski, T.; Wooley, K. L. *J. Am. Chem. Soc* .1996, *118*, 7239-7240.
[109] Huang, H.; Kowalewski, T.; Remsen, E. E.; Gertzmann, R.; Wooley, K. L. *J. Am. Chem. Soc.* 1997, *119*, 11653-11659.
[110] Liu, S. Y.; Armes, S. P. *J. Am. Chem. Soc.* 2001, *123*, 9910-9911.
[111] Mbanez, M. V. D.; Robinson, K. L.; Butun, V.; Armes, S. P. *Polymer* 2001, *42*, 29-37.
[112] Almgren, M.; Brown, W.; Hvidt, S. *Colloid Polym. Sci.* 1995, *273*, 2-15.
[113] Booth, C.; Attwood, D. *Macromol. Rapid Commun.* 2000, *21*, 501-527.
[114] Alexandridis, P.; Spontak, R. J. *Curr. Opin. Colloid and Interface Sci.* 1999, *4*, 130-139.
[115] Wanka, G.; Hoffmann, H.; Ulbricht, W. *Colloid Polym. Sci.* 1990, *268*, 101-117.
[116] Alexandridis, P.; Hatton, T. A. *Colloids Surf.* A 1995, *96*, 1-46.
[117] Nolan, S. L.; Phillips, R. J.; Cotts, P. M.; Dungan, S. R. *J. Colloid and Interface Sci.* 1997, *191*, 291-302.
[118] Alexandridis, P. *Curr. Opin. Colloid and Interface Sci* 1997, *2*, 478-489.
[119] Liu, Y. C.; Chen, S. H.; Huang, J. S. *Macromolecules* 1998, *31*, 2236-2244.
[120] Goldmints, I.; von Gottberg, F. K.; Smith, K. A.; Hotton, T. A. *Langmuir* 1997, *13*, 3659-3664.
[121] King, S. M.; Heenan, R. K.; Cloke, V. M.; Washington, C. *Macromolecules* 1997, *30*, 6215-6222.
[122] Mortensen, K.; Pedersen, J. S. *Macromolecules* 1993, *26*, 805-812.
[123] Won, Y. Y.; Brannan, A. K.; Davis, H. T.; Bates, F. S. *J. Phys. Chem. B* 2002, *106*, 3354-3364.
[124] Won, Y. Y.; Davis, H. T.; Bates, F. S.; Agamalian, M.; Wignall, C. D. *J. Phys. Chem. B* 2000, *104*, 7134-7143.

[125] Won, Y. Y.; Davis, H. T.; Bates, F. S. *Science* 1999, *283*, 960-963.
[126] Jain, S.; Bates, F. S. *Science* 2003, *300*, 460-464.
[127] Yu, K.; Eisenberg, A. *Macromolecules* 1996, *29*, 6359-6361.
[128] Yu, K.; Zhang, L. F.; Eisenberg, A. *Langmuir* 1996, *12*, 5980-5984.
[129] Yu, K.; Bartels, C.; Eisenberg, A. *Langmuir* 1999, 15, 7157-7167.
[130] Mortensen, K.; Brown, W.; Almdal, K.; Alami, E.; Jada, A. *Langmuir* 1997, *13*, 3635-3645.
[131] Gohy, J. -F.; Lohmeijer, B. G. G.; Varshney, S. K.; Schubert, U. S. *Macromolecules* 2002, *35*, 7427-7435.
[132] Polik, W. F.; Burchard, W. *Macromolecules* 1983, *16*, 978-982.
[133] Rolland, A.; O'Mullane, J. E.; Goddard, P.; Brookman, L.; Petrak, K. *J. Appl. Polym. Sci.* 1992, *44*, 1195-1203.
[134] Savic, R.; Luo, L.; Eisenberg, A.; Maysinger, D. *Science* 2003, *300*, 615-618.
[135] Luo, L.; Tam, D.; Maysinger, D.; Eisenberg, A. *Bioconjug. Chem.* 2002, *13*, 1259-1265.
[136] Verdonck, B.; Goethals, E. J.; Du Prez, F. E. *Macromol. Chem. Phys.* 2003, *204*, 2090-2098.
[137] Nuopponen, M.; Ojala, J.; Tenhu, H. *Polymer* 2004, *45*, 3643-3650.
[138] Moffitt, M.; Khougaz, K.; Eisenberg, A. *Acc. Chem. Res.* 1996, *29*, 95-102.
[139] Tuzar, Z. *Copolymer Micelles in aqueous media.* In: Webber, S. E.; Munk, P.; Tuzar, Z. (Eds), NATO ASL Series, Series E: Applied Science, Vol 327, Kluwer, Dordrecht.
[140] Shen, H. W.; Eisenberg, A. *J. Phys. Chem. B.* 1999, *103*, 9473-9487.
[141] Yu, Y. S.; Zhang, L. F.; Eisenberg, A. *Langmuir* 1997, *13*, 2578-2581.
[142] Shen, H. W.; Zhang, L. F.; Eisenberg, A. *J. Phys. Chem. B* 1997, *101*, 4697-4708.
[143] Selb, J.; Gallot, Y. In: Goodman, I. (Ed), *Developments in block copolymers, Applied Sciences*; Publishers: London, 1985, vol. 2, p27.
[144] Tuzar, Z.; Prochazka, K.; Zuskova, I.; Munk, P. *ACS Polym. Prepr. (Div. Polym. Chem.)* 1993, *34*, 1038-1039.
[145] Solomatin, S. V.; Broonich, T. K.; Bargar, T. W.; Eisenberg, A.; Kabanov, V. A.; Kabanov, A. V. *Langmuir* 2003, *19*, 8069-8076.
[146] Burke, S. E.; Eisenberg, A. *Langmuir* 2001, *17*, 6705-6714.
[147] Burke, S. E.; Eisenberg, A. *Polymer* 2001, *42*, 9111-9120.
[148] Chen, L.; Shen, H. W.; Eisenberg, A. *J. Phys. Chem. B* 1999, *103*, 9488-9497.
[149] Rager, T.; Meyer, W. H.; Wegner, G. *Macromol. Chem. Phys.* 1999, *200*, 1672-1680.
[150] Guenoun, P.; Muller, F.; Delsanti, M.; Auvray, L.; Chen, Y. J.; Mays, J. W.; Tirrell, M *Phys. Rev. Lett.* 1998, *81*, 3872-3875.
[151] Förster, S.; Hermsdorf, N.; leube, W.; Schnablegger, H.; Regenbrecht, M.; Akari, S.; Lindner, P.; Böttcher, C. I. *J. Phys. Chem. B* 1999, *103*, 6657-6668.
[152] Förster, S.; Hermsdorf, N.; Böttcher, C.; Lindner, P. *Macromolecules* 2002, *35*, 4096-4105.
[153] Gohy, J. -F.; Antoun, S.; Jérôme, R. *Polymer* 2001, *42*, 8637-8645.
[154] Raviv, U.; Giasson, S.; Kampf, N.; Gohy, J. -F.; Jérôme, R.; Klein, J. *Nature* 2003, *425*, 163-165.
[155] Kötz, J.; Kosmella, S.; Beitz, T. *Prog. Polym. Sci.* 2001, *26*, 1199-1232.
[156] Shen, H.; Zhang, L.; Eisenberg, A. *J. Am. Chem. Soc.* 1999, *121*, 2728-2740.

[157] Streitwieser, A., Jr. *Inorganic to Organic Chemistry*, 4th Ed.; Macmillan: New York, 1992; p1116.
[158] Zhu, J. T.; Yu, H. Z.; Jiang, W. *Macromolecules* 2005, *38*, 7492-7501.
[159] Jiang, Y.; Zhu, J. T.; Jiang, W.; Liang, H. J. *J. Phys. Chem. B.* 2005, *109*, 21549-21555.
[160] Zhu, J. T.; Yu, H. Z.; Jiang, W. *Polymer* 2005, *46*, 11962-11968.
[161] Selb, J.; Gallot, Y. *Macromol. Chem.: Macromol. Chem. Phys.* 1980, *181*, 809-822.
[162] Liu, S.; Armes, S. P. *Curr. Opin. Colloid. and Interface Sci.* 2001, *6*, 249-256.
[163] Vamvakaki, M.; Billingham, N. C.; Armes, S. P. *Polymer* 1998, *39*, 2332-2337.
[164] Cölfen, H. *Macromol. Rapid Commun.* 2001, *22*, 219-252.
[165] Tauer, K.; Khrenov, V.; Shirshova, N.; Nassif, N. *Macromol. Symp.* 2005, *226*, 187-201.
[166] Martin, T. J.; Procházka, K.; Munk, P.; Webber, S. E. *Macromolecules* 1996, *29*, 6071-6073.
[167] Gohy, J.-F.; Mores, S.; Varshney, S. K.; Zhang, J. X.; Jérôme, R. *e-polymers* 2002, paper No.21
[168] Borchert, U.; Lipprandt, U.; Bilang, M.; Kimpfer, A.; Rank, A.; Peschka-Suss, R.; Schubert, R.; Lindner, P.; Förster, S. *Langmuir* 2006, *22*, 5843-5847.
[169] Sidorov, S. N.; Bronstein, L. M.; Kabachii, Y. A.; Valetsky, P. M.; Soo, P. L.; Maysinger, D.; Eisenberg, A. *Langmuir* 2004, *20*, 3543-3550.
[170] Bronstein, L. M.; Sidorov, S. N.; Valetsky, P. M.; Hartmann, J.; Cölfen, H.; Antonietti, M. *Langmuir* 1999, *15*, 6256-6262.
[171] Chernyshov, D. M.; Bronstein, L. M.; Boerner, H.; Berton, B.; Antonietti, M. *Chem. Mater.* 2000, *12*, 114-121.
[172] Lee, A. S.; Buetuen, V.; Vamvakaki, M.; Armes, S. P.; Pope, J. A.; Gast, A. P. *Macromolecules* 2002, *35*, 8540-8551.
[173] Vamvakaki, M.; Palioura, D.; Spyros, A.; Armes, S. P.; Anastasisdis, S. H. *Macromolecules* 2006, *39*, 5106-5112.
[174] Cai, Y. L.; Armes, S. P. *Macromolecules* 2004, *37*, 7116-7122.
[175] Bütün, V.; Bennett, C. E.; Vamvakaki, M.; Lowe, A. B.; Billingham, N. C.; Armes, S. P. *J. Mater. Chem.* 1997, 7, 1693-1695.
[176] Nishi, T.; Kwei, T. K. *Polymer* 1975, *16*, 285-290.
[177] Sugihara, S.; Kanaoka, S.; Aoshima, S. *J. Polym. Sci. Part A: Polym. Chem.* 2004, *42*, 2601-2611.
[178] Topp, M. D. C.; Dijkstra, P. J.; Talsma, H.; Feijen, J. *Macromolecules* 1997, *30*, 8518-8520.
[179] Forder, C.; Patrickios, C. S.; Armes, S. P.; Billingham, N. C. *Macromolecules* 1996, *29*, 8160-8169.
[180] Szczubialka, K.; Nowakowska, M. *Polymer* 2003, *44*, 5269-5274.
[181] Arotçarena, M.; Heise, B.; Ishaya, S.; Laschewsky, A. *J. Am. Chem. Soc.* 2002, *124*, 3787-3793.
[182] Pergushov, D. V.; Remizova, E. V.; Gradzielski, M.; Linder, P.; Feldthusen, J.; Zezin, A. B.; Müller, A. H. E.; Kabanov, V. A. *Polymer* 2004, *45*, 367-378.
[183] Kamachi, M.; Kurihara, M.; Stille, J. K. *Macromolecules* 1972, *5*, 161-167.
[184] Golub, T.; de Keizer, A.; Cohen-Stuart, M. A. *Macromolecules* 1999, *32*, 8441-8446.
[185] Gohy, J. -F.; Creutz, S.; Garcia, M.; Mahltig, B.; Stamm, M.; Jérôme, R. *Macromolecules* 2000, *33*, 6378-6387.

[186] Lowe, A. B.; Billingham, N. C.; Armes, S. P. *Chem. Comm.* 1997, 1035-1036.
[187] Mahltig, B.; Gohy, J.-F.; Antoun, S.; Jérôme, R.; Stamm, M. *Colloid Polym. Sci.* 2002, *280*, 495-502.
[188] Förster, S.; Zisenis, M.; Wenz, E.; Antonietti, M. *J. Chem. Phys.* 1996, *104*, 9956-9970.
[189] Förster, S.; Antonietti, M. *Adv. Mater.* 1998, *10*, 195-217.
[190] Spatz, J. P.; Sheiko, S.; Möller, M. *Macromolecules* 1996, *29*, 3220-3226.
[191] Spatz, J. P.; Röscher, A.; Sheiko, S.; Krausch, G.; Möller, M. *Adv. Mater.* 1995, *7*, 731-735.
[192] Spatz, J. P. *Angew Chem., Int. Ed.* 2002, *114*, 3359-3362.
[193] Spatz, J. P.; Möβmer, S.; Möller, M. *Chem. Eur. J.* 1996, *2*, 1552-1555.
[194] Antonietti, M.; Wenz, Bronstein, L. M.; Seregina, M. S. *Adv. Mater.* 1995, *7*, 1000-1005.
[195] Klingelhofer, S.; Heitz, W.; Greiner, A.; Oestreich, S.; Förster, S.; Antonietti, M. *J. Am. Chem. Soc.* 1997, *119*, 10116-10120.
[196] Mössmer, S.; Spatz, J. P.; Möller, M.; Aberle, T.; Schmidt, J.; Burchard, W. *Macromolecules* 2000, *33*, 4791-4798.
[197] Kiriy, A.; Gorodyska, G.; Minko, S.; Stamm, M.; Tsitsilianis, C. *Macromolecules* 2003, *36*, 8704-8711.
[198] Gorodyska, A.; Kiriy, A.; Minko, S.; Tsitsilians, C.; Stamm, M. *Nano. Lett.* 2003, *3*, 365-368.
[199] Samii, A. A.; Karlstrom, G.; Lindman, B. *Langmuir*, 1991, *7*, 1067-1071.
[200] Samii, A. A.; Lindman, B.; Karlstrom, G. *Prog. Colloid Polym. Sci.* 1990, *82*, 280
[201] Alexandridis, P.; Yang, L. *Macromolecules* 2000, *33*, 3382-3391.
[202] Vagberg, L. J. M.; Cogan-Farinas, K. A.; Gast, A. P. *Macromolecules* 1991, *24*, 1670-1677.
[203] Lin, E. K.; Gast, A. P. *Macromolecules* 1996, *29*, 4432-4441.
[204] Gervais, M.; Gallot, B. *Macromol. Chem.: Macromol. Chem. Phys.* 1977, *178*, 2071-2078.
[205] Wu, G.; Chu, B. *Macromolecules* 1994, *27*, 1766-1773.
[206] Guo, C.; Liu, H. Z.; Che, J. Y. *Colloids Surf. A* 2000, *175*, 193-202.
[207] Kovacs, A. J.; Manson, J. A. *Kolloid Z. Z. Polym.* 1966, *214*, 1-8.
[208] Gast, A. P.; Vinson, P. K.; Cogan-Farinas, K. A. *Macromolecules* 1993, *26*, 1774-1776.
[209] Zhang, L. F.; Khougaz, K.; Moffitt, M.; Eisenberg, A. *Self-assembly of block polyelectrolytes.* In: Alexandridis, P.; Lindman, B. (Eds), *Amphiphilic block copolymers: self-assembly and applications.* Elsevier, Amsterdam. 2000.
[210] Moffitt, M.; Zhang, L. F.; Khougaz, K.; Eisenberg, A. *Micellization of ionic block copolymers in three dimentions.* In: Webber, S. E.; Munk, P.; Tuzar, Z. (Eds), NATO ASI series, series E: Applied sciences, vol 327, Kluwer, Dordrecht. 1996.
[211] Bang, J.; Jain, S.; Li, Z.; Lodge, T. P *Macromolecules* 2006, *39*, 1199-1208.
[212] LaRue, I.; Adam, M.; Sheiko, S. S.; Rubinstein, M. *Polym. Mater. Sci. Eng.* 2003, 88, 236.
[213] LaRue, I.; Adam, M.; da Silva, M.; Sheiko, S. S.; Rubinstein, M. *Macromolecules* 2004, *37*, 5002-5005.
[214] Canham, P. A.; Lally, T. P.; Price, C.; Stubbersfield, R. B. *J. Chem. Soc., Faraday Trans. 1* 1980, *76*, 1857.

[215] Liu, G. J.; Yan, X. H.; Li, Z.; Zhou, J.; Duncan, S. *J. Am. Chem. Soc.* 2003, *125*, 14039-14045.
[216] Ding, J.; Liu, G. J. *Macromolecules* 1997, *30*, 655-657.
[217] Zheng, R. H.; Liu, G. J.; Yan, X. H. *J. Am. Chem. Soc.* 2005, *127*, 15358-15359.
[218] Zhou, J.; Li, Z.; Liu, G. J. *Macromolecules* 2002, *35*, 3690-3696.
[219] Hu, J.; Liu, G. J. *Macromolecules* 2005, *38*, 8058-8065.
[220] Londono, J. D.; Dharmapurikar, R.; Cochran, H. D.; Wignall, G. D.; Mcclain, J. B.; Betts, D. E.; Canelas, D. A.; Desimone, J. M.; Samulski, E. T.; Chilluramartino, D.; Triolo, R. *J Appl. Crystallogr.* 1997, *30*, 690-695.
[221] Buhler, E.; Dobrynin, A. V.; DeSimone, J. M.; Rubinstein, M. *Macromolecules* 1998, *31*, 7347-7355.
[222] Zhou, S. Q.; Chu, B. *Macromolecules* 1998, *31*, 7746-7755.
[223] Walther, M.; Faulhammer, H.; Finkelmann, H. *Macromol. Chem. Phys.* 1998, *199*, 223-237.
[224] Choucair, A.; Eisenberg, A. *Eur. Phys. J. E.* 2003, *10*, 37-44.
[225] Soo, P. L.; Eisenberg, A. *J. Polym. Sci. Part B: Polym. Phys.* 2004, *42*, 923-938.
[226] Choucair, A.; Lavigueur, C.; Eisenberg, A. *Langmuir* 2004, *20*, 3894-3900.
[227] Shen, H. W.; Eisenberg, A. *Macromolecules* 2000, *33*, 2561-2572.
[228] Zhang, W. Q.; Shi, L. Q.; An, Y. L.; Gao, L. C.; Ma, R. J.; He, B. L. *Phys. Chem. Chem. Phys.* 2004, *6*, 109-115.
[229] Svensson, B.; Olsson, U.; Alexandridis, P. *Langmuir* 2000, *16*, 6839-6846.
[230] Yu, Y.; Eisenberg, A. *J. Am. Chem. Soc.* 1997, *119*, 8383-8384.
[231] Zhang, W. Q.; Shi, L. Q.; An, Y. L.; Gao, L. C.; Wu, K.; Ma, R. J. *Macromolecules* 2004, *37*, 2551-2555.
[232] Talingting, M. R.; Munk, S. E.; Webber, S. E.; Tuzar, Z. *Macromolecules* 1999, *32*, 1593-1601.
[233] Jorgensen, E. B.; Hvidt, S.; Brown, W.; Schillén, K. *Macromolecules* 1997, *30*, 2355-2364.
[234] Lei, L.; Gohy, J. -F.; Zhang, J.-X.; Varshney, S.; Jérôme, R. *Polymer* 2004, *45*, 4375-4381.
[235] [235] Zhang, L. F.; Eisenberg, A. *J. Polym. Sci. Part B: Polym. Phys.* 1999, *37*, 1469-1484.
[236] Ouati, N.; Viville, P.; Lazzaroni, R.; Minatti, E.; Schappacher, M.; Deffieux, A.; Borsali, R. *Langmuir* 2005, *21*, 1180-1186.
[237] Desbaumes, L.; Eisenberg, A. *Langmuir* 1999, *15*, 36-38.
[238] Terreau, O.; Batels, C.; Eisenberg, A *Langmuir* 2004, *20*, 637-645.
[239] Terreau, O.; Luo, L. B.; Eisenberg, A. *Langmuir* 2003, *19*, 5601-5607.
[240] Laschewsky, A. *Curr. Opin. Colloid and Interface Sci.* 2003, *8*, 274-281.
[241] Stahler, K.; Selb, J.; Landau, F. *Langmuir* 1999, *15*, 7565-7576.
[242] Lei, L.; Gohy, J.-F.; Willet, N.; Zhang, J. X.; Varshney, S.; Jérôme, R. *Polymer* 2006, *47*, 2723-2727.
[243] Li, Z.; Hillmyer, M. A.; Lodge, T. P. *Macromolecules* 2004, *37*, 8933-8940.
[244] Liu, Y.; Abetz, V.; Müller, A. H. E. *Macromolecules* 2003, *36*, 7894-7898.
[245] Fustin, C. A.; Abetz, V.; Gohy, J. -F. *Euro. Phys. J. E* 2005, *16*, 291-302.
[246] Zhou, Z.; Li, Z.; Ren, Y.; Hillmyer, M. A. Lodge, T. P. *J. Am. Chem. Soc.* 2003, *125*, 10182-10183.

[247] Erardt, R.; Zhang, M.; Bocker, A.; Zettl, H.; Abetz, C.; Frederik, P.; Krausch, G.; Abetz, V.; Müller, A. H. E. *J. Am. Chem. Soc.* 2003, *125*, 3260-3267.
[248] Lodge, T. P.; Hillmyer, M. Z.; Zhou, Z.; Talmon, Y. *Macromolecules* 2004, *37*, 6680-6682.
[249] Li, Z. B.; Kesselman, E.; Talmon, Y.; Hillmyer, M. A.; Lodge, T. P. *Science* 2004, *306*, 98-101.
[250] Lutz, J. F.; Laschewsky, A. *Macromol. Chem. and Phys.* 2005, *206*, 813-817.
[251] Erhardt, R.; Bocker, A.; Zettl, H.; Kayz, H.; Pyckhout-Hintzen, W.; Krausch, G.; Abetz, V.; Müller, A. H. E. *Macromolecules* 2001, *34*, 1069-1075.
[252] Xu, H.; Erhardt, R.; Abetz, V.; Müller, A. H. E.; Goedel, W. A. *Langmuir* 2001, *17*, 6787-6793.
[253] Jain, S.; Bates, F. S. *Macromolecules* 2004, *37*, 1511-1523.
[254] Zhu, J. T.; Yu, H. Z.; Jiang, W. *To be submitted.*
[255] Li, Z. B.; Hillmyer, M. A.; Lodge, T. P. *Macromolecules* 2006, *39*, 765-771.
[256] Prochazka, K.; Martin, T. J.; Webber, S. E.; Munk, P. *Macromolecules* 1996, *29*, 6526-6530.
[257] Tsitsilianis, C.; Voulgaris, D.; Stepanek, M.; Podhajecka, K.; Prochazka, K.; Tuzar, Z.; Brown, W. *Langmuir* 2000, *16*, 6868-6876.
[258] Stuart, M. A. C.; Hofs, B.; Voets, I. K.; de Keizer, A. *Curr. Opion. Colloid and Inter. Sci.* 2005, *10*, 30-36.
[259] Lysenko, E. A.; Chelushkin, P. S.; Bronich, T. K.; Eisenberg, A.; Kabanov, V. A.; Kabanov, A. V. *J. Phys. Chem. B* 2004, *108*, 12352-12359.
[260] Zhang, W. Q.; Shi, L. Q.; Gao, L. C.; An, Y. L.; Li, G. Y.; Wu, K.; Liu, Z. *Macromolecules* 2005, *38*, 899-903.
[261] Liu, S.; Zhu, H.; Zhao, H.; Jiang, M.; Wu, C. *Langmuir* 2000, *16*, 3712-3717.
[262] Chen, D. Y.; Jiang, M. *Acc. Chem. Res.* 2005, *38*, 494-502.
[263] Duan, H.; H. W.; Chen, D. Y.; Jiang, M.; Gan, W. J.; Li, S. J.; Wang, M.; Gong, J. *J. Am. Chem. Soc.* 2001, *123*, 12097-12098.
[264] Peng, H.; Chen, D.; Jiang, M. *J. Phys. Chem. B* 2003, *107*, 12461-12464.
[265] Peng, H.; Chen, D.; Jiang, M. *Langmuir* 2003, *19*, 10989-10992.
[266] Schrage, S.; Schlaad, H.; Antonietti, M. IUPAC Int. *Symp. Ionic. Polym. Creta. Abstract* AP20 2001.
[267] Yan, X. H.; Liu, G. J.; Hu, J. W. *Macromolecules* 2006, *39*, 1906-1912.
[268] Klok, H. A.; Lecommandoux, S. *Adv. Mater.* 2001, *13*, 1217-1229.
[269] Cornelissen, J. J. L. M.; Rowan, A. E.; Nolte, R. J. M.; Sommerdijk, N. A. J. M. *Chem. Rev.* 2001, *101*, 4039-4070.
[270] Douy, A.; Gallot, B. *Polymer* 1987, *28*, 147-154.
[271] Massey, J.; Power, K. N.; Manners, I.; Winnik, A. M. *J. Am. Chem. Soc.* 1998, *120*, 9533-9540.
[272] Massey, J.; Temple, K.; Cao, L.; Rharbi, Y.; Raez, J.; Winnik, W. A.; Manners, I. *J. Am. Chem. Soc.* 2000, *122*, 11577-11584.
[273] Wang, X. S.; Winnik, M. A.; Manners, I. *Macromol. Rapid Commun.* 2002, *23*, 210-213.
[274] Cao, L.; Massey, J.; Winnik, M. A.; Manners, I.; Riethmüller, S.; Banhart, F.; Möller, M. *Adv. Funct. Mater.* 2003, *13*, 271-276.

[275] Gohy, J.-F.; Lohmeijer, B. G. G.; Schubert, U. S. *Macromolecules* 2002, *35*, 4560-4563.

[276] Gohy, J.-F.; Lohmeijer, B. G. G.; Schubert, U. S. *Macromol. Rapid Commun.* 2002, *23*, 555-560.

[277] Branna, A. K.; Bates, F. S. *Macromolecules* 2004, *37*, 8816-8819.

[278] Determan, M. D.; Guo, L.; Thiyagarajan, P.; Mallapragada, S. K. *Langmuir* 2006, *22*, 1469-1473.

[279] Stepanek, M.; Matejicek, P.; Humpolickova, J.; Havrankova, J.; Podhajecka, K.; Spirkova, M.; Tuzar, Z.; Tsitsilianis, C.; Prochazka, K. *Polymer* 2005, *46*, 10493-10505.

[280] Felali, M.; Meier, M. A.; Schubert, U. S.; Gohy, J. -F. *Langmuir* 2005, *21*, 7995-8000.

[281] Cong, Y.; Li, B. Y.; Han, Y. C.; Li, Y. G.; Pan, C. Y. *Macromolecules* 2005, *38*, 9836-9846.

[282] Ma, Y.; Cao, T.; Webber, S. E. *Macromolecules* 1998, *31*, 1773-1778.

[283] Hattori, H. *Adv. Mater.* 2001, *13*, 51-54.

[284] Alivisatos, A, P. *Science* 1996, *271*, 933-937.

[285] Minko, S.; Kiriy, A.; Gorodyska, G.; Stamm, M. *J. Am. Chem. Soc.* 2002, *124*, 10192-10197.

[286] Glass, R.; Möller, M.; Spatz, J. P. *Nanotechnology* 2003, *14*, 1153-1160.

[287] Djalali, R.; Li.; S. Y.; Schmidt, M. *Macromolecules* 2002, *35*, 4282-4288.

[288] Fahmi, A. W.; Braun, H. G.; Stamm, M. *Adv. Mater.* 2003, *25*, 1201-1204.

[289] Nie, W.; Zhu, J. T.; Zhao, N. N.; Xie, R. G.; Chen, Y. M.; Jinag, W.; Ji, X. L. *Nanotechnology* 2006, *17*, 3313-3318.

[290] Zhang, D.; Qi, L.; Ma, J.; Cheng, H. *Chem. Mater.* 2001, *13*, 2753-2755.

[291] Qi, L.; Cölfen, H.; Antonietti, M *Angew. Chem., Int. Ed.* 2000, *39*, 604-607.

[292] Tjandra, W.; Yao, J.; Ravi, P.; Tam, K. C.; Alamsjah, A. *Chem. Mater.* 2005, *17*, 4865-4872.

[293] Cölfen, H.; Antonietti, M. *Langmuir* 1998, *14*, 582-589.

[294] Rösler, A.; Vandermeulen, G. W. M.; Klok, H.-A. *Adv. Drug Deliv. Rev.* 2001, *53*, 95-108.

[295] Kwon, G. S.; Okano, T. *Adv. Drug Deli. Rev.* 1996, *21*, 107-116.

[296] Trubetskoy, V. S. *Adv. Drug Deli. Rev.* 1999, *37*, 81-88.

[297] Bader, H.; Ringsdorf, H.; Schmidt, B. *Angew. Macromol. Chem.* 1984, *123/124*, 457-485.

[298] Yokoyama, M.; Inoue, S.; Kataoka, K.; Yui, N.; Sakurai, Y. *Macromol. Chem. Rapid Commun.* 1987, *8*, 431-435.

[299] Lodge, T. P.; Rasdai, A.; Li, Z.; Hillmyer, M. A. *J. Am. Chem. Soc.* 2005, *127*, 17608-17609.

[300] Sveson, M.; Johansson, H. O.; Tjerneld, F. *Appplication of amphiphilic block copolymers in separations.* In: Alexandridis, P.; Lindman, B. (Eds), *Amphiphilic block copolymers: self-assembly and applications.* Amsterdam: Elsevise; 2000. p305-318.

[301] Xu, R.; d'Oliveira, J. M. R.; Winnik, M. A.; Riess, G.; Croucher, M. D. *J. Appl. Polym. Sci: Apppl. Polym. Symp.* 1992, *51*, 135-149.

[302] Yan, D. Y.; Zhou, Y. F.; Hou, J. *Science* 2004, *303*, 65-67.

[303] Li, Z.; Zhang, Y.; Fullhart, P.; Mirkin, C. A. *Nano. Lett.* 2004, *4*, 1055-1058.

[304] Velonia, K.; Rowan, A. E.; Nolte, R. J. M. *J. Am. Chem. Soc.* 2002, *124*, 4224-4225.

[305] Boerakker, M. J.; Hannik, J. M.; Bomans, P. H. H.; Frederik, P. M.; Nolte, R. J. M.; Meijer, E. M.; Sommerdijk, N. A. J. M. *Angew. Chem., Inter. Ed.* 2002, *41*, 4239-4241.
[306] Du, J. Z.; Chen, Y. M.; Zhang, Y.; Han, C. C.; Fischer, K.; Schmidt, M. *J. Am. Chem. Soc.* 2003, *125*, 14710-14711.
[307] Du, J. Z.; Chem, Y. M. *Macromolecules* 2004, *37*, 5710-5716.
[308] Du, J. Z.; Chen, Y. M. *Angew. Chem., Int. Ed.* 2004, *43*, 5084-5087.
[309] Sevink, G. J. A.; Zvelindovsky, A. V. *Macromolecules* 2005, *38*, 7502-7513.
[310] He, X.; Liang, H. J.; Huang, L.; Pan, C. Y. *J. Phys. Chem. B* 2004, *108*, 1731-1735.
[311] Szostak, J. W.; Bartel, D. P.; Luisi, P. L. *Nature* 2001, *409*, 387-390.
[312] Lasic, D. D. *Nature* 1991, *351*, 613-613.
[313] Leng, J.; Egelhaaf, S. U.; Cates, M. E. *Europhys. Lett.* 2002, *59*, 311-317.

In: Chemical Physics Research Trends, Vol. 252
Editor: Samuel V. Arnold, pp. 117-164
ISBN: 978-1-60021-554-4

Chapter 3

PHOTOLUMINESCENCE OF SOME POLY(PARA-PHENYLENE) TYPE π- CONJUGATED POLYMERS DOPED WITH PHOSPHORESCENT AND FLUORESCENT DYES

S. A. Bagnich
Institute of Molecular and Atomic Physics,
National Academy of Sciences of Belarus,
Nezavisimosti Ave. 70, 220072 Minsk, Belarus

ABSTRACT

This paper presents the results of the investigation of the luminescence of ladder-type methyl-, phenyl-poly(para-phenylene) (Me-, PhLPPP), and poly(9,9-di(ethylhexyl)fluorine) (PF2/6) activated by phosphorescent and fluorescent dyes obtained under their optical excitation. It is shown that the use of molecules of the type of diketones (benzil and its derivatives) makes it possible to realize conversion of singlet excitations of the polymer to its triplet state and thus upgrade the efficiency of its population. This process results from the singlet-singlet energy transfer from the polymer to the dopant and the effective intersystem crossing in it followed by the triplet-triplet energy transfer back to the polymer. The use of these converters has made it possible to investigate in detail the spectral-kinetic properties of the phosphorescence of MeLPPP. It is shown that the main channel of polymer fluorescence quenching is the dipole-dipole controlled single step transfer from an excited donor to an acceptor. In the case of a planar structure of the polymer and the activator, along with the energy transfer, the charge transfer makes a great contribution to the polymer fluorescence quenching. The latter process is accompanied by the formation of an intermolecular complex (exciplex). The introduction of volume substituents into the molecule of both the polymer and the activator decreases the efficiency of this process. The use, as a singlet-triplet converter, of organometallic complexes containing heavy atoms leads to an increase in the phosphorescence intensity and in its decay rate, which is due to the remote resonantly enhanced external heavy atom effect. The investigation of the system MeLPPP/ Pt(II)octaethylporphyrin (PtOEP) has shown that the main channel of population of the activator triplet state is the singlet-singlet energy transfer, while both the singlet and triplet levels of the dopant are located below the respective levels of the polymer. The reason for this effect is the low population efficiency of the polymer triplet state. The addition of benzil leads to its appreciable increase and, as a consequence, to an increase in the contribution of the triplet channel to the excitation of the PtOEP phosphorescence. When the activator is excited to the absorption band proper, delayed fluorescence of the polymer is observed, in other words, up-conversion takes place. On the basis of the data on the quenching of delayed fluorescence by an electric field it has been shown that its source is the exciplex between PtOEP and MeLPPP generated by excitation of the S_1 state of PtOEP. On the basis of PF2/6 activated by fluorescent dyes two systems with white color luminescence have been proposed.

INTODUCTION

Along with small molecules, conjugated polymers exhibiting remarkable electronic and luminescent properties have a large potential in terms of their application in optoelectronic devices [1, 2]. Investigations in this field were stimulated by the discovery in 1977 of electrical conduction in polymers [3], for which the authors of these investigations were awarded in 2000 a Nobel Prize in the field of chemistry. In 1990 it was also discovered that conjugated polymers display electroluminescence [4]. These results gave impetus to investigations in many scientific groups that led to the development in 2004 of a color display based on polymer LEDs [5].

In principle, such major characteristics of LEDs based on conjugated polymers as brightness, efficiency, stability and lifetime are rapidly approaching the level required for their commercial application [6]. At the same time, there is a whole number of questions that have to be answered for gaining a deeper insight into the processes proceeding in conjugated polymers upon absorption of light by them and injection of electric charges into them from electrodes [7].

A large number of works on spectroscopy of conjugated polymers concern singlet excitation with a typical lifetime of 10^{-10} –10^{-8} s. However, it is well known that triplet excitations also play an important role in many photophysical and photochemical processes. Thus, in terms of application of conjugated polymers in light-emitting diodes the interest in triplet states is due to the fact that they inevitably arise in the process of recombination of charges into an active medium [8-10]. Consequently, such processes as triplet-triplet energy transfer and triplet-triplet annihilation can strongly influence the luminescent properties of OLEDs.

The existence of triplet excitations in conjugated polymers is confirmed by various indirect methods, e.g., such as optically detected magnetic resonance [11], electroluminescence [12], delayed luminescence [13] and transient absorption [14]. However, phosphorescence of these polymers is practically not observe a except for ladder-type methyl poly(para-phenylene) (MeLPPP) [15] and poly(9,9-di(ethylhexyl)fluorene) (PF2/6) [16].

The main reason for this is the weak spin-orbit coupling [17]. This leads, first, to a low efficiency of triplet state population as a result of the intersystem-crossing (ISC) and, second, to a low value of the quantum yield of phosphorescence. Moreover, at high temperatures effective migration of triplet excitations that is due to the long lifetime may lead to their quenching as a result of the triplet-triplet annihilation or energy transfer to the energy acceptor. The mobility of excitations can be reduced by introducing a polymer into a solid matrix. For instance, in investigating a solid solution of MeLPPP [18], it was found that its phosphorescence decays practically exponentially at a rate of 1.1 s^{-1}. However, the dynamics of the triplet state of this polymer in the film is absolutely different [15]. It was found that phosphorescence decays by the power law and in 1 ms its intensity decreases 1000 times. Such a behavior was explained by the effective quenching of triplet excitations in the sample under investigation. However, the nature of the quencher, whose concentration should be unjustifiably large for the realization of the observed effect, is still unclear.

There exists a point of view, according to which the observation of MeLPPP phosphorescence in the first batch of this polymer, as in the case of the phenyl derivative of this polymer [19], is due to the presence in it of heavy metal (palladium) atoms. As a result of the heavy atom effect, the forbidding on the radiative decay of the triplet state is lifted. Moreover, there is an increase in the efficiency of triplet state population as a result of intersystem crossing. As a consequence, the migration of triplet excitations in their lifetime to the centers including palladium atoms leads to emission of light with a larger quantum yield and a shorter decay time. And in the case of solution the main contribution to the phosphorescence is made by molecules containing no metal atoms, since the concentration of the latter is very small [19]. Within the framework of such an approach the difference in the decay rate of MeLPPP phosphorescence in the solution and the film can be explained without the assumption about the presence of a large number of quenchers in the bulk sample. The absence of phosphorescence in the latter samples is likely to be due to their high purity. The answer to this question can be given by the investigation of the sensitized phosphorescence observed upon introduction into the polymer of an appropriate activator.

Energy transfer to the activator molecules introduced into the active layer is a conventional way of upgrading the efficiency and improving the chromacity coordinates of polymer LEDs [20-22]. For this purpose fluorescent dyes with a high yield of luminescence were usually used [23]. However, according to the simple spin statistics, only 25% of recombination acts of charges injected into the active layer from electrodes lead to the formation of singlet excitations, thus setting the upper limit to the efficiency of these devices. And while in the case of polymer LEDs the relation between singlet and triplet excitations formed in the process of charge recombination is still under discussion [24-26], it is obvious that the use of luminescence of both kinds can considerably upgrade the efficiency of LEDs. A whole number of systems, into whose active layer a phosphorescent dye was introduced, were instigated [27-30]. The most interesting results were obtained for devices based on platinum (II) porphyrins (Pt(II)OEP) and iridium (III) tris(2-phenylpyridine) ($Ir(ppy)_3$). These materials incorporate heavy metal atoms. The strong spin-orbit interaction realized in such atoms leads to a mixing of singlet and triplet states, which in turn lifts the forbidding for transitions between states of a different multiplicity. As a result, the efficiency of population of the triplet state in the ISC process and the probability of its radiative decay sharply increase. However, there is one substantial difference between the LEDs using fluorescent and phosphorescent dyes. In the latter case, the efficiency sharply decreases at fairly large current densities [22]. First, the large lifetime of triplet states causes a saturation of emitting sites, and, second, the triplet-triplet annihilation markedly increases losses. Thus, it is obvious that before using such devices for commercial purposes, it is necessary to study well and understand all aspects of this phenomenon, both positive and negative ones. Earlier such attempts were already made for individual polymer/dopant systems [28, 29, 31].

This paper presents the results of the investigation of the photoluminescence of Me-, Ph-LPPP and PF2/6 doped with phosphorescent molecules of different types. These results make it possible to gain a deeper insight into the above problems. Moreover, two PF2/6-based systems are proposed, which can be used to obtain radiation of equi-energy white color and white light with allowance for the specific features of the spectral sensitivity of the human eye.

EXPERIMENTAL

The structural formulas of the polymers used in the present work are given in Figure 1. To prepare films, MeLPPP, PhLPPP, and PF2/6 were dissolved in toluene in a concentration of 10 mg/ml. The activator was added to the obtained solution in required quantities. The films were obtained by applying the solution on a quartz substrate rotating at a rotation rate of 1000 rpm. Such a concentration provides the formation of a MeLPPP film with an optical density of 0.25 at the first absorption maximum (456 nm), which corresponds to a thickness of 35 nm [32], and a PF2/6 film with a value of the optical density at the absorption spectrum maximum of the order of 0.4 corresponding to a thickness of 70 nm [33].

The absorption spectra of the films and solutions were recorded on a Cary 500 Scan (Varian) spectrophotometer. The stationary fluorescence spectra were measured on a Fluoromax (Spex) spectrofluorimeter. To register the quasi-stationary delayed luminescence, a disk phosphoroscope cutting 1 ms after the exciting pulse was used.

The time-resolved prompt fluorescence, delayed fluorescence, and phosphorescence spectra were recorded by an optical multichannel analyzer with a time gated intensified diode array detector with varied gate-width and delay time after the exciting pulse. A Lambda Physik FL 2001 dye laser pumped by a Lambda Physik XeCl-excimer laser (λ = 308 nm, repetition rate 10 Hz) was used for excitation.

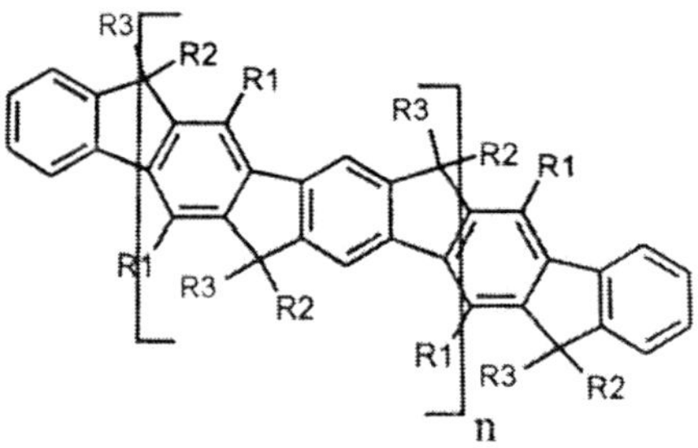

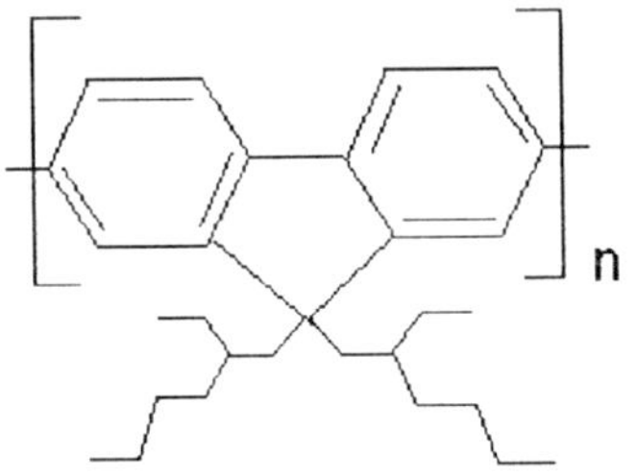

PF2/6

R1=n-C_6H_{13}, R2=1,4-(C_6H_4)-n-$C_{10}H_{21}$

R3=CH_3 - MeLPPP, R3=C_6H_5 – PhLPPP

Figure 1. Structural formulas of Me-, PhLPPP, and PF2/6.

The luminescence from an illuminated sport of size 3 mm in diameter was passed through a monochromator with a spectral resolution of 2 nm onto an optical multichannel analyzer. Prompt fluorescence was recorded with a gate length of 100 ns and a zero delay time. For measurements of delayed fluorescence and phosphorescence the detection window of up to 70 ms in width of the intensified diode array detector was operated synchronously but with a delay with respect to the laser pulse by 0.5 μs as a minimum. This allowed detection of a weak delayed luminescence after the intense prompt fluorescence. To increase the signal-to-noise ratio, the spectra were accumulated by averaging over 100-200 pulses. Except for the intensity-dependent measurements, the laser intensity was kept below 10 μJ/pulse to avoid damages of the polymer film. The sample was mounted either in a cryostat or in a quartz Dewar with liquid nitrogen.

For the fluorescence decay measurements a Ti:Sapphire laser producing 150 fs pulses at a repetition rate of 80 MHz, followed by a frequency doubling by the BBO crystal, was used to obtain a typical 400 nm (± 30 nm) excitation line. The detection system consisted of a streak camera with a maximal time resolution of about 4 ps, combined with a 0.25 m monochromator.

For the investigation of the electric field effect, a sandwich type LED structure ITO/SiO_x/MeLPPP/Al was used. Reverse bias was applied to the LED structure. The pulse generator produced 10 ms voltage pulses at variable delay times (-1 μs ... +0.6 μs with respect to the optical excitation) by means of an electrical trigger that was provided by the laser prior to the optical light pulse. The samples were attached to a holder of a home-built cryostat and kept under dynamic vacuum of 10^{-5} mbar.

RESULTS AND DISCUSSION

1. Sensitized Phosphorescence of LPPP

In [34, 35] it was shown that to solve the problem of a very low population of the triplet state of PF2/6 in the ISC process upon optical excitation of the polymer to the singlet state, benzil molecules can be used. Important properties of this substance are a high efficiency of going to the triplet state (0.92) and the absence of fluorescence ($\phi \sim 0$), which is due to the $n\pi^*$ nature of its lower exited states [36]. The scheme of the energy levels of the MeLPPP / benzil system is given in Figure 2. As in the case of PF2/6, in this system the singlet level of the polymer (22026 cm^{-1}) is lying higher than the respective level of benzil (20800 cm^{-1}) [37]. Consequently, the singlet excitation of the polymer will be quenched by the dopant molecules as a result of the singlet-singlet energy transfer. Effective ISC in benzil will lead to an almost 100% transition to the triplet state. However, since the triplet level of the polymer (16500 cm^{-1} in the case of MeLPPP) is lying below the dopant triplet state (18560 cm^{-1}) [38], back triplet-triplet energy transfer to the polymer will take place. Since as a result of the above processes a fairly high population of the triplet level of the polymer can be attained, it may be expected that in this case direct observation of MeLPPP phosphorescence will become possible. Unfortunately, in [34, 35] the investigations were carried out only for one concentration of benzil. In this case, it is rather difficult to judge all details of the first stage of the process under investigation, namely the fluorescence quenching. Below the results of the investigation of the MeLPPP and PF2/6 fluorescence quenching by benzil and 4-dimethylaminobenzil (DMAB) molecules are presented.

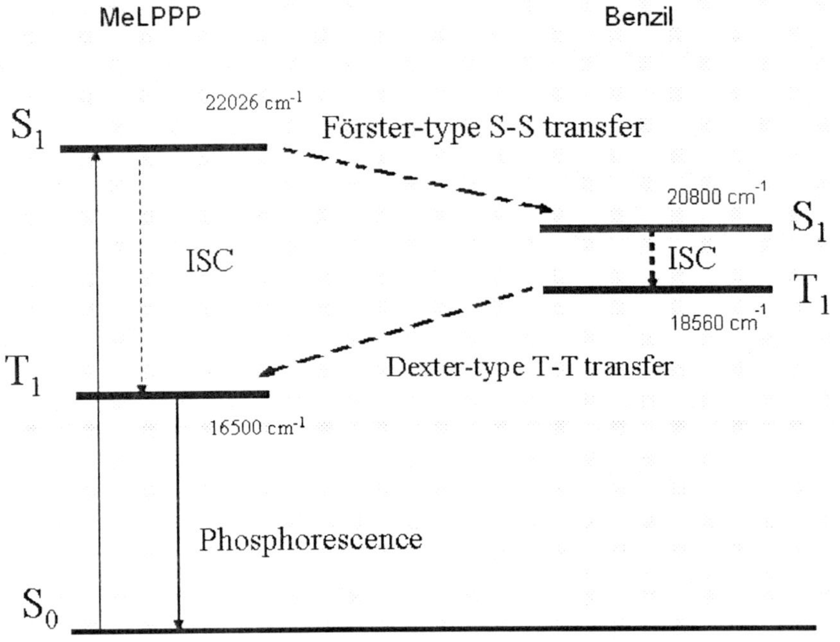

Figure 2. Scheme of the energy levels of the MeLPPP/benzil system.

1.1. Quenching of the Prompt Fluorescence of Polymers

The absorption and fluorescence spectra of the MeLPPP и PF2/6 films are given in Figure 3. Despite the above advantages of benzil as a singlet-triplet converter, it has one important disadvantage. Figure 4 shows its absorption spectrum. Analysis of Figures 3 and 4 points to the fact that in the region of overlap of the fluorescence spectrum of the polymer (energy donor) with the absorption spectrum of benzil (energy acceptor) the latter is characterized by extremely low values of the extinction coefficient. This effect is especially pronounced in the case of MeLPPP. At the same time this factor plays an extremely important role from the point of view of the energy transfer efficiency. According to the Förster theory, the transfer rate in the case of the dipole-dipole donor-acceptor interaction is defined by the following expression [39, 40]:

$$k_{D^*\to A} = \frac{9000(\ln 10)\kappa^2 \phi_D^0}{128\pi^5 n^4 N_A \tau_D^0 R^6} \int_0^\infty F_D(\lambda)\varepsilon_A(\lambda)\lambda^4 d\lambda \qquad (1)$$

where τ_D^0 and ϕ_D^0 are, respectively, the decay time and the quantum yield of the donor fluorescence in the absence of the acceptor, n is the refractive index of the medium, N_A is the Avogadro number, κ is the molecular orientation factor (2/3 for random orientation), λ is the wavelength , $F_D(\lambda)$ is the corrected fluorescence intensity of the donor with the total intensity normalized to unity, and $\varepsilon_A(\lambda)$ is the extinction coefficient of the acceptor at λ, and R is the mutual distance between the donor and acceptor molecules. Eq. (1) can be expressed more conveniently as

$$k_{D^*\to A} = \frac{1}{\tau_D^0}\left(\frac{R_0}{R}\right)^6 \tag{2}$$

where R_0 is the critical distance at which the excitation transfer and the fluorescence decay of the donor contribute equally and can be expressed as

$$R_0^6 = \frac{9000(\ln 10)\kappa^2\phi_D^0}{128\pi^5 n^4 N_A}\int_0^\infty F_D(\lambda)\varepsilon_A(\lambda)\lambda^4 d\lambda = \frac{9000(\ln 10)\kappa^2\phi_D^0}{128\pi^5 n^4 N_A}J \tag{3}$$

where J is the spectral overlap integral between the donor fluorescence and acceptor absorption spectra.

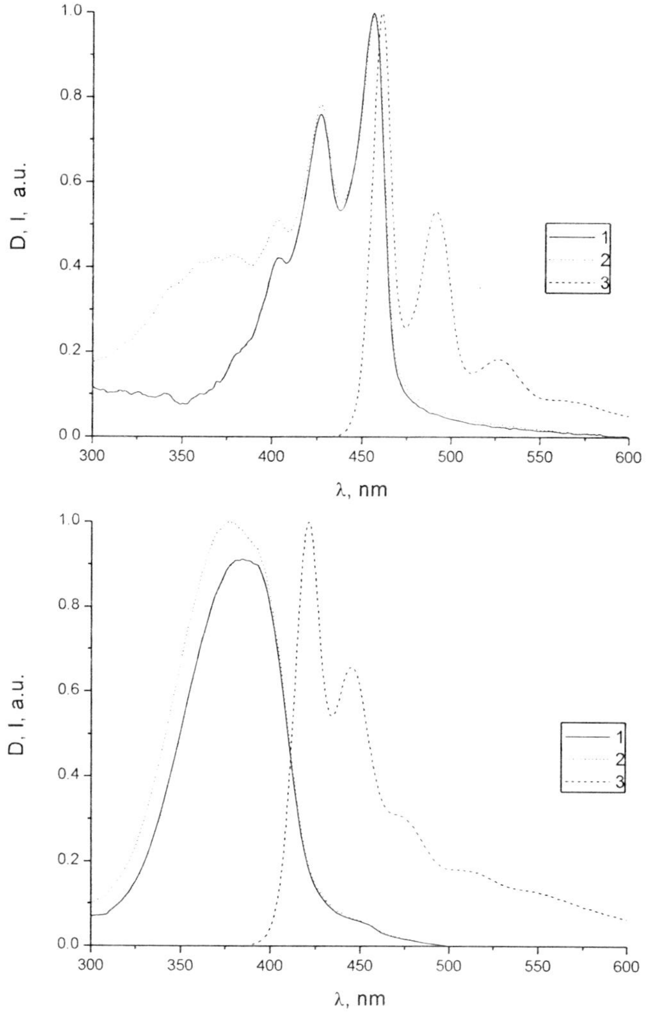

Figure 3. Absorption (1, 2) and fluorescence (3) spectra of MeLPPP (upper figure) and PF2/6 (lower figure) at T=293 K. Concentration of 4-dimethylaminobenzil: 1- 0; 2 – 20 wt %.

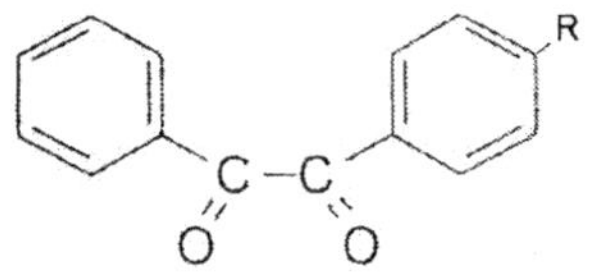

R=H – benzyl, $R=N(CH_3)_2$ - 4-dimethylaminobenzil

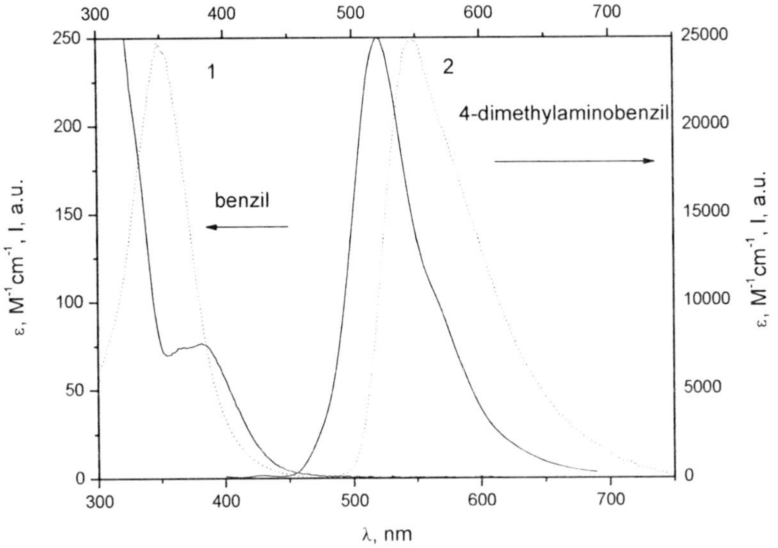

Figure 4. Absorption (1) (at 293 K) and phosphorescence spectra (2) (at 77K) of benzil (solid line) and 4-dimethylaminobenzil (dashed line) in ethanol. Arrows point to the scales for the extinction coefficient of the corresponding compounds. At the top the structural formulas of the molecules are given.

As is seen, the energy transfer rate is proportional to the spectral overlap integral *J*, whose value depends in turn on the value of the acceptor extinction coefficient. To increase the quenching efficiency, as an acceptor, DMAB molecules, whose spectral-luminescent data are given in Figure 4, were also used. Introduction of benzil into the polymer in no way affects the absorption spectra of the films. This is due to the small values of the extinction coefficient of this compound. In the case of DMAB, for high-concentration samples in the absorption spectra of activated films there appears a band with a maximum in the region of 350 nm corresponding to the absorption of the activator molecules (see Figure 3). The data on the DMAB absorption in the films and on the film thicknesses have made it possible to determine its molar concentration in the samples. For instance, the concentration of DMAB of 20% by weight corresponds to the molar concentration C=0.7 M for MeLPPP and C=0.43 M for PF2/6. The respective values for the benzil concentration were determined from the molecular mass ratio between these compounds.

Figure 5 shows the MeLPPP fluorescence spectra for various concentrations of benzil. As would be expected, the introduction of activator molecules into the polymer is followed by the latter's fluorescence quenching. Figure 6 presents the dependences of the relative fluorescence intensity of the polymer on the activator concentration.

On the basis of the data on the activator absorption and the polymer fluorescence the values of the critical radius R_0 given in Table 1 were calculated. For the calculations, the values of *n*(MeLPPP)=2 [32], *n*(PF2/6)=1.75 [41], ϕ_D^0 (MeLPPP)=45% и ϕ_D^0 (PF2/6)=60%.

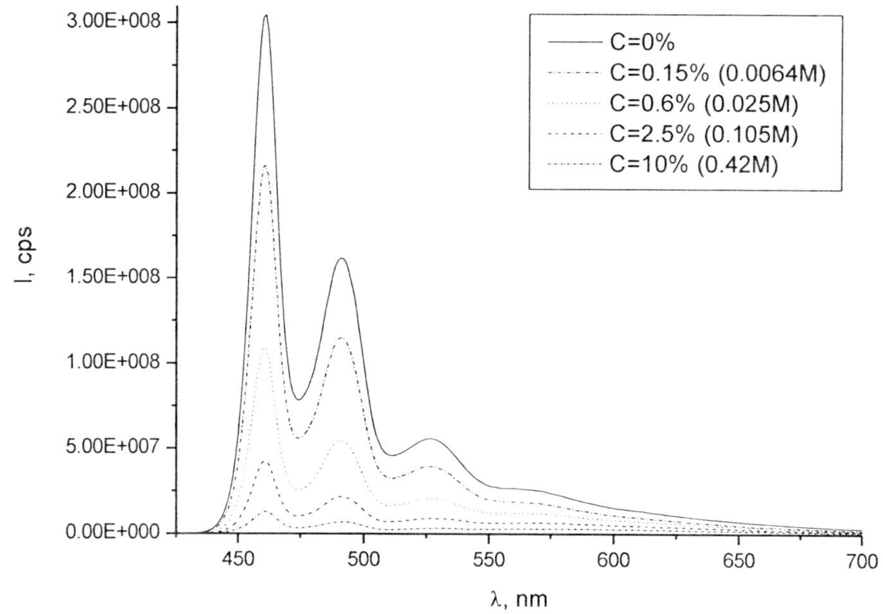

Figure 5. Prompt fluorescence spectra of the MeLPPP film with different concentrations of benzil at 293 K in air. (λ_{exc}=430 nm).

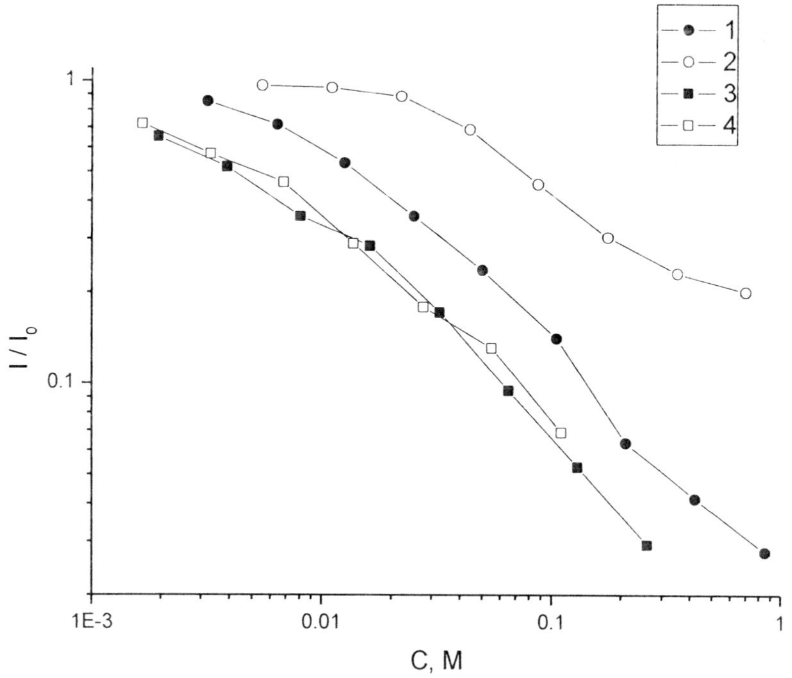

Figure 6. Relative fluorescence intensity of the polymers versus the activator concentration: 1 – MeLPPP/benzil, 2 - MeLPPP/DMAB, 3 – PF2/6 /benzil , 4 - PF2/6 / DMAB.

Table 1. Theoretical and experimental data for the critical radius of energy transfer, R_0 (Å), and the critical concentration of the acceptor, C_0 (M)

System	R_{0theor} / R_{0exp}	C_{0theor} / C_{0exp}
MeLPPP/benzil	6 / 15	0.24 / 0.014
MeLPPP/DMAB	12 / 8,5	0.029 / 0.079
PF/benzil	10 / 22,5	0.05 / 0.004
PF/DMAB	19 / 21	0.009 / 0.005

Unfortunately, there are no exact data for the quantum yields of fluorescence for these polymers. So they were estimated for 30% weight solutions of these polymers in the polystyrene matrix by the relative method [42] using, as a standard, a polystyrene film activated by molecules of fac-tris(2-phenylpyridyl)iridium(III) having a 100% quantum yield of phosphorescence [43, 44]. In determining the experimental values of the critical radius, the fact that the theory was developed within the framework of notions about spatially limited point dipoles was taken into account. In our case, the donor is a continuous medium (matrix). Therefore, as in [45], as an experimental value for the critical radius of energy transfer, we took half of the mean distance between acceptor molecules at which the fluorescence intensity was equal to half of the fluorescence intensity of the nonactivated sample. This fact was taken into account in calculating the theoretical values for the critical acceptor concentration (see Table 1) on the basis of the data for the critical radius.

From Table 1 it is seen that the most intensive quenching should be observed for polymers activated by DMAB molecules, which is due to the large value of the integral of spectral overlap between the donor fluorescence and the acceptor absorption and, as a result, to the large value of the critical radius of energy transfer. However, the experimental data show that for MeLPPP a more effective quenching takes place in the case of benzil, and for PF2/6 the curves of the relative fluorescence intensities versus the activator concentration for the two compounds practically coincide. Another very important result is that upon introduction of DMAB into the polymer the experimental values for the critical concentration and the critical radius are close to the theoretical values. This makes it possible to draw the conclusion about the quenching mechanism. In classical molecular crystals, the process determining the quenching rate is the migration of excitons inside the crystal. In amorphous media, as in conjugated polymers, such a process is dispersive transport, since they are characterized by a wide distribution of the energy of excited states. As a result, in disordered media upon an excitation act energy relaxation occurs and the mobility of excitons decreases with time [46, 47]. Nevertheless, equilibrium migration of donor excitations before they are trapped by acceptor molecules should lead to an increase in thevalue of the critical radius of energy transfer, which is practically not observed in the experiment with the use of DMAB for the acceptor. Proceeding from these data, it may be concluded that in this case the main quenching mechanism is dipole-dipole controlled single step transfer from an excited donor to an acceptor. The fact of practical temperature independence of the quenching efficiency also argues in favor of this conclusion. Indeed, since energy transport in disordered media is a thermally activated jumping process, its efficiency should be strongly influenced by temperature.

In the case of benzil, the observed quenching is much more effective than one would expect proceeding from the data of the theoretical consideration of the process under investigation. On the basis of the data obtained it may be concluded that in the case of benzil, besides singlet-singlet energy transfer, there is an additional channel of matrix fluorescence quenching. From Table 1 it is seen that the theoretical value of the critical radius of energy transfer in the case of the polymer activated by benzil is extremely small, especially for MeLPPP. Apparently, at distances between donor and acceptor molecules at which energy transfer can compete with the processes of intramolecular deactivation of the excited state of the donor the formation of an exciplex is possible. It is known [7, 48], that in this case donor-acceptor charge transfer occurs on a time scale of less than 0.1 ps, so that this process, along with the energy transfer, can make a great contribution to the quenching of the singlet state of the donor. For MeLPPP molecules with a rigid planar structure this process can be particularly effective. The twist-structure of the PF2/6 molecule makes it less effective. The presence in the DMAB molecule of a volume side substitution impedes the formation of an exciplex because of the impossibility to attain the donor-acceptor distance needed for this process.

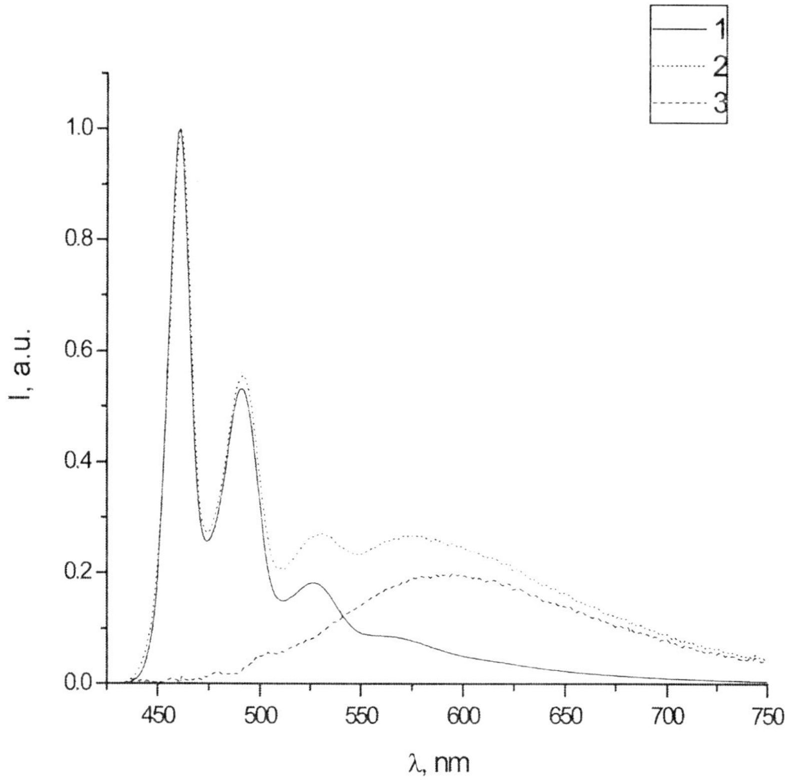

Figure 7. Normalized luminescence spectra of pure MeLPPP (1) and MeLPPP containing 20 wt % of benzil (2). Spectrum 3 represents the difference between spectra 2 and 1.

The data of Figure 7 also argue in favor of the quenching mechanism proposed above. It is well seen that the introduction of benzil into the polymer leads to not only quenching of its fluorescence, but also the appearance of a new luminescence characterized by a wide structureless spectrum shifted into the red region. At first glance the appearance of this band in the spectrum of the sample with a strongly quenching fluorescence can be due to the luminescence of aggregates or keto-defects [49] that is not quenched by benzil molecules. However, the data obtained for PhLPPP disprove this suggestion.

Figure 8 shows the fluorescence spectra of a pure sample and benzil- and DMAB-activated samples. First of all, it is seen from Figure 8 that the transition from MeLPPP to PhLPPP leads to a decrease in the quenching efficiency of polymer fluorescence. For DMAB, however, this effect is much weaker than for benzil-doped films. This is explained by the larger, compared to benzil, value of the critical radius for the polymer/DMAB system. In this case, there is no need for such a close approach of molecules for the act of energy transfer to be realized, which could be inhibited by the replacement of the methyl group by a more voluminous phenyl group. Importantly, the polymer fluorescence quenching efficiency for both molecules in PhLPPP is almost the same. It is seen that also for PhLPPP in the luminescence spectrum of the film with benzil a wide structureless band that is absent from the DMAB-containing sample is observed.

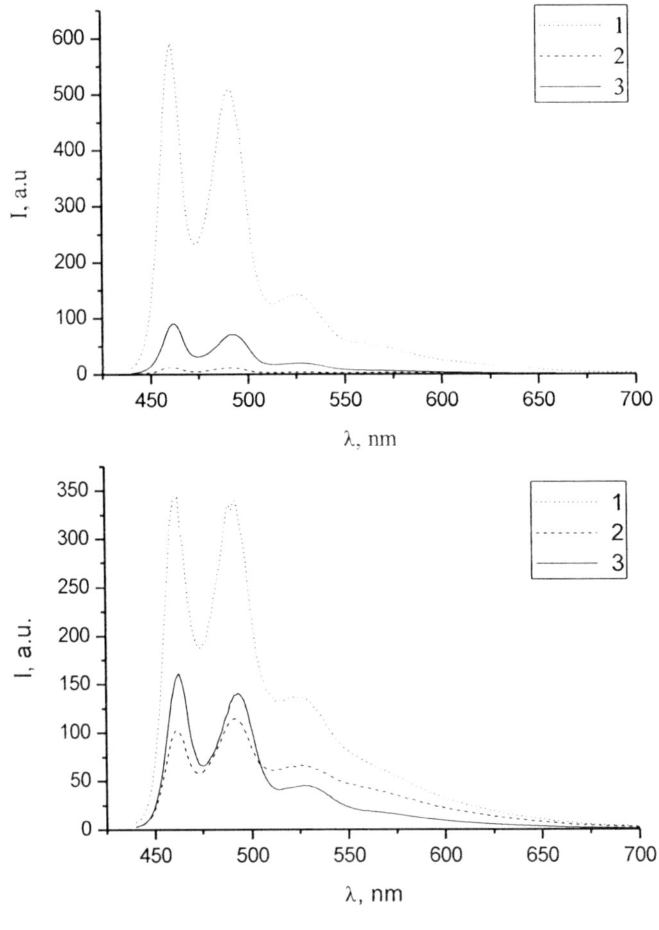

Figure 8. Luminescence spectra of pure (1) MeLPPP (top) and PhLPPP (bottom) and of those activated by 20 wt % benzil (2) and DMAB (3).

This permits concluding that the observed band is exactly due to the introduction into the polymer of benzil molecules and corresponds to the exciplex luminescence. It is necessary to point out that in the luminescence spectrum of highly concentrated benzil-activated PF2/6 samples a wide structureless band with a maximum at 550 nm is observed too. The change in the position of the band on going from PF2/6 to MeLPPP permits the conclusion that it is not an attribute of benzil molecules, e.g., the luminescence of its aggregates or excimers.

1.2. Delayed Luminescence of MeLPPP

As was shown in [34, 35], the molecules of benzil and, as may be expected, of its derivatives can serve as a converter of singlet excitations of a polymer into triplet excitations. At the same time the above data point to the presence in the case of polymer activation by benzil molecules, along with singlet-singlet energy transfer, of another mechanism of their fluorescence quenching including the formation of an exciplex. And although recombination of charges in the process of their deactivation can also lead, with a high probability, to a population of the triplet state of the polymer [50], it may be suggested, however, that the efficiency of singlet-triplet conversion in benzil-containing samples should be lower than in samples with DMAB , since part of the exciplexes decay through the radiative channel. To elucidate this question, quasi-stationary spectra of delayed luminescence of samples containing benzil and DMAB in such concentrations at which the quenching of polymer fluorescence is equal for both substances were obtained by means of a disk phosphoroscope. Figure 9 shows the spectra of such samples at 77 K. It is well seen that the introduction of the activator leads to a strong increase in the intensity of the band with a maximum at 610 nm. Taking into account the results of the previous investigations [15, 51], it may be argued that the observed band is a polymer phosphorescence sensitized by the activator molecules. From Figure 9 it is seen that the phosphorescence and delayed fluorescence intensity for the samples containing DMAB as an activator is much higher than in the samples with benzil. This points to the fact that in the case of benzil-containing polymers some singlet excitations of the polymer are deactivated by a different mechanism instead of conversion to the triplet state.

Despite the fact that DMAB is most likely a more effective converter than benzil, to investigate the properties of time-resolved delayed luminescence, the latter was used due to its better solubility in MeLPPP and the possibility of obtaining a more significant quenching of the polymer fluorescence (see Figure 6). Figure 10 shows the delayed luminescence spectra of the pure MeLPPP film and the film containing 20 wt % of benzil recorded with a delay of 1 ms after the excitation pulse and a 10 ms recording window. As in the case of fluorescence, in the benzil-containing sample a wide band that is absent from the pure polymer spectrum is observed.

An important difference from the data of [15] is the fact that in our case a phosphorescence decay close to exponential at the initial stage with a decay time of the order of 15 ms is observed (Figure 11). This result is in good agreement with the data of [52]. However, as will be shown below, the decay rate decreases with time, and in general the decay is nonexponential.

An interesting observation is the fact that the introduction of benzil into the polymer leads to an increase in the delayed fluorescence intensity rather than to its decrease, which would be expected proceeding from the fact of polymer fluorescence quenching by benzil.

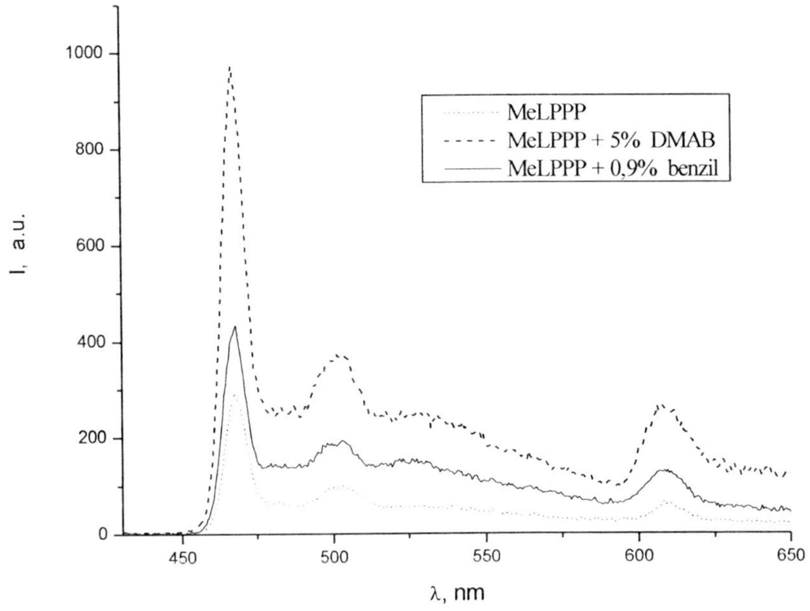

Figure 9. Quasi-stationary spectra of delayed luminescence of MeLPPP obtained by means of a disk phosphoroscope at 77 K.

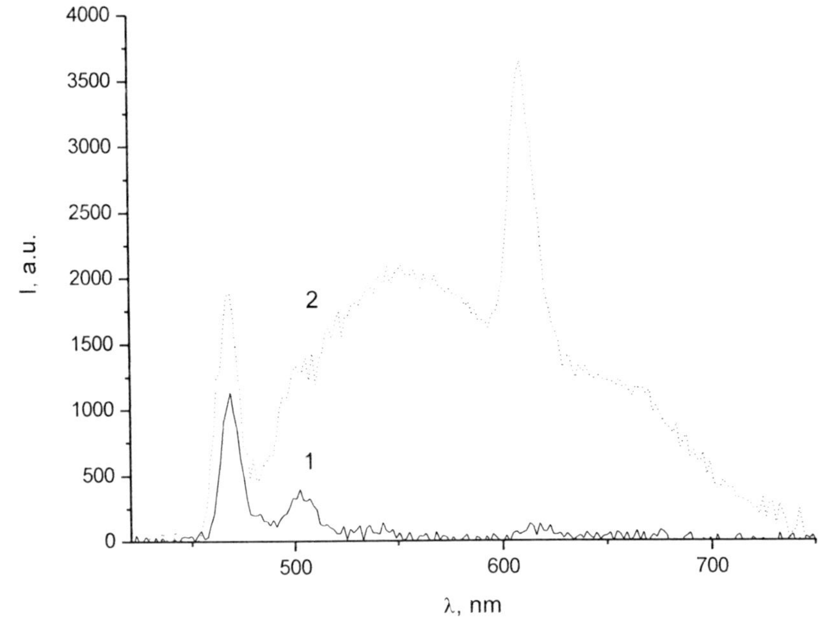

Figure 10. Delayed luminescence spectra of the film of pure MeLPPP (1) and MeLPPP doped with 20 wt % of benzil (2) at 77 K. λ_{exc}=440 nm. Both spectra are observed with a 1 ms delay after excitation and a 10 ms gate width.

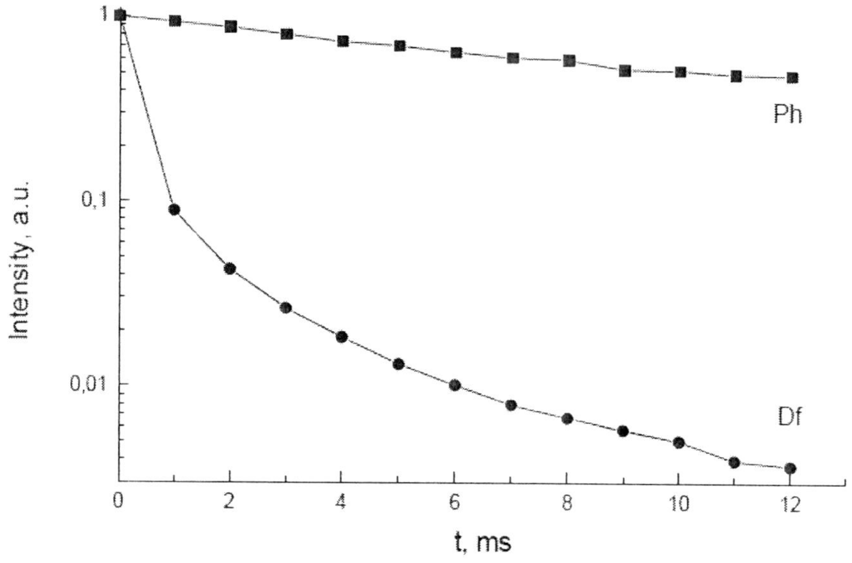

Figure 11. Decay of the phosphorescence and delayed fluorescence of MeLPPP with 20 wt % of benzil at 77 K. The curves were recorded with a gate width of 1 ms.

This effect can be explained as follows. It is well known that for a pure polymer a DF is caused by the recombination of geminate electron-hole pairs generated by the polymer excitation [51, 53]. In all likelihood the absence of the DF caused by the annihilation of the triplet excitons in the given polymer is due to the low efficiency of the intersystem crossing to the triplet state and, as a consequence, its small population. The introduction of benzil strongly increases the population of the triplet state of the polymer, as a result of which the number of annihilation events sharply increases leading to an increase in the annihilation delayed fluorescence (ADF) intensity. To test this hypothesis, the dependence of the phosphorescence and delayed fluorescence intensity on the exciting light intensity was investigated. Figure 12 presents the dependences obtained at two values of the time delay of registration after the exciting pulse. In both cases, in the range of values of the excitation intensity below 10 $\mu J/cm^2$ the phosphorescence intensity varies by the following law: $I_{Ph} \propto I_{exc}^{1/2}$. This dependence can take place in two cases: first, if the triplet-triplet annihilation is the main deactivation channel of the triplet state and, second, in the case where the main way of decay of singlet excitations of the polymer is their annihilation. The second process, however, can be neglected, since the data on the fluorescence quenching indicate that in a doped sample the polymer fluorescence has been quenched by a factor more than 30. Moreover, from the data of [53] it follows that even in the pure sample in the above range of values of excitation intensities the main deactivation channel of the singlet stat is the monomolecular decay. For elucidating the nature of the observed DF, the following result is important. In the case where the delayed fluorescence is caused by the annihilation of triplet excitations, the following relation between the phosphorescence, I_{Ph}, and delayed fluorescence, I_{DF}, should take place: $I_{DF}(I_{exc}) \propto I_{Ph}^2(I_{exc})$, at any intensity of the exciting light, I_{exc}, and whatever the character of the dependence $I_{Ph}(I_{exc})$. From Figure 12 it is seen that the above relation between the phosphorescence and the delayed fluorescence holds in a wide range of values of the excitation intensity with delayed luminescence registration in the time interval from 2 to 12 ms after the laser pulse.

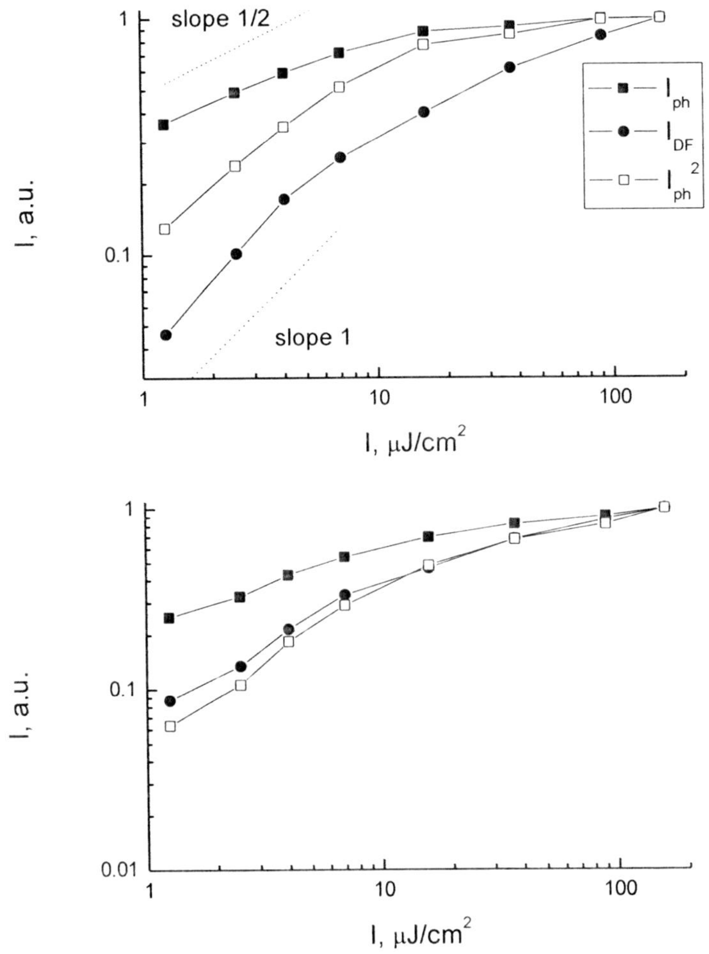

Figure 12. Dependence of the phosphorescence and delayed fluorescence intensity of MeLPPP with 20 wt % of benzil on the laser pulse energy at 77 K. The top curves are obtained with a 0.1 ms delay after excitation, the bottom ones with a 2 ms delay. All data were recorded with a gate width of 10 ms.

When the delay time is decreased to 0.1 ms, this relation holds only at excitation intensities lower than 10 $\mu J/cm^2$. At the same time it is known [53] that for a pure polymer a linear dependence of the DF intensity registered in the range from 1 ps to 10 ms on the laser radiation intensity in the same range of values of the latter is observed. On the basis of the data obtained it may be concluded that in the investigated system the delayed fluorescence of the polymer registered in the millisecond range arises from the annihilation of triplet excitons. In the microsecond range, the registered DF has a mixed character. The result obtained is another argument in favor of the statement that the delayed fluorescence of pure MeLPPP is not due to the TTA-process, but is rather associated with the recombination of charges generated in the process of its excitation.

If the delayed fluorescence is due to the annihilation of triplet excitations, then upon termination of the laser excitation it should decay according to the following formula

$$I_{DF}(t) = \frac{1}{2} f\gamma_{TTA}\left[I_{Ph}(t)\right]^2, \tag{4}$$

where f is the fraction of triplet encounters leading to the formation of singlets. However, from Figure 11 it is well seen that it decays much faster. Such a behavior cannot be described in terms of this simple relation characteristic of organic molecular crystals [54]. In conjugated polymers, as in the other amorphous materials, a more complex situation takes place. This is due to the statistical distribution of the excited state energy, whose optical manifestation is the inhomogeneous broadening of the spectrum. As a result, at low temperatures spectral diffusion that shows up as a shift of the luminescence spectrum and dispersive excitation transport are observed [55]. A typical manifestation of dispersive transport is the power character of the time dependence $R(t) \propto t^{-m}$, with a value of m close to unity. Indeed, Monte Carlo simulations in disordered media show that the rate coefficient for the TTA reaction γ is no longer constant and depends on time according to the law $\gamma_{TTA}(t) \sim t^{-n}$ with n close to unity [56, 57]. The time dependence of the rate coefficient for the process of annihilation of triplet excitations can be obtained from the expression

$$\gamma_{TTA}(t) \propto I_{DF}(t) / I^2_{Ph}(t). \tag{5}$$

The plot $\log(I_{df} / I_{ph}^2)$ versus $\log t$ should be linear with a $-n$ slope. As is seen from Figure 13, such a dependence really takes place with a slope close to the theoretical predictions.

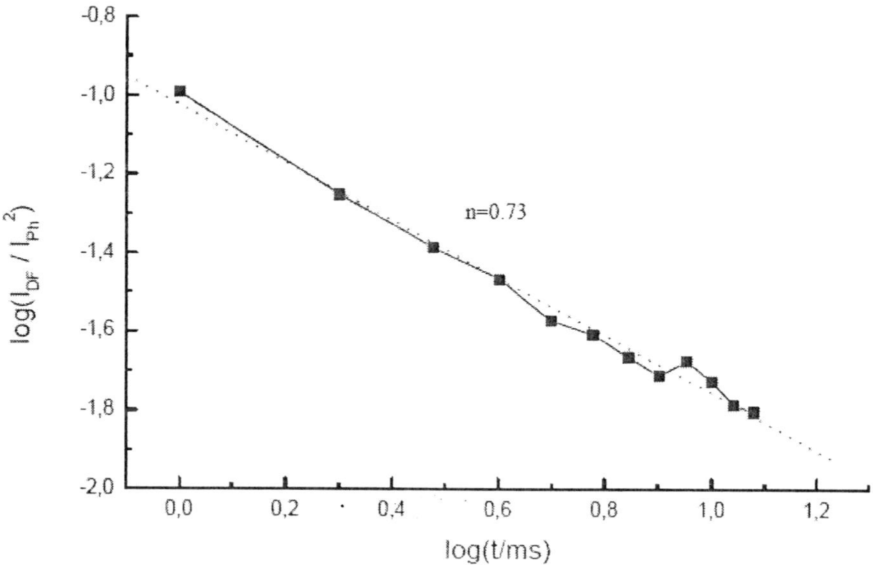

Figure 13. Dependence of $\log(I_{df} / I_{ph}^2)$ on $\log t$ for the MeLPPP film with 20 wt % of benzil at 77 K.

1.3. Delayed Luminescence of PhLPPP

The introduction of an atom of heavy metals into each monomeric unit of the polymer molecule markedly increases the probability of intersystem crossing, thus making it possible to investigate fairly easily the phosphorescence of such polymers [58, 59]. At the same time, however, it leads to a strong depopulation of the singlet state. Recently a new polymer (PhLPPP (its structure formula is given in Figure 1)) containing only an insignificant number of heavy metal atoms - of the order of 1 atom per 17000 monomeric units has been presented [19, 60]. As shown in [19, 60], such a low concentration produces a weak effect on the efficiency of triplet state population, but at the same time provides an effective channel of its radiative decay due to the migration of triplet excitons to the site containing metals. This permits direct observation of the PhLPPP phosphorescence in the photo- and electroluminescence spectra even at room temperature. The above-mentioned papers present the results of detailed studies of the PhLPPP luminescence. Below data that make it possible to gain a deeper insight into the mechanisms proceeding in such systems will be given.

Figure 14 shows the delayed luminescence spectrum of the pure polymer film obtained at room temperature in vacuum. It is seen that it consists of delayed fluorescence and phosphorescence. The decay of both types of luminescence is shown in Figure 15. Since the registration window considerably exceeds the luminescence decay time, the curves presented correspond to the integral kinetics. The true decay curves can be obtained by calculating the derivative of the integral curves. The result of this operation is given in Figure 16 on a double logarithmic scale. Obviously, both the DF and the phosphorescence decay by the power law $I(t) \propto t^{-n}$ with a value of n=1.2. A similar picture was obtained in [53] for MeLPPP at 80 K. The same paper gives a detailed justification of the power law decay for both types of delayed luminescence. Apparently, the same decay kinetics of both types of luminescence contradicts formula (4) describing the relation between them for the case where the DF source is the annihilation of triplet excitations. As the authors of [60] believe, at room temperature the polymer chain segments containing a palladium atom are responsible for the phosphorescence. The extremely low concentration of triplet excitations and the high rate of their nonadiative deactivation as a result of the interaction with phonons make the annihilation process ineffective, which shows up as a linear dependence of the phosphorescence intensity on the exciting light intensity [60]. As in the case of MeLPPP, delayed fluorescence is most likely due to the recombination of geminate electron-hole pairs generated in the process of polymer excitation.

An interesting result is observed upon introduction into the polymer of benzil molecules. From Figure 14 it is seen that in the sample containing 20% of the activator phosphorescence is practically absent, and the DF intensity somewhat increases. The latter effect has been already discussed for MeLPPP. As for the absence of phosphorescence, it may be suggested that an increase in the population of triplet excitons leads to an increase in the efficiency of their bimolecular interaction. As result of such interaction, the excitations localized at the sites containing an atom of heavy metals are also quenched.

Figure 17 shows the temperature dependence of the delayed fluorescence intensity (intensity of the first band with a maximum at 460 nm) and the phosphorescence intensity (intensity of the band with a maximum at 600 nm). Intensity measurements were made with a 1 μs delay after the exiting pulse and an integration window of 10 ms. It is seen that both types of luminescence demonstrate a different behavior in the pure and activated samples.

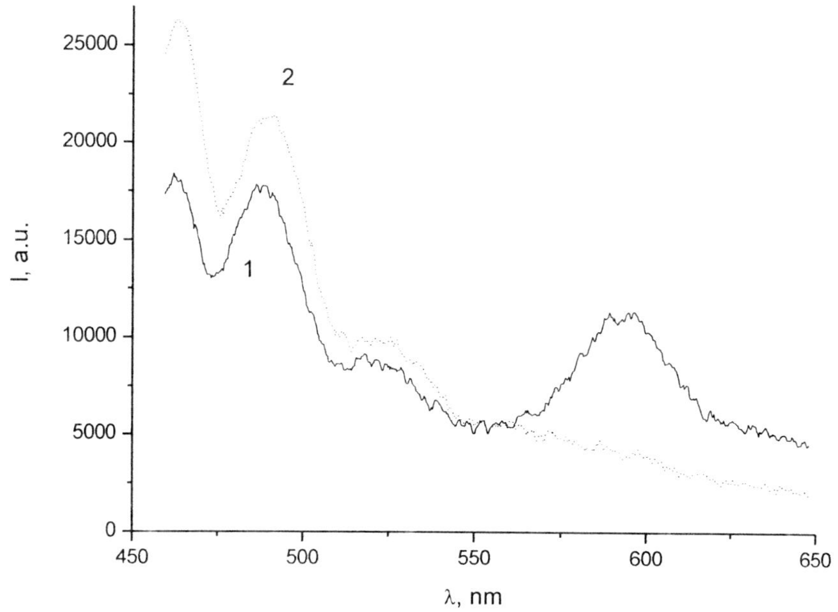

Figure 14. Delayed luminescence spectra of pure PhLPPP (1) and PhLPPP containing 20 wt % benzil (2) at room temperature in vacuum. Both spectra are observed with a 1 μs delay after excitation and a 10 ms gate width.

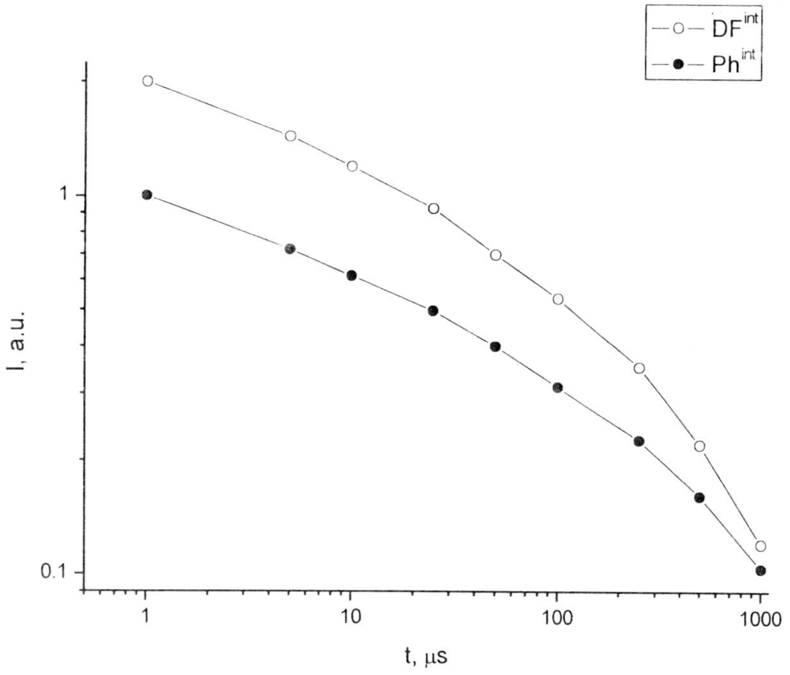

Figure 15. Decay of the time-integrated delayed fluorescence and phosphorescence of PhLPPP at 293 K in vacuum.

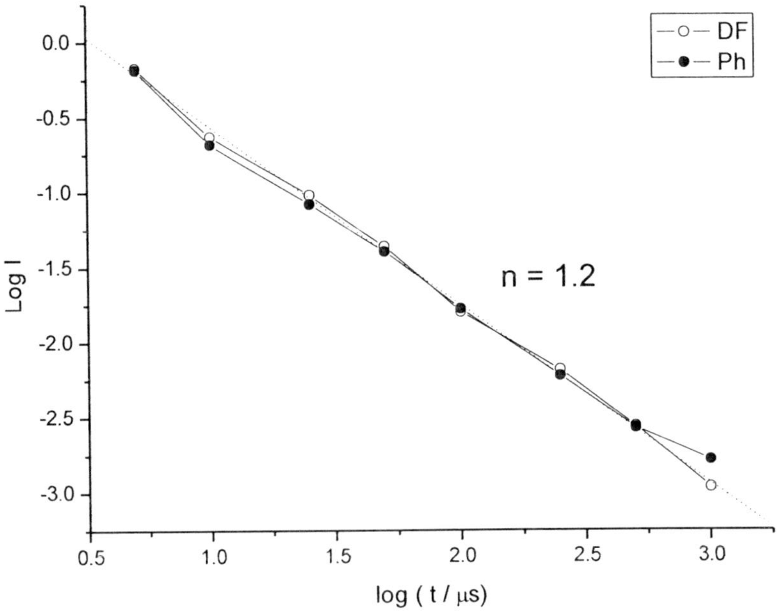

Figure 16. Derivative of the data from Figure 15.

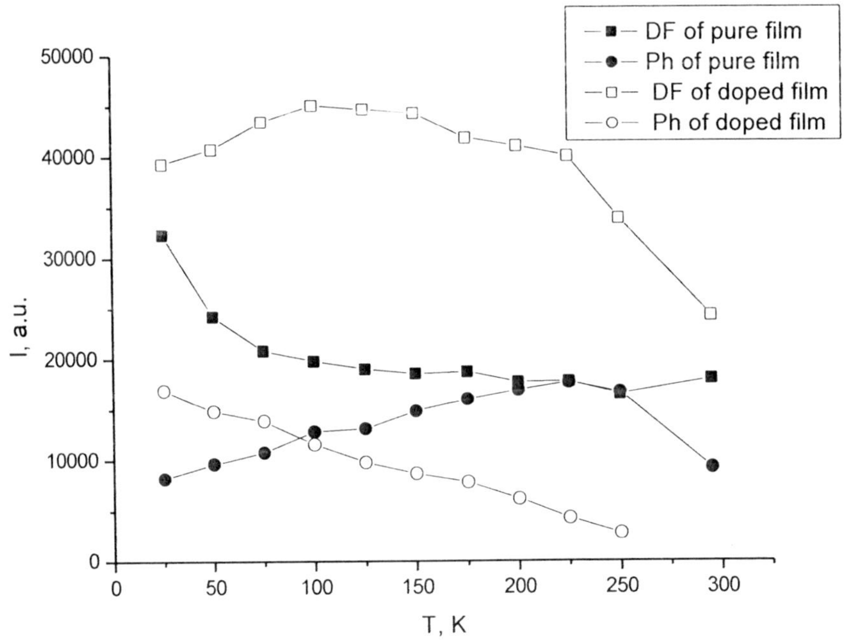

Figure 17. Temperature dependence of the phosphorescence and delayed fluorescence of pure PhLPPP and PhLPPP containing 20 wt % benzil. The spectra are observed with a 1 μs delay after excitation and a 10 ms gate width.

The pure polymer phosphorescence shows a nonmonotonic temperature dependence. This effect can be due the competition between two processes. The first process is the migration of triplet excitons. Obviously, its efficiency increases with temperature. As a consequence, there is an increase in the number of excitations reaching, as the result of migration, the sites containing heavy metal atoms, which leads to an increase in the phosphorescence intensity. The low density of triplet excitations makes their bimolecular quenching ineffective. The second factor is associated with the triplet excitation lifetime. It is well known that the rate of nonradiative deactivation of the triplet state increases with increasing temperature, which leads to a reduction in its lifetime. As a result, at high temperatures excitations have not time to reach the sites with a palladium atom and decay in a nonradiative way. The joint action of these factors leads, apparently, to the observed dependence. As for the delayed fluorescence, the increase in it intensity as the temperature decreases to below 100 K can be explained in terms of the accepted conception of its appearance through the recombination of geminate electron-hole pairs. A decrease in the temperature leads to a decrease in the mobility of charges upon their generation and, as a result, to an increase in the probability of their subsequent recombination.

The addition of benzil to the polymer leads to a change in the character of the dependence for both the phosphorescence and the DF. It is seen that it leads to a dramatic enhancement of delayed fluorescence, but not phosphorescence. This result permits concluding that in the activated film most triplet excitations interact with one another, which is followed by the production of annihilation delayed fluorescence. Indeed, an increase in the population of the triplet state in the doped sample also increases, as a consequence, the probability of their annihilation. In this case, both factors, namely the migration efficiency of triplet excitations and the rate of their nonradiative deactivation are acting in one direction, which leads to a monotone decrease in the phosphorescence intensity with increasing temperature. The dependence for the DF can be explained if we assume that it is to a greater extent due to the annihilation of triplet excitations. Then the increase in its intensity with temperature is associated with the migration efficiency of triplet excitations, and the sharp drop observed for temperatures above 200 K is due to the increase in the probability of their nonradiativ decay. The latter statement is corroborated by the good correlation with the data for the pure polymer phosphorescence.

Of particular interest are the data on the decay kinetics of the PhLPPP phosphorescence at low temperatures. For these measurements, the sample was placed in a quartz Dewar with liquid nitrogen. From Figure 18 it is seen that the decay curve contains a fast and a slow component. Since the registration window in recording this curve was 10 ms, the fast component does not reflect the real decay rate. As shown above in analyzing the data for room temperature, in fact the decay rate of the fast component is much higher. In the case of the slow component, the data given in Figure 18 can be used for analysis. Analysis sows that it decays almost exponentially with a decay time of 150 ms. This results completely coincides with the data of [19] on the lifetime of the triplet state of PhLPPP investigated by the method of transient absorption. As the authors of [19] suppose, the fast phosphorescence decay observed by them in the region of 100 μs is associated with the decay of the triplet excitations that have reached, as result of the migration, the centers including palladium atoms. Then it may be suggested that the slow component corresponds to the monomolecular decay of immobile triplet excitations localized on shallow traps. To test this hypothesis, the phosphorescence spectra recorded with a different time delay were investigated (Figure 19).

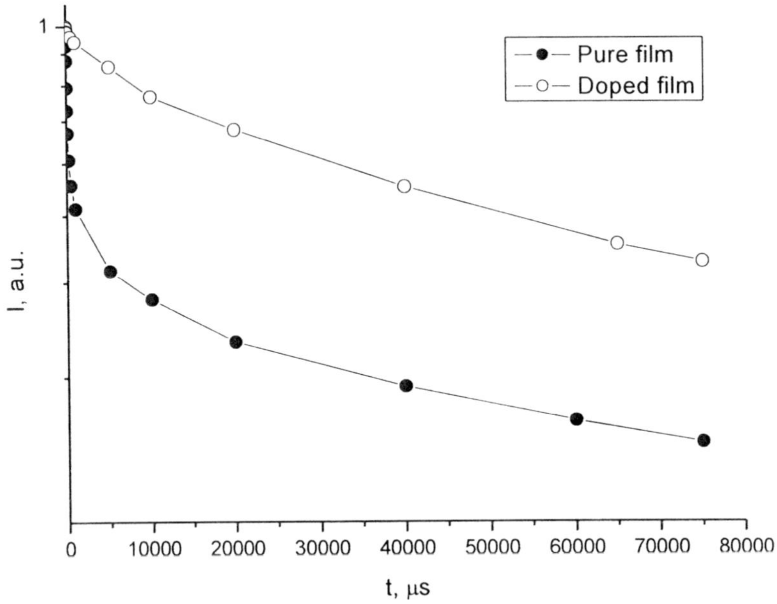

Figure 18. Decay of the phosphorescence of pure PhLPPP and PhLPPP with 20 wt % of benzil at 77 K. The curves were recorded with a gate width of 10 ms.

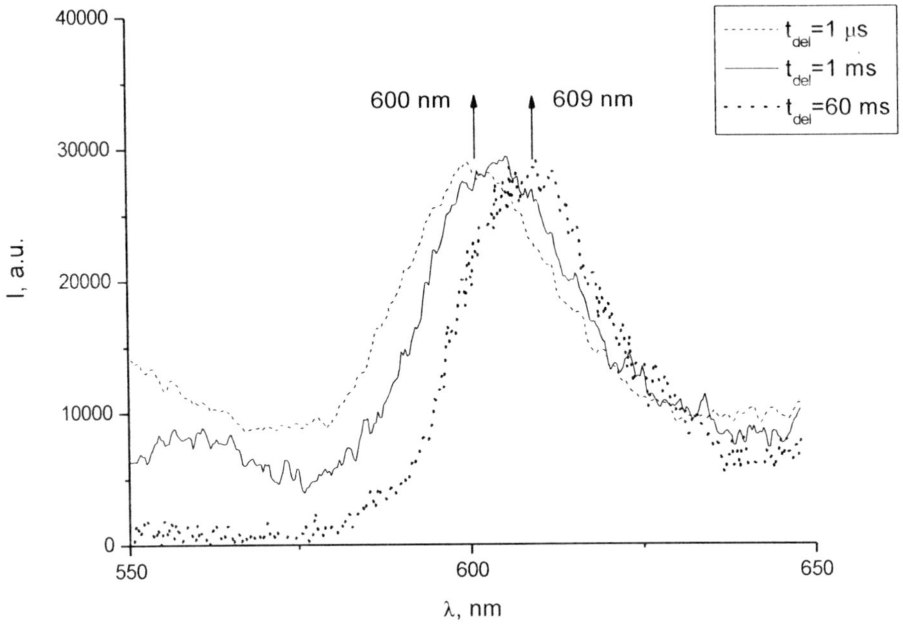

Figure 19. PhLPPP phosphorescence spectra obtained with a different time delay after the excitation pulse. The spectra are observed with a 10 ms gate width.

It is well seen that in the investigated system at low temperatures spectral diffusion is observed, which shows up as a shift of the phosphorescence band into the long wavelength region of the spectrum, and the value of this shift is 250 cm^{-1}. The introduction of benzil into the polymer leads to an almost complete disappearance of the fast component in the decay kinetics of its phosphorescence. However, this does not mean that it is really absent. With the measuring technique used this result points to a considerable decrease in its relative intensity, which agrees with the above data and their analysis.

1.4. Phosphorescence of Polymers Sensitized by an Iridium Complex

Recently in [61] it has been shown that the introduction into the polymer of molecules containing heavy atoms leads to an increase in the population of the triplet state of the matrix by a factor of more than one order of magnitude. And the process efficiency thereby does not depend on the relative position of the triplet levels of the polymer and the activator. It was shown that the basis of this phenomenon is formed by the remote resonantly enhanced external heavy atom effect, whose model was developed in [62, 63]. In this case, the increase in the rate of intersystem crossing is induced by the intermolecular spin-orbital interaction. Interestingly, in [61] an increase in the population efficiency of the polymer triplet state without a change in its lifetime was observed. It may be concluded that the above effect takes place only in the $S_1 \rightarrow T_1$ channel, without affecting the triplet-ground states transition $T_1 \rightarrow S_0$. At the same time in [64] the influence of the external heavy atom on the decay rate of the investigated compound phosphorescence was demonstrated.

Below the results of the investigation of the phosphorescence of MeLPPP and PF2/6 sensitized by the methyl derivative of fac-tris(2-phenylpyridyl)iridium(III) (Ir(meppy)$_3$ are presented. Figure 20 shows the structure of this compound and its absorption and phosphorescence spectra. It is well seen that in this case a situation analogous to the case of polymer activation by benzil molecules is realized. The spectral overlap between the complex absorption and the polymer fluorescence leads to an effective quenching of the latter as a result of the singlet-singlet energy transfer. The theoretical values of the critical radius of energy transfer equal to 21 and 26 Å for MeLPPP and PF2/6, respectively, point to a more effective quenching that in the case of their activation by benzil and DMAB. The back triplet-triplet energy transfer from Ir(meppy)$_3$ to the polymer should lead to a population of its triplet state. To part the effects caused by the external heavy atom from the simple sensitization of phosphorescence, samples with Ir(meppy)$_3$ and DMAB in such quantities at which equal quenching of the polymer fluorescence occurs were investigated. In this case, to obtain high-concentrated samples, instead of toluene, methyltetrahydrofuran as a solvent was used. In the case of MeLPPP, samples containing 2.5% Ir(meppy)$_3$ and 20% DMAB were studied. For such concentrations the fluorescence intensity of the polymer decreases five times compared to the sample containing no activator. For PF2/6 samples incorporating 5% of the activator were used. In this case, a decrease in the fluorescence intensity by a factor of 9 was attained.

Figure 21 shows the delayed luminescence spectra of pure and activated MeLPPP obtained with a 100 μs delay after the exciting pulse and a registration window of 70 ms. As is seen, the introduction of the activator leads to an increase in the phosphorescence and DF intensity. In so doing, the effect is practically the same for both types of activator. At first glance, no heavy atom effect is revealed here. However, an increase in the concentration of Ir(meppy)$_3$ to 5% is followed by a further increase in the phosphorescence intensity by a factor of 1.3, but the DF intensity thereby decreases twice.

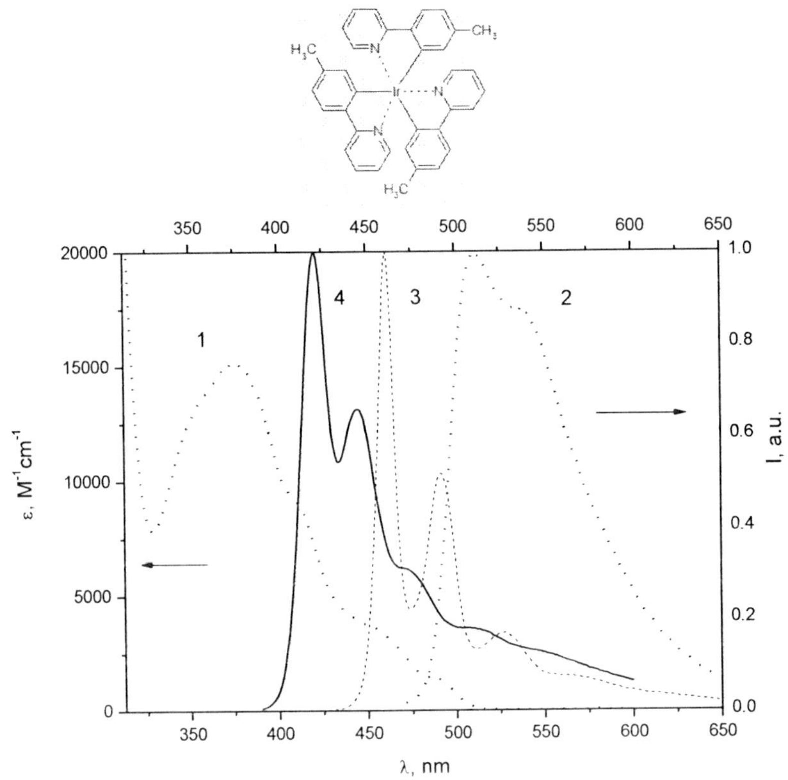

Figure 20. Absorption spectra (1) (at 293 K) and phosphorescence spectra (2) (at 77 K) Ir(meppy)$_3$ in MTHF. The fluorescence spectra of MeLPPP (3) and PF2/6 (4) are also given.

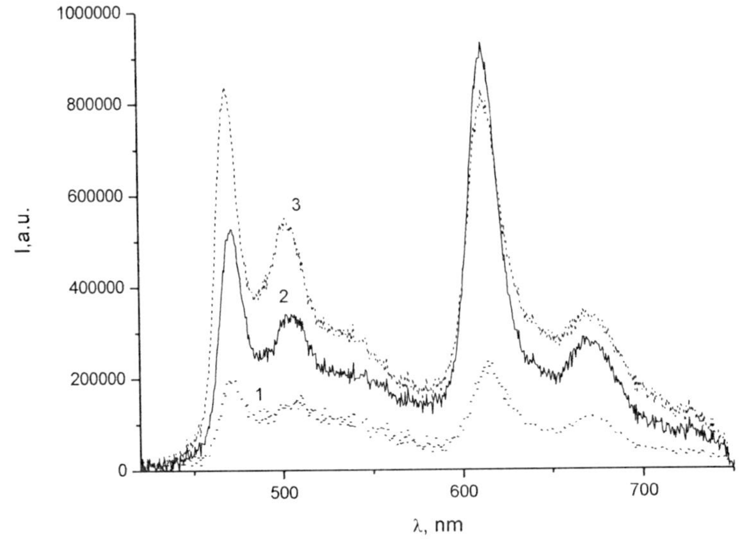

Figure 21. Delayed luminescence spectra of the film of pure MeLPPP film (1)and MeLPPP doped with 20 wt % of DMAB (2) and with 2.5% of Ir(meppy)$_3$ (3) at 77 K. λ_{exc}=440 nm. All spectra are observed with a 0.1 ms delay after excitation and a 70 ms gate width.

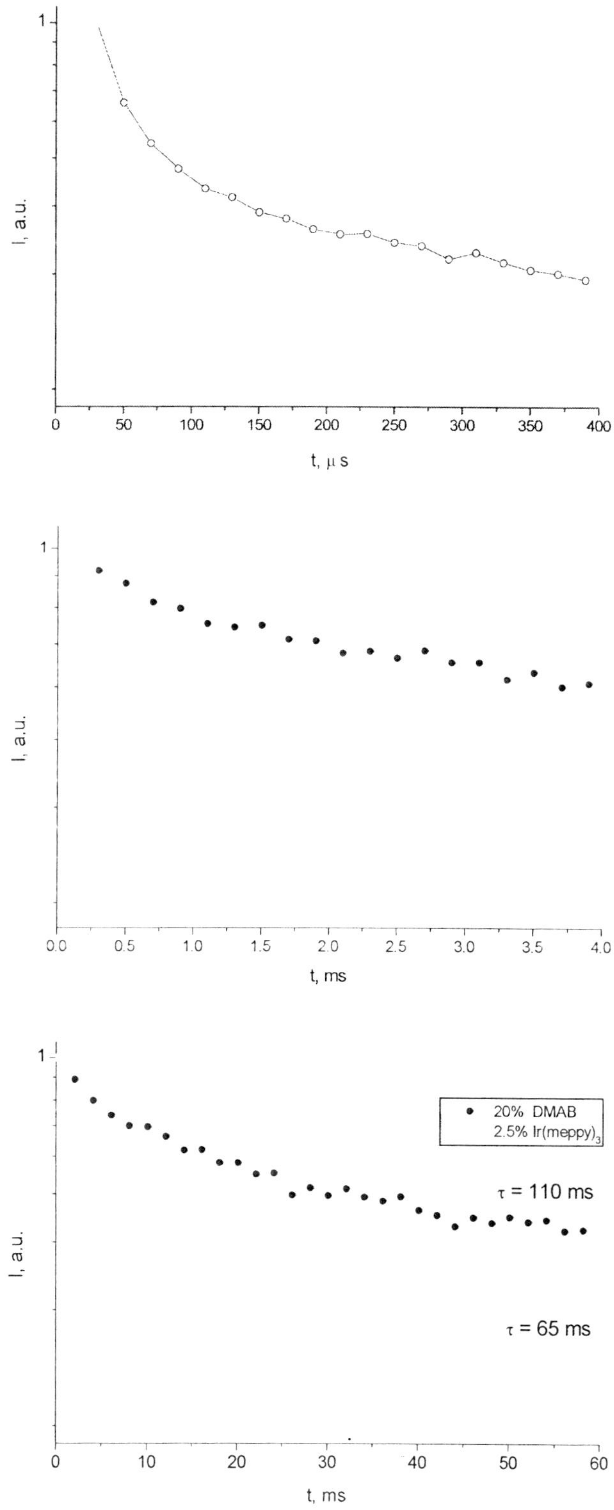

Figure 22. Decay of the phosphorescence of MeLPPP with 20 wt % of DMAB and 2.5% of $Ir(meppy)_3$ at 77 K. The curves were recorded with a gate width of 0.02 ms (top), 0.2 ms (middle), and 2 ms (bottom).

The decrease in the DF intensity is likely to be due to the decrease in the lifetime of triplet excitations caused by the external heavy atom of iridium. To elucidate this question, the decay kinetics of the sample was investigated.

From Figure 22 it is clearly seen that the phosphorescence of the sample activated by a organometallic complex decays much faster than the phosphorescence of the sample incorporating DMAB molecules. Since for each time interval the registration window was shorter than the decay time (except for the initial stage), it may be suggested that the given curves reflect the real dynamics. It is seen that the replacement of DMAB by $Ir(meppy)_3$ leads to not only the appearance of a fast component, but also to an increase in the decay rate of the long-lived component. It also should be taken into account that at the initial instant of time for the middle figure the real phosphorescence intensity for the sample with $Ir(meppy)_3$ is twice higher than for the DMAB-activated sample. The sample containing 5 % of $Ir(meppy)_3$ demonstrates a decay analogous to that given in the figure. It should be noted that in using a registration window of 20 μs for obtaining decay curves in the microsecond range, we failed to observe a polymer phosphorescence in the sample with a benzil derivative, which points to the absence of a fast component. Apparently, the presence of a fast component cannot be due to the quenching processes, since in this case an increase in the phosphorescence decay rate should be accompanied by an equally substantial decrease in its intensity, which is not observed in the experiment.

In the case of PF2/6 a more pronounced effect is observed. Figure 23 presents the delayed luminescence spectra for two activated samples obtained under the same conditions as for MeLPPP. At equal efficiencies of the polymer fluorescence quenching the delayed luminescence spectra of these systems differ widely.

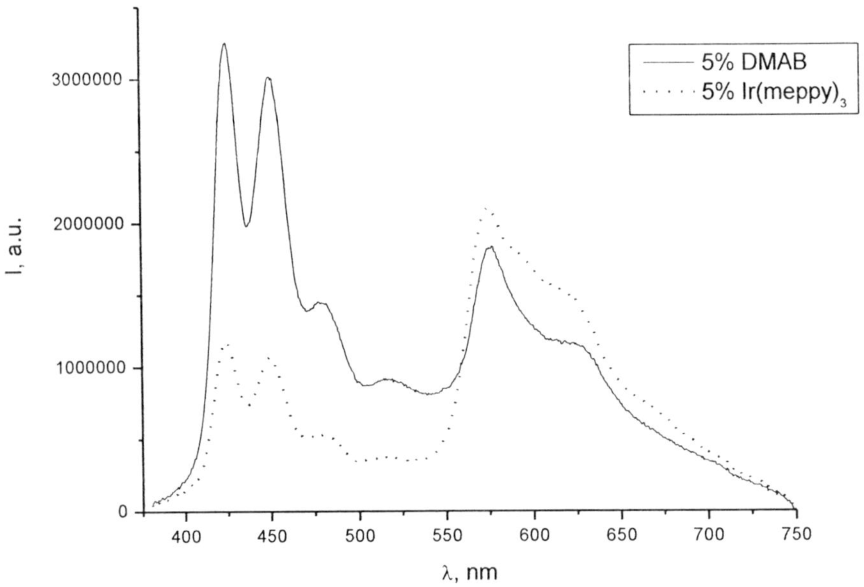

Figure 23. Delayed luminescence spectra of the PF2/6 film doped with 5 wt % of DMAB and 5% of $Ir(meppy)_3$ at 77 K. λ_{exc}=390 nm. Both spectra were observed with a 0.1 ms delay after excitation and a 70 ms gate width.

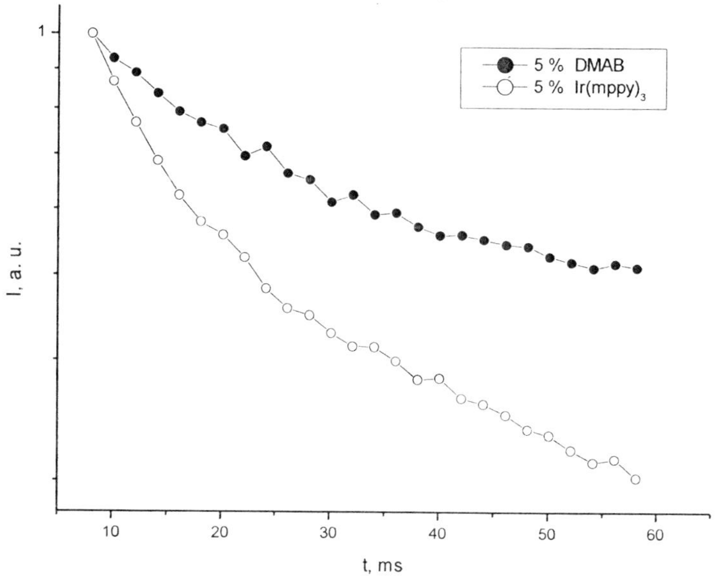

Figure 24. Decay of the phosphorescence of PF2/6 with 5 wt % of DMAB and 5% of Ir(meppy)$_3$ at 77 K. The curves were recorded with a gate width of 2 ms.

The phosphorescence of the sample with Ir(meppy)$_3$ is twice higher than the phosphorescence of the sample activated by DMAB. And its delayed fluorescence intensity is much lower. From Figure 24 it is seen that in so doing the phosphorescence of the sample incorporating a organometallic complex decays faster. All this points to the fact that the external heavy atom effect really takes place in Ir(meppy)$_3$-containing systems.

2. Luminescence of MeLPPP Doped with Pt(II)Octaethylporphyrin

As mentioned above, phosphorescence dyes containing atoms of heavy metals are widely used for upgrading the efficiency of OLEDs. For instance, in [30] for this purpose Pt(II)octaethylporphyrin was used. In the case of the MeLPPP/PtOEP system, the singlet and triplet levels of the activator are located lower than the respective levels of the matrix (Figure 25). Therefore, the population of the triplet state of the former is possible in principle as a result of the following processes: i) direct transfer of the singlet excitation of MeLPPP to the singlet state of the dopant followed by intersystem crossing to the triplet excited state and ii) generation of triplet excitations of MeLPPP via ISC of the singlet excited state followed by Dexter transfer to form a triplet exciton on the PtOEP guest. Below the results of the investigation of the photoluminescence of the MeLPPP/PtOEP system are presented. These results permit estimating the contribution of these processes to the creation of the population of activator triplet excitations.

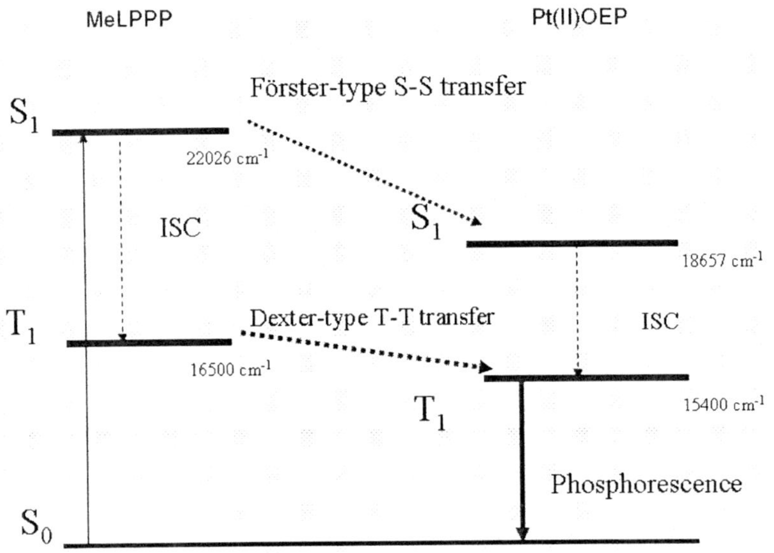

Figure 25. Scheme of the energy levels of MeLPPP/PtOEP.

2.1. Energy Transfer in MeLPPPp Doped with PtOEP

The structural formula of PtOEP and its absorption and phosphorescence spectra are given in Figure 26. The introduction of porphyrin into the polymer leads to a decrease in the fluorescence intensity of the latter and an increase in the rate of its decay, as well as to the appearance of phosphorescence of the former. For rather highly concentrated samples (>1% PtOEP) in their absorption spectra there appear distinct absorption bands of the activator with maxima at 536 and 382 nm, which permits determining the molar concentration of PtOEP in the samples (5% by weight corresponds to 0.05 M).

Looking at Figure 26, at first glance one may conclude that for this system the energy transfer efficiency should be fairly low, which is due to the small spectral overlap between the donor fluorescence and the acceptor absorption. However, the large value of the extinction coefficient at 536 nm (64000 $M^{-1}см^{-1}$) provides a fairly large value of the critical radius for the singlet-singlet energy transfer. For instance, the calculations performed with the use of formula (3) give the values of R_0=30 Å, which corresponds to the critical concentration C_{0theor} = 0.0019 M. The latter was determined taking into account the fact that the matrix acts as a donor (see Section 1.1). Figure 27 shows the dependences for the fluorescence intensity and its decay time on the concentration of PtOEP obtained at room temperature. In so doing, taking into account the nonexponential character of the decay, we took for the decay time the time interval during which the fluorescence intensity decreased e times. The analysis of these dependences gives an experimental value of the critical concentration of the acceptor, at which the fluorescence intensity of the donor decreases by one half, C_{0exp} = 0.0025 M, which is very close to the calculated value. The investigation of the influence of temperature on the transfer efficiency has revealed its complete independence of the temperature in the 4.2 – 300 K range.

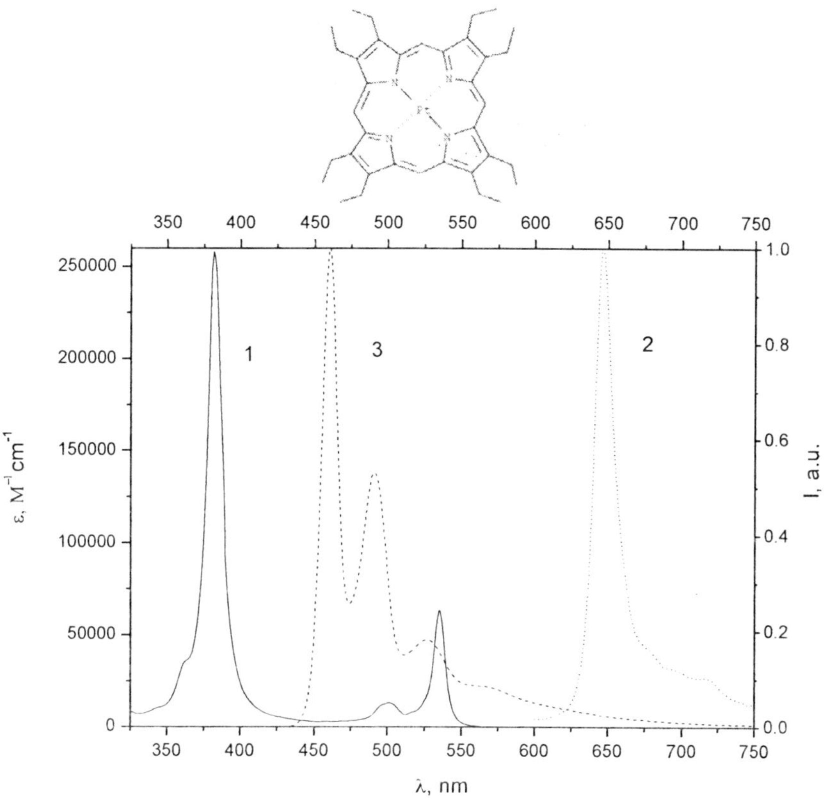

Figure 26. Absorption (1) and phosphorescence spectra of Pt(II)EOP (2) (λ_{exc}=536 nm) and prompt fluorescence spectrum of MeLPPP (3) at 293 K.

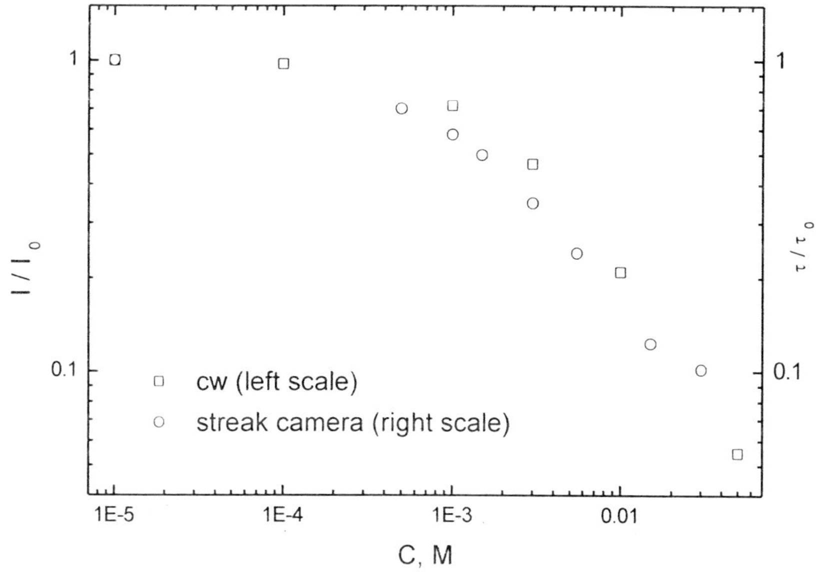

Figure 27. Dependence of the normalized cw intensity (left scale) and the normalized decay time (right scale) of the prompt fluorescence of MeLPPP on the PtOEP concentration at 293 K. The decay time is defined as the time in which the fluorescence has dropped to 1/e of its peak values.

These two results completely agree with the data of Section 1.1 and corroborate the conclusion that in the case of MeLPPP the polymer –activator singlet energy transfer is controlled by the single step Förster type mechanism.

To elucidate the role of the triplet-triplet energy transfer from the polymer to PtOEP in the population of the triplet state of the latter, we investigated the influence of oxygen on the activator phosphorescence. It is well known that oxygen is an excellent quencher of phosphorescence of organic compounds. Figures 28 and 29 show the phosphorescence spectra and decay curves of PtOEP obtained in vacuum and in air under its excitation through the matrix and direct excitation to the 536 nm absorption band.

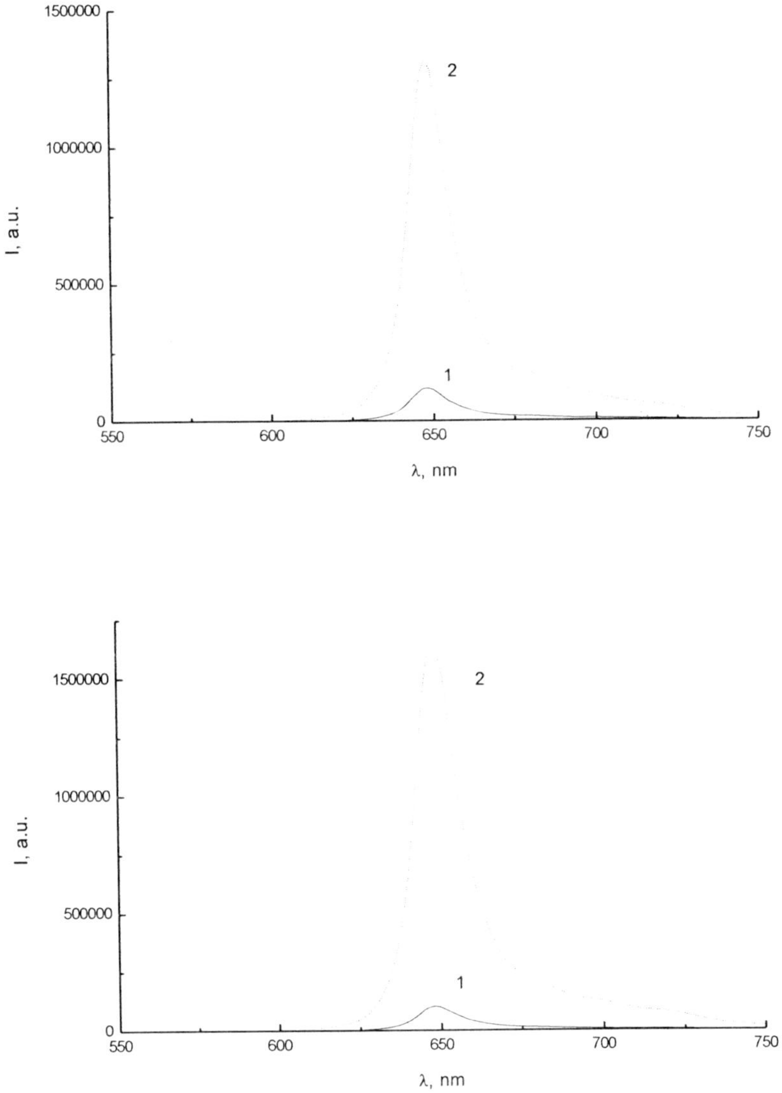

Figure 28. Phosphorescence of the sample containing 1 wt % PtOEP in MeLPPP. Measurements were performed at room temperature in air (1) or in vacuum (2). The excitation wavelength was 450 nm (top) and 536 nm (bottom).

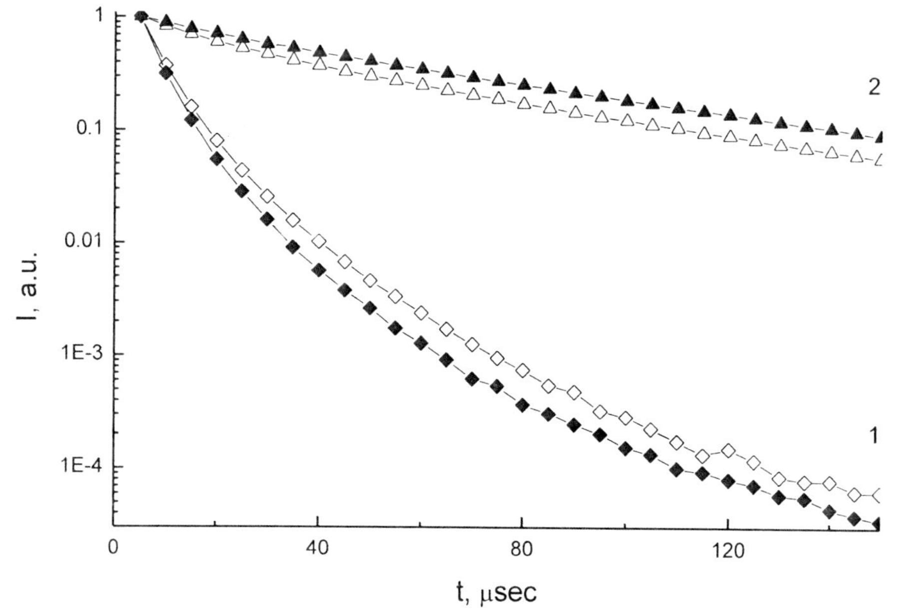

Figure 29. Decay curves of the Pt(II)EOP phosphorescence in the MeLPPP film in air (1) and in vacuum (2) at 293 K. Open characters correspond to λ_{exc}=450 nm, solid ones- to λ_{exc}=536 nm. C=1 wt%.

It is well seen that the switch from vacuum to air leads to a decrease in the phosphorescence intensity by a factor of 10-15 and a drastic increase in its decay rate. On the face of it, such a strong effect observed for a solid medium points to the participation in this process of mobile triplet excitations of the polymer. However, from Figures 28 an 29 it is seen that the quenching efficiency is independent of the way of excitation. This gives reason to exclude the triplet state of the polymer from the process under consideration. Otherwise the PtOEP phosphorescence quenching efficiency under excitation through the matrix should be higher than under its direct excitation. As for the quenching mechanism, the most probable fact is the formation in air of a weak complex between porphyrin and oxygen molecules.

The result obtained is not surprising taking into account the data of Section 1.2 that point to a low efficiency of formation of triplet states of MeLPPP in the process of intersystem crossing. To change this situation, 20% of benzil were additionally introduced into the system. As follows from Figure 6, the introduction of such a quantity of benzil leads to a decrease in the fluorescence intensity by a factor more than 30. Taking into account the fact that the main excitation source of porphyrin phosphorescence is the S_1 state of the polymer, one would expect a similar decrease for it as well. However, from Figure 30 it is seen that the phosphorescence intensity of the porphyrin introduced into the matrix in a concentration of the order of 0.001% has decreased by a factor of only 3.5. This points to the presence in the given system of additional ways of population of the triplet state of porphyrin. These are the following channels: singlet-singlet energy transfer from benzil to porphyrin; triplet-triplet energy transfer in the same pair, and energy transfer from the triplet level of the polymer to the triplet level of the porphyrin.

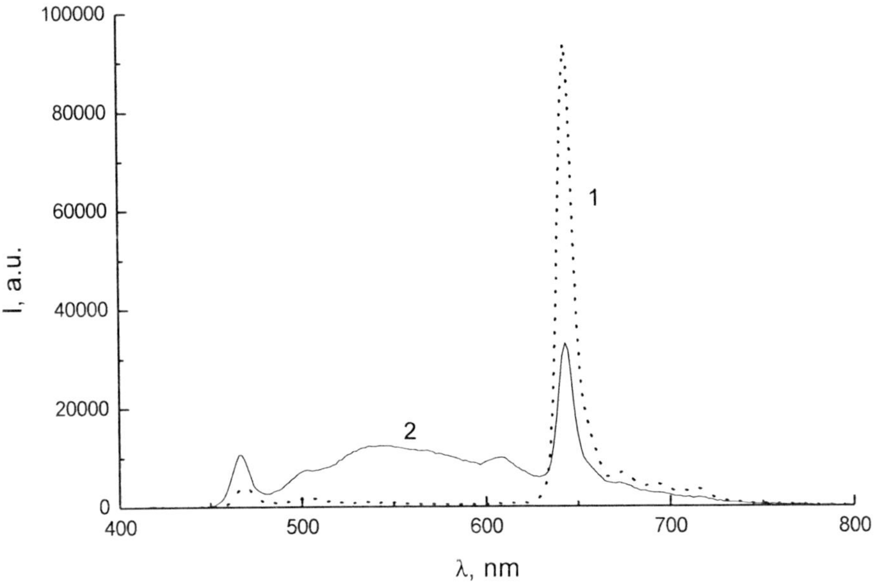

Figure 30. Delayed luminescence spectra of the MeLPPP film doped with 0.001 wt % of Pt(II)OEP without (1) and with 20 wt % of benzil (2) at 77 K. λexc=450 nm. Both spectra were observed with a 1 μs delay after excitation and a 10 ms gate width.

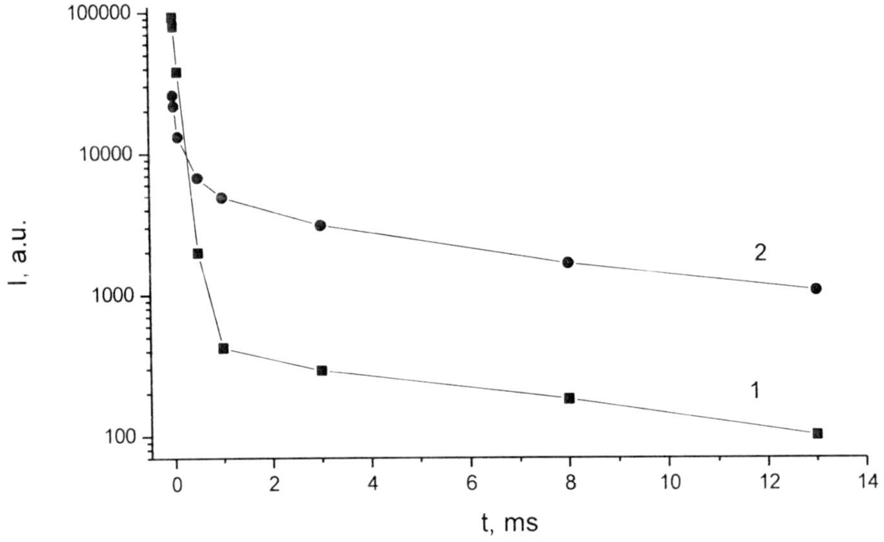

Figure 31. Decay of the Pt(II)OEP phosphorescence in the MeLPPP film without and with 20 wt % of benzil at 77 K. The curves were recorded with a gate width of 10 ms.

Taking into account the large quantity of benzil, the basic possibility of energy transfer by the first mechanism cannot be excluded. However, since the quantum yield of fluorescence of benzil is close to zero, it can definitively be stated that this channel plays no role in the population of the excited state of PtOEP [39, 40]. Since the phosphorescence decay times of the polymer (15 ms at the initial stage) and nonquenched benzil (4 ms) are much larger than the lifetime of the triplet state of porphyrin (0.1 ms) and differ from each other, the investigation of the decay kinetics of PtEOP permits revealing the presence of triplet-triplet transfer in the system and a donor for the porphyrin phosphorescence. Figure 31 shows the phosphorescence decay curves of phorphyrin in the polymer without benzil and with it. It is well seen that in both cases a fast and a slow components are observed. Roughly these curves can be described by the two-exponential law with times of 0.11 and 10 ms and weight factors of 99.5 and 0.05% for the matrix without benzil and 82 and 18% in its presence. Obviously, the contribution to the fast component is made by the PtOEP molecules excited as a result of the singlet-singlet energy transfer from the polymer. From Figure 31 it is well seen that the introduction of benzil dramatically decreases the number of acts of this type transfer. At the same time, there is also a marked increase in the contribution to the decay of the long-lived component. Proceeding from the decay time of the slow component, it may be concluded that the main energy donor for porphyrin is the triplet state of the polymer.

2.2. Sensitised Phosphorescence of PtOEP

Figure 32 shows the phosphorescence intensity of PtOEP versus the exciting radiation intensity for samples with its different concentration. It is seen that these dependences are linear for a low intensity of the incident light. For the sample containing 5% of porphyrin, a deviation from the linear law is observed at intensities larger than 1 μJ/pulse. One most probable mechanism of this effect is the bimolecular interaction of the triplet excitations of PtOEP. However, this explanation contradicts the fact that saturation of activator phosphorescence for samples with a small concentration begins sooner than for high-concentration samples, while the bimolecular quenching should decrease with decreasing concentration. This effect is not associated with any mechanism of bimolecular quenching in the matrix either. This is confirmed by the data of Figure 33. It is seen that the curve linearity is also disturbed in the case of direct excitation of the dopant into the $S_1 \leftarrow S_0$ transition independent of whether the matrix is a conjugated polymer, i.e., MeLPPP or optically neutral polystyrene.

The most plausible explanation of this effect is the bleaching of the ground state of PtOEP molecules in the process of population of their triplet state. At a 1% concentration of the activator the optical density at the wavelength of 536 nm is $D=3*10^{-3}$, which is equivalent to a fraction absorbance of $7*10^{-3}$, at an area density of dopant molecules of $3*10^{13}$ cm^{-2}. At a photon flux of 10 μJ/pulse, equivalent to $\approx 1.4*10^{-4}$ J/(pulse cm^2), the number of singlet states populating the triplet level is $3*10^{12}$ cm^{-2}, i.e., only one order of magnitude less than the number of dopant molecules. The fact that at this excitation intensity the dependence of the activator phosphorescence intensity begins to deviate from the linear law permits arguing that the effect of saturation in the population of the triplet state of PtOEP takes place. And it does not depend on how these excited states are formed: whether by direct excitation into the activator absorption band or through polymer excitation with subsequent energy transfer. Obviously, this effect is very important in using such a system as an active layer in organic LEDs and can restrict its efficiency, especially in the case of a low concentration of the activator [22].

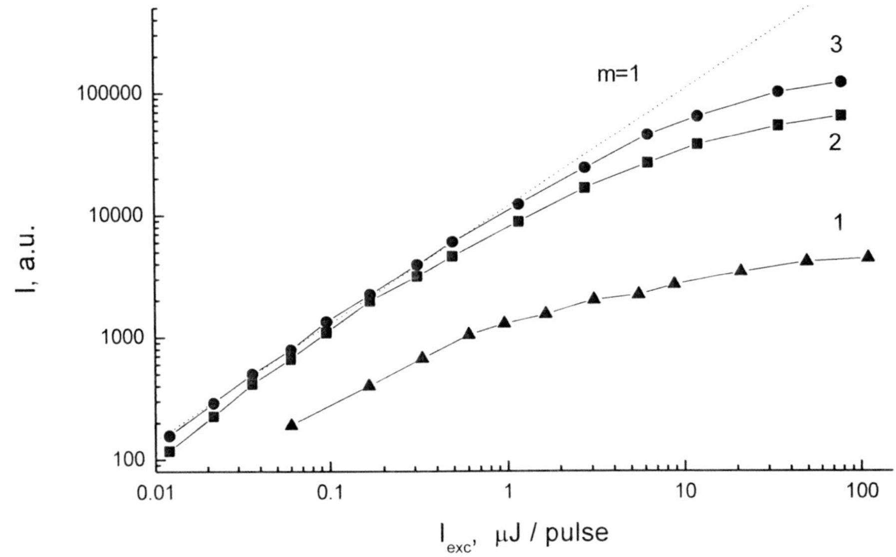

Figure 32. Dependence of the Pt(II)OEP phosphorescence intensity on the laser pulse energy at 293 K in air. λ_{exc}=450 nm. The concentration of PtOEP was 0.1 (curve 1), 1 (curve 2), and 5 wt % (curve 3).

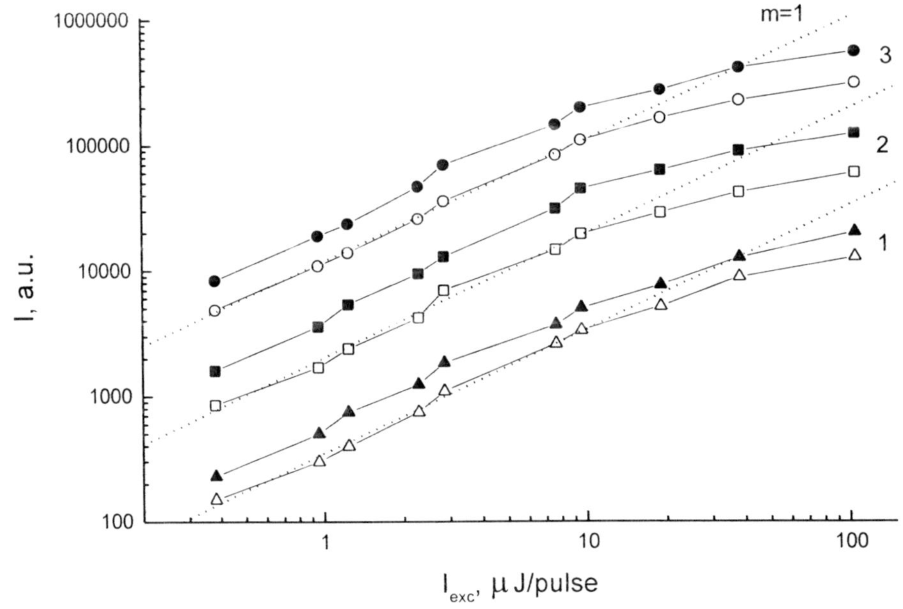

Figure 33. Dependence of the Pt(II)OEP phosphorescence intensity on the laser pulse energy in MeLPPP (1, 2) and polystyrene (3) films. λ_{exc}=536 nm. The concentration of PtOEP was 0.1 (curve 1), 1 wt % (curves 2 and 3). Open characters correspond to 293 K, solid ones - to 77 K. The curves are arbitrarily shifted on the ordinate scale.

2.3. Up-Conversion of Delayed Fluorescence of MeLPPP Doped with PtOEP

Figure 34 shows the delayed luminescence spectra of a pure polymer and a polymer containing 5% of PtOEP obtained at room temperature and at 77 K under excitation by light with a wavelength of 450 nm. It is well seen that a change in the temperature produces a different effect on the delayed fluorescence intensity of the polymer for both samples. For instance, for the pure polymer a decrease in the temperature leads to an increase in the DF intensity by a factor of 3, whereas for the activated sample its decrease by one and a half is observed. However, the most remarkable thing is that DF of MeLPPP can also be generated under excitation of the activated sample into the absorption band of PtOEP (536 nm). This effect is absent from the pure polymer under the same excitation conditions. Apparently, in the activated sample up-conversion of the delayed fluorescence of MeLPPP is realized.

As mentioned above, in the pure polymer the main contribution to the delayed fluorescence is made by the recombination of the geminately bound electron-hole pairs. It is essential to elucidate the DF source in the activated sample. One possible variant is the fact that at room temperature energy transfer takes place from the triplet state of PtOEP generated directly through its singlet state under excitation by light with a wavelength 536 nm or through the singlet state of the polymer under excitation by light with a wavelength 450 nm into the triplet state of MeLPPP. The subsequent annihilation of the polymer triplet states leads to the appearance of its DF of the annihilation type. The simple Bolzmann statistics predicts that under equilibrium conditions the fraction of triplet states of the polymer is defined by the following relation: $T_h/T_g = N_h/N_g \exp(- \Delta E_t /kT)$, where N_h/N_g is the ratio of host to guest chromophores and ΔE_t is the difference of the triplet energies.

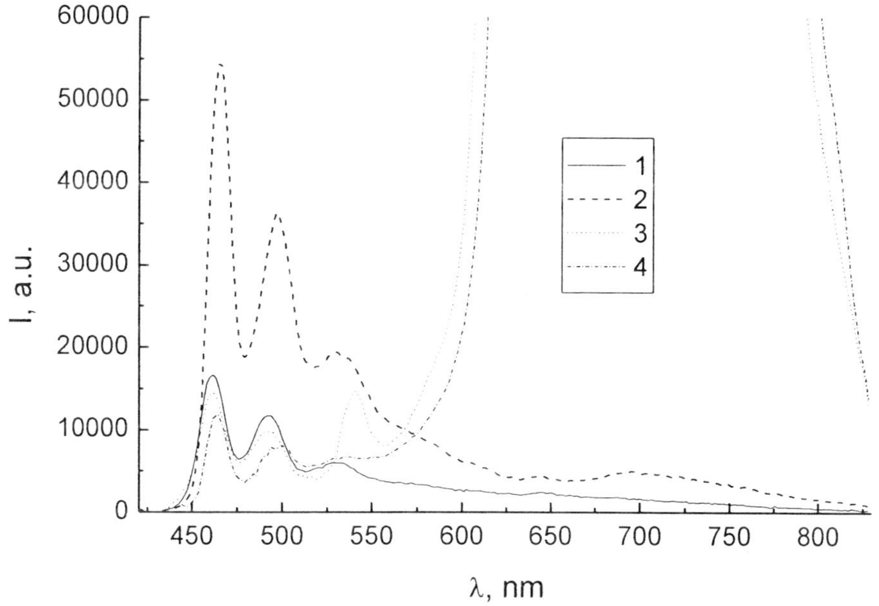

Figure 34. Delayed fluorescence spectra of the MeLPPP film without (1, 2) and with 5 wt % of Pt(II)EOP (3, 4). λ_{exc}=450 nm. Curves 1 and 3 were taken at 295 K, curves 2, 4 at 77 K.

Taking N_h/N_g as the reciprocal guest concentration and adopting the spectroscopically determined value ΔE_t = 1265 cm^{-1} yield $T_h/T_g \cong 0.04$ at 295 K. Under a laser excitation a concentration of polymer triplet excitations sufficient for their effective bimolecular annihilation can be attained indeed. It is obvious, however, that this process can work only at room temperature, but in no way at 77 K. At the same time from Figure 34 it is seen that the temperature weakly influences the DF intensity of the activated sample.

Another possible mechanism of the appearance of polymer DF under direct excitation of the dopant is the annihilation of the triplet excitations of the latter leading to the appearance of a highly excited singlet state of PtOEP followed by energy transfer to the matrix. In this case the phosphorescence of PtOEP, I_{Ph}, and the DF from MeLPPP, I_{DF}, had to be kinetically of the first and second order, respectively, implying that $I_{DF}(I_{exc})$ be proportional to $I_{Ph}^2(I_{exc})$ at any intensity I_{exc} of excitation. The pertinent experiment was done in air implying that there is O_2-quenching of triplet excitations. For checking that $I_{DF}(I_{exc})$ is proportional to $I_{Ph}^2(I_{exc})$ this is irrelevant, though, because DF and Ph would be similarly affected. The result of the experiment is negative anyhow (Figure 35). This confirms that DF from the polymer host is not caused by bimolecular fusion of two PtOEP triplets.

To elucidate the possible participation in the process of generation of the observed DF of charged particles, namely geminate electron-hole pairs, we investigated the influence of electric field on it. An electric field pulse was applied before or after the laser radiation pulse generating the registered DF. Figure 36 presents the dependence of the polymer DF intensity and the dopant phosphorescence intensity on the voltage applied before the exciting pulse.

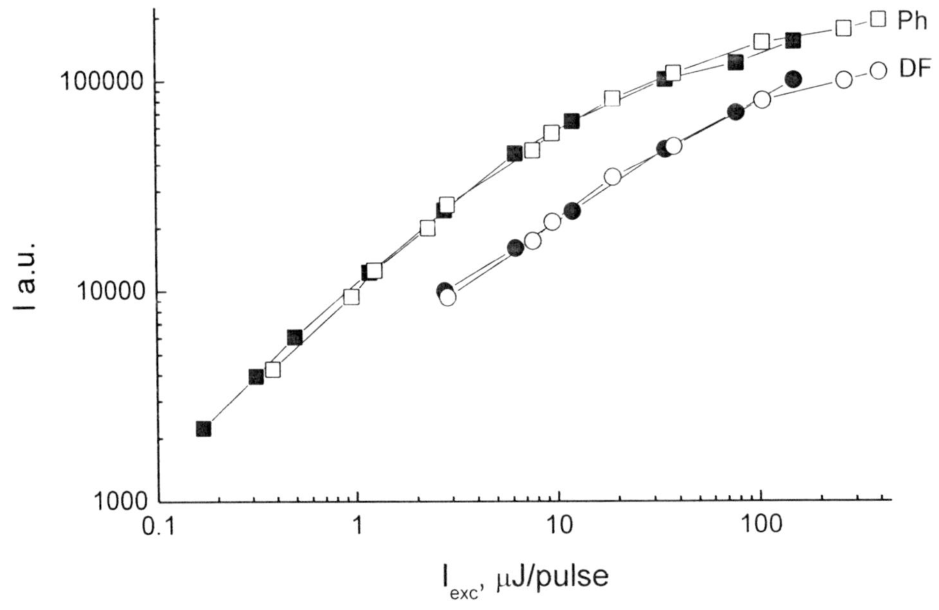

Figure 35. Dependence of the PtOEP phosphorescence and delayed fluorescence of MeLPPP on the laser pulse energy at 295 K in air. The concentration of the dopant is 5 % wt. Open characters correspond to the excitation into the $S_1 \leftarrow S_0$ 0-0 transition of PtOEP, solid characters - to the excitation into the polymer absorption band.

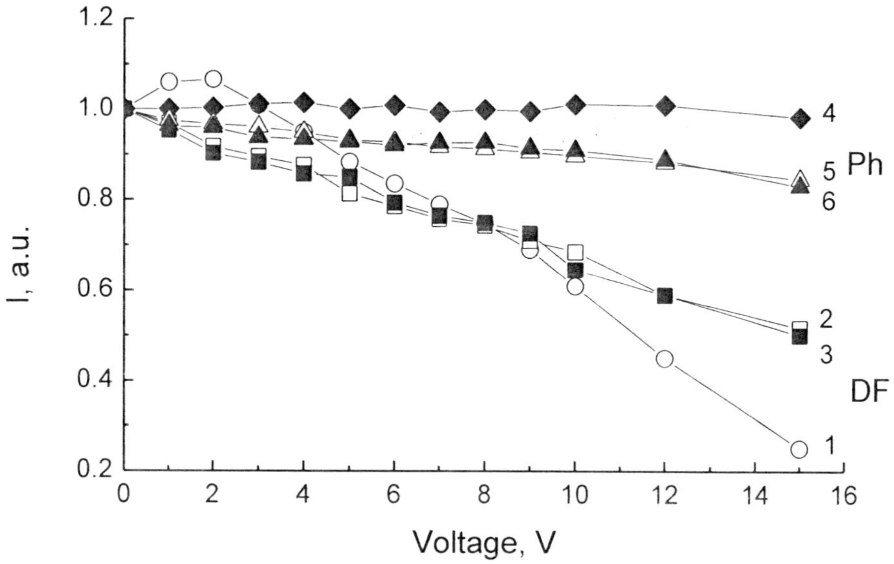

Figure 36. Dependence of the delayed fluorescence intensity of pure MeLPPP (curve 1) and MeLPPP doped with 5 wt % of PtOEP (curves 2 and 3) and of the PtOEP phosphorescence intensity (curves 4-6) on applied voltage. Open characters correspond to λ_{exc}=450 nm, solid ones - to λ_{exc}=536 nm. Curve 4 was measured with a polystyrene film doped with 5 wt % of PtOEP.

It is seen from Figure 36 that the DF of the pure polymer is strongly quenched by he electric field, which agrees with the previous results. The DF of the activated sample is also quenched by the electric field, although in this case the effect is somewhat weaker. This effect is independent of the method of dopant excitation, directly or through the polymer. There is an insignificant quenching, by about 15% at $E=3*10^6$ V/cm, of the phosphorescence of PtOEP in MeLPPP, which is absent when an inert polymer polystyrene is used. All this points to the fact that in the case of MeLPPP DF should be caused by charged particles, since neutral singlet and triplet excitations react to the external electric field rather weakly.

Figure 37 shows the decay curves of the polymer delayed fluorescence for various samples, temperatures, and excitation methods. It is seen that when an electric field is applied there is a sharp decrease in the intensity independent of whether PtOEP is excited directly or through MeLPPP. This indicates that as the DF source, charged couloumbically bound particles should act. To confirm this conclusion, we checked the presence of transient photocurrent upon excitation of the activated sample into the dopant absorption band. Since the energy of the singlet or triplet state of PtOEP is insufficient for photocurrent generation, its presence points to the fact that charge transfer from the excited state of the dopant to the polymer should take place. Figure 38 shows the presence of transient photocurrent. This means that direct excitation of the singlet state of PtOEP leads to photocurrent sensitization, in other words, to a charge transfer from the dopant to MeLPPP.

Taking into account the above results, it may be concluded that the most suitable candidate for the polymer DF source in the activated sample is the exciplex resulting from the charge transfer from the excited dopant to the polymer molecule. The planar structure of both the polymer and dopant molecules promotes their approach to a distance sufficient for the exciplex formation. And although the lifetime of the singlet state of PtOEP is only 1 ps [65], this is quite enough, since it is known [7, 48] that donor-acceptor charge transfer in such systems occurs on a time scale less than 0.1 ps.

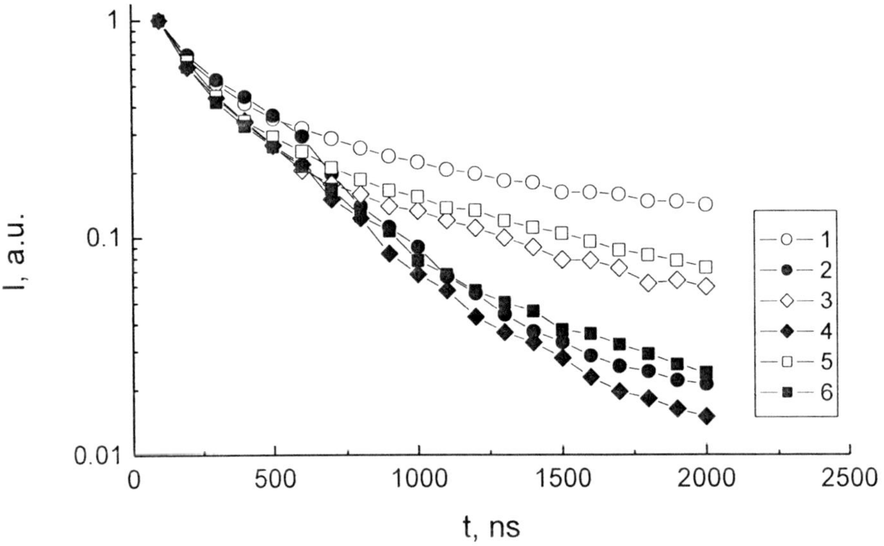

Figure 37. Decay of the DF intensity of MeLPPP at zero electric field (open characters) and E=3*10^6 V/cm (solid characters) at room temperature. The concentration of PtOEP was 0% (curves 1 and 2) and 5 wt % (curves 3-6). Curves 1-4 were measured upon excitation at 450 nm, curves 5 and 6 - at 536 nm.

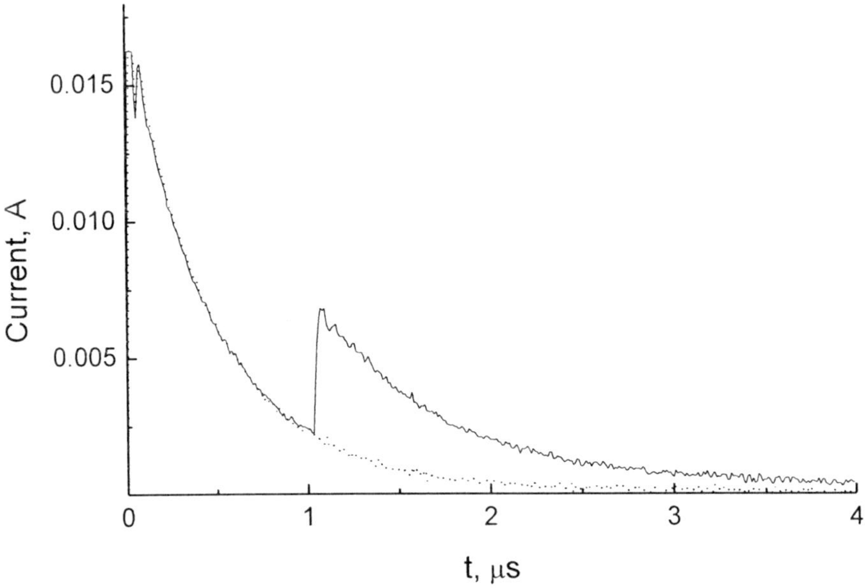

Figure 38. Transient current observed upon applying a step voltage of 15 V to the diode. The current spike at 1 μs is the instantaneous photocurrent response upon excitation of the $S_1 \leftarrow S_0$ transition of PtOEP by a laser flash. Dotted line corresponds to the signal without optical excitation.

Apparently, the energy of such an exciplex is insufficient for the generation of electron-hole pairs in the polymer. This means that DF cannot be caused directly by the exciplex. For this to happen, a bimolecular interaction of some particles, whose concentration is high enough for their interaction to be the main deactivation channel, is needed. One possible scenario of the development of such a situation is the exciplex deactivation leading to the formation of the triplet state of the polymer or the dopant. Subsequent heteroannihilation of the triplet excitations of the polymer and dopant is followed by the appearance of a highly excited singlet state of the polymer, from which electron-hole pairs can be formed. Their subsequent recombination leads to the appearance of delayed fluorescence, as is done in the pure polymer. Taking into account that the number of triplet excitations of the dopant is much larger than the number of triplet excitations of the polymer, their heneroannihilation can be considered as a quasi-monomolecular process. Since the migration efficiency of triplet excitations in the polymer depends on the temperature, the probability of annihilation should decrease with its drop and, as a result, the DF intensity should decrease, which is actually observed in the experiment.

2. White Luminescence of PF2/6 Doped with Fluorescent Dyes

The development of OLEDs emitting white color light is also a very important problem in terms of their practical application. One such application is the use of white-light LEDs for back illumination of liquid crystal displays. On the other hand, from white light one can always obtain radiation of required wavelength. Evidently, the development of a light source on the basis of only one substance is impossible, since at present no materials emitting white light are known. To solve this problem, several different methods are used: 1) in LEDs several active layer emitting different spectral ranges are used [66]; 2) as an active layer, a substance with a wide emission spectrum is used, which is brought up to white by means of special filters [67]; 3) to obtain white light, luminescence of excimers and excilpexes can be used [68], 4) white light can be obtained by connecting LEDs emitting in the blue, green, and red spectral region [69]; 5) use for an active layer of a mixture of polymers, whose luminescence covers the entire visible range [70]; and, finally, 6) addition to the emitting layer of several organic dyes with suitable luminescence spectra [71]. It was the last of the above methods that was used in the present paper to develop a PF2/6-based system emitting white light. Polyfluorene PF2/6 is widely used as an active layer in organic LEDs [21] and its basic spectroscopic properties have been studied fairly well [72].

To characterize the luminescence in terms of its color, the formalism and definitions established by Commission Internationale de l'Eclairage (CIE) in 1931 were used [73]. Within the framework of this approach the color quality (chromaticity) is characterized by its two CIE coordinates (x, y) as a point in a two-dimensional space (Figure 39). To obtain these coordinates, the color matching functions of the CIE (Figure 40) that are associated with the spectral sensitivity of three color receptors of the human eye are used. The coordinates are determined by means of the following formulas:

$$X = \int I(\lambda) X(\lambda) d\lambda$$
$$Y = \int I(\lambda) Y(\lambda) d\lambda$$
$$Z = \int I(\lambda) Z(\lambda) d\lambda$$
$$x = \frac{X}{X+Y+Z}$$
$$y = \frac{Y}{X+Y+Z} \quad (6)$$

where $I(\lambda)$ is the luminescence spectrum, $X(\lambda)$, $Y(\lambda)$ and $Z(\lambda)$ are the color matching functions for the CIE color coordinate system. The luminescence of any color can be given as a superposition of three main components:

$$I(\lambda)=X\,X(\lambda) + Y\,Y(\lambda) + Z\,Z(\lambda) \quad (7)$$

White light is characterized by coordinates x=1/3 и y=1/3. In other words, white color, from the point of view of the human eye, incorporates all the three components in equal quantities. Besides the given definition of white color, there also exists the conception of equi-energy white color, i.e., color of radiation with a uniform intensity distribution throughout the visible spectrum. In the present work, two systems emitting white light in accordance with the above two definitions are proposed.

The choice of PF2/6 as the basis for solving the problem posed is primarily due to its luminescence properties. From comparison of Figures 3 and 40 it is well seen that the fluorescence spectrum of this polymer is in the blue region and very well overlaps the $Z(\lambda)$ function. Apart from this, it is characterized by a large value of the quantum yield. Unfortunately, the low position of the triplet level of the polymer (see Figure 23) makes it impossible to use, as an activator, the phosphorescent organometallic complex emitting in the green region of the spectrum for solving the problem posed. Therefore, in the present work white luminescence was obtained by means of singlet-singlet energy transfer from the polymer matrix to the fluorescent dyes. Taking into account the fact that stability and a long service life are the basic requirements placed upon organic LEDs, in selecting adequate activators preference was given to laser dyes. And while at present there is a large number of dyes overlapping the entire visible range of the spectrum [74], the choice of a dyes suitable for solving the problem posed was difficult because of the fact that their spectral-luminescent properties strongly depend on the properties of the medium.

To obtain luminescence of an equi-energy white color, coumarin 6 and 2-thiophenamine were used as activators (Figure 41). Figure 42 shows the normalized absorption and fluorescence spectra of these compounds in toluene. The choice of toluene as a solvent for the dyes is due to two reasons. First, it is used, as a rule, to prepare polymer films and, second, the luminescent properties in it are close to those in polyfluorene. From Figure 42 it is well seen that the absorption spectra of the activator molecules well overlap the fluorescence spectrum of the polymer, which creates prerequisites for realizing singlet-singlet energy transfer.

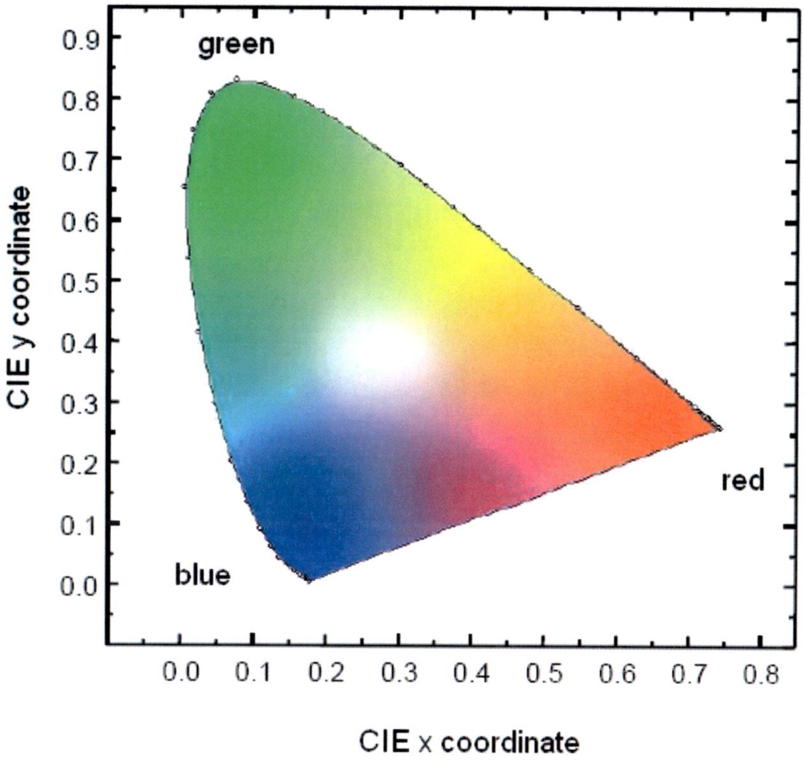

Figure 39. Chromaticity diagram for the CIE color coordinate system.

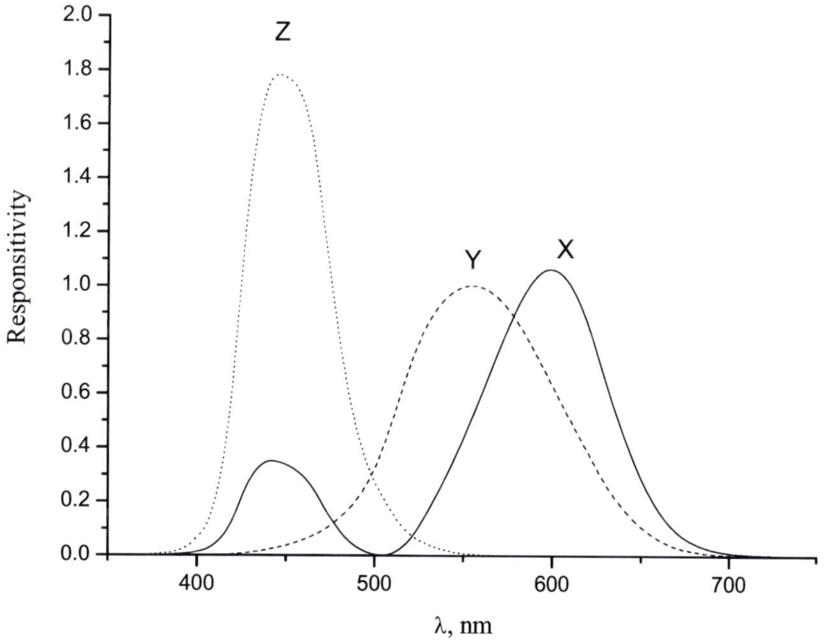

Figure 40. Spectral responsitivity of the CIE 1931 color matching functions.

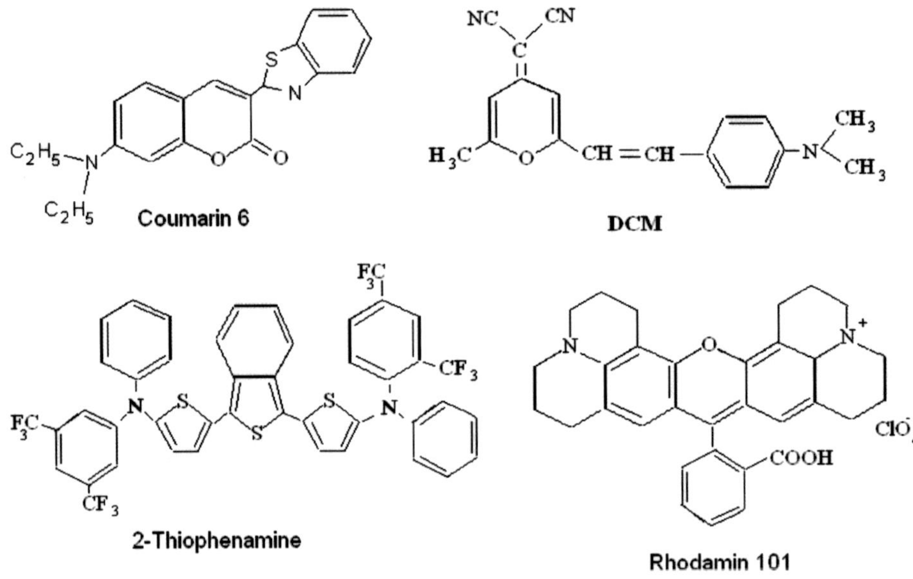

Figure 41. Structural formulas of the compounds used.

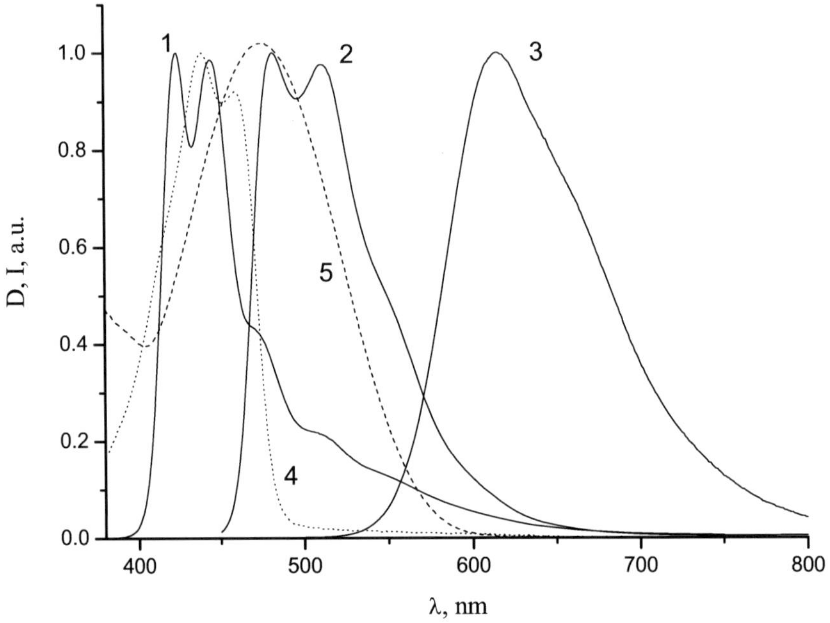

Figure 42. Normalized fluorescence (1-3) and absorption (4, 5) spectra of PF2/6 (1), coumarin 6 (2, 4), and 2-thiophenamine (3, 5) in toluene.

Figure 43 shows the fluorescence spectrum of the polyfluorene film containing 0.025 wt % of coumarine 6 and 2.2 wt % of 2-thiophenamine. It is seen that in general this spectrum can be considered as an equi-energy spectrum in the 415-610 nm range. Analysis of this spectrum from the point of view of the CIE approach gives values of the chromaticity coordinates x and y 0.32 and 0.33, respectively, which corresponds to practically white light.

Such a system can serve as an active layer in organic LEDs used for obtaining radiation of practically any color by means of spectral filtration.

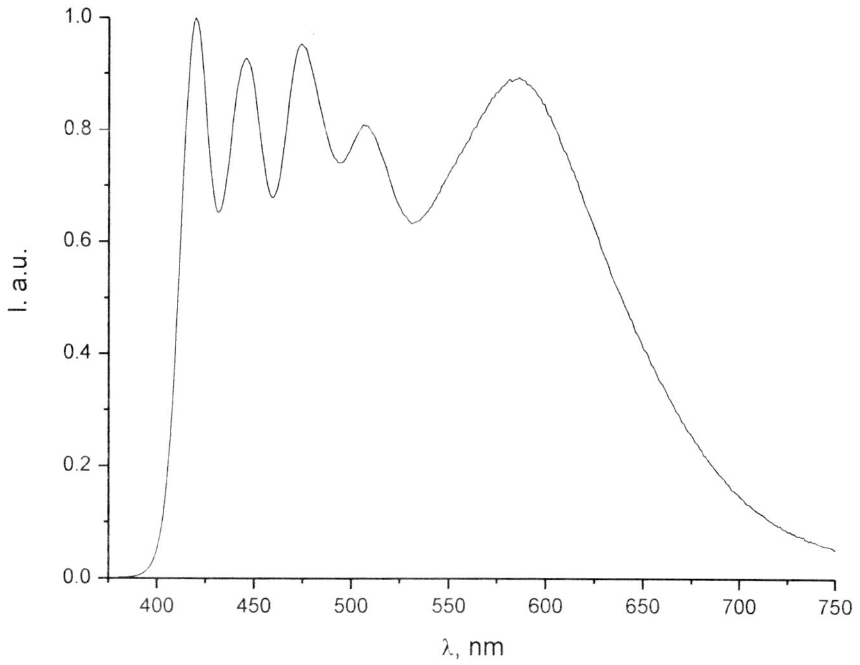

Figure 43. Normalized fluorescence spectrum of the PF2/6 film containing 0.025 wt % coumarine 6 and 2.2 wt % 2-thiophenamine.

Besides equi-energy white light, one can also obtain a system with emission of white color that takes into account the spectral sensitivity of the human eye receptors. To solve this problem, we used DCM molecules as the main dopant (Figure 41). The absorption and fluorescence of this dye in toluene are given in Figure 44. Its absorption spectrum overlaps the fluorescence spectrum of PF2/6. The fluorescence spectrum of DCM in toluene is shifted towards the blue region by 60 nm from the spectrum observed for the solution in methanol [75]. In principle, this permits using only one dye to obtain white color emission. Figure 45 shows the luminescence spectrum of the polymer film containing 0.2 wt % of DCM. It is seen that the fluorescence spectrum of DCM in the polymer somewhat differs from its spectrum in toluene. Nevertheless, even the luminescence of such a system has a color close to white. In this case, the chromaticity coordinates have values of x=0.31 and y=0.36. There is an excess of the green component. To correct this situation and improve the color, we added to the system 0.013 wt % of rhodamine 101. Its fluorescence maximum in the polymer is in the region of 595 nm, and the absorption spectrum overlaps the DCM fluorescence spectrum (the absorption spectrum maximum in the region of 570 nm) and practically does not overlap the fluorescence spectrum of the polymer. Thus, in this case cascade energy transfer is realized. The luminescence spectrum of such system given in Figure. 45 is characterized by the chromaticity coordinates x=0.33 and y=0.35, which is very close to white.

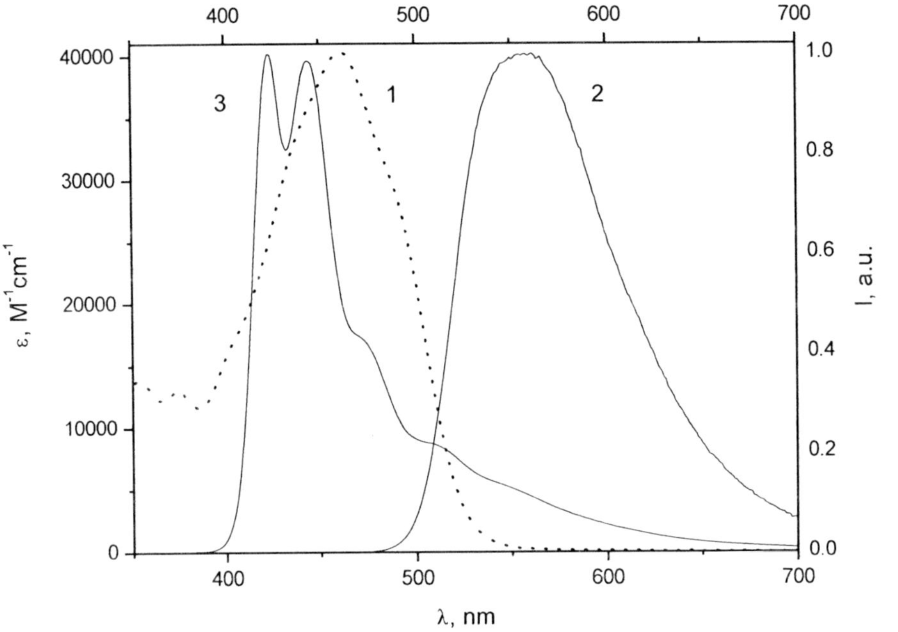

Figure 44. Normalized fluorescence (1, 2) and absorption (3) spectra of PF2/6 (1) and DCM (2, 3) in toluene.

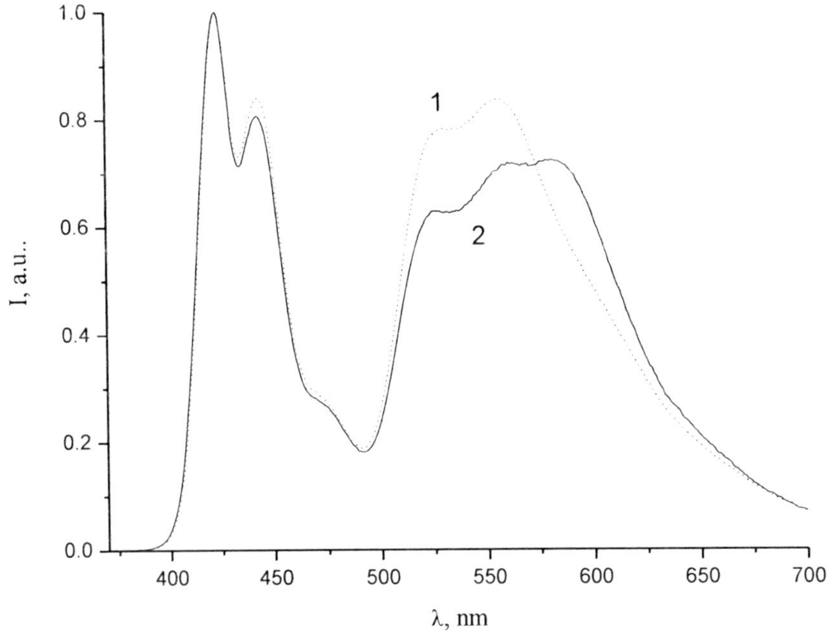

Figure 45. Normalized fluorescence spectrum of the PF2/6 film containing 0.2 wt % DCM (1, 2) and 0.013 wt % rhodamine 101 (2).

CONCLUSIONS

The results of the investigations presented in this paper show that the absence of phosphorescence or its low quantum yield, that take place for the majority of conjugated polymers, is due to the very low population efficiency of the triplet state as a result of the intersystem crossing. The use of molecules of the type of diketones characterized by a practically 100 % transition to the triplet state makes it possible to considerably increase it as a result of the triplet excitation transfer to the polymer molecule, which enabled us to investigate the spectral-kinetic properties of the phosphorescence of the polymer used in the work. It has been revealed that in activated samples of MeLPPP, as opposed to the pure polymer, delayed fluorescence is generated as a result of the bimolecular interaction of the triplet excitations of the polymer. It has been shown that for the investigated polymers the main mechanism of their fluorescence quenching is the dipole-dipole controlled single step transfer from an excited donor to an acceptor. In using benzil molecules as an activator, a considerable contribution to the florescence quenching is made by the formation of complexes with charge transfer between the polymer and activator molecules - exciplexes. The presence of volume side substituents in both the polymer and the dopant markedly decreases the efficiency of this mechanism. When organometallic complexes containing heavy metal atom are used as a singlet-triplet converter, the phosphorescence intensity and its decay rate increase compared to analogous systems containing no heavy metal atom. This effect is attained as a result of the remote resonantly enhanced external heavy atom effect.

The low probability of intersystem crossing leads to the fact that in the polymer activated by molecules whose singlet and triplet levels are lower in energy than the respective levels of the polymer (MeLPPP/PtOEP), the main channel of population of dopant excited states is the singlet-singlet energy transfer from the polymer. The introduction of the singlet-triplet converter considerably increases the contribution of the triplet channel to the population of the triplet state of the activator. The investigation of the delayed luminescence of the MeLPPP/PtOEP system has revealed that direct excitation of the activator molecules into their proper absorption band leads to the appearance of delayed fluorescence of the polymer matrix (up-conversion). On the basis of the data on the influence on it of an external electric field it has been concluded that the source of the observed effect is the exciplex between the polymer and activator molecules.

The use of fluorescent dyes permits obtaining on the basis of PF2/6 white light sources by realizing a cascade energy transfer. The choice of laser dyes as dopants is due to their high photostability and large values of the extinction coefficient, which leads to a high efficiency of energy transfer and, as a consequence, to their insignificant concentration in the sample. The latter fact is important from the point of view of retaining the electrical properties of conjugate polymers.

The author whishes to thank Prof U. Sherf for supplying Me-, and PhLPPP, Dr M. Pfeiffer for supplying 2-thiophnamine, Prof. H. Bässler, Dr J. Lupton, and Dr. D. Hertel for providing the possibility to carry out investigations of time-resolved luminescence. The author also expresses special gratitude to Prof. H. Bässler for support in conducting part of the investigations and for fruitful discussions of the results obtained.

Part of this work was supported by the Belarusian Republican Foundation for Basic Research (project № Ф05-204).

REFERENCES

[1] Skotheim T. *Handbook of Conducting Polymers*, Marcel Dekker: New York, 1986.

[2] *Handbook of Organic Conductive Molecules and Polymers*, Nalwa H.S.; Ed.; Wiley: New York, 1997.

[3] Shirakawa H.; Louis E.J.; MacDiarmaid A.G.; Alan G.; Chiang C.K.,; Heeger A.J. *Chem. Commun.* 1977, *16*, 578–580.

[4] Burroughes J.H.; Bradley D.D.C. *Nature*, 1990, *347*, 539-540.

[5] Overwijk M. *Philips Research Password*, 2004, *20*, 10-13.

[6] Ziemelis K. *Nature*, 1999, *399*, 408-410.

[7] *Primary Photoexcitations in Conjugated Polymers: Molecular Exciton Versus Semiconductor Band Model*, Sariciftci N.; Ed.; World Scientific: Singapore, 1997.

[8] Sinha S.; Rothe C.; Guentner R.; Scherf U.; Monkman1 A. P. *Phys. Rev. Lett.* 2003, *90*, 127402.

[9] Shuai Z.; Beljonne D.; Silbey R.J.; Bredas J.L. *Phys. Rev. Lett.* 2000, *84*, 131-134.

[10] Wohlgenannt M.; Tandor K.; Mazumdar S.; Ramasesha S.; Vardeny Z.V. *Nature*, 2001, *409*, 494-497.

[11] Dyakonov V.; Rosler G.; Schwoerer M.; Frankevich E. L. *Phys. Rev. B* 1997, *56*, 3852-3862.

[12] Karg S.; Dyakonov V.; Meier M.; Ries W.; Paasch G. *Synth. Met.* 1994, *67*, 165-168.

[13] Partee J.; Frankevich E. L.; Uhlhorn B.; Shinar J.; Ding Y.; Barton T. J. *Phys. Rev. Lett.* 1999, *82*, 3673-3676.

[14] Pichler K.; Halliday D.A.; Bradley D.D.C.; Burn P. L.; Friend R. H.; Holmes A.V. *J. Phys. Condens. Matter* 1993, *5*, 7155-7172.

[15] Romanovskii Yu.; Gerhard A.; Schweitzer B.; Scherf U.; Personov R.I.; Bässler H. *Phys. Rev. Lett.* 2000, *84*, 1027-1030.

[16] Rothe C.; Monkman A.P. *Phys. Rev. B* 2002, *65*, 073201.

[17] McGlynn S.P.; Azumi T.; Kinoshita M. *Molecular spectroscopy of the triplet state*, Prentice-Hall, Inc., Englewood Cliffs: New Jersey, 1969.

[18] Romanovskii Yu.; Bässler H. *Chem. Phys. Lett.* 1999, *326*, 51-57.

[19] Lupton J.M.; Pogantsch A.; Piok T.; List E.J.W.; Patil S.; Scherf U. *Phys. Rev. Lett.* 2002, *89*, 167401.

[20] Wang X.; Ogino K.; Tanaka K.; Usui H. *IEEE Journal of Selected Topics in Quantum Electronics* 2004, *10*, 121-126.

[21] Higgins R.W.; Monkman A.P.; Nothofer H.-G.; Scherf U. *Appl. Phys. Lett.* 2001, *79*, 857-859.

[22] Higgins R.W.; Monkman A.P.; Nothofer H.-G.; Scherf U. *J. Appl. Phys.* 2002, *91*, 99-105.

[23] Shoustikov A.A.; You Y.; Thompson M.E. *IEEE J. Selected Topics Quantum Electron.* 1998, *4*, 3-13.

[24] Cao Y.; Parker I.D.; Yu G.; Zhang C.; Heeger A.J. *Nature,* 1999, *397*, 414-417.

[25] Baldo M.A.; O'Brien D.F.; Thompson M.E.; Forrest S.R. *Phys. Rev. B.* 1999, *60*, 14422-14428.

[26] Wohlgenannt M.; Tandor K.; Mazumdar S.; Ramasesha S.; Vardeny Z.V. *Nature*, 2001, *411*, 617.

[27] Kwong R. C.; Lamansky S.; Thompson M.E. *Adv. Mater.* 2000, *12*, 1134-1138.
[28] Cleave V.; Yahioglu G.; Le Barny P.; Hwang D.-H.; Holmes A.B.; Friend R.H.; Tessler N. *Adv. Mater.* 2001, *13*, 44-47.
[29] Lane P.A.; Palilis L.C.; O'Brien D.F.; Giebeler C.; Cadby A.J.; Lidzey D.G.; Campbell A.J.; Blau W.; Bradley D.D.C. *Phys. Rev. B.* 2001, *63*, 235206.
[30] Yang X.H.; Neher D.; Scherf U.; Bagnich S.A.; Bässler H. *J. App. Phys.* 2003, *93*, 4413-4419.
[31] Cleave V.; Yahioglu G.; Le Barny P.;. Friend R.H.; Tessler N. *Adv. Mater.* 1999, *11*, 285-288.
[32] Bagnich S.A.; Im C.; Bässler H.; Neher D.; Scherf U. *Chem. Phys.* 2004, *299*, 11-16.
[33] Hertel D.; Bassler H.; Guentner R.; Scherf U. *J. Chem. Phys.* 2001, *115*, 10007-10013.
[34] Rothe C.; Guentner R.; Scherf U.; Monkman A.P. *J. Chem. Phys.* 2001, *115*, 9557-9562.
[35] Rothe C.; Palsson L.O.; Monkman A.P. *Chem. Phys.* 2002, *285*, 95-101.
[36] Lamola A. A.; Hammond G. S. *J. Chem. Phys.* 1965, *43*, 2129-2135.
[37] Murov L.; Carmichael I.; Hug G. L. *Handbook of Photochemistry,* Dekker: New York, 1993.
[38] Monkman A. P.; Burrows H. D.; Miguel M. D. G.; Hamblett I.; Navaratnam S. *Synth. Met.* 2001, *116*, 75-79.
[39] Foerster Th. *Ann. Phys.* 1948, *2*, 55-75.
[40] Lakowicz J.R. *Principles of Fluorescence Spectroscopy*, Plenum Press: New York, 1983.
[41] Rothe C.; Hintschich S.I.; Palsson L.-O.; Monkman A.P.; Guenter R.; Scherf U. *Chem. Phys. Lett.* 2002, *360*, 111-116.
[42] Williams A. T. R.; Winfield S. A.; Miller J. N. *Analyst* 1983, *108*, 1067-1072.
[43] Goushi K.; Kawamura Y.; Sasabe H.; Adachi C. *Jap. J. Appl. Phys.* 2004, *43*, L 937–L 939.
[44] Kawamura Y.; Sasabe H.; Adachi C.; *Jap. J. Appl. Phys.* 2004, *43*, 7729–7730.
[45] Suljovrujic E.; Ignjatovic A.; Srdanov V. I.; Mitsumori T.; Wudl F. *J. Chem. Phys.* 2004, *121*, 3745-3750.
[46] Elschner A.; Mahrt R.F.; Pautmeier L.; Bässler H.; Stolka M.; McGrane K. *Chem. Phys.* 1991, *150*, 81-91.
[47] Meskers S.C.J.; Hübner J.; Oestreich M.; Bässler H. *Chem. Phys. Lett.* 2001, *339*, 223-228.
[48] Basko D.M.; Conwell E.M. *Phys. Rev. B* 2002, *66*, 155210.
[49] Lupton J. *Chem. Phys. Lett.* 2002, *36*, 366–368.
[50] Morteani A. C.; Dhoot A. S.; Kim J.-S.; Silva C.; Greenham N.C.; Murphy C.; Moons E.; Cinb S.; Burroughes J.H.; Friend R.H. *Adv. Mater.* 2003, *15*, 1708-1712.
[51] Romanovskii Yu.V.; Gerhard A.; Schweitzer B.; Personov R.I.; Bässler H. *Chem. Phys.* 1999, *249*, 29-39.
[52] Piok T.; Gadermaier C.; Wenzl, F.P.; Patil S.; Montenegro R.; Landfester K.; Lanzani G.; Cerullo G.; Scherf U.; List E.J.W. *Chem. Phys. Lett.* 2004, *389*, 7–13.
[53] Hertel D.; Romanovskii Yu.V.; Schweitzer B.; Scherf. U.; Baessler H. *Synth. Met.* 2001, *116*, 139-143.
[54] Pope M; Swenberg C. S. *Electronic Processes in Organic Crystals and Polymers*, 2nd ed. Oxford Science Publications: New York, 1999.

[55] Richert R.; Bässler H. *J. Chem. Phys.* 1986, *84*, 3567-3572.

[56] Scheidler M.; Cleve B.; Bässler H.; Thomas P. *Chem. Phys. Lett.* 1994, *225*, 431-436.

[57] Nikitenko V. R.; Hertel D.; Bässler H. *Chem. Phys. Lett.* 2001, *348*, 89-94.

[58] Wilson J.S.; Koehler A.; Friend R.H.; Al-Suti M.K.; Al-Mandhary M.R.A.; Khan M.S.; Raithby P.R. *J. Chem. Phys.* 2000, *113*, 7627-7634.

[59] Koehler A.; Wilson J.S.; Friend R.H.; Al-Suti M.K.; Khan M.S.; Gerhard A.; Baessler H. *J. Chem. Phys.* 2002, *116*, 9457-9463.

[60] Reufer M.; Schindler F.; Patil S.; Scherf U.; Lupton J.M. *Chem. Phys. Lett.* 2003, *381*, 60-66.

[61] Rothe C.; King S.; Monkman A. *Nature* 2006, in press.

[62] Petelenz P.; Funfschillinf J. *Chem. Phys.* 1990, *145*, 333-342.

[63] Basu G.; Kubasik M.; Anglos D.; Kuki A. *J. Phys. Chem.* 1993, *97*, 3956-3967.

[64] Romanova Z.S.; Deshayes K.; Piotrowiak P. *J. Am. Chem. Soc.* 2001, *123*, 2444-2445.

[65] Bulovic V.; Balbo M.A..; Forest S.R. In: *Organic Electronic Materials,* Farchioni R.; Grosso G. Eds.; Springer: 2001, 391-441.

[66] Shen Z.; Borrows P.E.; Bulvic V.; Forrest S.R.; Thompson M.E. *Science* 1997, *276*, 2009-2011.

[67] Jordan R. H.; Dodabalapur A.; Strukelj M.; Miller T. M. *Appl. Phys. Lett.* 1996, *68*, 1192-1194.

[68] Wang Y.Z.; Sun R.G.; Meghdadi F.; Leising G.; Epstein A.J. *Appl. Phys. Lett.* 1999,*74*, 3613-3615.

[69] Tasch S.; Brandstatter C.; Meghdadi F.; Leising G.; Athouel L.; Froyer G. *Adv. Mater.* 1997, *9*, 33-36.

[70] Tasch S.; List E. J. W.; Ekstroem O.; Graupner W.; Leising G.; Schlichting P.; Rohr U.; Geerts Y.; Scherf U.; Mullen K. *Appl. Phys. Lett.* 1997, *71*, 2883-2885.

[71] Jiang X.Y.; Zhang Z.L.; Zhang B.X.; Zhu W.Q.; Xu S.H. *Synth. Met.* 2002, *129*, 9-13.

[72] Ariu M.; Lidzey D.C.; Bradley D.D.C. *Synth. Met.* 2000, *111*, 607-610.

[73] *Recommendations on Uniform Colour Spaces,* 2nd ed. CIE Pub. No. 15.2, CIE: Vienna, 1978.

[74] Brackmann U. *Laser Dyes*, 3d ed. Lambda Physik AG: Goettingen, · Germany, 2000.

[75] Lee S.-H.; Jung K.-H.; Sung J. H.; Hong K.-H.; Nam C. H. *J. Chem. Phys.* 2002, *117*, 9858-9861.

In: Chemical Physics Research Trends, Vol. 252
Editor: Samuel V. Arnold, pp. 165-213

ISBN: 978-1-60021-554-4

Chapter 4

EXPLORING BASIC CHEMICAL CONCEPTS WITH THE QUANTUM THEORY OF ATOMS IN MOLECULES

Ricardo A. Mosquera, María J. González Moa, Ana M. Graña, Marcos Mandado and Antonio Vila

Department of Physical Chemistry, University of Vigo,
36310-Vigo, Galicia, Spain

ABSTRACT

The development of the Quantum Theory of Atoms in Molecules (QTAIM) by Bader's research group has provided chemists with a rigorous physical basis for getting insight about many problems. This paper reviews results obtained (mostly by our research group) in the topological analysis of the electron density of a large set of organic molecules with QTAIM. The results are connected and discussed in relation to several basic concepts of Chemistry. We concentrate on those concepts where QTAIM results are not keeping in line with widespread traditional explanations that are still common in textbooks and research papers. In particular, we deal with some recent findings about approximate transferability (one of the first subjects attacked by the author of this theory), strain energy, anomeric effect, hydrogen bonding, and resonance structures. Our results on approximate transferability indicate the need for defining at least four specific methylene groups in the vicinity of a functional group when the chain is in antiperiplanar arrangement. Nevertheless, we can only distinguish two specific methylene groups when the chain is in gauche arrangement. Also atomic energies cannot be used as a transferable property unless they are computed from an electron density that verifies the virial theorem very strictly. QTAIM study of heterocycles reveals that homodesmotic strain energies comprise energy terms that are not related to ring opening. These terms are not included in a QTAIM based definition of strain energy. The partitioning of molecular energy into one- and two-center atomic terms, with atoms defined in terms of the QTAIM, gives rise to a new interpretation of the anomeric effect, with results that are not in line with either the electrostatic or the stereoelectronic models developed for it. Also, classical ideas about hydrogen bonding have to be re-examined according to QTAIM results. Finally, the study of several protonation and hydride addition process has revealed several trends that are not in line with the resonance model.

1. INTRODUCTION

As the title of this chapter indicates, we have applied the Quantum Theory of Atoms in Molecules (QTAIM) to study some basic chemical problems. For some of them, we have got results that question traditional chemical interpretations.

In spite of QTAIM, also known as AIM or Bader's electron density analysis, has become very popular in the last years and has been detailed reviewed in diverse papers and monographs [1-6], epigraph 2 contains a brief overview of this theory. Most of the applications detailed in forthcoming epigraphs have been selected, taking care of including the most provocative ones. Nevertheless, we will start speaking about approximate transferability, because, this concept is, in one way or another, involved in many other topics [7,8].

From transferability we will move to a nearly antonym that is strain energy. We will see that traditional schemes for computing strain energy are including something more than ring opening energy and that QTAIM provides alternative values. They coincide with the traditional ones for hydrocarbons, as shown more than 20 years ago by Wiberg and Bader [9], but they are different for heterocycles as we have found recently [10-12].

Another problem, where QTAIM results are in conflict with traditional ideas based on the molecular orbital paradigm is the origin of the anomeric effect [13-16]. We will see that QTAIM does not support any kind of stereoelectronic model [17-20] but, in our opinion, points to a very simple explanation based on electron repulsions, specially those between the electron density contained in the basins of the central hydrogens and non bonded charge concentrations [21], which are more or less what most of chemists are used to call lone pairs.

We also deal with the lack of hydrogen bond critical points in what are reported as typical examples for this interaction like 1,2-ethanediol, or orto-dihydroxybenzene and how these points appear when the geometry is slightly distorted from equilibrium [22]. Finally, we review how the results obtained for protonation of several systems and hydride addition to various molecules differ from the predictions of the resonance model [23-26].

2. OVERVIEWING QTAIM

QTAIM can be viewed as a topological analysis of the electron density function, $\rho(\vec{r})$. As in any topological analysis, the localization of singular points plays a basic role. In this case we have singular points in the real space spanned by the 3 coordinates representing the position of any electron, $\vec{r}$. Looking at the relief map of $\rho(\vec{r})$ in the plane of pirrole (figure 1) that contains all its nuclei, we can observe ten local maxima, also called electron density attractors, whose coordinates correspond very approximately to those of the ten nuclei in the molecule. Along every bond there is a saddle point with two negative eigenvalues of the Hessian matrix of $\rho(\vec{r})$. These points are also called bond critical points or BCPs. Finally, inside the ring, we observe another saddle point, whose Hessian matrix presents two positive eigenvalues. It is called a ring critical point (or RCP) and it is characteristic of cyclic structures. Finally in molecules like cubane, where there is a cage structure, we observe the presence of a relative minimum, one per cage, which is named cage critical point (CCP) and is surrounded by ring critical points, six in this case. It this context, a recent paper by Castillo et al. has proved that in a highly twisted system (1,12-difluorobenzo[c]phenanthrene) a cage structure can be delimited by only two ring surfaces [27].

The topological analysis also looks at the gradient paths of the electron density, which are shown for the main plane of pirrole in figure 2. We observe they form a vector field where every group of field lines ends at a different nucleus. These groups of lines are delimited by surfaces given by what is known as the zero flux condition (1), which is a mathematical condition rigorously derived [28] from Schwinger's principle of stationary action [29]. These surfaces are represented in the pirrole plane as lines that intersect a certain vanishing limit of the molecular electron density defining the atomic basins, that are disjoints regions of the space (figure 3). In this context, an atom Ω can be defined as the joint of a nucleus and its electron basin. The integration of the proper density function within the atomic basin provides the atomic properties, like the atomic electron population (2), the atomic electron kinetic energy (3) or the atomic volume (4).

$$\nabla\rho(\vec{r})\cdot\vec{n}(\vec{r})=0 \qquad (1);$$

$$N(\Omega)=\int_{\Omega}\rho(\vec{r})d\vec{r} \qquad (2);$$

$$K(\Omega)=\frac{1-\gamma}{4}\int_{\Omega}\left\{2[\nabla\nabla'\rho(\vec{r},\vec{r}')]_{\vec{r}=\vec{r}'}-\nabla^{2}\rho(\vec{r})\right\}d\vec{r} \qquad (3);$$

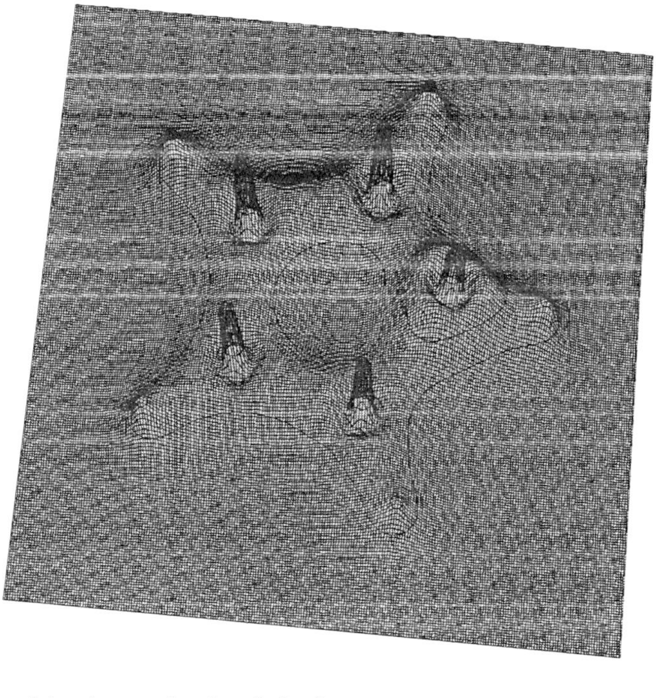

Figure 1. Relief plot of the electron density of pirrole.

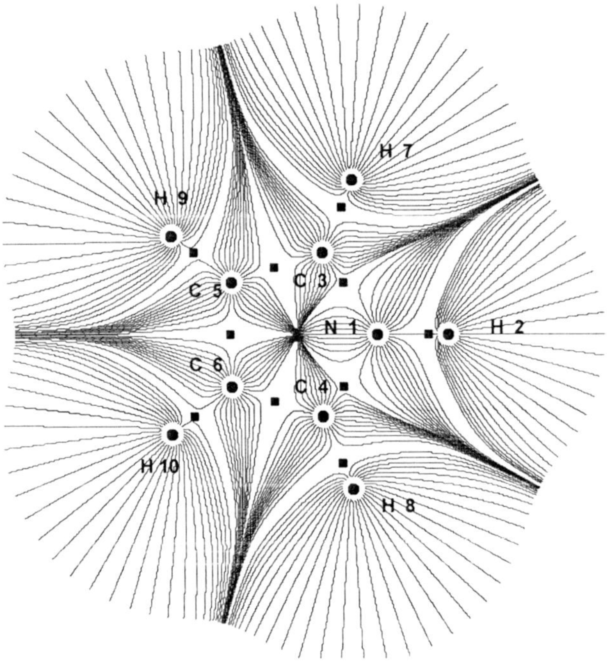

Figure 2. $\nabla \rho(\vec{r})$ plot in the main plane of pirrole.

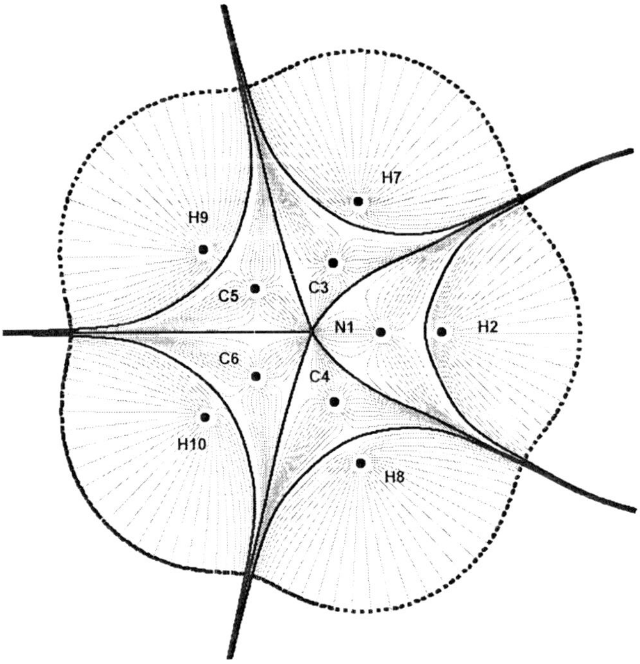

Figure 3. Intersection of the zero flux surfaces (solid lines) and vanishing limit of $\rho(\mathbf{r})$ (dots) with the main plane of pirrole. $\nabla \rho(\vec{r})$ lines in grey.

$$v(\Omega) = \int_{\Omega} d\vec{r} \tag{4}$$

Turning back to critical points, it has to be said that the conduction of the eigenvector associated to the positive eigenvalue of every BCP gives rise to the atomic interaction lines or bondpaths (figure 4). According to Bader's original formulation of QTAIM, bondpaths are the physical representation of chemical bonds [1], being a necessary and sufficient condition for the existence of bondpaths. Nevertheless, the interpretation of bondpaths in several systems (biphenyl, inclusion complexes of He in adamantane, etc.) has risen significant controversies [30-36].

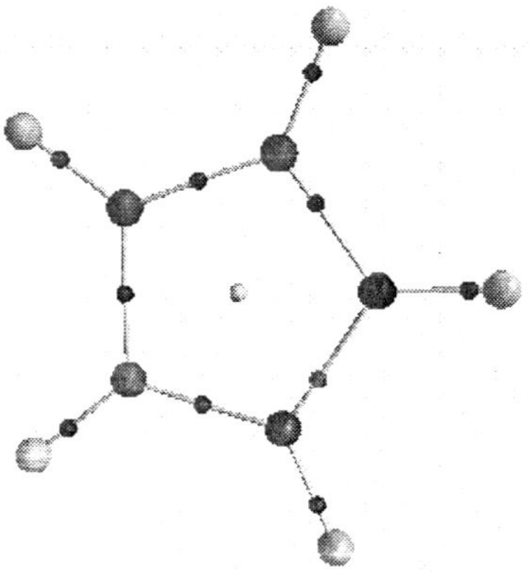

Figure 4. AIM2000 [37] plot showing the BCPs (between every pair of nuclei), the RCP (inside the ring) and bondpaths of pyrrole.

3. Approximate Transferability

One of the first ideas that should be known about transferability is that, as a consequence of the Hohenberg and Kohn's theorem [38] we can only speak of approximate transferability [39]. This leads to define transferability limits to compare geometries and atomic and bond properties of similar compounds. On first thoughts these limits should be close to typical experimental errors or computational accuracy. This is quite easy to establish for geometries, but not so for atomic properties.

In principle, 0.001 au could be a reasonable transferability limit for the atomic electron population. In fact, this limit is achieved perfectly in many cases, like comparison of oxygens of aldehydes [40] or fluorines of fluroalkanes [41]. As we can see, looking at the corresponding averaged electron populations and maximum deviation shown in parenthesis for the least significant digit (figure 5). Nevertheless, the situation is different for other series like carbonyl carbon of aldehydes or oxygen of methylethers (figure 6).

To explain these cases we looked at L(Ω), which is the integrated value of the laplacian of rho, and which should be zero for an atom that is perfectly delimited. We observe the absolute values of this property are much larger when the transferability limit is not achieved. This allows introducing a condition for L(Ω) values to study transferability. Moreover, if we plot the atomic populations versus L(Ω) for every series of compounds we observe that, within a certain interval of L(Ω) values, there is a linear relation (figure 7). This allows us to represent a set of atoms obtained with not so low L(Ω) values by the fitted intercept of the electron population, written as $N^0(\Omega)$, which is the electron population obtained with no integration error.

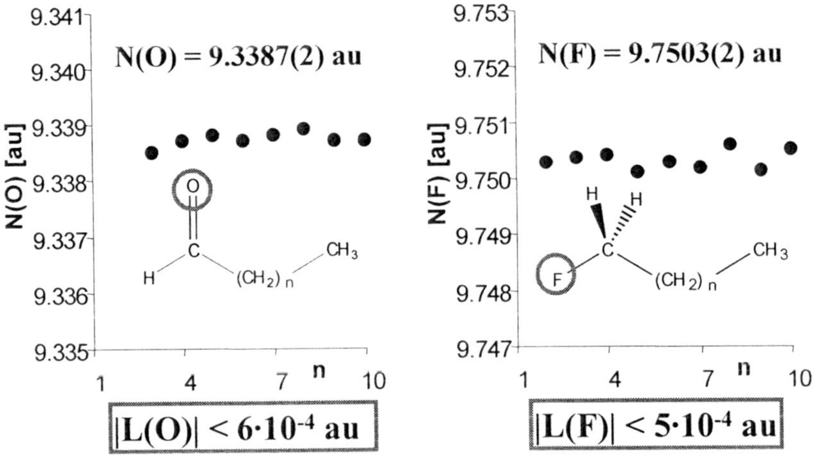

Figure 5. Electron atomic populations of oxygen in a series of $HCO(CH_2)_nCH_3$ aldehydes and fluorine in a series of $FCH_2(CH_2)_nCH_3$ 1-fluoroalkanes. All the values are computed for completely antiperiplanar conformers.

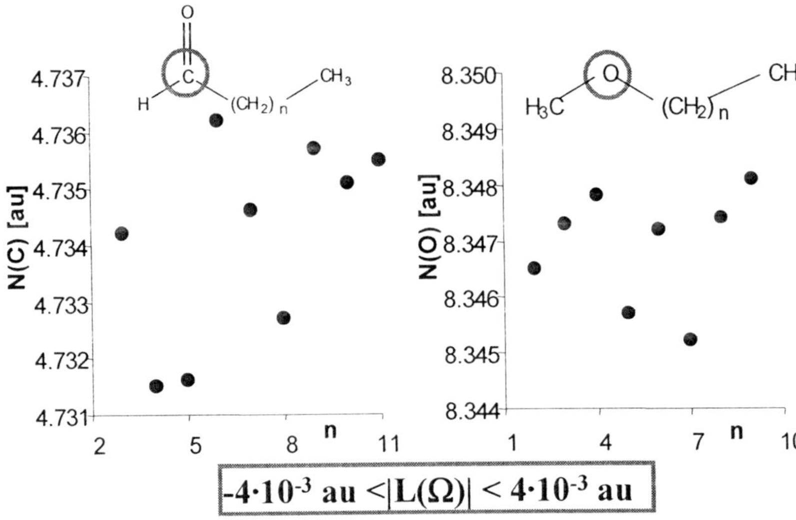

Figure 6. Electron atomic populations of carbonyl carbon in a series of $HCO(CH_2)_nCH_3$ aldehydes and oxygen in a series of $CH_3O(CH_2)_nCH_3$ methylethers. All the values are computed for completely antiperiplanar conformers.

Atomic energy was a more problematic quantity for establishing transferability limits. Looking again at the atomic energy of the oxygen atom, E(O), of a homologous series of aldehydes [40], we observe that so similar atoms like the oxygen of dodecanal and nonanal differed by almost 30 kJ mol^{-1}. The situation was completely similar for all the ketones studied, with the oxygen atom becoming more destabilized when the size of the molecule increases, and giving rise to unexpected differences of the oxygen energy in position isomers, like 5-undecanone and 6-undecanone (figure 8).

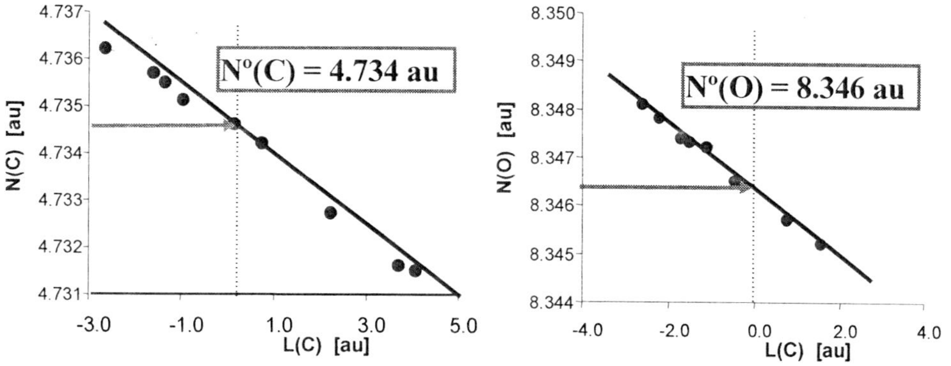

Figure 7. N(Ω) vs. L(Ω) plot for the carbonyl carbon in a series of $HCO(CH_2)_nCH_3$ aldehydes and oxygen in a series of $CH_3O(CH_2)_nCH_3$ methylethers. All the values are computed for completely antiperiplanar conformers.

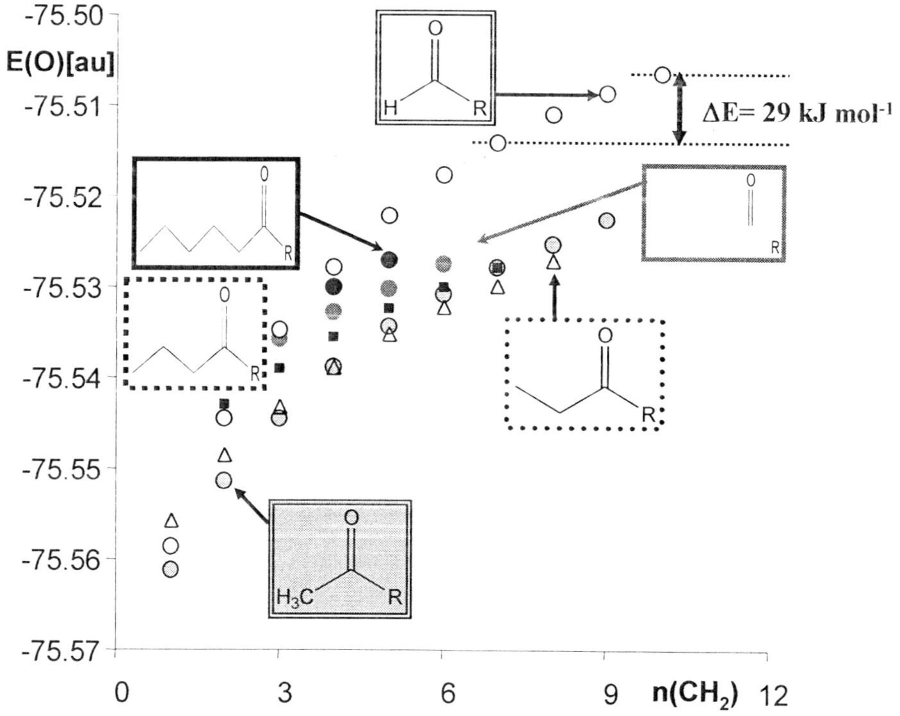

Figure 8. E(O) values for the series of aldehydes (open face circles) and several groups of linear ketones including molecules with 1 to 12 carbon atoms. All values computed with HF/6-31++G(d,p) electron densities.

This trend also holds for alkanols [42], methylethers and other kinds of monoethers [43] and diethers [44] (figure 9a). It is not only a characteristic of oxygens, but we also find it in the carbonyl carbons of aldehydes and ketones [45] and in the carbons of linear alkanes [46], with HF or DFT levels (figure 9b). The same situation was also observed in every series of homologous compounds we have studied, like cyanoalkanes [47] or fluoroalkanes [41]. Therefore, we inferred this trend is common to any atom in any series and computational level and prevents atomic energy transferability. Nevertheless, when the atomic energies of every series are plotted versus the inverse of the total number of electrons in the molecule, Z, we get an empirical linear relation (figure 10). It allows representing each series of nearly transferable atoms by the intercept of this plot, which would represent the energy for an infinity number of electrons, $E^{\infty}(\Omega)$. This relationship also provides a method for classifying atoms into different types as shown for aldehydes, methylketones and dialkylketones (figure 10). We have observed this relation holds for the rest of homologous series we have studied: alkanes [46], alkanols [42], cyanoalkanes [47], fluoralkanes [41], ethers [43,48] and, diethers [44,49].

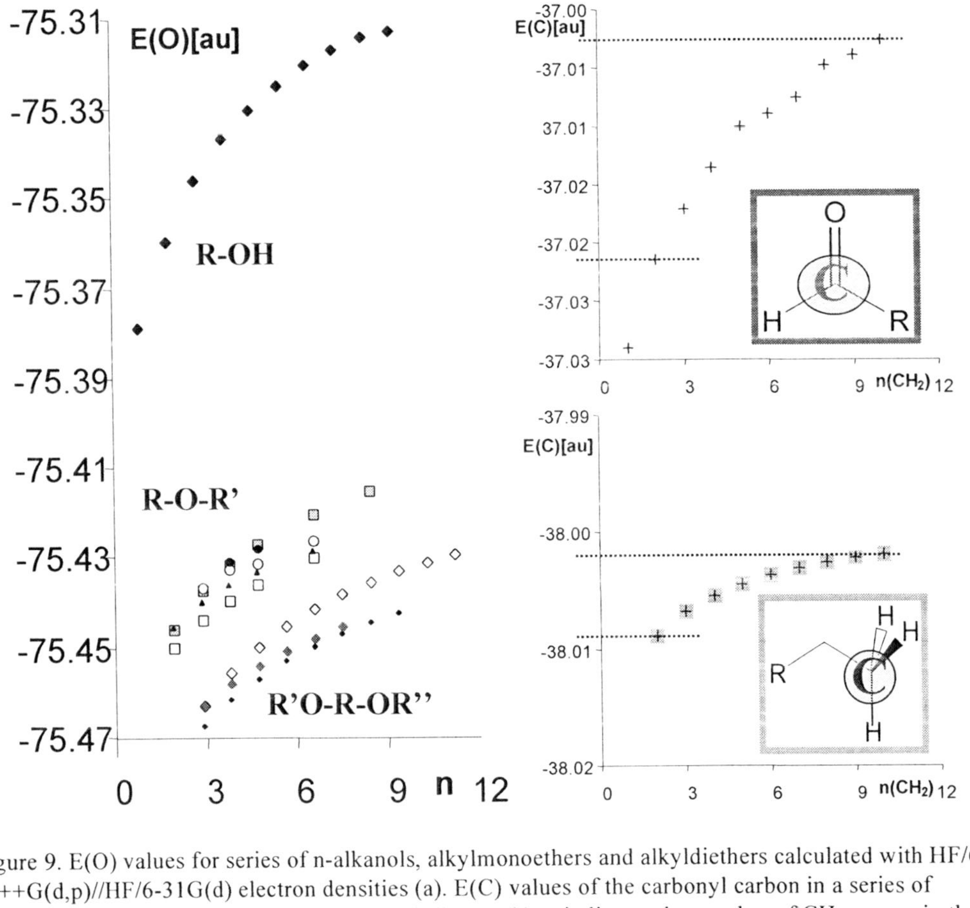

Figure 9. E(O) values for series of n-alkanols, alkylmonoethers and alkyldiethers calculated with HF/6-31++G(d,p)//HF/6-31G(d) electron densities (a). E(C) values of the carbonyl carbon in a series of aldehydes and of methyl carbon of a series of alkanes (b). n indicates the number of CH_2 groups in the molecule.

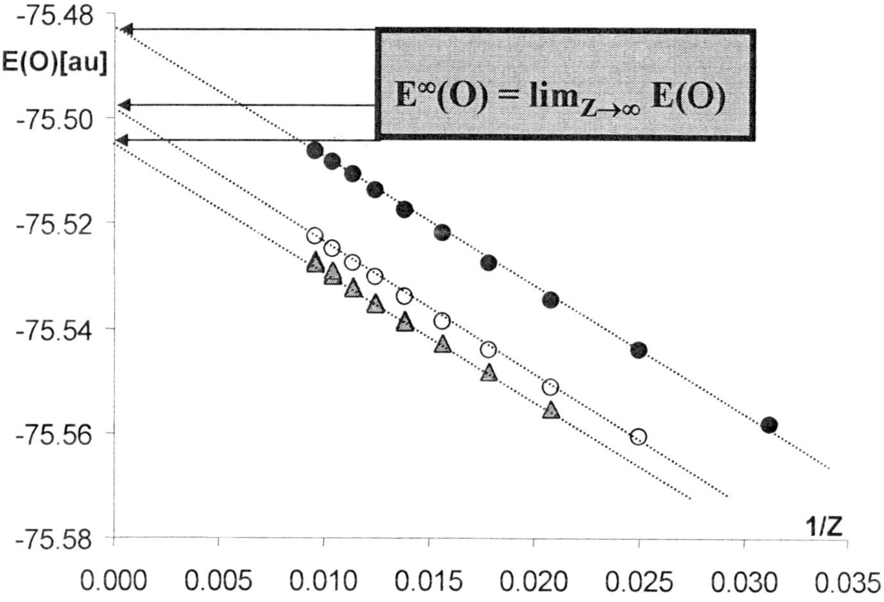

Figure 10. E(O) vs. Z^{-1} plots for aldehydes (propanal to dodecanal),methylketones (2-butanone to 2-dodecanone) and dialkylketones ($CH_3(CH_2)_nCO(CH_2)_mCH_3$) (n,m > 0, n+m < 10).

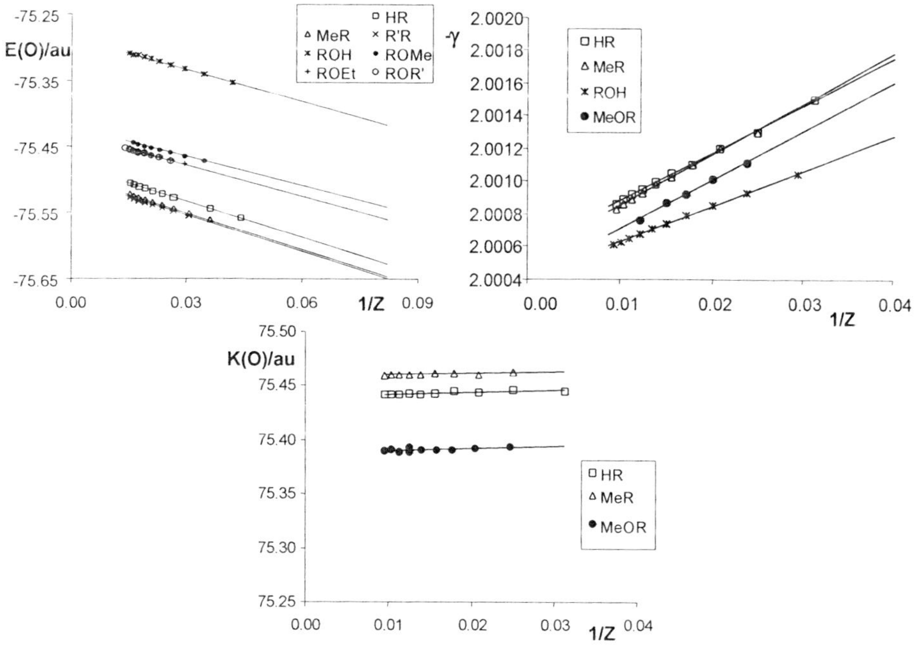

Figure 11. Plots of E(Ω), K(Ω) and γ vs. the inverse of the total number of electrons in the molecule.

Overall, the atomic energy was not found as a transferable property in any homologous series. This problem comes from the fact QTAIM atomic energies are usually computed correcting the atomic electron kinetic energy, K(Ω), with the molecular virial ratio γ (equation {1}). This method has computational advantages, because it avoids computing two-electron interactions. It can be observed that, probably because of the convergence criteria employed to obtain the molecular electron density, the molecular virial ratio is linearly correlated with the inverse of the number of electrons. In contrast, K(Ω) is nearly transferable for each homologous series (figure 11) [50]. Therefore, E(Ω) linear dependence on Z^{-1} is just a shortcoming due to virial ratio correction of K(Ω) values. In fact, Cortés-Guzmán and Bader have demonstrated that nearly transferable atomic energies can be obtained for a certain homologous series when γ approaches its ideal value with high accuracy [51]. This situation can be achieved using self consistent virial scaling (SCVS) optimizations [52].

$$E(\Omega) = (1+\gamma)K(\Omega) \quad \{1\}$$

No significant problems were found for obtaining accurate transferability limits for other atomic properties like the first moment of the electron density, dipole moment, volume, or Shannon entropy, as well as for bond properties.

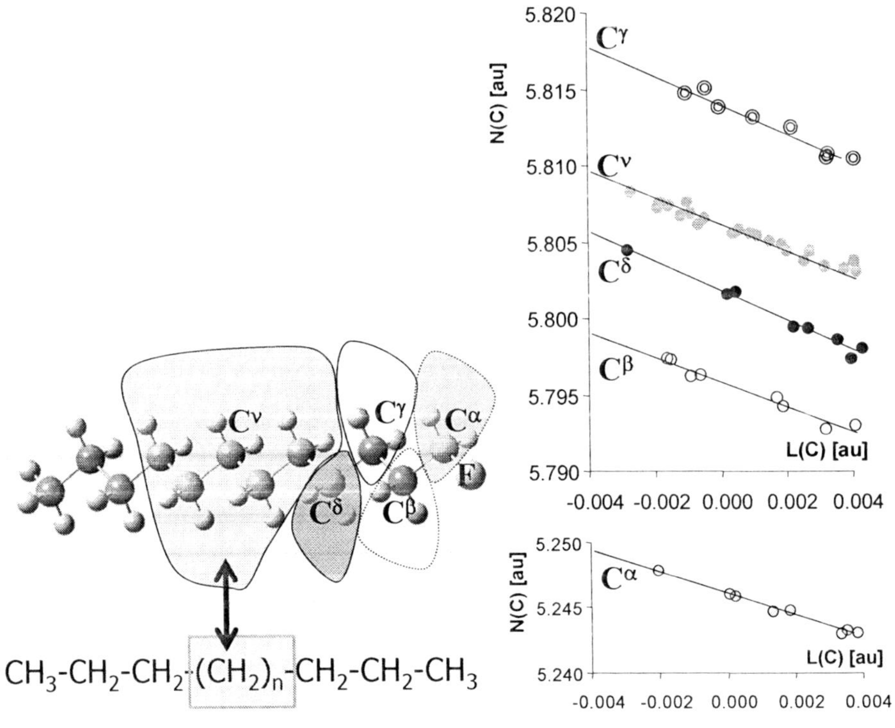

Figure 11. N(C) vs. L(C) plots for nearly transferable carbons in 1-fluoroalkanes in completely antiperiplanar conformer.

Once the transferability limits were established we investigated how transferability is affected by diverse factors. One of the simplest is the conformational change. If we take a series of functionalized linear alkanes, like 1-fluoralkanes, in completely antiperiplanar arrangement, we observe (using the above commented N(Ω) vs. L(Ω) plots shown in figure 11), that α carbons form a nearly transferable group, the same is true for β, γ, and δ atoms, whereas all the rest of the carbons (excluding the terminal groups) belong to a common nearly transferable group [41]. They are called by us normal or C^{υ} carbons, and present the same properties obtained for the inner methylenes of unfuctionalized *n*-alkanes.

Rotating from anti to gauche conformer, we observe that the population of α and β carbons is different, but they continue as different types. In contrast, C^{υ} cannot be distinguished from γ and δ carbons, which are not specific groups in gauche conformers (figure 12) [41]. In summary, α, β, γ and δ carbons of an alkyl chain are significantly influenced by a functional group in antiperiplanar arrangement, but this influence reduces to α and β in gauche. This trend also holds for other series and computational levels [53].

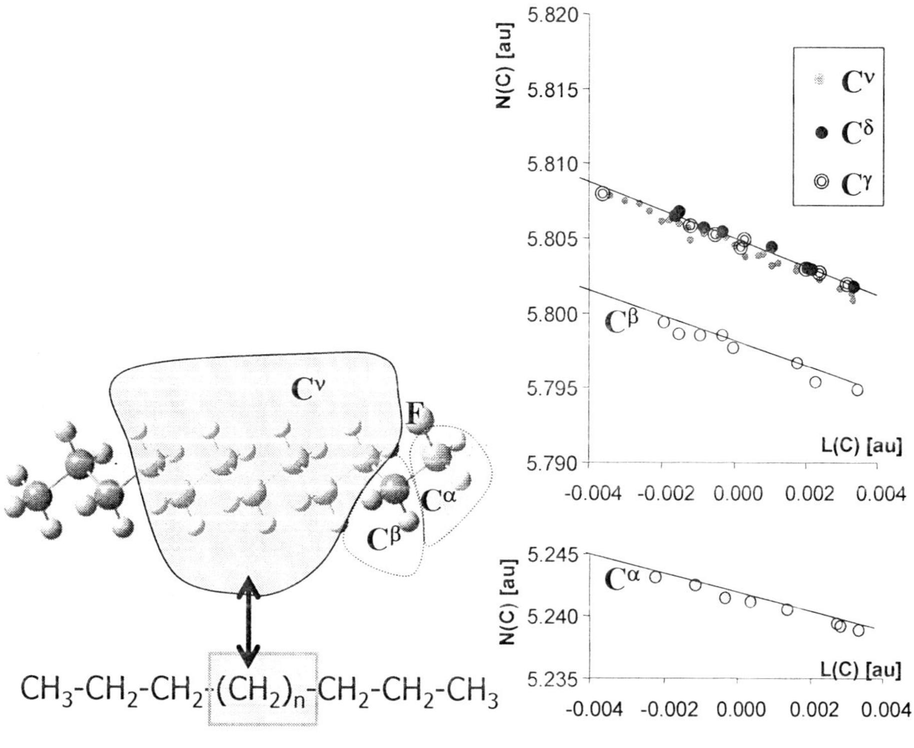

Figure 12. N(C) vs. L(C) plots for nearly transferable carbons in 1-fluoroalkanes in gauche conformers.

4. Strain Energy

Strain energy (SE) is one of the basic concepts generally invoked in the Chemisry of cyclic compounds [54-56]. Many definitions of strain energy have appeared in the chemical literature, and none of them can be considered as unique [57]. In what follows we will assume SE is defined as the difference between the enthalpy of formation of a certain cyclic compound and that of a strainless reference. This reference can be a real molecule, a hypothetic system (in fact, this is what we do when we use group contribution models), or a set of compounds related by a reaction that can be isodesmic [58], homodesmotic [59], hiperomodesmotic [60], etc. In these cases, the strain energy is given by the opposite of the reaction enthalpy.

In the case of cycloalkanes, we can compute the strain energy using the homodesmotic process represented by reaction {R1}, where n ≥ 3 and we assume that cyclohexane is strainless. Obviously, the number of methylene groups in the cycle, n, should be at least three, and the strain energy is given by equation {2}.

$$6\ C_nH_{2n}\ (gas) \rightarrow n\ C_6H_{12}\ (gas) \qquad \{R1\}$$

$$SE^1(n) = \frac{6\Delta H_f^0(C_nH_{2n}) - n\Delta H_f^0(C_6H_{12})}{6} \qquad \{2\}$$

Alternatively, we can use the homodesmotic processes {R2}, where n is still the number of mehtylenes in the cycle. This process includes two linear compounds and m refers to the number of methylenes in the linear reactive (m ≥ 2). The variation of m produces different reactions. The strain energy is given by equation {3}, which is in fact a set of equations depending on the particular value of m.

$$C_nH_{2n}\ (gas) + C_mH_{2m+2}\ (gas) \rightarrow C_{n+m}H_{2(n+m)+2}\ (gas) \qquad \{R2\}$$

$$SE^{2,m}(n) = \Delta H^0_f(C_nH_{2n}) + \Delta H^0_f(C_mH_{2m+2}) - \Delta H^0_f(C_{n+m}H_{2(n+m)+2}) \qquad \{3\}$$

Table 1 lists the strain energies obtained from 3 to 6 membered rings using the first reaction, and the set of reactions contained in the second expression for 2 ≤ m ≤ 7. All the values have been calculated using the most reliable experimental formation enthalpies [61]. We observe the different values obtained for the strain energy of cyclopropane span within 2 kJ mol^{-1}. This is not too much, and can be assigned to experimental errors. Differences even smaller are obtained for larger cycles.

Table 1. Strain energies (in kJ mol^{-1}) computed with processes R1 and R2 (2 ≤ m ≤ 7) for C_nH_{2n} cycloalkanes (3 ≤ n ≤ 6) making use of experimental formation enthalpies listed in reference 61

n	{R1}	{R2}						QTAIM
		m=2	m=3	m=4	m=5	m=6	m=7	
3	114.9	113.8	115.2	115.4	114.7		115.0	115.8
4	110.7		111.7	111.3		111.2		109.2
5	26.5		27.4		26.4			29.7
6	0.0			0.8				0.3

Using QTAIM, the strain energy is obtained as the summation of the differences between the atomic energy in the cycle and that of an atom in a linear nearly transferable compound (equation {4}). The values obtained in this way, using HF/6-31++G(d,p) electron densities, are also in close agreement with the homodesmotic ones. The largest discrepancy does not reach 4 kJ mol^{-1} (Table 1).

$$SE^{QTAIM} = \sum_{\Omega} E(\Omega) - E^{0}(\Omega) \qquad \{4\}$$

QTAIM also provides an interpretation for the origin of strain energy, as was discusses by Wiberg et al two decades ago [9]. In fact, the electron population of carbons is larger in the cycle than in linear compounds, whereas the reverse is true for hydrogens. This indicates the transference of electron density from hydrogens to carbons when a cycle is formed, what is in line with the increased electronegativity of carbons in small rings predicted by hybridization. It can be also observed that hydrogens become destabilized when they lose electron density, whereas the energies of the carbons become more negative. As, the stabilization of carbons does not compensate hydrogen destabilization the strain energy of the cycloalkane is positive.

4.1. Strain Energy in Heterocycles

The above presented explanation for the origin of the strain energy of cycloalkanes is among the first successes of QTAIM. But, what happens when we move from cycloalkanes to heterocycles? To get an answer for this question we have considered the oxirane molecule (C_2H_4O). The first point to be noticed is the diversity of homodesmotic reactions that can be defined for this case. Thus, we can open the cycle breaking just one bond to get diethyl ether, or break two bonds reacting with ethane and methanol to produce ethylmethyl ether and ethanol, or breaking three bonds to give dimethyl ether and two molecules of ethanol (reactions R3 to R5, respectively). Also we can use, what is called the diagonal reaction (R6), where two molecules of oxirane form one of 1,4-dioxocyclohexane, or the semidiagonal one (R7), where oxirane reacts with cyclopropane to give oxacyclohexane, both of them are also homodesmotic. In this point, we just indicate the possibility of defining other reactions that are just isodesmic, like R8. Also, the number of reactions becomes much larger if we replace ethane and methanol by other alkanes and alkanols. Moreover, the number of homodesmotic processes rises for unsymmetrically substituted oxiranes.

$$C_2H_4O + C_2H_6 \rightarrow C_2H_5OC_2H_5 \qquad \{R3\}$$

$$C_2H_4O + C_2H_6 + CH_3OH \rightarrow CH_3OC_2H_5 + C_2H_5OH \qquad \{R4\}$$

$$C_2H_4O + C_2H_6 + 2\ CH_3OH \rightarrow CH_3OCH_3 + 2\ C_2H_5OH \qquad \{R5\}$$

$$2\ C_2H_4O \rightarrow C_4H_8O_2 \qquad \{R6\}$$

$$C_2H_4O + C_3H_6 \rightarrow C_5H_{10}O \qquad \{R7\}$$

$$C_2H_4O + 3\ C_2H_6 \rightarrow CH_3OCH_3 + C_3H_8 \qquad \{R8\}$$

We calculated the homodesmotic strain energies using equations {R3} to {R7} at three different computational levels: HF/6-311++G(d,p), MP2/6-311++G(d,p) and B3LYP/6-311++G(d,p). Looking at the results presented in Table 2, we observe a dependence with the computational level (from 55 to 120 kJ mol^{-1}), which is quite large. But, worse than that, the strain energies obtained for oxirane using the same computational level on different processes never differ by less than 90 kJ mol^{-1} and can reach almost 270 kJ mol^{-1}.

Expressing reaction enthalpy as a function of atomic energies and taking into account the vibrational corrections (expression {5}), we observe the strain energy for a certain homodesmotic process *r*, SE_r, contains more terms than that due to ring opening, ΔE_r^{RO}.

$$
\begin{aligned}
-SE_r &= \Delta H_r = \Delta E_r - RT = \left[\sum_{\Omega} \nu_i E(\Omega_i)\right] - RT \\
&= \Delta E_r^{RO} + \Delta E_r^{NC} + \Delta E_r^{ZPVC} - RT
\end{aligned} \qquad \{5\}
$$

Table 2. Strain Energies computed for oxirane using homodesmotic processes {R3} to {R7} and three computational levels

SE [kJ mol^{-1}]	{R3}	{R4}	{R5}	{R6}	{R7}
HF/6-311++G(d,p)	53.5	197.4	53.3	213.9	321.3
MP2/6-311++G(d,p)	115.6	85.5	114.6	103.6	223.7
B3LYP/6-311++G(d,p)	118.6	118.9	118.7	109.0	201.3

The vibrational corrections (ΔE_r^{ZPVC}) cannot be the reason for the big differences listed in Table 2 as they are always less than 15 kJ mol^{-1}. Another term present in {5} is the energy change experienced by the non cyclic compounds, ΔE_r^{NC}. Analyzing reaction {R3}, we see that two specific methyls of ethane are transformed into the specific methyls of diethyl ether (figure 13). This represents 213.6 kJ mol^{-1}, which are included inside the strain energy. Looking at reaction {R4}, we observe this term involves interconversions among different methyl groups and that due to transforming the hydroxyl of methanol into that of ethanol (figure 13). This results in a different value (146.2 kJ mol^{-1}) for this non cyclic energy term.

Turning back to the ring opening energy term, ΔE_r^{RO}, we observe that its components and total amount in the {R3} reaction differ significantly from those in the {R4} one (figure 14). Thus, although the difference in the estabilization gained by the oxygen atom in the linear ether is only 3.4 kJ mol^{-1} less in reaction {R4} than in {R3}, there are very important differences in the destabilization experienced by α CH_2 in linear compounds with regard to oxirane. This results in a difference of more than 90 kJ mol^{-1} when comparing ΔE_r^{RO} values for {R3} and {R4}.

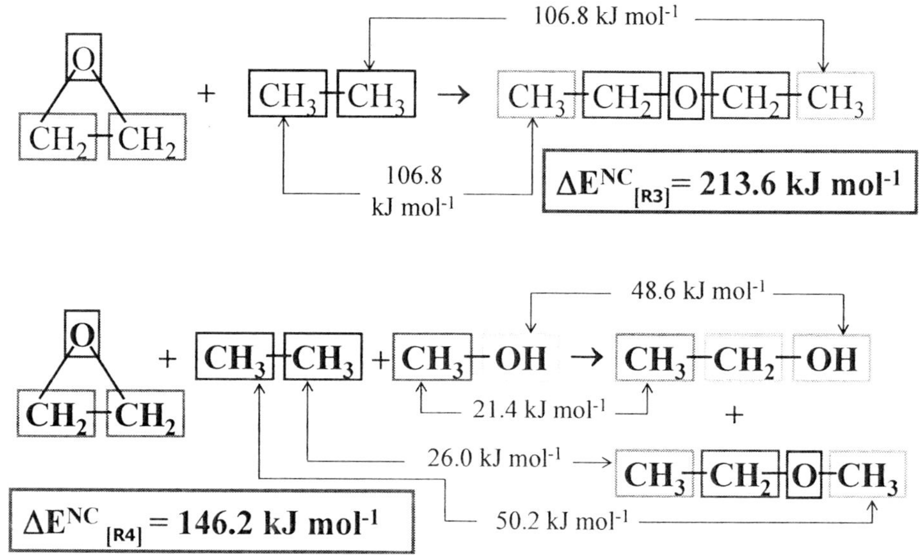

Figure 13. Atomic groups scheme for reactions {R3} and {R4} indicating the diverse contributions to the ΔE_r^{NC} term in these processes.

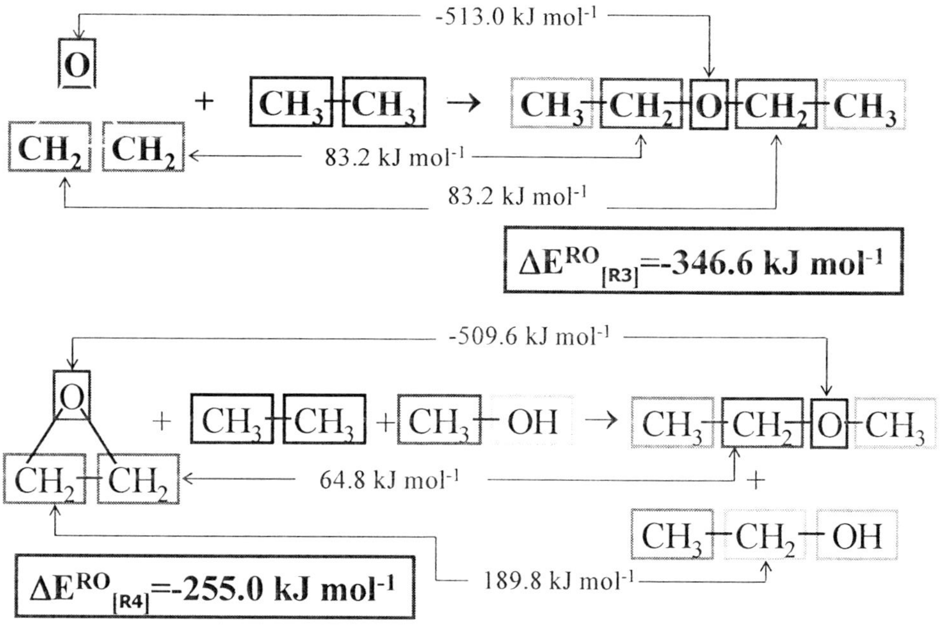

Figure 14. Atomic groups scheme for reactions {R3} and {R4} indicating the diverse contributions to the ΔErRO term in these processes.

The calculation of these terms for the five processes at the B3LYP level making use of QTAIM atomic energies (Table 3), shows the already commented very small variations of the vibrational correction, large changes in the non cyclic term, and also in the ring opening one. The compensation between them at this computational level gives rise to the pretty good constancy of SE_r obtained for reactions {R3}-{R5} with B3LYP, which is not obtained with HF and MP2 levels.

Table 3. Components of the SE of homodemostic processes {R3} to {R7}. All values in kJ mol^{-1} and computed at the B3LYP/6-311++G(d,p) level

	{R3}	{R4}	{R5}	{R6}	{R7}
SE_r	118.6	118.9	118.7	109.0	201.3
$-\Delta E_r^{ZPVC}$	-14.4	10.1	-13.1	-9.7	-14.3
$-\Delta E_r^{NC}$	-213.6	-146.2	-52.0	0.0	-96.6
$-\Delta E_r^{RO}$	346.6	255.0	173.8	118.7	312.2

The variation displayed by the ring opening term is quite large. As previously shown it has two components: the energy implied in transforming the cyclic oxygen into an oxygen atom of a strainless compound, and that due to transforming the methylenes of oxirane into strainless CH_2^{α} of linear ethers or alkanols. We observe there are four different transformations for the oxygen atom in the five reactions here considered, yielding: the specific oxygen of ethoxyethane in {R3}, the specific oxygen of methoxyethane in {R4}, the specific oxygen of methoxymethane in {R5}, an oxygen γ to other oxygen in {R6}, and the oxygen of tetrahydropirane (nearly equivalent to that of a dialkyl ether) in {R6}. Also, the second component of ΔE_r^{RO} (the linearization of a cyclic methylene that is simultaneously α and β to an oxygen atom) represents a different trasnformation for every reaction (figure 15). Therefore, the diversity of ΔE_r^{RO} values in Table 3 is reasonable.

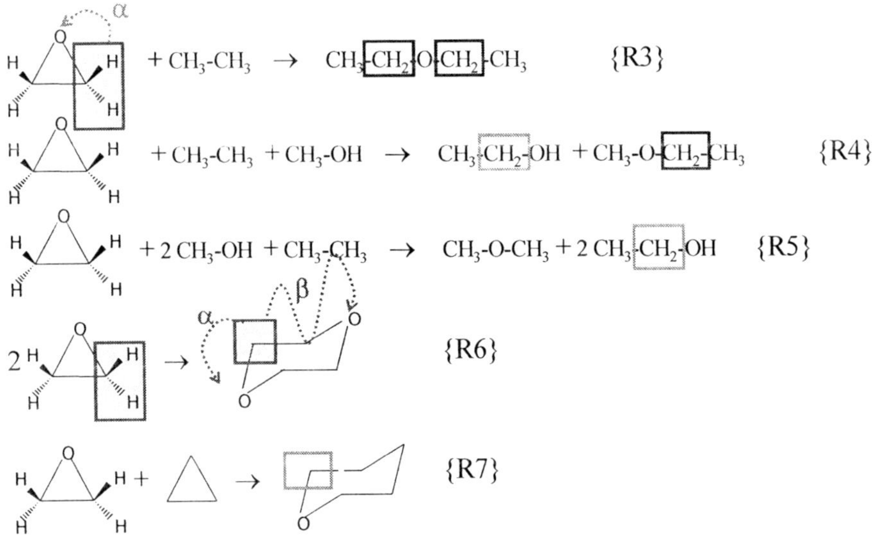

Figure 15. Transformations experienced by the cyclic methylene of oxirane in processes {R3} to {R7}. It can be observed that the slightest change corresponds to process {R6} where the methylene group is α and β to oxygen atoms.

In contrast, the QTAIM definition of strain energy provides a unique value. The nearly transferable atoms to include in equation {4} correspond to the oxygen and inner methylenes of long polyethers of formula $C_2H_5(OC_2H_5)_nOC_2H_5$, where $n \geq 4$. $[E(\Omega)-E^0(\Omega)]$ terms indicate that carbon and hydrogens are estabilized in oxirane (by 128.2 and 23.7 kJ mol^{-1}, respectively), whereas the oxygen experiences a big destabilization in the cycle (462.1 kJ mol^{-1}). The QTAIM obtained value for the strain energy of oxirane (110.9 kJ mol^{-1}) is only near the ring opening energy obtained in the {R7} process (Table 3), which is the case where transformations of methylene and oxygen are not affected by other chemical influences but the formation of a strainless compound (figure 15).

4.2. Strain Energy in Perfluorocycloalkanes

To properly include only the energy due to ring formation in SE, we should eliminate changes in the energy due to electron transfers from groups that are not present in the cycle. Whereas this is possible for cycloalkanes, where the CH_2 electron population (8.000 au) coincides with its value in the central methylenes of a *n*-alkane [9], it cannot be done for heterocycles. For instance, the methylenes of oxirane are at the same time α and β to an oxygen that is only bonded to two methylene groups. The most similar situation in an acyclic compound is that of the inner methylenes in 1,2-diethoxyethane. Nevertheless, comparing the electron populations of these methylene fragments (respectively, 7.405 and 7.340 au [49] at the HF/6-31++G(d,p) level) we conclude they are different indeed.

The above discussion indicates that the extension of the QTAIM-treatment on the origin of SE of cycloalkanes [9] to heterocycles is not a straightforward task [11]. Thus, the SE computed as summation of differences between atomic energies in the cycle and in strainless molecules, SE^{QTAIM}, does not recover the SE computed with homodesmotic, ultradiagonal, or other processes, SE_r, because of: i) their different physical meaning, ii) shortcomings that can affect to both procedures [10,11]. To analyse the origin of these differences we have considered the series formed by the five smallest cyclic perfluorocycloalkanes, *c*-PFAs, (*c*-C_nF_{2n}; $3 \leq n \leq 7$) [12], carring out B3LYP/6-31++G(d,p) full optimizations and frequency calculations on the most stable conformers of *c*-PFAs, and on the completely antiperiplanar conformer of the *n*-PFAs of formula *n*-C_nF_{2n+2} with $4 \leq n \leq 8$, using the virial scaling option [52] in the GAMESS program [62]. All the conformers were optimized until the virial ratio differed from 2 by less than 10^{-5}. The expected planar, puckered, and chair conformers were obtained for, respectively, *c*-C_3F_6, *c*-C_4F_8, and *c*-C_6F_{12}. Their main geometrical features [12] allow to distinguish between axial (or pseudoaxial), F_a, and equatorial (or pseudoequatorial), F_e, fluorine atoms in these compounds. The twist-boat conformer of *c*-C_6F_{12} displays D_2 symmetry with two kinds of carbon atoms (C_a and C_b) and three of fluorine atoms (pseudoaxial, F_a, pseudoequatorial, F_e, and those bonded to C_b, F_b) The conformer with nearly envelope disposition (but with C_1 symmetry) is obtained for *c*-C_5F_{10}, and a C_2 twist-chair one for *c*-C_7F_{14}. The symmetry of these conformers led us to refer to their atoms by the numbering shown in figure 16.

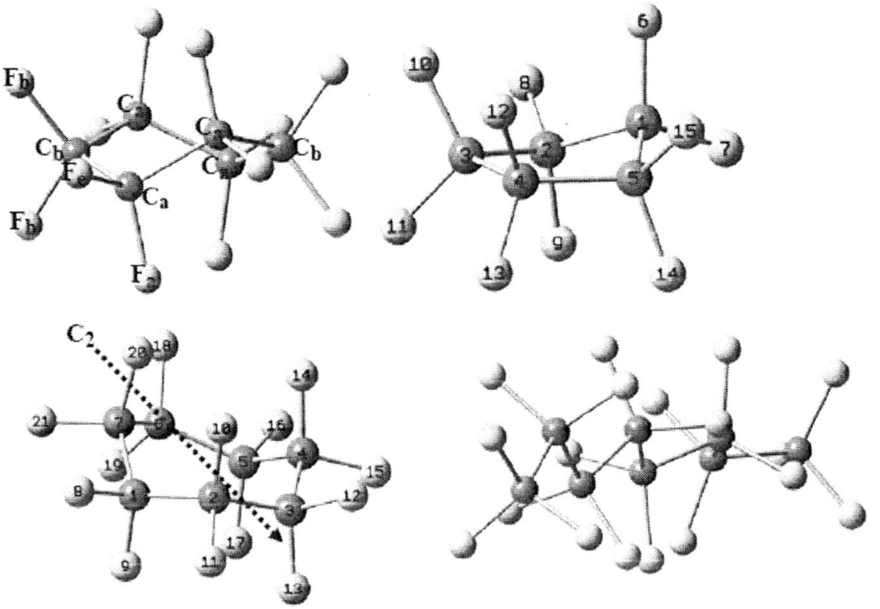

Figure 16. Twist-boat conformer of c-C_6F_{12}, and most stable conformers of c-C_5F_{10}, c-C_7F_{14}, and n-C_8F_{18} perfluoroalkanes.

It has to be noticed that the completely antiperiplanar conformers of n-PFAs, which are their global minima, present main dihedral angles that differ from 180° significantly, displaying a helical structure previously found at several computational levels [63-64] and rationalized as due to F···F destabilizing 1,5-diaxial interactions [65]. Thus, these conformers present C_2 symmetry instead of the C_{2h} or C_{2v} observed, respectively, for n-alkanes with an even or odd number of carbons.

The SE^{QTAIM} calculation of c-PFAs is untouched by an important limitation affecting usual heterocycles: the different electron populations in cyclic and nearly transferable linear fragments. For these molecules CF_2 cyclic fragments of symmetric conformers and the nearly transferable central CF_2 of linear perfluoroalkanes, n-PFAs, display a total population of 24.000 au. However, SE^{QTAIM} and SE_r values computed for c-PFAs disagree strongly.

The total B3LYP/6-31++G(d,p) electronic energy (and also those corrected with ZPVE after frequency calculations) of n-PFAs are linearly correlated with the number of CF_2 groups in the compound following fitting equation {6}. The corresponding fitting of ZPVE uncorrected total electronic molecular energies provides –237.70242 a.u. for $E(CF_2)$ and –337.52732 a.u. for $E[CF_3]$. This fitted $E[CF_2]$ energy differs from that obtained by summation of the QTAIM integrated energies of the atoms in the group (Table 4) by values that range from -35 to -93 kJ mol^{-1} for CF_2 and from 111 to 181 kJ mol^{-1} for CF_3. We remark that the fitted CF_2 energy is also clearly different (-89.0 kJ mol^{-1}) from that of the nearly transferable CF_2 fragment (central CF_2 of n-C_8F_{18} as commented below). In contrast, molecular electronic energies are reproduced by those obtained with equation {6} and these parameters within 0.8 kJ mol^{-1}. Also, the fitted group energies are more in line with the conventionally accepted practice in the classical organic chemistry context of strain energy than the QTAIM group energies of nearly transferable groups. Nevertheless, nothing assures that all the molecules fitted actually contain the parameterized atomic groups.

$$E[CF_3(CF_2)_nCF_3]^{n\text{-PFA}} = 2\,E[CF_3] + n\,E[CF_2] \qquad \{6\}$$

Atomic electron populations computed for C_nF_{2n+2} *n*-PFAs from n=5 to 8 (Table 4) indicate, at least, the presence of four different groups in these compounds: i) terminal CF_3 groups; ii) CF_2 groups bonded to CF_3 (CF_3^{α}) characterized by the smallest electron population and are positively charged; iii) CF_2 in β to a CF_3 (CF_2^{β}), which display a slight negative charge; and iv) the central CF_2 groups in *n*-perfluoroheptane (CF_2^{CC}) and *n*-perfluorooctane (CF_2^{C}), which are neutral or nearly neutral. It has to be stressed that the group definition we are employing here follows the general rules shown in previous studies on approximate group transferability using QTAIM [9,40-43,45-48,66-67]. This is a much stricter definition of atomic group than those used by Benson [68] or Liebman [55,56], where the four different CF_2 groups here defined for *n*-C_8F_{18} could be considered as four different environments of the same group. The different criterion for group definition can be considered as one of the factors that give rise to obtain different values of the strain energies of heterocycles using QTAIM and group contribution models.

Looking at the corresponding E(Ω) values (Table 4) we observe that: i) this property varies significantly for CF_3, CF_2^{α}, and CF_2^{β} groups along the series of molecules here considered; ii) The energy gap between two equivalent fragments of molecules that differ in one CF_2 becomes progressively smaller as the number of carbon rises; and iii) the only atomic groups that differ by less than 6 kJ mol^{-1} (which is close to the limits of the energy accuracy obtained when summing the atomic energies of a molecule) are the CF_2^{CC} of *n*-C_7F_{16} and the CF_2^{C} of *n*-C_8F_{18}. Therefore, we conclude that the CF_2^{C} group of *n*-C_8F_{18} can be considered as a good approximation for the CF_2 nearly transferable group of *n*-PFAs. In this context, we remind that perfect transferability has been demonstrated to be an unattainable goal [39]. In contrast, approximate transferability has not been reached for the remaining groups yet. Also, the three CF_2 groups defined above present energies that are significantly different in the same molecule, CF_2^{α} being the most stabilized (if we exclude *n*-C_5F_{12} where $CF_2^{\beta\beta}$, which is clearly specific as it is β to two CF_3 groups is the most stable) and CF_2^{C} the least.

Two groups of cyclic conformers can be considered according to their atomic properties: those presenting common values for each kind of atoms (*c*-C_3F_6, *c*-C_4F_8, and chair conformer of *c*-C_6F_{12}) and those where several CF_2 groups have to be considered for each ring (*c*-C_5F_{10}, twist-boat conformer of *c*-C_6F_{12}, and *c*-C_7F_{14}). The formers display neutral charge, whereas positive, negative and neutral CF_2 groups are present in the latter. Anyway, all of them present either E(Ω) or N(Ω) values that differ significantly from those of the above introduced nearly transferable linear CF_2 group. Overall, cyclic CF_2 groups become more stable as the ring size increases, the average stability of CF_2 groups of *c*-C_7F_{14} being less than those in *c*-C_6F_{12} and *c*-C_5F_{10} though.

The irregular geometry of *c*-C_5F_{10} conformer gives rise to different values of atomic electron population and energies for each atom. We notice that C atoms with the largest atomic populations are the most stabilized (C1 and C2). The twist-boat conformer of *c*-C_6F_{12} presents two different CF_2 groups (Table 4). The one with pseudoaxial and pseudoequatorial fluorines being the most stabilized and electron populated. Finally, four different CF_2 groups are present in the optimized conformer of *c*-C_7F_{14} as some are equivalent because of its C_2 symmetry (figure 16). The largest destabilization and smallest electron population corresponds to the CF_2 group whose carbon atom (C6) is on the C_2 axis.

Several homodesmotic processes can be employed to calculate the SE_r of *c*-PFAs here considered using the total molecular energies and ZPVEs of the *n*-PFA shown in Table 5. Some of these processes are summarized by equations {R9} - {R11}. The SE_r computed with $2 \leq m \leq 5$ (when all the linear compounds involved in the process belong to the series n-C_4F_{10} – n-C_8F_{18}) differ for a same cycle as much as 6 to 14 kJ mol^{-1} (Table 5). These differences are larger than those originated by ZPVE corrections, ΔSE_r^{vib}, which are always less than 2 kJ mol^{-1} (Table 5).

$$c\text{-}(CF_2)_n + n\ CF_3(CF_2)_mCF_3 \rightarrow n\ CF_3(CF_2)_{m+1}CF_3 \quad \{R9\}$$

$$c\text{-}(CF_2)_n + CF_3(CF_2)_mCF_3 \rightarrow CF_3(CF_2)_{m+n}CF_3 \quad \{R10\}$$

$$c\text{-}(CF_2)_n + CF_3(CF_2)_mCF_3 + (n\text{-}2)\ CF_3(CF_2)_{m+1}CF_3 \rightarrow (n\text{-}1)\ CF_3(CF_2)_{m+2}CF_3 \quad \{R11\}$$

It is also noticeable that c-C_5F_{10} and both conformers of c-C_6F_{12} display negative SE_r values, with the chair conformer of c-C_6F_{12} presenting the most negative ones. A plausibly negative strain energy has been discussed for perchlorocycloalkanes by Mahavadi et al. [69] bearing in mind that SEs are defined as relative to some strainless species, which need not have the lowest energy of species composed by the groups contained therein.

Table 4. QTAIM atomic electron population, N(Ω), and energies, E(Ω), computed for linear and cyclic perfluoroalkanes. All data in au

Compound	Group	N(C)	N(F)[a]	-E(C)	-E(F)[a]	N(CF_2)	-E(CF_2)
n-C_8F_{18}	CF_2^{C}	4.788	9.606[b]	37.21143	100.22854[b]	24.000	237.66851
n-C_7F_{16}	CF_2^{CC}	4.787	9.606[b]	37.21402	100.22652[b]	23.998	237.66706
n-C_8F_{18}	CF_2^{A}	4.794	9.606[b]	37.22165	100.22946[b]	24.006	237.68057
n-C_7F_{16}	CF_2^{A}	4.794	9.606[b]	37.22281	100.22774[b]	24.006	237.67830
n-C_6F_{14}	CF_2^{A}	4.793	9.606[b]	37.21950	100.22604[b]	24.004	237.67157
n-C_5F_{12}	CF_2^{AA}	4.798	9.606[b]	37.22308	100.22686[b]	24.010	237.67680
n-C_8F_{18}	CF_2^{P}	4.778	9.604[b]	37.21904	100.23497[b]	23.987	237.68898
n-C_7F_{16}	CF_2^{P}	4.779	9.604[b]	37.22278	100.23117[b]	23.987	237.68511
n-C_6F_{14}	CF_2^{P}	4.777	9.604[b]	37.21992	100.22993[b]	23.985	237.67978
n-C_5F_{12}	CF_2^{P}	4.776	9.604[b]	37.21849	100.22753[b]	23.985	237.67355
n-C_8F_{18}	CF_3	4.181	9.610[b]	36.76560	100.27692[b]	33.009	337.59635
n-C_7F_{16}	CF_3	4.180	9.610[b]	36.76398	100.27432[b]	33.010	337.58692
n-C_6F_{14}	CF_3	4.183	9.610[b]	36.76506	100.27217[b]	33.012	337.58157
n-C_5F_{12}	CF_3	4.181	9.610[b]	36.76294	100.26886[b]	33.011	337.56953
c-C_3F_6	CF_2	4.788	9.606	37.23883	100.22026	24.000	237.67935
c-C_4F_8	CF_2	4.780	9.611[c] 9.609[d]	37.23889	100.23020[c] 100.22766[d]	24.000	237.69675
c-C_5F_{10}	$C(1)F_2$	4.799		37.26039		24.007	237.71931
c-C_5F_{10}	$C(2)F_2$	4.796		37.25319		24.004	237.71179
c-C_5F_{10}	$C(3)F_2$	4.789		37.22777		23.997	237.69254
c-C_5F_{10}	$C(4)F_2$	4.787		37.21815		23.996	237.69152
c-C_5F_{10}	$C(5)F_2$	4.791		37.23779		23.999	237.69899

Table 4. Continued

c-C_6F_{12}	CF_2	4.801	9.598[c] 9.601[d]	37.24707	100.23563[c] 100.22127[d]	24.000	237.70397
c-C_6F_{12}[e]	CF_2	4.803	9.600[c] 9.602[d]	37.24463	100.24079[c] 100.22148[d]	24.005	237.70690
c-C_6F_{12}[e]	C(t)F_2	4.791	9.600	37.21505	100.23975	23.991	237.69455
c-C_7F_{14}	C(1)F_2	4.814		37.24678		24.014	237.70331
c-C_7F_{14}	C(2)F_2	4.798		37.23139		23.992	237.70023
c-C_7F_{14}	C(3)F_2	4.798		37.23120		23.992	237.70030
c-C_7F_{14}	C(4)F_2	4.814		37.24659		24.014	237.70334
c-C_7F_{14}	C(5)F_2	4.803		37.22918		24.000	237.70210
c-C_7F_{14}	C(6)F_2	4.794		37.20784		23.989	237.69893
c-C_7F_{14}	C(7)F_2	4.807		37.22944		24.000	237.70194

[a]For *c*-C_5F_{10} and *c*-C_7F_{14} see figure 16. [b]Average value. [c]Equatorial or pseudoequatorial F. [d]Axial or pseudoaxial F. [e]Twist boat conformer.

The sequence of SE_r values obtained for equivalent processes (same equation and same m value) follows the same order obtained for cycloalkanes [54,70]. Thus, SE_r(*c*-C_3F_6) > SE_r(*c*-C_4F_8) > SE_r(*c*-C_7F_{14}) > SE_r(*c*-C_5F_{10}) > SE_r(*c*-C_6F_{12}, chair). Nevertheless, the SE_r obtained for *c*-C_3F_6 is much larger than that for *c*-C_4F_8 (114-116 kJ mol^{-1}), whereas the SE_r generally accepted values of cyclopropane and cyclobutane [10,54] differ by less than 5 kJ mol^{-1}. This result agrees with previous calculations of SE_r carried out for *c*-C_3F_6 and *c*-C_4F_8 [71-72], where it is also found that perfluorination increases the SE_r of cyclopropane and decreases that of cyclobutane [71-73].

Table 5. Molecular energies, vibrational corrections, virial ratios, γ, and integration errors of the cyclic and linear perfluoroalkanes as obtained from B3LYP/6-31++G(d,p) calculations. $L_{max}(\Omega)$ indicates the maximum absolute value achieved for $L(\Omega)$ along each molecule. Values in parenthesis refer to the atoms in the central fragment[a]

	*E*au	ZPVE kJ mol^{-1}	10^6 (2-γ)	ΣE(Ω)-E kJ mol^{-1}	10^3[ΣN(Ω)-N] au	$10^3 L_{max}(\Omega)$ au
c-C_3F_6	-713.03876	89.4	0.97	1.8	1.5	1.9
c-C_4F_8	-950.78601	124.0	5.61	-2.6	0.8	1.9
c-C_5F_{10}	-1188.51282	158.9	0.20	-3.5	2.2	1.6
c-C_6F_{12} (chair)	-1426.22222	192.0	0.15	-4.1	4.4	2.2
c-C_6F_{12} (twist)	-1426.21569	192.1	0.06	-2.7	1.9	1.5
c-C_7F_{14}	-1663.91036	224.9	1.07	0.6	2.4	1.7
n-C_4F_{10}	-1150.45945	141.8	0.55	-	-	-
n-C_5F_{12}	-1388.16186	174.1	0.16	-2.9	0.9	2.6 (2.6)
n-C_6F_{14}	-1625.86425	206.2	0.29	-4.1	2.6	2.6 (0.1)
n-C_7F_{16}	-1863.56702	238.5	0.76	-1.9	2.9	1.5 (1.0)
n-C_8F_{18}	-2101.26896	270.4	3.72	0.3	2.6	1.9 (0.3)

[a] CF_2^C in *n*-C_8F_{18} and *n*-C_7F_{16} and CF_2^{β} in *n*-C_6F_{14} and *n*-C_5F_{12}.

Thus values of 226 kJ mol^{-1} for *c*-C_3F_6 and from 61 to 75 kJ mol^{-1} for C_4F_8 were reported by Liebman *et al.*[71] on the basis of Benson's group increments and experimental ΔH_f(gas) values for both molecules. HF/6-311G** optimizations lead to SE_r values of 472 and 65 kJ mol^{-1} for *c*-C_3F_6 depending on the linear compounds used as references for the homodesmotic process [72]. Finally, diagonal and ultradiagonal SE_r values of 259 kJ mol^{-1} were computed at the same computational level assuming a zero SE for perfluorocyclohexane [72], which is not confirmed by our calculations. Nevertheless, as commented above, the diverse SE_r are relative to different reference strainless species.

Moreover, the difference between SE_r values of c-C_7F_{14} and c-C_5F_{10} is significantly larger than that between the SE_rs of c-C_7H_{14} and c-C_5H_{10} (0.4 kJ mol^{-1} from experimental data [54] and 1.2 kJ mol^{-1} using MP2/6-31+G* calculations [70]). Finally, SE_r values calculated for the twist-boat conformer of c-C_6F_{12} are approximately equal to those obtained for c-C_5F_{10}.

In contrast with the multiple SE_r values associable to a given cyclic compound, QTAIM provides a unique SE^{QTAIM} value for each ring. It is defined as the difference between the summation of group energies in the cycle and those of the nearly transferable fragments of the corresponding linear groups, which are taken as the strainless reference. Another way to obtain a unique value of SE for each ring consists in using fitted values for the group energies of the strainless reference. The energy of each group is obtained by fitting a series of electron molecular energies of linear compounds to the number of groups of each kind presented by them. Hereafter, the values computed with this reference are named SE^f.

We observe that the election of the strainless reference is not meaningless. In fact, SE^f values display similar evolution and values to SE_r ones for the series here considered (Table 6). In contrast, SE^{QTAIM} values differ a lot from SE_r ones (hundreds of kJ mol^{-1}), they even display opposite signs in several cases (c-C_3F_6, c-C_4F_8, and c-C_7F_{14}), and follow a different sequence of relative values for the series (Table 6). On the contrary, SE^{QTAIM} values only contain the energy involved in ring opening.

Why are SE^{QTAIM} values so different from SE_r and SE^f? The calculation of SE_r using the group energies provided by QTAIM (Table 4) gives insight about this question. Let's consider, for instance, one molecule of c-C_3F_6 experiencing {R11} with m=5 (figure 17). We observe that reagents are made up by: 3 specific CF_2 3-membered ring groups, 6 specific CF_3 groups of n-C_7F_{16}, 6 specific CF_2^{α} groups of n-C_7F_{16}, 6 specific CF_2^{β} groups of n-C_7F_{16}, and 3 specific CF_2^{C} group of n-C_7F_{16}. The products are made up by: 6 specific CF_3 groups of n-C_8F_{18}, 6 specific CF_2^{α} groups of n-C_8F_{18}, 6 specific CF_2^{β} groups of n-C_8F_{18}, and 6 nearly transferable CF_2^{C} groups. Therefore, none of the groups is common to reagents and products. Also, only six groups are strictly involved in the ring opening process (the 3 specific CF_2 3-membered ring groups of reagents and 3 of the 6 nearly transferable CF_2^{C} groups). The difference between the energy of these groups (85.4 kJ mol^{-1}) is the one implied in the ring opening process, ΔE_r^{RO}, and in this process (but not in all, see below) is just the opposite to SE^{QTAIM}. The difference between the energies of the remaining groups (-256.7 kJ mol^{-1}) is an energy term included in SE_r and due to the presence of linear compounds in reagents, ΔE_r^{NC}. The summation of ΔE_r^{NC} and ΔE_r^{RO} differs only from $-SE_H$ in an error term, ε_r, originated in the numerical computation of $E(\Omega)$ in each molecule. This error term is given by the summation of the errors in the molecular energy listed in Table 5 once they are multiplied by the stoichiometric coefficients in equation {R9}.

The process represented by equation {R9} with m=4 is an example where ΔE_r^{RO} (96.8 kJ mol^{-1}) and $-SE^{QTAIM}$ (85.4 kJ mol^{-1}) do not coincide. It is easy to demonstrate that the energy difference (11.4 kJ mol^{-1}) should be, as it is, three times the difference between the energy of the nearly transferable CF_2^{C} group and that of the corresponding specific group in n-C_7F_{16} (3.8 kJ mol^{-1}, Table 4).

Table 6. SEs computed for *c*-PFAs with processes {R9}-{R11} and their components due to ZPVE, SE_r^{vib}. Maximum variations observed for both quantities in every compound (ΔSE_r and ΔSE_r^{vib}), and SE^f values as well as the components of SE_r^{QTAIM} are also shown.[a] All values in kJ mol^{-1}

	process	m	SE_r	ΔSE_r	SE_r^{vib}	ΔSE_r^{vib}	SE^{QTAIM}	SE^f	ΔE_r^{RO}	ΔE_r^{NC}	ε_r
c-C_3F_6	{R9}	2	172.4	6.0	-7.4	0.8	-85.4	179.9			
		3	172.5		-7.1				61.3	-246.6	-5.8
		4	175.4		-7.3				96.8	-274.3	5.2
		5	169.4		-6.6				85.4	-256.7	4.7
	{R10}	2	173.4		-7.3						
		3	172.4		-7.0				53.7	-231.8	1.4
	{R11}	2	172.4		-7.2						
		3	174.4		-7.2				67.3	-247.4	1.5
		4	171.4		-6.8				85.4	-258.7	4.9
c-C_4F_8	{R9}	2	57.1	7.9	-4.9	1.0	-296.6	62.2			
		3	57.2		-4.6				264.4	-328.8	-2.6
		4	61.1		-4.8				311.8	-365.7	12.0
		5	53.1		-3.9				296.6	-342.3	11.4
	{R10}	2	57.1		-4.6						
	{R11}	2	57.2		-4.7						
		3	60.1		-4.7				282.3	-338.8	8.4
		4	55.1		-4.1				296.6	344.3	11.5
c-C_5F_{10}	{R9}	2	-4.3	9.9	-2.3	1.3	-450.4	-1.9			
		3	-4.3		-1.9				410.4	-411.0	-3.0
		4	0.6		-2.2				469.6	-457.1	15.2
		5	-9.3		-1.1				450.5	-427.8	14.4
	{R11}	2	-4.3		-2.0						
		3	-0.4		-2.1				440.1	-430.2	11.6
		4	-7.4		-1.3				450.5	-429.9	14.6
c-C_6F_{12}[a]	{R9}	2	-21.8	11.9	-1.4	1.5	-558.6	-20.2			
		3	-21.7		-0.9				510.4	-493.2	-3.6
		4	-15.9		-1.2				581.4	-548.5	18.3
		5	-27.8		0.1				558.6	-513.4	17.3
	{R11}	2	-21.7		-1.0						
		3	-16.8		-1.2				551.9	-521.6	14.7
		4	-25.8		-0.1				558.6	-515.4	17.5
c-C_6F_{12}[b]	{R9}	2	-4.6	11.9	-1.4	1.5	-539.9	-3.1			
		3	-4.5		-0.9				491.7	-493.2	-5.1
		4	1.3		-1.2				562.7	-548.5	16.8
		5	-10.6		0.1				539.9	-513.4	15.8
	{R11}	2	-4.5		-1.0						
		3	0.3		-1.2				533.2	-521.6	13.1
		4	-8.6		-0.1				539.9	-515.4	15.9
c-C_7F_{14}	{R9}	2	16.3	13.9	-0.8	1.8	-605.4	17.3			
		3	16.4		-0.2				549.1	-575.4	-9.6
		4	23.2		-0.6				632.0	-639.9	15.9
		5	9.3		1.0				605.4	-599.0	14.8
	{R11}	2	16.4		-0.3						
		3	22.2		-0.5				602.5	-613.1	12.2
		4	11.3		0.8				605.4	-601.0	14.9

[a]Chair conformer. [b]Twist conformer. [c]SE^Q and SE^f values do not dependent on the process and are common for each compound.

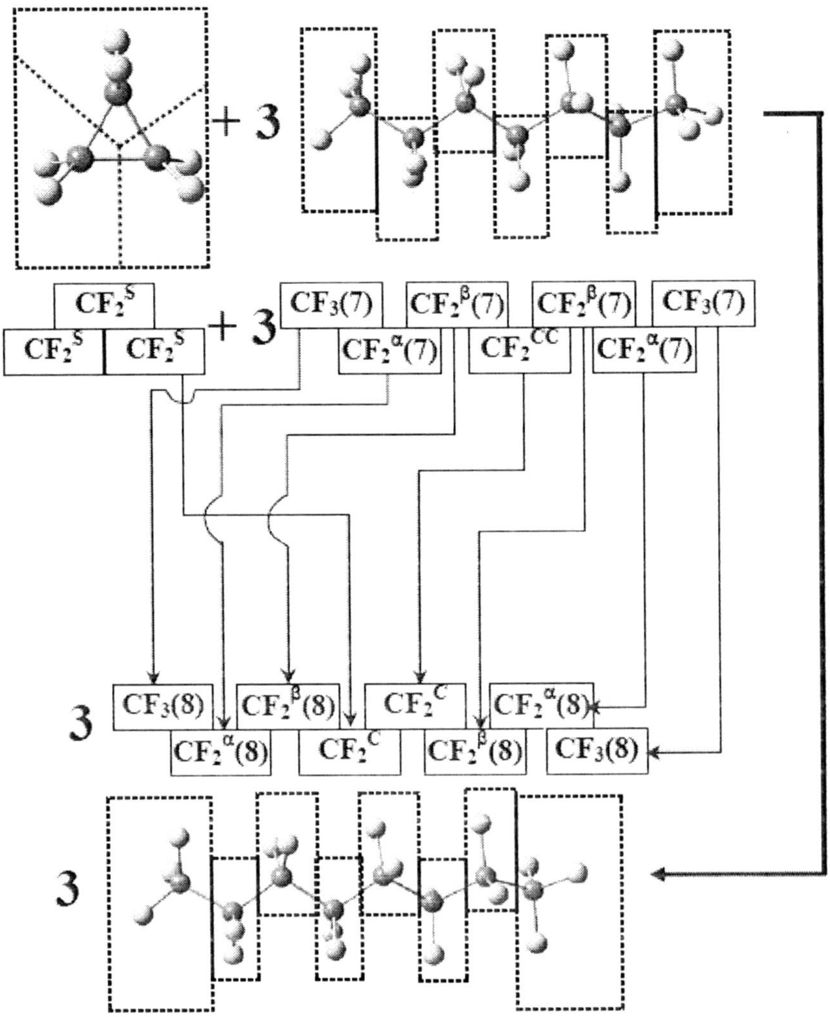

3 [$CF_2^S \rightarrow CF_2^C$]	3·(28.5)	ΔE^{RO} = 85.4
3 [$CF_2^{CC} \rightarrow CF_2^C$]	3·(-3.8)	ΔE^{NC} = -256.7
6 [CF_2^{α} (C_7F_{14})→ CF_2^{α}(C_8F_{16})]	6·(-10.2)	
6 [CF_2^{β} (C_7F_{14})→ CF_2^{β}(C_8F_{16})]	6·(-6.0)	
6 [CF_3 (C_7F_{14})→ CF_3 (C_8F_{16})]	6·(-24.8)	

Figure 17. Scheme indicating the transformation of molecular groups experienced within homodesmotic process {2} for perfluorocyclopropane with m=5 and which of them belong to ΔE^{RO} and ΔE^{NC}. The energy variation associated to each change is shown in parenthesis (in kJ mol^{-1}).

The values obtained for SE^{QTAIM} are negative for all the cycles here studied (Table 6), as the specific CF_2 cyclic groups are always more stabilized than the near transferable linear one (Table 4). The larger the cycle the more negative SE^{QTAIM} is. This does not keep in line with the evolution sequence of SE_r. Nevertheless, when SE^{QTAIM} values are divided by the number of CF_2 groups in the molecule we obtain the largest stabilization per CF_2 groups corresponds to the chair conformer of *c*-C_6F_{12} (-93.1 kJ mol^{-1}) that is close to the average found in *c*-C_5F_{10} (-90.1 kJ mol^{-1}) and *c*-C_7F_{14} (-86.5 kJ mol^{-1}). Finally, the stabilization gained by a CF_2 group in *c*-C_4F_8 (-74.1 kJ mol^{-1}) is much larger than in *c*-C_3F_6 (-28.5 kJ mol^{-1}).

Looking at the atomic components of CF_2 stabilization in the cycles and at the differences between the atomic populations in the cycle and the nearly transferable linear CF_2 (Table 4), we observe that: i) all the carbon atoms gain stability in the cycles (C6 in *c*-C_7F_{14} being the only exception); ii) Most of the fluorine atoms of *c*-C_nF_{2n} compounds with n>4 are more stabilized than in the nearly transferable linear CF_2, and they lose electron population that is gained by the carbons; iii) Perfluorocyclobutane presents a specific behavior with pseudoequatorial fluorines stabilized more than the destabilization displayed by axial fluorines. In this case the carbon atoms lose electron population (with regard to the CF_2^C fragment) gained by the fluorines (more by those in pseudoequatorial disposition); iv) CF_2 groups of perfluorocyclopropane display also specific characteristics, as their atomic electron populations coincide with those of the CF_2^C fragment. In contrast, carbons result stabilized whereas fluorines are destabilized giving rise to the lowest value of ΔE^{RO} in the series.

4.3. Theoretical Approach to the Relations between Diverse SEs

Strain energies, SE, are computed using a homodesmotic, diagonal, or other chemical process, SE_r, using equation {7}, where ν_i^r refers to the stoichiometric coefficients of reactant i in process r (positive for products and negative for reagents) and E_i^{el} and E_i^{vib} denote, respectively, the electronic (including nuclear-nuclear repulsion) and vibrational molecular energy of reactant i. According to QTAIM the electronic molecular energy can be obtained by adding the atomic energies, $E_i(\Omega)$. Because of the numerical procedure employed to compute $E_i(\Omega)$ values, their summation differs from the electronic molecular energy in term ε_i, usually employed as one of the parameters that estimate the quality assignable to the QTAIM computed atomic properties, and calculated by equation {8}. Replacing E_i^{el} in {7} we obtain {9}. {9} becomes {13} when we introduce: the vibrational contribution to strain energy, SE_r^{vib} (defined by {10}); the total error affecting the computation of SE_r with atomic energies, ε_r, (defined by {11}); and the homodesmotic strain energy computed with QTAIM atomic energies, SE_r^{QTAIM}, (defined by {12}).

$$SE_H = - \Sigma_i \, \nu_i^r \, (E_i^{vib} + E_i^{el}) \qquad \{7\}$$

$$\varepsilon_i = [\Sigma_\Omega \, E_i(\Omega)] - E_i^{el} \qquad \{8\}$$

$$SE_r = - \Sigma_i \, \nu_i^r \, \{[\Sigma_\Omega \, E_i(\Omega)] + E_i^{vib} - \varepsilon_i\} \qquad \{9\}$$

$$SE_r^{vib} = - \Sigma_i \, \nu_i^r \, E_i^{vib} \qquad \{10\}$$

$$\varepsilon_r = \Sigma_i \, \nu_i^r \, \varepsilon_i \qquad \{11\}$$

$$SE_r^{QTAIM} = - \Sigma_i \, \nu_i^r \, \Sigma_\Omega \, E_i(\Omega) \qquad \{12\}$$

$$SE_r = SE_r^{vib} + SE_r^{QTAIM} + \varepsilon_r \qquad \{13\}$$

The SE_r^Q term can be split considering different criteria. One of them is distinguishing between atoms involved (Ω_{RO}) and not involved (Ω_{NC}) in ring opening as indicates equation {14}. This allows to define the ring opening energy, ΔE_r^{RO}, and the term due to other non-cyclic groups, ΔE_r^{NC}, involved in the homodesmotic process (equations {15} and {16} respectively). Using these terms equation {13} transforms into {17}.

$$SE_r^{QTAIM} = - \Sigma_i \nu_i^r [\Sigma_{\Omega RO} E_i(\Omega_{RO}) + \Sigma_{\Omega NC} E_i(\Omega_{NC})] \quad \{14\}$$

$$\Delta E_r^{RO} = \Sigma_i \nu_i^r \Sigma_{\Omega RO} E_i(\Omega_{RO}) \quad \{15\}$$

$$\Delta E_r^{NC} = \Sigma_i \nu_i^r \Sigma_{\Omega NC} E_i(\Omega_{NC}) \quad \{16\}$$

$$SE_r = SE_r^{vib} + \varepsilon_r - [\Delta E_r^{RO} + \Delta E_r^{NC}] \quad \{17\}$$

Depending on the homodesmotic process, ΔE_r^{RO} defined by {15} may contain in the products atoms that belong to the nearly transferable linear groups, Ω_{RO}^t, or to specific linear groups, Ω_{RO}^s. So, in general, we can write {18}, where the first summation extends to products, p, and the second to reagents, q. As the only reagent involved in ring opening is the cycle, no atom included in the second summation belongs to a nearly transferable linear group. Also, the summation can be reduced to the atoms in the cycle, Ω_c, and ν_q^r is -1. Considering a certain homodesmotic process, HT, where all the Ω_{RO} atoms of the products belong to nearly transferable linear groups, ΔE_r^{RO} coincides with the definition given for a SE computed using nearly transferable linear groups as the strainless reference, SE^{QTAIM}, as is written in equation {19}. The energy of a specific atom in a linear compound can be related to that of the corresponding nearly transferable atom by equation {20}. Then the ΔE_r^{RO} term of any process can be related to SE^{QTAIMt} by equation {21}.

$$\Delta E_r^{RO} = \Sigma_p \nu_p^r [\Sigma_{\Omega ROt} E_p(\Omega_{RO}^t) + \Sigma_{\Omega ROs} E_p(\Omega_{RO}^s)] - \Sigma_q |\nu_q^r| \Sigma_{\Omega ROs} E_q(\Omega_{RO}^s)] \quad \{18\}$$

$$\Delta E_{HT}^{RO} = [\Sigma_p \nu_p^r \Sigma_{\Omega ROt} E_p(\Omega_{RO}^t)] - [\Sigma_{\Omega c} E(\Omega_c^s)] = - SE^{QTAIM} \quad \{19\}$$

$$E_p(\Omega_{RO}^s) = E_p(\Omega_{RO}^t) + \delta E_p(\Omega_{RO}) \quad \{20\}$$

$$\Delta E_H^{RO} = -SE^{QTAIM} + [\Sigma_p \nu_p^r \Sigma_{\Omega ROs} \delta E_p(\Omega_{RO})] \quad \{21\}$$

The second term on the right hand side of {21} will be named in what follows as non-transferability contribution to ring opening energy, ΔE_r^{RONT}. This term is different for each process and may be positive or negative depending on the groups considered. Thus, {17} becomes {22}.

$$SE_r = SE_r^{vib} + \varepsilon_r + SE^{QTAIM} - [\Delta E_r^{RONT} + \Delta E_r^{NC}] \quad \{22\}$$

Finally, we should look for a relationship between SE_r and the strain energy obtained using fitted group energies, SE^f. The term SE^f is defined as the difference between the energy of the cycle, E_c, and the summation of the fitted energies of the corresponding groups resulting from ring opening, E_G^f. It is given by equation {23}, where n_G is the number of G groups in a cycle. In this case, the balance of molecular electronic energies included in {7} is given by {24}, where ε_f represents the error introduced by the fitted value in this calculation. Therefore, {25} confirms that SE_r and SE^f only differ by small terms and should be very similar as shown in Table 5. In fact, the largest difference between SE^f and SE_r is 10.5 kJ mol^{-1} (process {R9} with m=5 for c-C_3F_6), whereas it does not exceed 4.3 kJ mol^{-1} for the chair conformer of c-C_6F_{12}.

$$SE^f = E_c - \Sigma_G\, n_G\, E_G^f \qquad \{23\}$$

$$\Sigma_i\, \nu_i^r\, E_i^{el} = SE^f + \varepsilon_f \qquad \{24\}$$

$$SE_r = SE^f + SE_r^{vib} - \varepsilon_f \qquad \{25\}$$

As commented in a previous section, group energies obtained by fitting equations like {6} correspond neither to specific nor to nearly transferable groups. They are a certain average of the diverse group energies in the series of molecules. Comparing the equations obtained by inserting {23} in {25} and {19} in {22}, and taking into account {8} and {11}, we get {26}. The summation in the first term on the right hand side represents the nearly transferable energy of the groups resulting from ring opening, $\Sigma_G\, n_G\, E_G^t$. Thus, we finally get {27}, which indicates the average character of fitted group energies and how they include the main causes (energy differences due to transformation of non cyclic reagents, non-transferability of groups obtained by ring opening) that make SE^{QTAIMt} and SE_r so different. The average character is even more clear considering molecules, like perfluorocycloalkanes, where all the linear fragments obtained by ring opening are equivalent and we can write {28}.

$$\Sigma_G\, n_G\, E_G^f = [\Sigma_p\, \nu_p^r\, \Sigma_{\Omega ROt}\, E_p(\Omega_{RO}^t)] + \Delta E_r^{RONT} - \Delta E_r^{NC} - \varepsilon_f - \Sigma_{i\neq c}\, \varepsilon_i \qquad \{26\}$$

$$\Sigma_G\, n_G\, E_G^f = [\Sigma_G\, n_G\, E_G^t] + \Delta E_r^{RONT} - \Delta E_r^{NC} - \varepsilon_f - \Sigma_{i\neq c}\, \varepsilon_i \qquad \{27\}$$

$$E_G^f = E_G^t + [(\Delta E_r^{RONT} - \Delta E_r^{NC} - \varepsilon_f - \Sigma_{i\neq c}\, \varepsilon_i)/n_G] \qquad \{28\}$$

Homodesmotic strain energies SE_r include an energy term, ΔE_r^{NC}, that QTAIM atomic energies show as due to the balance between non-cyclic fragments and that should not be present in SE. Moreover, this term varies depending on the particular homodesmotic process selected to compute SE_r and gives rise to the multiplicity of SE_r values that can be obtained for a same cycle. For example, six different values of SE_r (ranging from 9.3 to 23.2 kJ mol^{-1}) are obtained in this work for the seven homodesmotic processes considered for c-C_7F_{14}. In contrast, SE computed by using the exclusively energy of the cyclic molecule and the energy of nearly transferable linear fragments, SE^{QTAIM}, provides a unique value for each cycle that only contains energy due to ring opening. The computation of SE replacing the energy of the nearly transferable linear fragment with the energy obtained by fitting the molecular energies to the number of CF_2 groups, SE^f, introduces an averaged ΔE_r^{NC} term as a mean of different groups included in the fitting. Thus, SE^f value is unique for each cycle but includes energy that is not due to ring opening. It has to be remarked that SE^{QTAIM} values span a different range and display different trends to those computed with homodemostic, diagonal, or other processes. So, on first view SE^{QTAIM} values can be considered surprising in spite of their physical meaning.

5. Anomeric Effect

The anomeric effect has been defined as the preference for a gauche arrangement of the dihedral angles centered in the R-Y-C-Z fragment, where Y is an atom bearing at least one lone pair of electrons and Z is an electronegative atom [13-16]. Electrostatic interactions between bond dipoles were early invoked to explain this effect [74,75]. Nevertheless, this framework, called the electrostatic model, was unable to rationalize the anomeric effect in many compounds [76,77]. A different explanation (referred as hyperconjugative model) considers the stabilization arises from the delocalization of one lone pair into the anti-bonding C-Z orbital that takes place when the *lp*-Y-C-Z unit adopts an antiperiplanar arrangement [17], what is frequently represented by the resonance structure shown in figure 18. However, NBO [78-81] studies have shown that other orbital and bond-antibond interactions, as well as steric and electrostatic effects have to be taken into account, with relative importance that depends on the system under consideration [18-20,76]. Nevertheless, the NBO picture of the anomeric effect is practically the same of the hyperconjugative model for the most well known anomeric units, like OCO, NCN, or FCO, which are those we have currently studied [21,82].

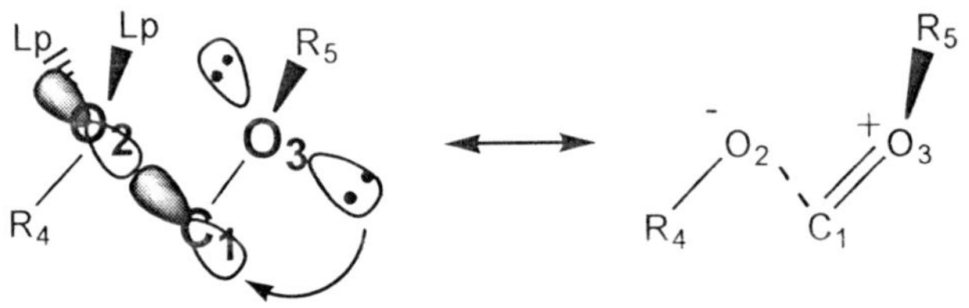

Figure 18. Resonance structure representing $\sigma_n \rightarrow \sigma^*_{C-O}$ electron transference.

For the sake of concision, we will focus in one molecule, methanediol [21], and computational level, B3LYP, Results obtained at other levels for diverse molecules follow the same trends [21,82]. The anomeric effect of this molecule is revealed when comparing the relative energies of the ***gg*** conformer (with two gauche units) and the ***tt*** with two antiperiplanar units (figure 19).

It has to be noticed that a previous QTAIM study on the origin of the anomeric effect was carried out [83]. The results obtained were not keeping in line with the view that anomeric effect derives from a differential interaction of a lone pair with an antiperiplanar CH_2-O bond. Apart from using higher computational levels and considering more compounds, the current study differs from that of Werstiuk *et al.* [83] by calculating deloclaization indices, DIs, and by a discussion more based upon integrated properties rather than upon bond properties and singular points of the laplacian of the electron density. Although Werstiuk et *al.* had already noted the differences in the atomic electron population and energies of conformers of dimethoxymethane, our discussion, combining these differences with geometrical distortions, lead us to propose a new interpretation of the anomeric effect.

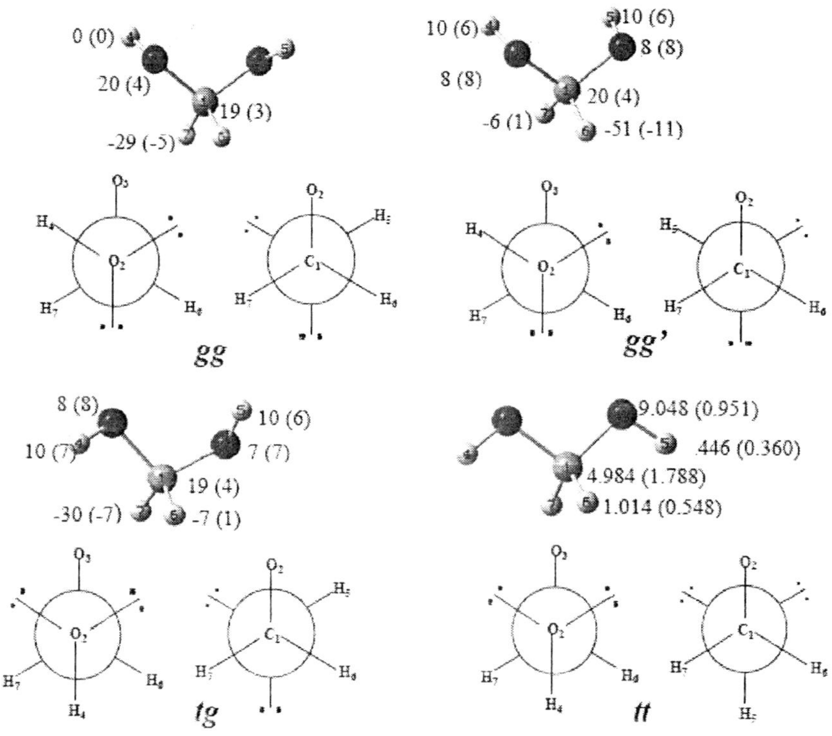

Figure 19. Structure and atom numbering of the conformers of methanediol. Absolute N(Ω) values (for ***tt***) in au. Relative atomic electron populations, ΔN(Ω), (for ***tg***, ***gg***, and ***gg'***) in au and multiplied by 10^3. Electron populations due to delocalizations over other atoms, $N^d(\Omega)$, are shown in parenthesis (absolute values for ***tt*** in au, relative values for the remaining forms multiplied by 10^3). All values as obtained from B3LYP/6-311++G(2d,2p) calculations. ***tt***, ***gg*** and ***gg'*** B3LYP/6-311++G(2d,2p) full optimized geometry display, respectively C_{2v}, C_2, and C_S symmetries with O_3-C_1-O_2-H_4 dihedral angles of 180, 62.5, and 79.4 degrees. ***tg*** was not found as a conformer at this level, the structure (with C_1 symmetry) here presented is the result of the B3LYP/6-311++G(2d,2p) restricted optimization with O_3-C_1-O_2-H_4 and H_5-O_3-C_1-O_2 dihedral angles fixed at the values in the HF/6-311++G(2d,2p) optimized conformer: -165.7° and 53.8° respectively.

Some results obtained with QTAIM in our study (Table 7) are not in line with the hyperconjugative model. Thus, in ***gg***, autolocalization, λ(O), and electron population, N(O), of oxygens are larger than in ***tt***, whereas the delocalization between both oxygens is depleted in ***gg*** with regard to ***tt***.

The relative atomic electron populations of ***gg***, suggest that an electron transfer from the central hydrogens to weighty atoms accompanies the conformational change from ***tt*** (figure 20). This electron population variation induces variations of atomic energies that describe the gauche stabilization as due to the increased stabilization gained by the central carbon (figure 20). We can say that the electron transference from hydrogens to better attractors (that is atoms with larger nuclear charge) stabilizes the conformer.

Table 7. Main relative (with regard to conformer *tt*) properties of the *gg* conformer of methanediol

	ΔE kJ mol^{-1}	$\Delta\theta_{O\text{-}C\text{-}O}$ degrees	ΔN(O) au	ΔN(C) au	ΔN(H) au	Δλ(O) au	Δδ(O,O) au
B3LYP	-32.0	8.5	0.019	0.018	-0.028	0.002	-0.006
HF	-29.7	7.1	0.019	0.019	-0.032	0.011	-0.001
MP2	-32.4	8.3	0.010	0.036	-0.029	0.003	-0.007

6-311++G(2d,2p) 6d basis set used with the three computational levels.

Which is the origin of the electron transference? We believe it is the increase experienced in the ***gg*** conformer by the repulsions with oxygen lone pairs. Why this? Well, if we look at Newman projections, we see that every one of the two central hydrogens goes from 2 to 3 gauche interactions with oxygens lone pairs. And consequentely, from 2 to 1 dihydrogen gauche interactions (figure 21). So, basically, we are replacing one gauche interaction with a hydrogen by one with a lone pair. Representing the lone pairs by the maxima of non bonded charge concentration (NBCC) around the oxygen we observe the distance between both centers of charge diminishes in GG. Also taking the value of the electron density as a rough estimation of the charges involved in the repulsion, we observe the electron density at the lone pair is substantially larger than at hydrogen nuclei (Table 8).

Table 8. Relevant indicators for the main repulsive energies in the diverse conformers of methanediol as computed from B3LYP/6-311++G(2d,2p) optimized electron density. Distances, R, in Å and charge densities, ρ, in a.u

	tt			*tg*			*gg*		
	$R_{Hc\text{-}Ht}$	ρ_{Hc}	ρ_{Ht}	$R_{H\text{-}NBCC}$[c]	ρ_H	ρ_{NBCC}[d]	$R_{H\text{-}NBCC}$[c]	ρ_H	ρ_{NBCC}[d]
a	2.323	0.4584	0.4328	2.020	0.4668	0.9511	2.085	0.4588	0.9516
b				2.228	0.4501	0.9537	2.396	0.4322	0.9562

a and **b** refer, respectively, to an interaction between an oxygen NBCC and a central hydrogen (Hc) and a terminal hydrogen (Ht). [c]Distance to the NBCC are computed as distances to the maximum of $-\nabla^2\rho(\mathbf{r})$ that represents the oxygen lone pair involved in the interaction. [d]Electron density computed at that point.

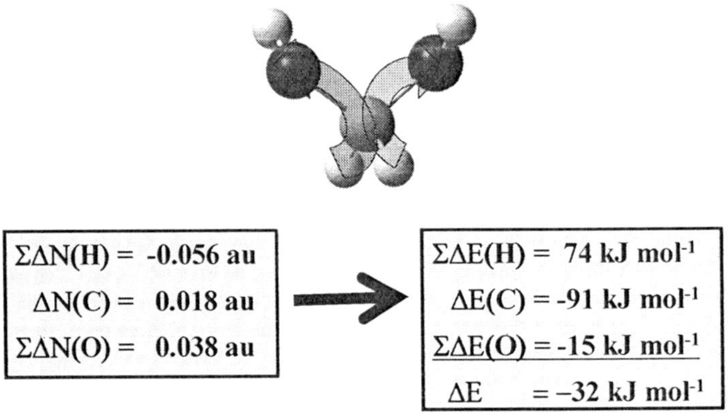

Figure 20. Electron transference from hydrogens of central CH_2 group to C and O atoms resulting from comparing ***gg*** and ***tt*** atomic properties. The relative energy of ***gg*** with regard to ***tt*** can be explained as the balance among hydrogens destabilization and C and O estabilization.

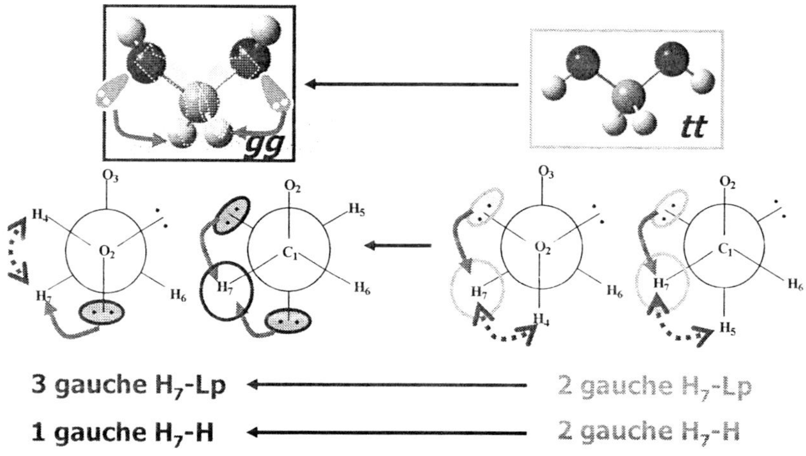

Figure 21. Gauche interactions in ***gg*** and ***tt*** conformers of methanediol.

Looking for more definite evidences, we computed the Coulomb-exchange and core attraction interactions between every pair of basins. These are really expensive calculations as they involve integrations over two atomic basins. The results seemed to be discouraging because we got a reduction of the O-H Coulomb-exchange interactions in ***gg*** with regard to ***tt***, both for those including two gauche lone pairs (-210.0 kJ mol^{-1}) and those including one (-40.1 kJ mol^{-1}) (Table 9). Nevertheless, this is not the calculation we need, because we are computing the interaction with the final electron density, which has been reduced. Therefore, what we need is to compare the repulsion experienced by the same amount of electron density in the central hydrogens, this can be done averaging the potential energy function (Table 10).

Table 9. Relative values[a] of total core attractions $E^A(\Omega)$, sum of Coulomb and exchange interactions, $E^{Q+EX}(\Omega,\Omega')$, and internuclear repulsions, $E^{NR}_{n,n'}$ experienced by each atomic basin, Ω, in the *gg* conformer of methanediol with regard to those in the *tt* one. All values in kJ mol^{-1}

Ω	C1	O2	O3	H4	H5	H6	H7
$\Delta E^A(\Omega)$	-1026.2	402.7	402.7	-271.4	-271.4	615.2	615.2
$\Delta E^{Q+EX}(\Omega,C1)$	222.5						
$\Delta E^{Q+EX}(\Omega,O2)$	357.7	195.3					
$\Delta E^{Q+EX}(\Omega,O3)$	357.7	-1378.9	195.3				
$\Delta E^{Q+EX}(\Omega,H4)$	-17.6	-14.3	359.5	-1.2			
$\Delta E^{Q+EX}(\Omega,H5)$	-17.6	359.5	-14.3	16.5	-1.2		
$\Delta E^{Q+EX}(\Omega,H6)$	-121.9	-210	-40.1	-6.8	-52.7	-34.9	
$\Delta E^{Q+EX}(\Omega,H7)$	-121.9	-40.1	-210	-52.7	-6.8	-48.2	-34.9
$\Delta E^{NR}_{O2,n(\Omega)}$ [b]	-81.9						
$\Delta E^{NR}_{O3,n(\Omega)}$	-81.9	-1813.6					
$\Delta E^{NR}_{H4,n(\Omega)}$	3.0	-19.7	756.9				
$\Delta E^{NR}_{H5,n(\Omega)}$	3.0	756.9	-19.7	123.4			
$\Delta E^{NR}_{H6,n(\Omega)}$	53.4	-21.4	170.0	3.5	-103.5		
$\Delta E^{NR}_{H7,n(\Omega)}$	53.4	170.0	-21.4	-103.5	3.5	3.5	
$\Delta E(\Omega)$	-91.3	-7.0	-7.0	3.5	3.5	33.6	33.6

[a]Values computed using HF/6-311++G(2d,2p) electron densities. [b]n(Ω) represents the nucleus of Ω electron basin.

Table 10. Relative electrostatic potential,[a] V(Ω), averaged for each atomic basin, Ω, in *gg* with regard to those in *tt* with its terms: core attraction potential with every nucleus n, $V^A_n(\Omega)$, Coulomb, $V^Q(\Omega,\Omega')$, and exchange, $V^{EX}(\Omega,\Omega')$, potentials with Ω' basin, All values in au multiplied by 10^3. Potential terms due to the interactions within the same basin and with the corresponding nucleus are not listed

Ω	C1	O2	O3	H4	H5	H6	H7
$\Delta V(\Omega)$	-18.9	-12.1	-12.1	5.3	5.3	10.2	10.2
$\Delta V^A_{C1}(\Omega)$		4.4	4.4	-1.3	-1.3	-15.9	-15.9
$\Delta V^A_{O2}(\Omega)$	-2.3		71.3	1.9	-315.3	6.6	-64.8
$\Delta V^A_{O3}(\Omega)$	-2.3	71.3		-315.3	1.9	-64.8	6.6
$\Delta V^A_{H4}(\Omega)$	4.2	1.2	-36.8		-46.5	-1.3	40.3
$\Delta V^A_{H5}(\Omega)$	4.2	-36.8	1.2	-46.5		40.3	-1.3
$\Delta V^A_{H6}(\Omega)$	-2.9	1.1	-6.0	-1.5	41.7		-0.7
$\Delta V^A_{H7}(\Omega)$	-2.9	-6.0	1.1	41.7	-1.5	-0.7	
$\Sigma_n \Delta V^A_n(\Omega)$	-1.9	35.3	35.3	-321.0	-321.0	-35.8	-35.8
$\Delta V^Q(\Omega,C1)+\Delta V^{EX}(\Omega,C1)$		11.3	11.3	-11.9	-11.9	14.3	14.3
$\Delta V^Q(\Omega,O2)+\Delta V^{EX}(\Omega,O2)$	6.5		-61.1	5	374.2	-3	60.1
$\Delta V^Q(\Omega,O3)+\Delta V^{EX}(\Omega,O3)$	6.5	-61.1		374.2	5	60.1	-3
$\Delta V^Q(\Omega,H4)+\Delta V^{EX}(\Omega,H4)$	-2.1	-0.8	14.6		17	0.1	-17
$\Delta V^Q(\Omega,H5)+\Delta V^{EX}(\Omega,H5)$	-2.1	14.6	-0.8	17		-17	0.1
$\Delta V^Q(\Omega,H6)+\Delta V^{EX}(\Omega,H6)$	-13	-9.2	-2.1	-5.8	-52.4		-8.6
$\Delta V^Q(\Omega,H7)+\Delta V^{EX}(\Omega,H7)$	-13	-2.1	-9.2	-52.4	-5.8	-8.6	
$\Sigma_{\Omega'} [\Delta V^Q(\Omega,\Omega')+\Delta V^{EX}(\Omega,\Omega')]$	-17.0	-47.4	-47.4	326.3	326.3	46.0	46.0

[a]Values computed using HF/6-311++G(2d,2p) electron densities.

When this was done, we got that, in fact, the Coulomb-exchange potential originated by the oxygen with two gauche interactions over the central hydrogen in ***gg*** was larger than that in ***tt*** (58.8 10^{-3} au), and larger enough to compensate the depletion of the Coulomb-exchange potential due to the other atom (-46.1 10^{-3} au). This is a direct evidence for the steric origin of the electron density transfer that produces the so called anomeric effect.

Looking at the ***tg*** conformer, where one OCOH unit is antiperiplanar and the other is gauche, we see H7 presents two lone pair gauche interactions, whereas H6 presents 2. Consequentely with the explanation for ***gg***, H7 loses more electron population than H6 (figure 22). This electron population is transferred mainly to better attractors and, contrary to what could be expected according to the hyperconjugative model; both oxygens gain the same amount of electron density. This, again, supports our explanation of the anomeric effect.

Finally the conformation ***gg'*** is a limit case where H6 displays four gauche interactions with lone pairs, and H7 only two. The variations of electron population reflect this fact and once more the electron density is gained by the best attractors and even a little part by hydroxylic hydrogens (figure 23).

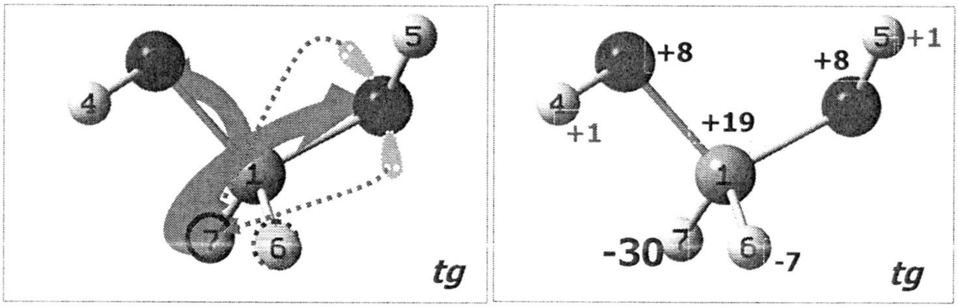

Figure 22. Lone pair-hydrogen interactions in the *tg* conformer of methanediol (dotted lines), relative electron transferences and relative atomic electron populations (with regard to *tt* conformer, in au, and multiplied by 10^3).

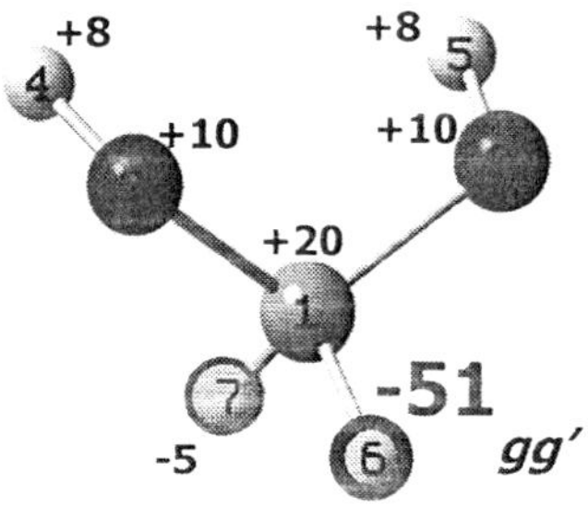

Figure 23. Relative atomic electron population in the gg' conformer of methanediol (with regard to the tt conformer, in au, and multiplied by 10^3).

The evolution of the electron density along the conformational change of fluoromethanol [82] is a further evidence for our interpretation. When going from the anti to the gauche conformer H5 experience more gauche interactions, whereas H6 presents the same number (figure 24). Consequentely the electron population at H5 is reduced whereas that of H5 remains practically unchanged (figure 24). The population of the fluorine atom experiences very small changes as its electron density is retained by a large nuclear charge. Carbon and oxygen populations increase substantially.

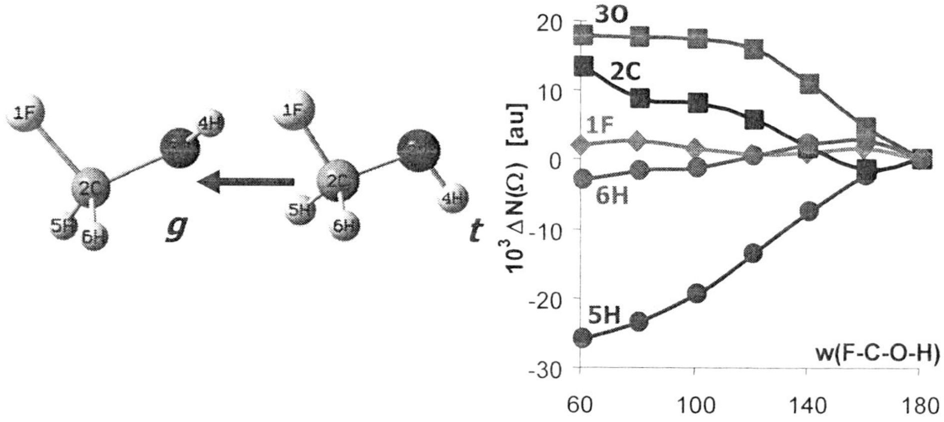

Figure 24. Evolution of atomic electron populations during the internal rotation in fluoromethanol. Values computed with B3LYP/6-311++G(2d,2p) 6d electron densities.

6. Hydrogen Bonding

Traditional ideas about hydrogen bonding (HB) include assuming its presence in the most stable conformers of molecules that contain 1,2-diol moieties and considering the R-O···H-X hydrogen bonding is formed by transferring electron density of the oxygen lone pair (HB acceptor) to the hydrogen (HB donor). Both points can be examined in the light of QTAIM. The first point is related clearly to intramolecular hydrogen bonding (IHB). In contrast, the second one should be studied better in adducts displaying intermolecular hydrogen bonding, like methoxymethane dimers [84], or the adducts of water and O-containing compounds [85].

6.1. IHB Bondpaths in 1,2-Diols

Frequency shifts observed in IR and NMR studies, abnormally long O-H bond lengths (experimental or calculated) and interatomic distances, or stabilization of gauche rotamers have been employed as "evidences" of IHB [86]. Nevertheless, all these facts could also be explained as the result of diverse nonbonding interactions. On the basis of these facts, IHB and a cyclic five-membered ring structure have been assumed elsewhere in the chemical literature to explain diverse features displayed by molecules containing the 1,2-diol structural unit.

The presence of IHB in 1,2-ethanediol was investigated in several structural studies (experimentally [87] and at diverse computational levels [88-97]). The stability sequence of 1,2-ethanediol conformers has been reasoned assuming (from geometrical considerations) that one IHB takes place in two conformers [96-97], named tGg′ and gGg′ (figure 1) with the nomenclature used by Radom *et al.*[88]. Also, an assumed IHB in 1,2-dihydroxybenzene (catechol) has been considered crucial for the enhanced antioxidant activity of flavonoids [98-99]. These computational works claim the existence of IHB in catechol on the basis of geometrical considerations. The red shift displayed by the O-H stretching bands was also used as a proof of IHB in this molecule [100].

R.A. Klein, using QTAIM, has stated recently, that contrary to what had been assumed [101-104], IHB between adjacent hydroxyls does not occur either in glucopyranose [105] or in saturated 1,2-diols [106]. Almost at the same time, Rozas et al. examined, also using the QTAIM, the IHB in several benzene derivatives, including catechol. They concluded the existence of a BCP and, consequently, IHB in catechol [107]. Both Klein's and Rozas' conclusions about IHB presence are based on accepting that two atoms are bonded when there is a bondpath connecting them. This is one of the most controverted items of QTAIM (see section 2 and references 30-36).

Aiming to understand when vicinal OH groups present a BCP associable to an IHB we have re-examined with some detail molecules that contain the 1,2-diol moiety on sp^3-C and sp^2-C backbones: 1,2-ethanediol and catechol [22]. Thus, the geometry and the electron density of the ***tGg′*** and ***gGg′*** conformers of 1,2-ethanediol and those of the main conformer catechol were calculated employing B3LYP/6-311++G**, MP2/6-311++G**, and QCISD/6-311++G** levels of theory. The electron density obtained for restricted optimisations with the R distance (figure 25) fixed at diverse values, ranging from 0.2 Å longer than the conformer to 0.5 Å shorter, were also considered for each conformer of 1,2-ethanediol, as well as intermediate structures calculated to determine the R distance where the IHB is formed with a precision of ± 0.005 Å. The conformer and the results of ten restricted optimisations where the interatomic R distance or the ω dihedral angle (figure 25) was fixed were considered for catechol. Again, the calculation of intermediate structures allowed us to determine the R distance where the IHB is formed with a precision of ± 0.005 Å, and the ω dihedral angle within ± 2 degrees.

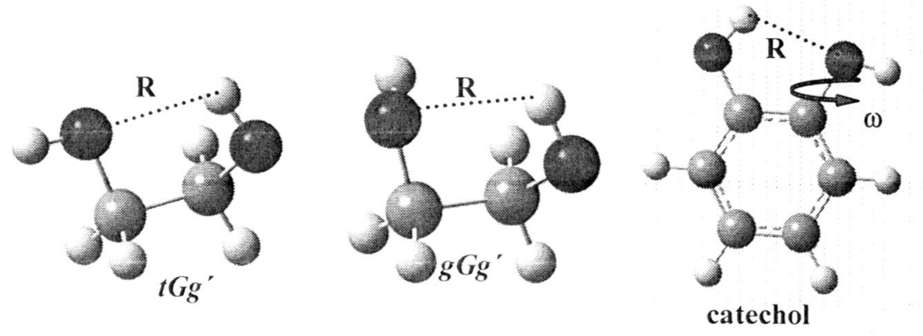

Figure 25. ***tGg'*** and ***gGg'*** conformers of 1,2-ethanediol, and most stable conformer of catechol, showing interatomic distance, R, and dihedral angle, ω, fixed in restricted optimizations.

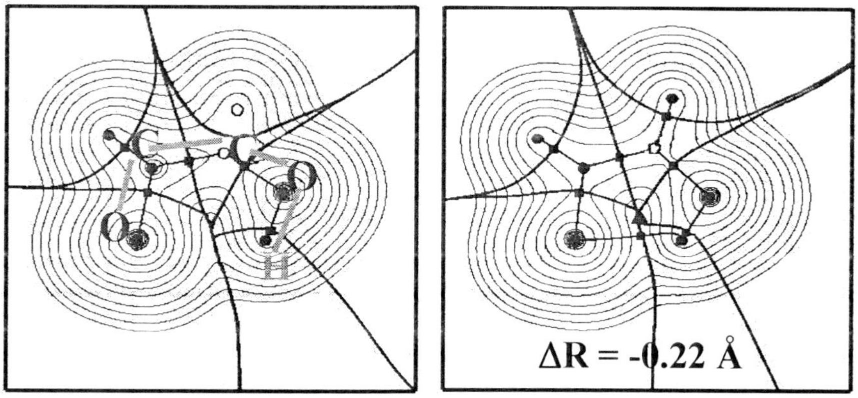

Figure 26. MORPHY [108,109] Plots of the QCISD/6-311++G** electron density in the plane containing IHB donor and acceptor atoms of ***tGg'*** conformer of 1,2-ethanediol (left), and at a ***tGg'*** distorted conformation (right) with R = 1.940 Å. It has to be noticed that the BCP is already observed for R = 2.171 Å. Squares denote BCPs, triangles RCPs, open face circles are projections of nuclei and bold face ones correspond to in-plane nuclei.

The QCISD/6-311++G** and the B3LYP/6-311G** electron densities of the ***tGg′*** and ***gGg′*** conformers of 1,2-ethanediol do not present any bond path connecting donor H and accepting O (figure 26 shows this fact for ***tGg'***). This confirms the conclusion obtained by Klein with several DFT levels about the lack of IHB in this compound [106]. Nevertheless, we observe that a RCP, a BCP, and its associated bond path connecting the H donor and the O acceptor appears when R is reduced, 0.220 Å in ***tGg'*** and 0.305 Å in ***gGg'*** (figure 26). This geometry modification increases the molecular energy in only 1.4 and 2.7 kJ mol^{-1}, respectively, according to the B3LYP calculations.

In 1,2-ethanediol, the electron population of the donor hydrogen, N(H), the first moment of its electron density, μ(H), its volume, v(H), and energy, E(H), vary monotonously when R lengthens, the three former quantities increase and the latter decreases (figure 27). These trends agree with the Koch and Popelier criteria for IHB presence [110]. However, it has to be observed that these quantities do not show any singularity at the BCP formation. Thus, it is impossible to establish when the hydrogen bond is established using them. There is also no singularity in the behaviour displayed by delocalisation indices at the BCP formation, and a monotonous fall is observed [22], as expected, when R lengthens.

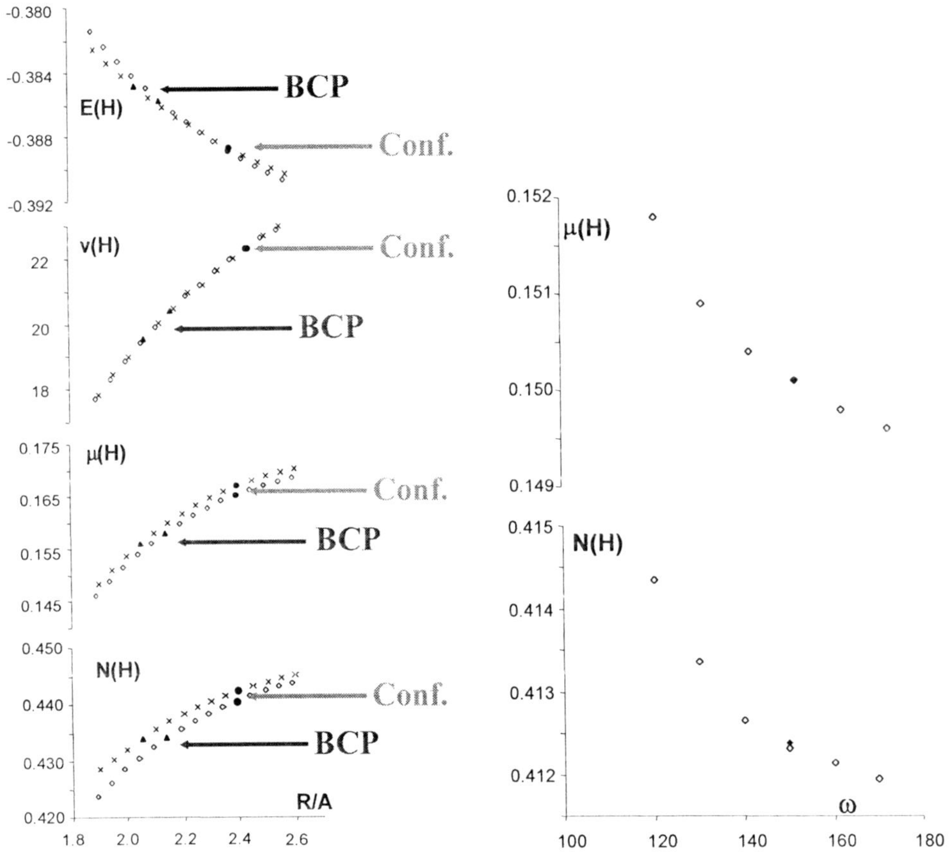

Figure 27. Evolution of N(H), μ(H), v(H), and E(H) with R (left). Crosses refer to ***gGg'*** and rhombuses to ***tGg'*** conformation of 1,2-ethanediol. Evolution of N(H) and μ(H) with ω for the conformer of catechol (right). All the values, but R and ω, in au. Notice that both representations are plotted with different scales. The optimised geometries are pointed out with a black dot and the geometries which correspond to the BCP formation are pointed out with a black triangle. All values in au but distances in Å and obtained from B3LYP/6-311++G** electron densities.

We started our study on 1,2-dihydroxybenzene repeating the QTAIM analysis of the B3LYP/6-31+G** electron density of the optimised conformer. There is really a BCP between H and O that are supposed to be connected by IHB, as Rozas et al. reported [107]. However, the corresponding bond path passes through the hydroxyl BCP of the hyptothetical H donor [22]. This means the optimised structure lays over the limit where the IHB vanishes.

In fact, the BCP disappears when the electron density is computed for geometries with slightly longer R distances, and it is observed again for slightly shorter R distances, although the corresponding bond path does not cross over the hydroxyl BCP any more. Optimising with diffuse functions on all atoms, B3LYP/6-311++G** we observe neither H···O BCP nor RCP for the conformer. The same trend is observed when more polarization functions are added (B3LYP/6-311++G(3d,3p) optimisation). Thus, we concluded [22] that the BCP observed by Rozas et al. for the conformer of catechol is due to the lack of diffuse functions over the hydrogens in the basis set used, what is inappropriate to study the IHB.

The study of integrated properties along the formation of IHB in 1,2-dihydroxybenzene was carried out in two different ways. The IHB can be formed by approaching H_9 and O_8, just as in 1,2-ethanediol, but it can be also obtained by rotating the ω dihedral angle between 180° and 120°, what approaches the lone pair of the acceptor atom closer to the donor atom, but keeping the distance between the two atoms nearly constant. We have observed that a BCP and a bond path interconnecting them appears when the interatomic distance is reduced from 2.165 to 2.115 Å (increasing the energy in 0.3 kJ mol^{-1}) or when the dihedral angle is rotated 36 degrees from its arrangement in the conformer (this increases the electronic molecular energy in 2.1 kJ mol^{-1}).

The results obtained for N(H), E(H), μ(H), and v(H) varying R follow the same trends observed in 1,2-ethanediol. On the contrary, the trends obtained for these properties varying ω are very different (figure 27). Thus, N(H) and μ(H) increase during the IHB formation. The variations experienced by E(H) and v(H) are very small and their representations versus ω are a non-significant dispersion of points. These results do not agree with the criteria reported by Koch and Popelier for the HB formation: N(H) and μ(H) should decrease with the IHB formation; nevertheless they only decrease when the IHB is formed by approaching the atoms. It has to be noticed that the variation of ω does not produce any significant variation in R (a rotation of 60° only reduces this distance in 0.009 Å). This could indicate that the previous reported trend for IHB (decrease in N(H) and μ(H)) [110] can be more related to the decrease of their interatomic distance than to IHB formation. Moreover, N(H) enhacement upon hydrogen bond formation was previously reported by Rozas et al. in HB complexes involving Li, Be, or B [111], what led them to coin the term "inverse hydrogen bond". Also, for a series of complexes of hydrides, we have observed that the electron population of hydrogen donor ehnaces whenever it is bonded to an atom that is much more electronegative than the hydrogen acceptor in the base (Table 10) [112].

In contrast with the lack of IHB associable BCPs in conformers of 1,2-diols (also in their radicals, like β-hydroxyethoxy radical [113]), they are present in the most stable conformers of 1,3-propanodiol (figure 28) and 1,4-butanediol [114]. In spite of this topological difference, the relative conformational energies of 1,2-ethanediol, 1,3-propanediol, and 1,4-propanediol can be split to a very good approximation into four different contributions: OH···O units (with or without BCP), $\Delta E_{OH\cdots O}$, 'gauche' stabilization of hydrogens and carbons due to the oxygen lone pairs, $\Delta E_{g(H)}$ and $\Delta E_{g(C)}$, and stabilizations due to the relative arrangement of hydroxyl and β-methylene groups, $\Delta E_{OH\cdots HC}$. An additional term has to be included when comparing conformers with different arrangement of the O-C-C-O unit in 1,2-ethanediol, ΔE'. $(n^C_1 - n^C_2)_{OH\cdots HC}$ represent the difference in the number of OH···HC interactions in conformation 1 (n^C_1) and 2 (n^C_2). Terms involving g(H) and g(C) in equation {29} have a similar meaning.

Table 10. Variation of the electron population (a.u. multiplied by 10^3) of the donor hydrogen, $\Delta N(H_d)$, of a series of hydrides adducts with the electronegativity difference between the heteroatom of the H donor molecule and that of the acceptor one, $\Delta\chi$, measured according to diverse scales (Allred, Pauling, and Pearson)

Adduct[a]	$\Delta N(H_d)$	$\Delta\chi$ (Allred)	$\Delta\chi$ (Pearson, eV)	$\Delta\chi$ (Pauling)
$H_2S...HF$	+22	1.7	4.19	1.40
$H_2O...HF$	-14	0.6	2.87	0.54
$H_2S...HCl$	-26	0.4	2.08	0.58
$H_2O...HCl$	-98	-0.7	0.76	-0.28
$H_3P...HF$	+28	2.0	4.79	1.79
$H_3N...HF$	+11	1.0	3.11	0.94
$H_3P...HCl$	-19	0.7	2.68	0.97
$H_3N...HCl$	-116	-0.3	1.00	0.12
HCl...HF	+18	1.3	2.11	0.82
HF...HF	-13	0.0	0.00	0.00
HCl...HCl	-17	0.0	0.00	0.00
HF...HCl	-14	-1.3	-2.11	-0.82
$H_2S...HOH$	-7	1.1	1.32	0.86
$H_2O...HOH$	-38	0.0	0.0	0.0
$H_2S...HSH$	-30	0.0	0.0	0.0
$H_2O...HSH$	-98	-1.1	-1.32	-0.86

[a] H_d is always contained in the second molecule.

$$\Delta E_{conf}^{C_1-C_2} \approx \Delta E_{OH..O} + (n^{C_1} - n^{C_2})_{OH..HC} \cdot \Delta E_{OH..HC} + (n^{C_1} - n^{C_2})_{g(H)} \cdot \Delta E_{g(H)} + (n^{C_1} - n^{C_2})_{g(C)} \cdot \Delta E_{g(C)} + \Delta E \quad \{29\}$$

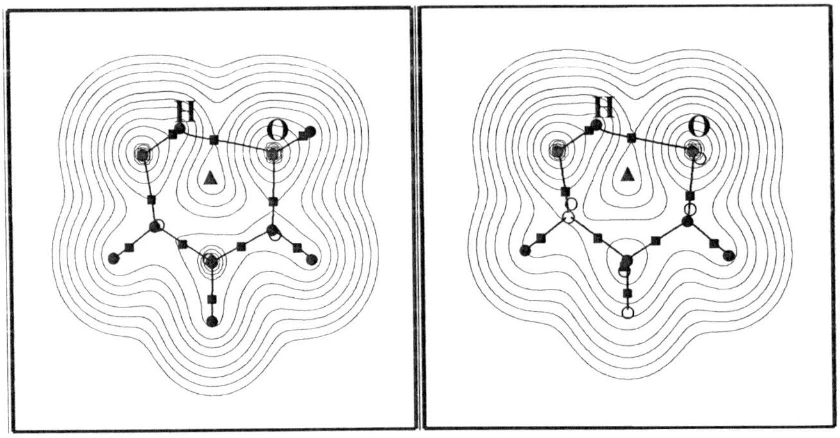

Figure 28. Morphy [108,109] plots of the B3LYP/6-311++G(2d,2p) electron density in the O-H···O plane for tGG'g (left) and gGG'g (right) conformers of 1,3-propanediol Bond critical points and ring critical points are represented by ■ and ▲, respectively.

This partitioning scheme allows to estimate and rationalize the energy due to the O-H···O interactions, $\Delta E_{OH..O}$, from the relative electronic energies of the conformers without neglecting important structural changes between them nor introducing arbitrary chosen reference conformations. The values of $\Delta E_{OH\cdots O}$ for IHB display a very good linear correlation with the $r(O\cdots H)^{-1}$ values. These linear correlations indicate a steeper decrease of $\Delta E_{OH\cdots O}$ when r(O···H) increases for the intramolecular case than for the intermolecular one, which can be attributed to ring-strain effects. Thus, the following IHB energies are obtained using this partitioning scheme: 24.2 kJ mol^{-1} (1,4-propanediol), 18.4 kJ mol^{-1} (1,3-propanediol), 9.2 kJ mol^{-1} (g'Gt and g'Gg conformers of 1,2-ethanediol), and 0.5 kJ mol^{-1} (g'Gg' conformer of 1,2-ethanediol) [114]. We observe that, although 1,2-propanediol IHB energies are less than those presented by diols with IHB BCPs, all of them display positive values.

In this context we also have to notice that BCPs associable to IHB are not found for the most stable conformers of oxalic acid at HF, B3LYP, and MP2 levels. Neverthelss, they appear applying slight distortions to the conformer geometry that do not give rise to any important modification of molecular energy or QTAIM atomic properties [115].

6.2. Electron Transferences in Intermolecular Hydrogen Bond

It is well known that hydrogen bonding has an eminent importance for the structure, function, and dynamics of chemical and biochemical systems. Nevertheless, and as this is an unfortunate rule affecting many fundamental chemical concepts, there is not a single definition of hydrogen bond. More than sixty years ago Linus Pauling stated "Under certain conditions an atom of hydrogen is attracted by rather strong forces to two atoms instead of only one, so that it may be considered to be acting as a bond between them. This is called a hydrogen bond" [116]. This definition is probably the only classical view of HB that has not been outdated by the whole palette of hydrogen bonds known nowadays [117]. However, most of textbooks still explain HB as due to the electron density deficit of H donor attached to a highly electronegative atom, which allows an approach of the electron density of one lone pair of the HB acceptor (usually O, F, or N).

This hypothesis can be tested directly using QTAIM electron density analysis. For instance, we can approach to monomers of neutral molecules until they form an stable dimer or adduct. When this is done for methoxymethane dimers [84], we observe that the atomic electron populations of the unsymmetric dimers (structures III and IV in figure 29) indicate there is an electron transference between the monomers that keeps not in line with the classical interpretation. Thus the electron density of the monomer containig the oxygen donor/s is increased (0.022 au in III and 0.012 au in IV) whereas the one containing H donor/s is reduced. It has to be remarked that O acceptor/s never experience the largest atomic electron population changes and that it is always positive (Table 11). In contrast, H donors lose electron density in the dimer with regard to the isolated monomer. In fact the electron population depletion of H donor basin is one of criteria established by Koch and Popelier for characterising HB [110].

The QTAIM study of a set of HB complexes formed by O- and S-containing monomers and water indicates that both O and S acceptors increase their electron population while that of H donors is depleted in O···H-O complexes and remains practically unchanged in S···H-O complexes (figure 30). In both cases, and contrary to what was found for the dimer of methoxymethane, the monomer that contains the H donor gains electron population. Another general trend that can be observed in figure 30 is that most of the electron density transferred from base to water monomer comes from the hydrogens of the base. This is in line with the important role played by hydrogens as electron density suppliers or accepto rs in many chemical processes, as pointed decades ago by Stuchbury and Cooper [118].

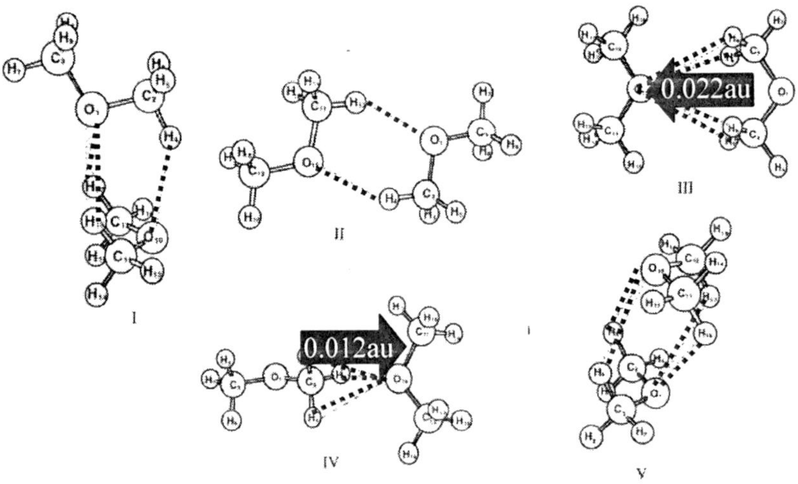

Figure 29. Structures for methoxymethane dimer indicating BCPs associable to O··H-C IHB and electron transference between monomers.

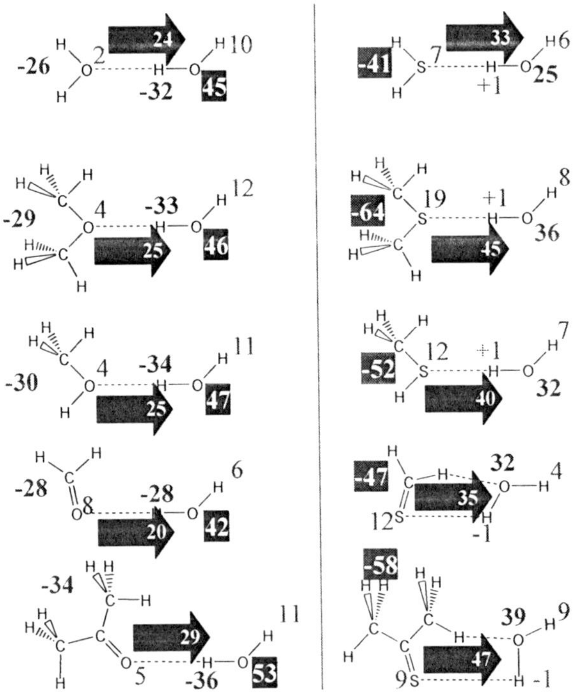

Figure 30. Main QTAIM atomic electron population variations and electron transferences between monomers for several HB complexes with water. All values in au multiplied by 10^3 and obtained from B3LYP/aug-cc-pVDZ electron densities.

Table 11. QTAIM variations of electron population (in au and multiplied by 10^3) of C, O, and Hs involved in HB in dimethoymethane dimmers

Dimer	ΔN(O1)	ΔN(C2)	ΔN(C3)	ΔN(O10)	ΔN(C11)	ΔN(C12)	-ΔN(H)
I	4	18	-1	3	8	10	25,25,26
II	5	23	-2	5	23	-2	38,38
III	1	0	1	2	3	3	5,5,5,5
IV	3	5	-7	3	2	2	5,3,3
V	1	11	1	1	11	1	5,5,5,5

7. On the Reliability of the Resonance Model in Protonation and Hydride Addition Processes

The resonance model is still one of the most often employed tools for explaining the mechanism of chemical processes or predicting their products. The study and application of this model consumes a significant amount of time to chemistry students. Nevertheless, diverse evidences found by several research groups (including ours) point to its inadequacy to describe the evolution of the electron density in various simple chemical processes. Even, the resonance model cannot explain some experimental facts like the evolution of pKas along certain series of organic compounds. Conformational equilibria, protonations or hydride additions are examples of simple processes where the resonance model leads to explanations that contradict those obtained from the modern quantum chemical methods for the electron density analysis, such as QTAIM.

Wiberg and Laidig's paper on the origin of the rotational barriers adjacent to doubles bonds can be considered as one of the first examples on the inadequacy of resonance model [119]. It shows that the electron population of amidic nitrogen is lower in any of the transition states for the internal rotation than in the planar conformer. In contrast, the electron population of the oxygen remains practically unchanged during the process. This does not keep in line with the traditional explanation that consideres the rotational barrier is due to the breakdown of the amidic resonance in the transition state. (scheme 1). Even, C=O and C-N bond lengths are nearly equal in conformer and transition states, in spite the resonance model predicts that C-N should shrink and C=O should lengthen in transition states. These results, obtained at the HF/6-31G(d,p) level were confirmer later at the MP2/6-31G(d) one [120].

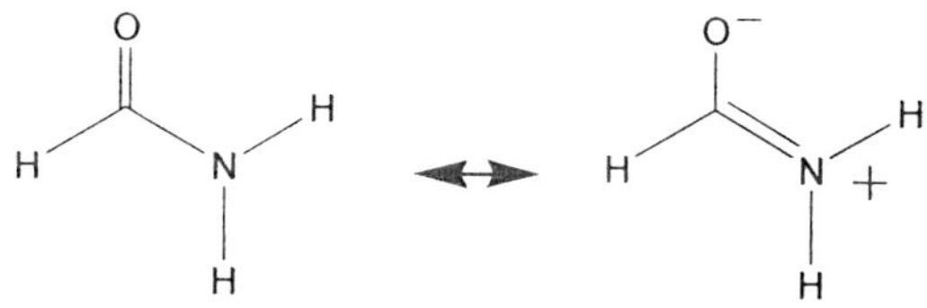

Scheme 1.

The first paper gave rise to a long controversy, where several researchers did not only criticize the paper, but also the basis of QTAIM [121]. Several rebuttals demonstrated some physical inconsistencies of these criticisms [122,123]. Later on, it was concluded that the same problem affected thioformamide [124]. Other authors found new evidencies where the descriptions provided by the resonance model were not compatible with the evolution experienced by electron densities. For example, the electron distribution of several XN_2^+ diazonium cations (X= F, OH, NH_2, CH_3) is not compatible with their traditional Lewis structure: $X\text{-}N^+{\equiv}N$, because the positive charge on the nitrogen β to X is more positive than that of the α nitrogen, which is even negative [125]. The pKas observed for the series: dimethyl sulfide, dimethyl sulfoxide, dimethylsulfone, follows the reverse trend to that expected using the resonance model. Nevertheless, it can be explained successfully with the results of a QTAIM electron density analysis carried out at the HF/6-31++G** level [126].

Our group has carried out a systematic study of the protonation processes in oxygenated compounds using QTAIM as the basic tool to analyse the evolution of the electron density obtained at several computational levels (HF, B3LYP, MP2, and even QCSID). This study has comprised hitherto carbonylic systems [127,128] and linear [129,130] and cyclic ethers [131,132]. Our main conclusion is most of the charge of the protonated species is basically present in the proton itself, whereas the electron population of the oxygen is not depleted in the cation (as it is assumed traditionally, according to scheme 2). On the contrary, the oxygen gains electron population in the protonated form, taking electron density, basically, from the remaining hydrogens in the molecule (figure 31), which also provide the electron density gained by the proton (around 0.35 au). In contrast, bonding of oxigenated compounds to Li^+, Na^+, or K^+ produces complexes where the molecule remains practically neutral. QTAIM properties (Table 12) reveal the stability of the latter complexes is due to ion-dipole interaction and not to covalent bonding, like that of the protonated species [133]. It is also noticeable that alkaline cations are placed in the main plane of the molecule, whereas the proton is bonded following one of the direction determined by oxygen nucleus and NBCC maxima [133].

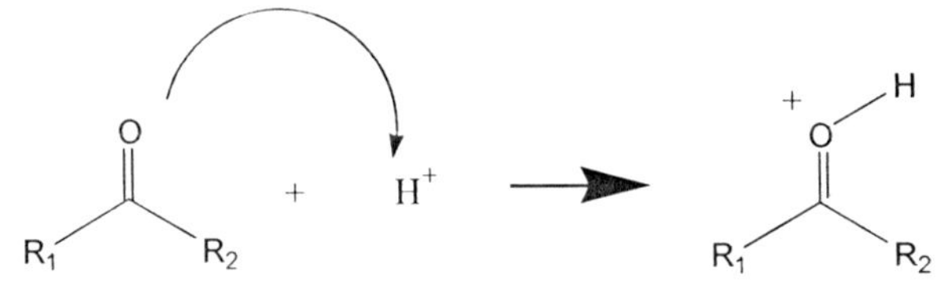

Scheme 2.

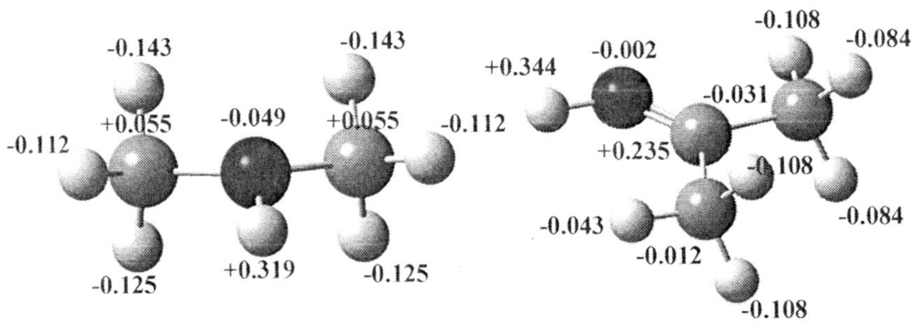

Figure 31. Variations of atomic electron populations in protonated methoxymethane (left) and propanone (right). All values in au with regard to the neutral molecule.

Table 12. Main atomic and bond properties (all values in au) of $X^{+}\cdots OC_2H_6$ cations, X^{+} = H^{+}, Li^{+}, Na^{+}, and K^{+}

	H^{+}	Li^{+}	Na^{+}	K^{+}
$\Delta N(X^{+})$	0.319	0.048	0.038	0.036
ρ_c	0.346	0.041	0.028	0.024
H_c	-0.481	0.014	0.008	0.004

In the last years we have extended our study in several ways: a) considering the protonation of other systems where the carbonyl group is conjugated with delocalized π systems that contain heteroatoms: uracile and cytosine [23,24,134]; b) analysing the protonation of delocalized π systems that do not contain oxygen: aniline [135] and indole [136]; c) comparing QTAIM results with those obtained employing other methods of electron density analysis [25,26,137] (Hirshfeld charges [138-139]); d) studing other processes like hidride addition [25-26], which showed that the resonance forms described in scheme 3 are not compatible with QTAIM and Hirshfeld results, or explaining the sequence of pKas of phenol derivatives [140]; e) studing electron delocalization in neutral systems [139,141,142].

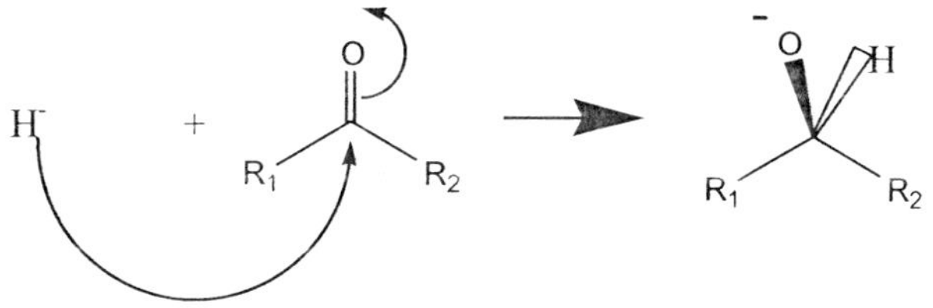

Scheme 3.

Atomic properties and delocalization indices of protonated forms of uracile and cytosine are not in line with the resonance model [23]. The changes of atomic electron populations and energies between neutral and protonated forms of in several model compounds: vinylketone, methyl formiate, and N-methylformamide were employed to propose an alternative model to explain the stability sequence of protonated forms of uracil [24] and cytosine [136]. These explanations were based upon a set of rules that can be extended to any protonation: i) The donation of electron population is facilitated by nucleus with small atomic number; ii) The closer the distance to the proton, the easier the electron density donation is; iii) The donation of electron population between bonded atoms follows the direction of the bond. The orientation of the bond with regard to the proton can make the electron transference easier (when the electron density approaches the proton), or more difficult (when the electron density moves away the proton). This explains why $\Delta N(H_6)$ is lower than $\Delta N(H_8)$ and $\Delta N(H_7)$ in protonated divinyl ketone (figure 32); and iv) π–transferences are generally easier than the σ ones, when both are possible.

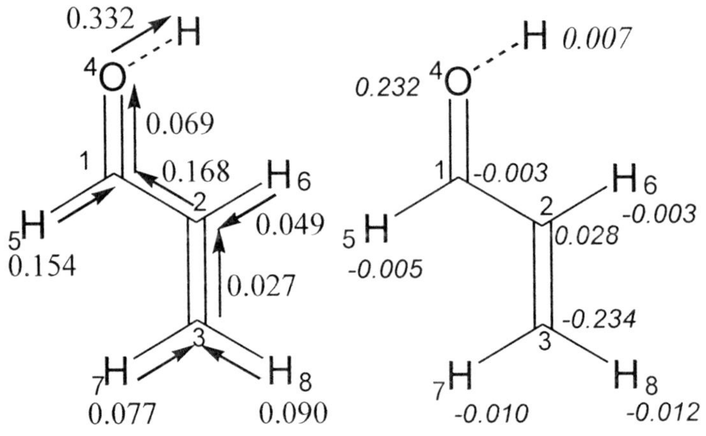

Figure 32. Variations experienced by σ (left) and π (right) atomic electron protonation in divinyl ketone upon protonation.

This can be expected considering that π–electron density is more polarizable than σ-electron density. In terms of energy, for similar variations of π and σ electron population, the former implies smaller variations of atomic energies. This suggests that H-X electron transferences (X representing any heavy atom), which are basically of σ character, should make an important contribution to the protonation energy, since electron transferences taking place between atoms with very different atomic numbers (like H-X) are accompanied by larger variations of core attraction than equivalent transferences between two heavy atoms.

8. Conclusion

The results reviewed in this chapter show that QTAIM electron density analysis allows getting insight about diverse basic chemical concepts. The conclusions obtained are in line with experimental facts but, in some cases do not support the explanations provided by classical models of Chemistry, such as the resonance model or the hyperconjugative model of anomeric effect.

QTAIM has been proved as a very useful tool to study group transferability. The results obtained with this topological electron density analysis indicate that functional groups modify the properties of methylene groups in α, β, γ, and δ dispositions when the alkyl chain is completely antiperiplanar. In constrast, the functional group effects are limited to α and β methylenes when it is in gauche disposition to the alkyl chain. Atomic energies only can be employed for transferability studies when they are obtained with very close verification of the virial theorem.

Strain energies computed with homodesmotic processes contain terms due to non cyclic reagents present in the reaction and to ring opening. Both terms differ from one homodesmotic process to another. In contrast, QTAIM strain energy, computed as the difference between the electronic molecular energy of the cycle and the summation of the atomic energies of the corresponding nearly transferable atoms, provides a unique value for this quantity.

The evolution of QTAIM computed atomic electron populations along conformations of anomeric compounds containing O-C-O, O-C-F, and N-C-N units (which are those hitherto studied), as well as those of delocalization indices and bond properties of anomeric compounds, are not in line with the predictions of the hyperconjugative model of anomeric effect. QTAIM interpretation explains the conformational preference for the R-O-C-O gauche units as due to the transference of electron density from the hydrogens of the central methylene to C and O atoms that takes place in H-C-O-Lp gauche arrangements. Atomic partitioning of the average potential experienced in each atomic basin supports this transference is due to the fact that hydrogen basin is affected by a larger potential when it is gauche to a lone pair than when it is gauche to another hydrogen basin.

Traditionally assumed intramolecular hydrogen bonds, such as those of the most stable conformers of 1,2-ethanediol and catechol are not represented by any bondpath when their electron density in analysed within the QTAIM framework. Nevertheless, these bondpaths are found when slight distortions are applied on the conformer geometry. No singularity is found for any atomic property or delocalization index for the geometry where the BCP associable to IHB appears. Moreover, the partitioning of conformational energy taking into account several features related to variations of atomic energies gives rise to non-zero IHB energy for these conformers.

Intermolecular hydrogen bond formation is accompanied by an increase of the electron population of the HB acceptor atom, whereas that of the donor hydrogen may be depleted or not depending on the electronegativity difference between the atom it is attached and the HB acceptor.

Finally, the resonance forms employed to explain several basical chemical processes are not suported by the evolution experienced by QTAIM atomic and bond properties, as well as by electron delocalization indices.

REFERENCES

[1] Bader, R. F. W. *Atoms in Molecules - A Quantum Theory*; International Series of Monographs on Chemistry, No 22; Oxford University Press: Oxford, 1990.

[2] Popelier, P. L. A. *Atoms in Molecules. An Introduction*, Pearson Education: Harlow, 2000.

[3] Bachrach, S. M. in *Reviews in Computational Chemistry*; Lipkowitz, K. B.; Boyd, D. B. (eds.), VCH: New York 1994; vol V.

[4] Bader, R. F. W. *Chem. Rev.* 1991, *91*, 893-928.

[5] Bader, R. F. W. *Monashefte fur Chemie* 2005, *136*, 819-854.

[6] Bader, R. F. W. *Acc. Chem. Res.* 1985, *18*, 9-15.

[7] Rouvray, D. H. *Top. Curr. Chem.* 1995, *173*, 1-30.

[8] Bader, R. F. W.; Popelier, P. L. A.; Keith, T. A. *Angew. Chem. Int. Ed. Engl.* 1994, *33*, 620-631.

[9] Wiberg, K. B.; Bader, R. F. W.; Lau, C. D. H. *J. Am. Chem. Soc.* 1987,*109*, 1001-1012.

[10] Vila, A.; Mosquera, R. A. *Chem. Phys.* 2003, *287*, 125-135.

[11] Vila, A.; de la Puente, E.; Mosquera, R. A. *Chem. Phys. Lett.* 2005,*405*, 440-447.

[12] Vila, A.; Mosquera, R. A. *J. Phys. Chem. A.* 2006, *110*, 11752-11759.

[13] Thacher, G. R. J. The Anomeric Effect and Associated Stereoelectronic Effects; ACS Symposium Series No. 539. American Chemical Society: Washington DC, 1993.
[14] Kirby, J. The Anomeric Effect and Related Stereoelectronic Effects at Oxygen; Springer Verlag: New York. 1983.
[15] Deslongchamps, P. Stereoelectronic Effects in Organic Chemistry; Wiley: New York. 1983.
[16] Juaristi, E.; Cuevas, G. The Anomeric Effect; CRC Press: Boca Ratón, 1995.
[17] Romers, C.; Altona, C.; Buys, H. R.;Havinga, E. *Top. Stereochem.* 1969, *4*, 39.
[18] Carballeira, L.; Pérez-Juste, I. *J. Comput. Chem.* 2000, *21*, 462-477.
[19] Salzner, U.; Schleyer, P. v. R. *J. Am. Chem Soc.* 1993, *115*, 10231-10236.
[20] Salzner, U.; Schleyer, P. v. R. *J. Org. Chem.* 1994, *59*, 2138-2155.
[21] Vila, A.; Mosquera, R. A. *J. Comput. Chem.* 2006, accepted.
[22] Mandado M.; Graña, A. M.; Mosquera, R. A. *Chem. Phys. Phys. Chem.* 2004, *6*, 4391-4396.
[23] González Moa, M. J.; Mosquera, R. A. *J. Phys. Chem. A* 2003, *107*, 5361-5367.
[24] González Moa, M. J.; Mosquera, R. A. *J. Phys. Chem. A* 2005, *109*, 3682-3686.
[25] Mandado, M.; Mosquera, R. A.; Van Alsenoy, C. J. Phys. Chem. A 2004, 108, 7050-7055.
[26] Mandado, M.; Van Alsenoy, C.; Mosquera, R. A. Chem.Phys. Lett. 2005, 405, 10-17.
[27] Castillo, N.; Matta, C. F.; Boyd, R. J. *Chem. Phys. Lett.* 2005, *409*, 265-269.
[28] Bader, R. F. W. *Pure and Appl. Chem.* 1988, *60*, 145-155.
[29] Schwinger, J. *Phys. Rev.* 1951, *82*, 914-927.
[30] Poater, J.; Solà, M.; Bickelhaupt, F. M. *Chem. Eur. J.* 2006, *12*, 2889-2895.
[31] Bader, R. F. W. *Chem. Eur. J.* 2006, *12*, 2896-2901.
[32] Haaland, A.; Shorokhov, D. J.; Tverdova, N. V. *Chem. Eur. J.* 2004, *10*, 4416-4421; *Chem. Eur. J.* 2004, *10*, 6210-6210 (Corrigendum).
[33] Bader, R. F. W.; De-Cai, F. *J. Chem. Theory Comput.* 2005, *1*, 403-414.
[34] Cioslowski, J,; Mixon, S. T. *Can. J. Chem.* 1992, *70*, 443-449.
[35] Cioslowski, J.; Mixon, S. T. *J. Am. Chem. Soc.* 1992, *114*, 4382-4387.
[36] Bader, R. F. W., *J. Phys. Chem. A* 1998, *102*, 7314-7323.
[37] Biegler-König, F. W.; Schönbohm, J.; Bayles, D. *J. Comp. Chem.* 2001, *22*, 545-559.
[38] Hohenberg, P.; Kohn, W. *Phys. Rev.* 1964, *136*, B864-B871.
[39] Bader, R. F. W.; Becker, P. *Chem. Phys. Lett.* 1988, *148*, 452-458.
[40] Graña, A. M.; Mosquera, R. A. *J. Chem. Phys.* 1999, *110*, 6606-6616.
[41] Quiñónez, P. B.; Vila, A.; Graña, A. M.; Mosquera, R. A. *Chem. Phys.* 2003, *287*, 227-236.
[42] Mandado, M.; Graña, A. M.; Mosquera, R. A. *J. Mol. Struct. (THEOCHEM)* 2002, *584*, 221-234.
[43] Vila, A.; Carballo, E.; Mosquera, R. A. *Can. J. Chem.* 2000, *78*, 1535-1543.
[44] Vila, A.; Mosquera, R. A. *Chem. Phys. Lett.* 2001, *345*, 445-452.
[45] Graña, A. M.; Mosquera, R. A. *J. Chem. Phys.* 2000, *113*, 1492-1500.
[46] Lorenzo, L.; Mosquera, R. A. *Chem. Phys. Lett.* 2002, *356*, 305-312.
[47] López, J. L.; Mandado, M.; Graña, A. M.; Mosquera, R. A. *Int. J. Quantum Chem.* 2002, *86*, 190-198.
[48] Vila, A.; Mosquera, R. A. *J. Chem. Phys.* 2001, *115*, 1264-1273.

[49] Vila, A; Carballo, E.; Mosquera, R. A. *J. Mol. Struct. (THEOCHEM)* 2002, *617*, 219-224.
[50] Mandado, M.; Vila, A.; Graña, A. M.; Mosquera, R. A.; Cioslowski, J. *Chem. Phys. Lett.* 2003, *371*, 739-743.
[51] Cortés-Guzmán, F.; Bader, R. F. W. *Chem. Phys. Lett.* 2003, *379*, 183-192.
[52] Lehd, M.; Jensen, F. *J. Comput. Chem.* 1991, *12*, 1089-1096.
[53] Mandado, M.; Graña, A. M.; Mosquera, R. A. *Chem. Phys. Lett.* 2002, *355*, 529-537.
[54] Wiberg, K. B. *Angew. Chem. Int. Ed. Engl.* 1986, *25*, 312-322.
[55] Liebman, J. F.; Greenberg, A. *Chem. Rev.* 1976, *76*, 311-365.
[56] Greenberg, A.; Liebman, J. F. Strained Organic Compounds; Academic Press: New York, 1978.
[57] Allinger, N. L.; Burkert, U. Molecular Mechanics; American Chemical Society: Washington, D. C., 1982.
[58] Hehre, W. J.; Ditchfield, R.; Radom, L.; Pople, J. A. *J. Am. Chem. Soc.* 1970, *92*, 4796-4801.
[59] George, P.; Trachtman, M.; Bock, C. W.; Brett, A. M. *Tetrahedron* 1976, *32*, 317-323.
[60] Hess, B. A.; Schaad, L. J. *J. Am. Chem. Soc.* 1983, *105*, 7500-7505.
[61] Cioslowski, J.; Schimeczek, M.; Liu, G.; Stoyanov, V. *J. Chem. Phys.*2000, *113*, 9377-9389.
[62] Schmidt, M.W.; Baldridge, K.K. ; Boatz, J.A.; Elbert, S.T.; Gordon, M.S.; Jensen, J.H.; Koseki, S.; Matsunaga, N.; Nguyen, K.A.; Su, S.J.; Windus, T.L.; Dupuis, M.; Montgomery, J.A. *J. Comput. Chem.* 1993, *14*, 1347-1363.
[63] Paul, A.; Wannere, C. S.; Kasalova, V.; Schleyer, P. v. R.; Schaefer III, H. F. *J. Am. Chem. Soc.* 2005, *127*, 15457-15469.
[64] Paul, A.; Wannere, C. S.; Schaefer III, H. F. *J. Phys. Chem. A* 2004, *108*, 9428-9434.
[65] Watkins, E. K. ; Jorgensen, W. L. *J. Phys. Chem. A* 2001, *105*, 4118-4125.
[66] Wiberg, K. B.; Bader, R. F. W.; Lau, C. D. H. *J. Am. Chem. Soc.* 1987, *109*, 985-1001.
[67] Bader, R. F. W.; Bayles, *J. Phys. Chem. A* 2000, *104*, 5579-5589.
[68] Benson, S. W.; Cruickshank, F. R.; Golden, D. M.; Haugen, G. R.; O'Neal, H. E.; Rodgers, A. S.; Shaw, R.; Walsh, R. *Chem. Rev.* 1969, *69*, 279-324.
[69] Mahavadi, M.; Zeiger, D. N.; Naqib, D,; Roux, M. V.; Notario, R.; Liebman, J. F. *Int. J. Quantum Chem.* 2003, *95*, 784-790.
[70] Dudev, T.; Lim, C. *J. Am. Chem. Soc.* 1998, *120*, 4450-4458.
[71] Liebman, J. F.; Dolbier, W. R., Jr.; Greenberg, A. *J. Phys. Chem.* 1986, *90*, 394-397.
[72] Zeiger, D. N.; Liebman, J. F. *J. Mol. Struct.* 2000, *556*, 83-94.
[73] Roux, M. V.; Notario, R.; Zeiger, D. N.; Liebman, J. F *J. Fluorine Chem.* 2001, *112*, 91-94.
[74] Anderson, C. B.; Sepp, D T. *J. Org. Chem.* 1967, *32*, 607-611.
[75] Perrin, C. L.; Armstrong, K. B.;Fabian, M. A. *J. Am. Chem. Soc.* 1994, *116*, 715-722.
[76] Suárez, D.; Sordo, T. L.; Sordo, J. A. *J. Am. Chem. Soc.* 1996, *118*, 9850-9854.
[77] Tvaroska, I.; Bleha, T. Advances in Carbohydrate Chemistry and Biochemistry, vol. 47; Academic Press: San Diego, 1989. p. 107.
[78] Foster, J. P.; Weinhold, F *J. Am. Chem. Soc.* 1980, *102*, 7211-7218.
[79] Reed, A. E.; Weinhold, F. *J. Chem. Phys.* 1985, *83*, 1736-1740.
[80] Carpenter, J. E.; Weinhold, F. *J. Mol. Struct. (THEOCHEM)* 1988, *169*, 41-62.

[81] Weinhold, F. Encylcopedia of computational chemistry, vol. 3, Schleyer, P. v. R.; Allinger, N. L.; Clark, T.; Gasteiger, J.; Kollman, P. A.; Schaefer III, H. F.; Schreiner, P. R. (eds.), Wiley: Chichester, 1988, p. 1792.
[82] Vila A.; Mosquera, R. A., *J. Comput. Chem.* in press.
[83] Werstiuk, N. H.; Laidig, K. E.; Ma, J. in G. J. R. Thatcher (Ed.), Anomeric Effect and Associated Stereoelectronic Effects, ACS Symp. Ser. 539, Washington DC, 1993. p. 176-.
[84] Vila, A.; Mosquera, R. A., Hermida-Ramón, J. M. *J. Mol. Struct. (THEOCHEM)*, 2001, *541*, 149-158.
[85] Vila, A.; Mosquera, R. A. *Chem. Phys.*, 2003, *291*, 73-80.
[86] Jeffrey, G. A. An Introduction to Hydrogen Bonding, Oxford University Press: New York, 1997.
[87] Tichy, M. *Adv. Organic Chem.* 1965, *5*, 117-.
[88] Radom, L.; Lathan, W. A.; Hehre, W. J.; Pople, J. A. *J. Am. Chem. Soc.* 1973, *95*, 693-698.
[89] Van Alsenoy, C.; Van den Emden, L.; Schäfer, L. *J. Mol. Struct. (THEOCHEM)* 1984, *108*, 121-128.
[90] Ikuta, S.; Nomura, O. *Chem. Phys. Lett.* 1989, *154*, 71-76.
[91] Cabral, B. J. C.; Alburquerque, L. M. P. C.; Fernández, F. M. S. S. *Theor. Chim. Acta* 1991, *78*, 271-280.
[92] Nagy, P. I.; Dunn III, W. J.; Alagona, G.; Ghio, C. *J. Am. Chem. Soc.* 1991, *113*, 6719-6729.
[93] Nagy, P. I.; Dunn III, W. J.; Alagona, G., Ghio, C. *J. Am. Chem. Soc.* 1992, *114*, 4752-4758.
[94] Murcko, M. A.; DiPaola, R. A. *J. Am. Chem. Soc.* 1992, *114*, 10010-10018.
[95] Alagona, G.; Ghio, C. *J. Mol. Struct (THEOCHEM)* 1992, *254*, 287-300.
[96] Reiling, S.; Brickmann, J.; Schlenkrich, M.; Bopp, P. A. *J. Comp. Chem.* 1996, *17*, 133-147.
[97] Manivet, P.; Masella, M. *Chem. Phys. Lett.* 1998, *288*, 642-646.
[98] Zhang, H-Y.; Sun, Y-M.; Wang, X-L. *Chem. Eur. J.* 2003, *9*, 502-508.
[99] Vedernikova, I.; Proynov, E.; Salahub, D.; Haemars, A. *Int. J. Quantum Chem.* 2000, *77*, 161-173.
[100] Kjaergaard, H. G.; Howard, D. L.; Schofield, D. P.; Robinson, T. W.; Ishiuchi, Sh-I.; Fujii, M. *J. Phys. Chem. A* 2002, *106*, 258-266.
[101] S. Mendonca, S.; Johnson, G. P.; French, A. D.; Laine, R. A. *J. Phys. Chem. A* 2002, *106*, 4115-4124.
[102] Lewis, B. E; Schramm, V. L. *J. Am. Chem. Soc.* 2001, *123*, 1327-1336.
[103] Schmidt, R. K.; Karplus, M.; Brady, J. W. *J. Am. Chem. Soc.* 1996, *118*, 541-546.
[104] Liu, Q.; Brady, J. W. *J. Am. Chem. Soc.* 1996, *118*, 12276-12286.
[105] Klein, R. A. *J. Am. Chem. Soc.* 2002, *124*, 13931-13937.
[106] Klein, R. A. *J. Comput. Chem.* 2002, *23*, 585-599.
[107] Rozas, I.; Alkorta, I.; Elguero, J. *J. Phys. Chem. A* 2001, *105*, 10462-10467.
[108] MORPHY98, a topological analysis program written by Popelier, P. L. A. with a contribution from Bone, R. G. A (UMIST, Engl., EU).
[109] Popelier, P. L. A. *Chem. Phys. Lett.* 1994, *228*, 160-164.
[110] Koch, U.; Popelier, P. L. A. *J. Phys. Chem.* 1995, *99*, 9747-9754.

[111] Rozas, I.; Alkorta, I; Elguero, J. *J Phys Chem A* 1997, *101*, 4236-4244.
[112] Vila, A.; Mosquera, R. A. *Int. J. Quantum Chem.* 2006, *106*, 928-934.
[113] Mandado, M.; Mosquera, R. A., Van Alsenoy, C. *Phys. Chem. Chem. Phys.* 2005, *7*, 3290-3296.
[114] Mandado, M.; Mosquera, R. A., Van Alsenoy, C. *Tetrahedron* 2006, *62*, 4243-4252.
[115] Hermida-Ramón, J. M.; Mosquera, R. A. *Chem. Phys.* 2006, *323*, 211-217.
[116] Pauling, L. The Nature of the Chemical Bond, Cornell University Press: Ithaca, 1939.
[117] Steiner, T. *Angew. Chem. Int. Ed.* 2002, *41*, 48-76.
[118] Stutchbury, N. C. J; Cooper, D. L. *J. Chem. Phys.* 1983, *79*, 4967-4972.
[119] Wiberg, K. B.; Laidig, K. E. *J. Am. Chem. Soc.* 1987, *109*, 5935-5943.
[120] Wiberg, K. B.; Breneman, C. M. *J. Am. Chem. Soc.* 1992, *114*, 831-840.
[121] Perrin,C. *J. Am. Chem. Soc.* 1991, *113*, 2865-2868.
[122] Laidig, K. E. *J. Am. Chem. Soc.* 1992, *114*, 7912-7913.
[123] Gatti, C.; Fantucci, P. *J. Phys. Chem.* 1993, *97*, 11677-11680.
[124] Laidig, K. E.; Cameron, L. M. *J. Am. Chem. Soc.* 1996, *118*, 1737-1742.
[125] Glaser, R.; Choy, G. S. K. *J. Am. Chem. Soc.* 1993, *115*, 2340-2347.
[126] 126 Speers, P.; Laidig, K. E.; Streitwieser, A. *J. Am. Chem. Soc.* 1994, *116*, 9257-9261.
[127] Graña, A. M.; Mosquera, R. A. *Chem. Phys.* 1999, *243*, 17-26.
[128] Mosquera, R. A.; Graña, A. M. *Rec. Res. Devel. Chem. Phys.* 2001, *2*, 23-36.
[129] Vila, A.; Mosquera, R. A. *J. Phys. Chem. A* 2000, *104*, 12006-12013.
[130] Vila, A.; Mosquera, R. A. *Chem. Phys. Lett.* 2000, *332*, 474-480.
[131] Vila, A.; Mosquera, R. A. *Tetrahedron.* 2001, *59*, 9415-9422.
[132] Vila, A.; Mosquera, R. A. *Chem. Phys. Lett.* 2003, *371*, 540-547.
[133] Vila, A.; Pardo, J. M.; Mosquera, R. A. *Chem. Phys. Lett.* 2003, *375*, 499-505.
[134] González Moa, M. J.; Mandado, M.; Mosquera, R. A *Chem. Phys. Lett.* 2006, *428*, 255-261.
[135] Graña, A. M.; Hermida-Ramón, J. M.; Mosquera, R. A. *Chem.Phys. Lett.* 2005, *412*, 106-109.
[136] Otero, N.; González Moa, M. J.; Mandado, M.; Mosquera, R. A. Chem.Phys.Lett. 2006, 428, 249-254.
[137] Mandado, M.; Van Alsenoy, C.; Mosquera, R. A. J. Phys. Chem. A 2005, 109, 8624-8631.
[138] Hirshfeld, F. L. *Theor. Chim. Acta* 1977, 44, 129-138.
[139] De Proft, F.; Van Alsenoy, C.; Peeters, A.; Langenaker, W.; Geerlings, P. *J. Comput. Chem.* 2002, *23*, 1198-1209.
[140] Mandado, M.; Mosquera, R. A.; Graña, A. M. *Chem. Phys. Lett.* 2004, *386*, 454-459.
[141] Vila, A.; Mosquera, R. A. *J.Phys.Chem. A* 2005, *109*, 6985-6989.
[142] González Moa, M. J.; Mosquera, R. A. *J. Phys. Chem. A* 2006, *110*, 5934-5941.

In: Chemical Physics Research Trends, Vol. 252
Editor: Samuel V. Arnold, pp. 215-246
ISBN: 978-1-60021-554-4

Chapter 5

NOVEL PYROPHEOPHORBIDE FULLERENE CONJUGATES AS PART OF MODULAR CARRIER SYSTEMS IN PDT AND AS PROMISING UNITS IN ARTIFICIAL PHOTOSYNTHESIS

***Eugeny A. Ermilov*[a], *Matthias Helmreich*[b], *Norbert Jux*[b] *and Beate Röder*[a*]**

[a]Humboldt-Universität, Institut für Physik, Photobiophysik, Newtonstr.15, 12489 Berlin, Germany

[b]Universität Erlangen-Nürnberg, Institut für Organische Chemie, Henkestr. 42, 91054 Erlangen, Germany

ABSTRACT

Dendrimers became of rising interest in the past decade in different fields of biophysical research due to their highly branched and well-defined architecture, which consists of a high number of functional end groups. On the other hand the unique photophysical and photochemical properties of fullerenes have led to the synthesis of a large number of compounds in which fullerenes are covalently linked to different kinds of photoactive groups. Some of these compounds could serve as artificial light harvesting or photosynthesis model systems due to the high capability of the fullerene moiety to act as an electron acceptor or even an electron accumulator. Additionally, C_{60} can be used as a versatile building block for the construction of globular dendrimers, which opens up ways for biomedical applications of fullerene based high branched molecular systems. One possibility is the use of the fullerene as functional core of an efficient multiplier in modular drug delivery systems for the photodynamic therapy (PDT) of tumors.

In this work we report on the photophysical investigations of dendritic fullerene-dye complexes with up to 24 pyropheophorbide *a* units attached to a fullerene core. Their electronic properties are discussed in detail. Depending on the addition pattern on the fullerene, after light absorption in these molecular systems either internal electron or/and energy transfer processes occur. Due to the length and high flexibility of the carbon

* E-mail: roeder@physik.hu-berlin.de

chains different energy transfer processes are possible. Beside Förster resonance transfer excitonic interactions occur due to intramolecular stacking effects of the dye molecules. The formation of different types of energy traps was postulated and proved experimentally. Their strong influence on the photophysical properties of fullerene-dye compounds in dependence of the number of appended dye molecules is discussed. A phenomenological model describing the different energy transfer processes in dendritic dye systems is given. Furthermore the potential of the fullerene-dye-complexes as a combined artificial light-harvesting and charge separation unit is discussed. Finally, their possible use for PDT applications is shown.

INTRODUCTION

Dendrimers are large, highly branched macromolecules, which have several distinctive architectural components, namely (i) an initial core; (ii) interior layers, called generations, composed of repeating units radially attached one after another; and (iii) the open ends on the periphery [1-4]. Dendrimers can contain selected functions in predetermined sites of their tree-like structure and also host molecules in inner cavities and deep ditches caused by their multi-branched structure. Since their first discovery by the group of Vögtle in 1978 [5] their unique properties are attracting increasing attention in many scientific fields, among them medicine [6, 7], guest-host chemistry [8, 9], optical electronics [10, 11], and physics [12–15].

Already years ago the photodynamic therapy (PDT) was proposed to be a powerful treatment against cancer and other diseases [16]. But still today one of the crucial problems is that no sensitizer is known, which actively and selectively accumulates in target tissues [17]. Several carrier systems have been tested in the past to overcome this problem [18] but it turned out, that only antibodies or their fragments show a sufficiently high selectivity for this purpose [17, 19]. Due to the fact, that such antibodies can carry only a few dye molecules covalently linked without losing their bioactivity [20], it was proposed to use linker-molecules (multipliers), which can carry a large number of photosensitizing molecules and are linked via one chemical bond to the antibody. Based on this the "Modular carrier system" was proposed [17] consisting of an addressator, a multiplier and lots of photosensitizers. It is not a surprise that as an efficient multiplier a dendritic structure was chosen [7] due to its highly branched and well-defined architecture. Already in 1991 Roberts [21] reported the possibility to couple dendrimers to antibodies and recently Patri [22] reported about antibody-dendrimer complexes carrying fluorescence markers, where the antibodies are still active and selective.

Since the initial discovery of the fullerene C_{60} and the development of a method for its preparation, fullerene based complex molecular systems have received great attention [e.g. 23–28]. The high capability of C_{60} to act as an electron acceptor or even as an electron accumulator has led to the synthesis of a large number of compounds in which the fullerene is covalently linked to photoactive groups serving as potential electron donors [29–33]. Furthermore, the special photophysical properties of C_{60} allow different possibilities for its biomedical applications [34–37]. For example, C_{60} is able to efficiently generate phototoxic singlet oxygen due to a practically unity value of intersystem crossing from the singlet excited state to the first excited triplet state. Unfortunately, the low solubility of unsubstituted fullerenes in different media appears to be a major problem for their direct biological applications.

It has been shown that C_{60} can be used as versatile building block for the construction of globular dendrimers [38–40]. Moreover, an antibody or an antibody fragment can be linked to the fullerene molecule [41–43]. These structural modifications increase the solubility of such C_{60} containing supramolecules. Taking advantage of these facts fullerenes can be used as functional core of an efficient multiplier in modular drug delivery systems for PDT.

Recently, many studies have been directed towards using dendrimers to mimic the photosynthetic light-harvesting antennae systems [44–49]. What makes these supramolecules promising candidates as artificial light-harvesting antennae is the large number of absorbing units (chromophores attached to the outermost branching units of a dendrite structure) that grows exponentially with their generation number, and the relatively short distance between the center (where a reaction center, or an energy trap, or a chemical sensor can be located) and the periphery of a dendrimer. It is expected, that dendrimers will capture the optical radiation and deliver it via energy transfer to an acceptor at the center with high efficiency. But it should be mentioned, that these systems are all based on directed energy transfer between donors and acceptors of different kind only and no other interaction between donor chromophores was reported. Besides our former work [50] only in a few papers the existence of a stronger interaction between identical dye-molecules in dendrimers was proposed [9, 15]. Yeow et al. observed this effect in fifth generation of porphyrin-DAB-dendrimers, but no further investigation was carried out. Nevertheless all results shown in [15] are in good agreement with the general model described below.

In biological photosynthesis, photoinduced electron transfer is one of the most important processes for conversion of solar energy into chemical energy. Molecular design of artificial photosynthetic systems involving restricted spatial arrangement between covalently or noncovalently linked electron donor and acceptor moieties has been one of the important goals during the past two decades [51–55]. Among the large number of electron-transfer assemblies characterized, C_{60}-fullerene based systems have demonstrated great potential as photosynthetic reaction centers due to above mentioned high capability of C_{60} to act as an excellent electron acceptor [29, 30, 56–60]. On the other hand, the prominent ground-state absorption in the visible region of the spectrum of π-conjugated macrocycles, such as tetrapyrroles [61–63], together with their rich redox chemistry, provides both excellent light-harvesting and electron-donor performances.

In the present paper we report on the photophysical studies of novel fullerene based dye complexes with up to 24 pyropheophorbide *a* units attached via dendrite branches to a fullerene C_{60} core. We focused on the development of two different types of complexes: one exhibiting efficient photoinduced electron transfer from initially excited chromophores to the fullerene, and another showing high probability of energy transfer between dye molecules. As a chromophore, the well-known photosensitizer pyropheophorbide *a* (pyropheo) (see Fig. 1) which can also act as an electron donator [53] was chosen.

In the first part we will discuss the photophysical properties of simple compounds consist of only two pyropheo molecules, and the influence of the fullerene core on the photoactivity of the chromophores in solutions with different polarity. Then we expand the monofullerene (*bis*)pyropheo motif to larger systems which carry six pyropheo chromophores utilizing dendrimers as multiplying units. Due to the high local concentration of the dye molecules in fullerene-pyropheo complexes the chromophores undergo Förster energy transfer between them as well as partly excitonic interaction. In the second part of the paper the influence of the fullerene core on the relative efficacy of the photoinduced transfer processes is investigated. Moreover, the relationship between spatial geometry of the fullerene-dendron-dye molecular systems and there photophysical properties is discussed. Last part of the paper deals with the highly branched and symmetrical fullerene-dendrimer supramolecules carrying 12 and 24 pyropheo chromophores.

1. Photophysical Properties of Fullerene-Based and Fullerene-Free Compounds with Two Pyropheophorbide *A* Chromophores

For all compounds presented in this article, the starting point was the isolation of chlorins from spinach or algae (*chlorella*, *spirulina*, or *urtica dioica*) and their subsequent modification according to literature procedures [64, 65] to give pyropheophorbide *a* (pyropheo), with its purity checked for each batch by ^{1}H NMR. Due to the inherent sensitivity of pyropheo to light and oxygen all reactions needed to be carried out in the dark and under an inert atmosphere. Therefore, it also seemed reasonable to introduce the dye at the very end of the syntheses to avoid loss of material. For the synthesis of monofullerene-(*bis*)pyropheophorbide *a* (FP2) C_{60} was reacted with a malonate unit connected to a tert-butyldimethylsilyl-protected octane-1,8-diol. Utilizing well-known fullerene chemistry this malonate was added to C_{60} and deprotected to give a fullerene diol compound. Pyropheo was subsequently attached to this material by a modified Sheehan-coupling [66, 67] to give compound FP2 (see Figure 1). The octyl chain serves as a spacer unit and does not play a vital role for the syntheses of FP2.

It was shown [28, 33] that addition of six malonates to the fullerene molecule causes a dramatic decrease of the electron-accepting properties of C_{60}. Due to the octahedral hexa-addition pattern, the fullerene π-system is broken up into a system of eight separate benzene rings which does not act as an electron acceptor anymore. Therefore, the new compound (*bis*)pyropheophorbide monofullerene hexaadduct – (FHP2, see Figure 1) was designed and synthesized. The aforementioned silyl-protected fullerene-diol was reacted with diethyl malonate and carbon tetrabromide under conditions optimized for a C_{60} hexaaddition to give the fullerene hexaadduct. After deprotection and coupling with pyropheo the desired FHP2 molecule was obtained.

As a reference compound, the (*bis*)pyropheophorbide malonate P2 (Figure 1) was synthesized using the abovementioned coupling conditions. All compounds were completely characterized by ^{1}H/^{13}C NMR, UV/Vis and IR spectroscopy and FAB mass spectrometry [59].

The absorption spectra of pyropheo, P2, FP2 and FHP2 in DMF are shown in Fig. 2 [68]. The shape of the absorption spectra of all samples is similar to that of metal free tetrapyrroles consisting of the Soret band and the split $Q_{x,y}$-bands (see Table 1), but in case of FP2 and FHP2 an additional shoulder at the blue edge of the absorption spectrum appears as a result of the fullerene chromophore contribution. However, the covalent coupling between P2 and C_{60} results in strong changes in the absorbance abilities of the FP2 compound. It is seen, that FP2 has practically two times lower absorbance than P2.

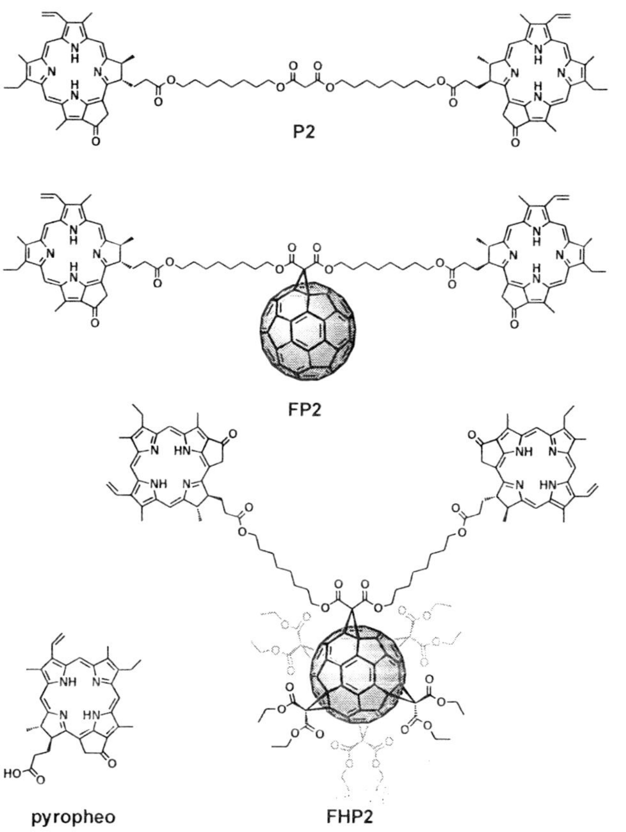

Figure 1. Structural formula of pyropheo, P2, FP2, and FHP2.

Table 1. Photophysical parameters of pyropheo, P2, FP2 and FHP2 in DMF

Sample	Soret [nm]	Q_y [nm]	λ_{fl}^{max} [nm]	τ_{fl} [ns]	Φ_{fl}	Φ_{ISC}	Φ_Δ
pyropheo	414	668	675	7.0	0.30	0.52	0.50
P2	414	668	675	5.7	0.23	0.49	0.43
FP2	414	669	677	0.5 (34 %) 5.7 (66 %)	0.025	0.04	0.03
FHP2	414	668	675	5.7	0.23	0.49	0.43

Intersystem crossing quantum yield, Φ_{ISC}, was estimated using picosecond transient absorption spectroscopy (ps-TAS) [68, 69]. To calculate the singlet oxygen quantum yield, Φ_Δ, the solution of H_2TPP in DMF was used as reference ($\Phi_\Delta = 0.69$ [70]).

The fluorescence of all samples originates from the pyropheo moiety (Figure 3). The fluorescence intensity of P2 is slightly lower than that of pyropheo. In contrast to pyropheo and P2, a strong quenching of the fluorescence signal was observed for FP2. The estimations have shown that the fluorescence quantum yield of FP2 is 8.6 and 11.2 times lower than the fluorescence quantum yields of P2 and pyropheo, respectively (Table 1).

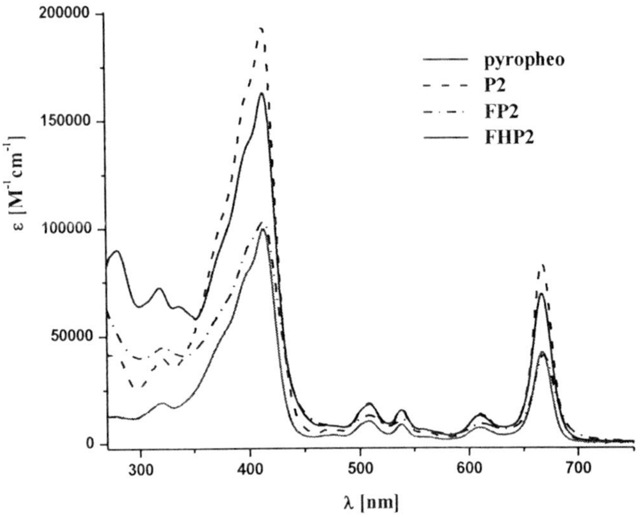

Figure 2. UV/Vis absorption spectra of pyropheo, P2, FP2 and FHP2 in DMF [after 68].

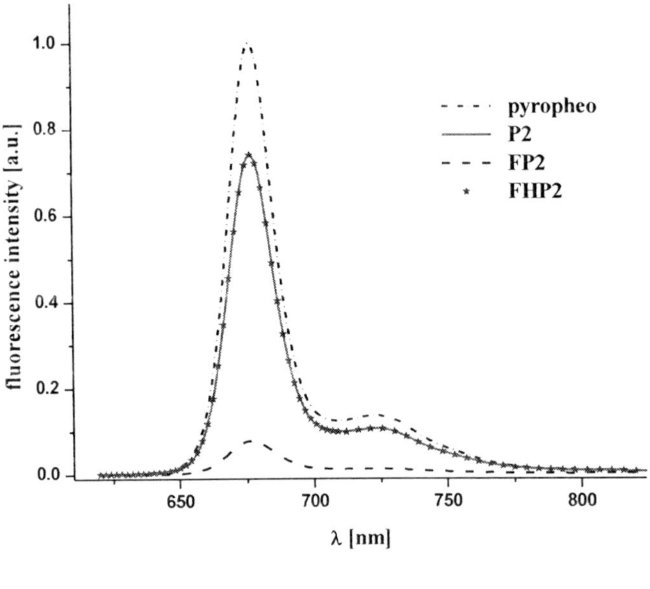

Figure 3. Steady-state fluorescence spectra of pyropheo, P2, FP2 and FHP2.

For pyropheo and P2 time-resolved measurements of the fluorescence decay have shown a single exponential depopulation of the first excited singlet state with a lifetime of 7.0 and 5.7 ns, respectively (see Table 1). The time-resolved fluorescence of FP2 has a double exponential decay with times of τ_1 = 5.7 ns and τ_2 = 0.5 ns. The amplitude of the fast component is approximately two times lower compared to that of the slow one.

As it was mentioned above, the photosensitized singlet oxygen generation is one of the important parameters in order to decide whether a photosensitizer is suitable for PDT or not. It is well-known that singlet oxygen ($^1O_2$, $^1\Delta_g$) is generated by energy transfer from the excited triplet state of a sensitizer molecule to molecular oxygen in its triplet ground state

($^3O_2$, $^3\Sigma_g^-$). Since the $^1O_2$ state lies 0.98 eV above the triplet ground state, the energy difference between T_1 and S_0 states of a sensitizer should be above or around this value to allow the generation of singlet oxygen [71–73]. This prerequisite holds for pyropheo with a triplet energy E_T of ~1.32 eV [74]. Furthermore, for pyropheo molecule the intersystem crossing (ISC) $S_1 \rightarrow T_1$ has a high probability with a quantum yield of 52 % in DMF.

As was to be expected, the fast deactivation of the first excited singlet state of FP2 should result in a dramatic decrease of the ISC quantum yield and a reduction of the photoinduced singlet oxygen generation. Using the time-resolved detection of the singlet oxygen luminescence [75] we obtained a quantum yield of the singlet oxygen generation of $\Phi_\Delta = 0.03$ for FP2 in DMF whereas for pyropheo and P2 compounds this value was 0.5 and 0.43, respectively. Weak interactions between the two pyropheo molecules linked to the one chain in P2 may explain the difference in the experimental data observed for P2 and pyropheo compounds. Surprisingly, in non-polar toluene the singlet oxygen quantum yield of FP2 has a value only two times lower compared to P2 (0.29 and 0.54, respectively [76]) whereas the quenching of the fluorescence in both solvents is similar (the quenching rate of the first excited singlet state of FP2, k_q, equals to 1.8×10^9 s^{-1} in DMF and 1.4×10^9 s^{-1} in toluene [76]). This means, that the population of the excited triplet state belonging to pyropheo or/and C_{60} moieties is much higher for FP2 solved in toluene than for that in DMF solution. To explain these photophysical properties of FP2 in polar and nonpolar solvents we investigated the nature of the fast quenching of the first excited singlet state of the pyropheo moiety and the way (or ways) of triplet state population of FP2 in toluene.

It should be remembered, that in FHP2 the fullerene π-system is broken up and the electron accepting abilities of C_{60} are strongly reduced. As result, the fullerene core acts as neutral tecton and does not affect the dissipation of electronic excitation energy within FHP2 (the photophysical parameters for FHP2 are the similar to the reference compound P2, see Table 1). Thus, one can conclude that the photoinduced electron transfer between initially excited pyropheo moiety and the fullerene core is the reason of the fast and strong quenching of the FP2 fluorescence.

In order to estimate the influence of an electron transfer (ET) process on the photophysical properties of FP2, the forward free enthalpy for charge separation, ΔG^0, has been calculated using the Rehm-Weller relation [77–79] both in polar DMF and in non-polar toluene

$$\Delta G^0 = E_{1/2}^{ox}\left(D/D^+\right) - E_{1/2}^{red}\left(A/A^-\right) - E_{0,0}^* - \frac{e^2}{4\pi\varepsilon_0\varepsilon_s R} - \frac{e^2}{8\pi\varepsilon_0}\left[\frac{1}{r_{d^+}} + \frac{1}{r_{a^-}}\right]\left[\frac{1}{\varepsilon_{ref}} - \frac{1}{\varepsilon_s}\right] \quad (1)$$

where $E_{1/2}^{ox}\left(D/D^+\right)$ is the half-wave oxidation potential of the donor molecules (pyropheo); $E_{1/2}^{red}\left(A/A^-\right)$ is half-wave reduction potential of the acceptor (fullerene); $E_{0,0}^*$ is $S_{0,0} \rightarrow S_{1,0}$ transition energy of the donor; e is the charge of the transferred electron; ε_0 is the vacuum permittivity; ε_s and ε_{ref} are the dielectric constants of the solvent used for photophysical studies (DMF or toluene) and for measurements of the redox potential, respectively; R is the separation distance between donor and acceptor molecules; r_{d^+} and r_{a^-} are the effective radii of the cation and the anion radical, respectively. According to eq. (1) and using the following data: $E_{1/2}^{ox}$ = 0.42 V [51], $E_{1/2}^{red}$ = -0.52 V [24, 80, 81], $E_{0,0}^{*}$ = 1.84 eV, r_{a^-} = 4.4 Å [81], r_{d^+} = 4 Å and R = 12 Å (the last two values were estimated using the HyperChem program [82]), ε_s = 37 (DMF) or ε_s = 2.38 (toluene) we have calculated the value of ΔG^0 = -0.9 eV for the sample solved in DMF and -0.1 eV in toluene. These negative values of the forward free enthalpy for charge separation, ΔG^0, show that the excited donor (pyropheo) undergoes the exothermic ET reactions with the fullerene.

The Marcus first order ET rate constant k_{ET} was computed from [24]

$$k_{ET} = \frac{2\pi}{\hbar}\frac{J^2}{\sqrt{4\pi k_B T\lambda}}\exp\left[-\frac{(\Delta G^0 + \lambda)^2}{4k_B T\lambda}\right] \tag{2}$$

where J is the transfer integral; k_B and $\hbar$ are the Boltzmann and Planck constants, respectively; T is the absolute temperature; $\lambda = \lambda_I + \lambda_S$ is the reorganization energy which can formally be divided into an intramolecular, λ_I, part and a contribution from inertial solvent reorganization, λ_S [52]. As shown in [83, 84] λ_I can be chosen as 0.3 eV, whereas the λ_S value was calculated by using the usual approximation [85]

$$\lambda_S = \frac{e^2}{4\pi\varepsilon_0}\left(\frac{1}{n^2} - \frac{1}{\varepsilon_s}\right)\left(\frac{1}{2r_{a^-}} + \frac{1}{2r_{d^+}} - \frac{1}{R}\right), \tag{3}$$

where n is the refractive index of the solvent. Using equations (1)-(3) the value of ET rate was calculated to be 2.5×10^9 s^{-1} in DMF and 1.7×10^9 s^{-1} in toluene.

Based on the data from time-resolved fluorescence measurements the rate of ET from the first excited singlet state of the pyropheo moiety, k_{ET}, was estimated to be 1.8×10^9 s^{-1} in polar solvent and 1.4×10^9 s^{-1} in toluene. It is seen that these values are in a good agreement with the results of theoretical calculations.

After being formed, the charge-separated (CS) state, having an energy of 0.94 eV (see Figure 4, top), undergoes in DMF charge recombination decay directly to the ground state with a time of 320 ps, as was found by analysis of the ps-TAS data [69]. These results can explain the very low value of photosensitized singlet oxygen generation quantum yield of FP2 in DMF.

In nonpolar toluene the CS state has a higher energy (E_{CS} = 1.74 eV) and lies above the pyropheo triplet state as well as the fullerene triplet state (Figure 4, bottom). According to the value of the quantum yield of photoinduced singlet oxygen generation and ps-TAS data we infer that in toluene a significant excitation of the pyropheo triplet state occurs due to relaxation of the CS state.

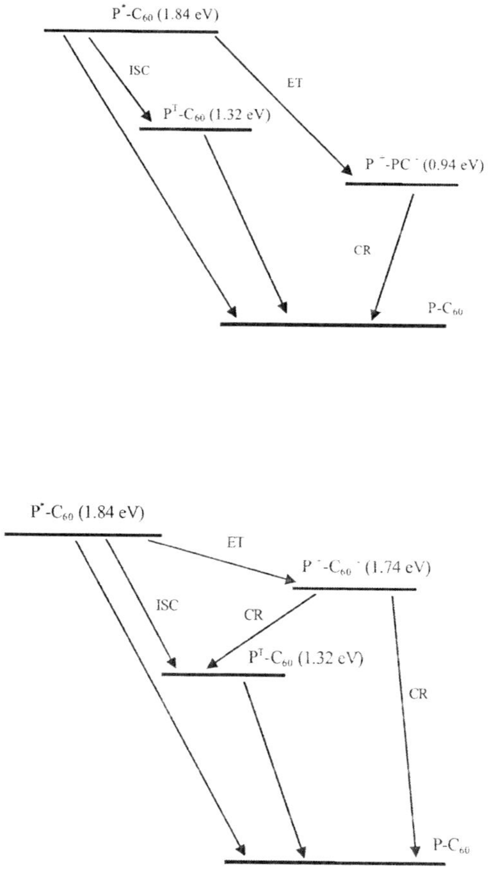

Figure 4. Scheme of energy levels of FP2 in DMF (top) and toluene (bottom) and transitions between them.

The photosensitized singlet oxygen quantum yield of FP2 in polar solvent is very low due to the fast and efficient depopulation of the first excited singlet state of pyropheo moiety. In order to use the fullerene-dye molecular systems for PDT, it is necessary to prevent the ET processes between dye and fullerene moieties as it is done in FHP2. In fact, it was demonstrated, that *in vitro* the phototoxic activity of FHP2 towards human leukemia T-lymphocytes (Jurkat cells) is much higher with respect to FP2 [86]. Moreover, the cell activity of FHP2 is even higher compared to P2 which has no fullerene core.

The obtained results show that fullerene hexakisadducts can be used as building blocks in modular carrier systems for photosensitizers, whereas for the mono-adduct FP2 the photoinduced electron transfer results in a dramatic reduction of singlet oxygen generation. Nevertheless, from observations made in the present investigations, it seems reasonable to assume that efficient photoconducting materials could be synthesized using FP2 motive and, moreover, it could be used as starting point for the construction of light-harvesting systems consisting of several separate pyropheo molecules and a charge-separating centre. In the next part we will discuss the photophysical properties of three different compounds carrying six pyropheo chromophores.

2. Electron and Energy Transfer in Dendritic Molecular Systems Carrying Six Pyropheophorbide Chromophores

The incorporation of a first-generation Newkome-type dendrimer allowed the synthesis of a series of three compounds, each carrying six pyropheo moieties: one without fullerene (P6), one with an electron accepting fullerene core (FP6, a [5:1]fullerene monoaddact) and one [5:1]fullerene hexaaduct (FHP6) which, due to its broken-up fullerene chromophore, does not accept electrons anymore (see Figure 5).

FP6 and FHP6 were obtained by using well-known organic chemistry applied to fullerenes and chlorin compounds. Pyropheophorbide *a* was connected to a 1,4-diamino butane spacer which was later on needed as the chlorin carrier. Malonic acid was modified by first attaching a C_6-spacer unit and subsequent connection to a first-generation Newkome-dendrimer. For the synthesis of FP6 this malonate unit was connected to the fullerene core utilizing a modified Bingel reaction. After liberation of the hexaacid the pyropheophorbide carrier was attached to the core which yielded the desired hexakis(pyropheophorbide *a*) fullerene FP6.

To obtain FHP6 five additional diethyl malonates were coupled to the fullerene core. For the preparation of the P6 compound the bis-dendrimer hexaacid was coupled with the aminylated pyropheophorbide system under Steglich conditions [87, 88]. The steady-state absorption spectra of pyropheo (as reference), P6, FP6 and FHP6 dissolved in DMF are presented in Figure 6. For FHP6 the absorbance increases by a factor of six with respect to pyropheo. This is not true for both P6 and FP6. The Q(0,0) absorption transition of pyropheo has an extinction coefficient of $\varepsilon = 4.2\times10^4$ $M^{-1}cm^{-1}$ whereas this value for P6 is $\varepsilon = 19.2\times10^4$ $M^{-1}cm^{-1}$. For FP6 the extinction coefficient in the maximum of the absorption Q(0,0) band is only $\varepsilon = 12.7\times10^4$ $M^{-1}cm^{-1}$ which is significantly lower than the value obtained for FHP6 and even for P6 [89]. The Q(0,0) band of FHP6, P6 and FP6 is bathochromic shifted compared to that one of pyropheo by about 2 nm (see also Table 2), while the Soret band of all three compounds carrying six pyropheo chromophores is slightly split and the second peak has its maximum around 403 nm.

Table 2. Selected photophysical parameters of FHP6, FP6, P6 and pyropheo in DMF

Sample	Soret [nm]	Q_y [nm]	λ_{fl}^{max} [nm]	τ_{fl}^{a} [ns]	Φ_{fl}	Φ_{ISC}	Φ_{Δ}
pyropheo	414	668	675	7.0 (1.0)	0.3	0.52	0.50
P6	413 403	670	680	0.14 (0.34) 0.7 (0.14) 3.3 (0.44) 5.7 (0.08)	0.058	0.19	0.15
FP6	413 403	670	678	0.11 (0.75) 0.6 (0.1) 1.7 (0.05) 5.7 (0.1)	$7\cdot10^{-4}$	0.04	0.02
FHP6	414 403	670	677	0.17 (0.27) 0.9 (0.14) 3.7 (0.51) 5.7 (0.08)	0.078	0.24	0.22

[a] Amplitudes of decay components are given in parenthesis.

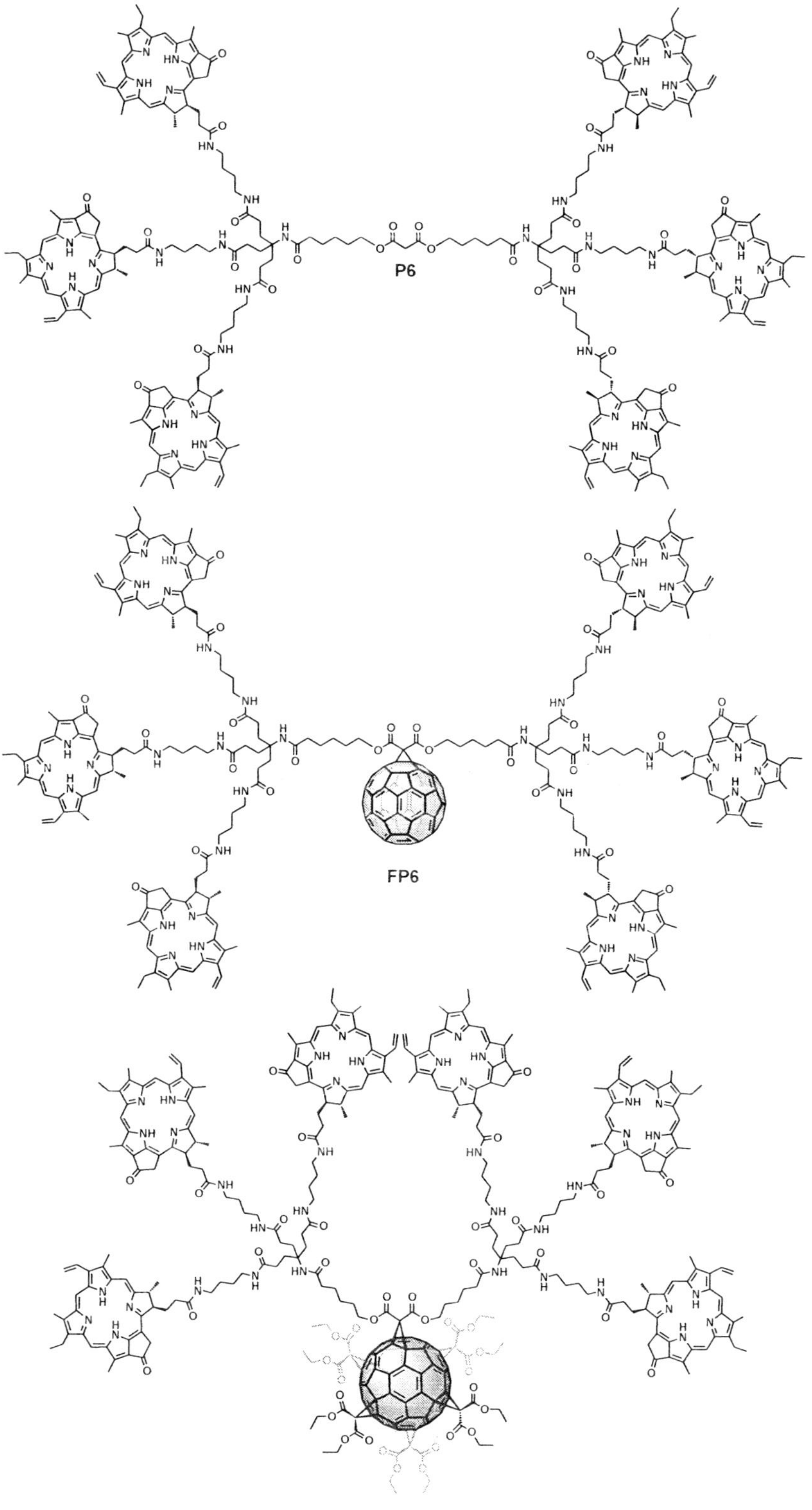

Figure 5. Structural formula of P6, FP6 and FHP6.

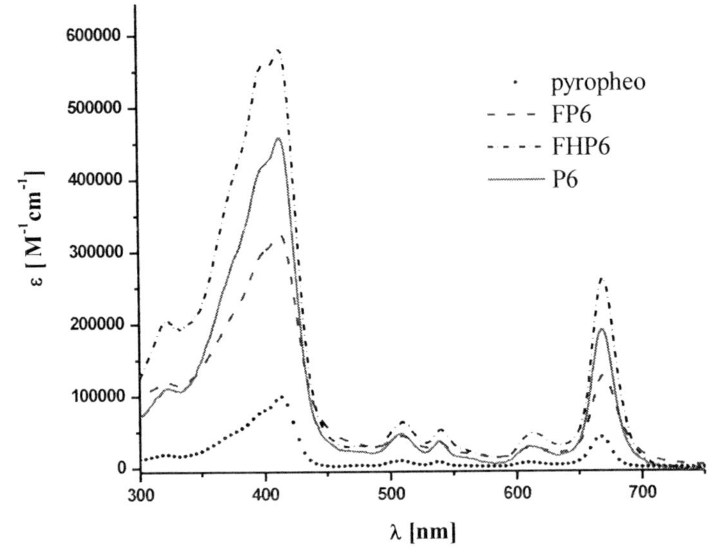

Figure 6. Steady-state absorption spectra of FP6, FHP6, P6 and pyropheo in DMF.

These effects already indicate an interaction between dye molecules in such (*hexa*)pyropheo molecular systems and the results of the interactions are more visible in the fluorescence measurements. The maximum of the emission spectrum of FHP6, FP6 and P6 is bathochromic shifted by 2, 3 and 5 nm with respect to that one of the reference pyropheo (Figure 7 and Table 2). Moreover, the shape of the fluorescence spectra of all three compounds under investigations is changed: it becomes broader and the gap between the main band and the vibronic shoulder is not so pronounced as for the pyropheo itself. The estimations have shown that the fluorescence quantum yield of FHP6 and P6 is 3.8 and 5.2 times lower than that of pyropheo, whereas for FP6 the quenching of the fluorescence is much stronger (see Table 2).

The fluorescence decay of FHP6, P6 and FP6 shows a complicate character and was fitted in all cases by a four exponential function. The fluorescence decay of FP6 exhibits a fast (τ_1 = 110 ps) and dominant deactivation process of the S_1-state (Table2). The intermediate times were determined to τ_2 = 0.6 ns and τ_3 = 1.7 ns, respectively. The longest decay time has a value of τ_4 = 5.7 ns similar to that observed for the fluorescence decay of P2 and FHP2 (see Table 1).

For FHP6 and P6 the short component was measured to be 0.17 ns and 0.16 ns, respectively. The two intermediate decay times had nearly similar values (about 0.8 ns and 3.5 ns), and the longest one was fitted to 5.7 ns. It should be mentioned, that all fluorescence decay times estimated for P6 are slightly shorter compared to FHP6. The long decay component had the lowest amplitude (below 10%) for these two compounds.

Using picosecond transient absorption spectroscopy (ps-TAS) an additional decay time was observed for recovery of the ground state population of FP6, FHP6 and P6. Decay components with values of 24 ps, 49 ps and 30 ps, respectively, were resolved. Moreover, it should be pointed that this short decay times were not observed in the time-resolved fluorescence experiments.

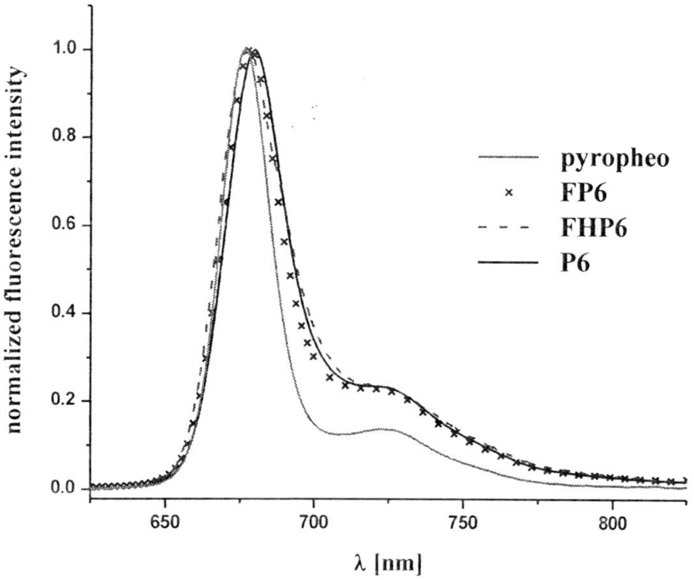

Figure 7. Steady-state fluorescence spectra of pyropheo, P6, FP6 and FHP6 in DMF.

Because of additional deactivation channels of the first excited singlet state of the (*hexa*)pyropheo compounds it is not surprising that their ISC quantum yields as well as their photoinduced singlet oxygen generation are reduced compared to those values of parent pyropheo. The intersystem crossing quantum yield of FP6 was determined to $\Phi_{ISC} = 0.04$ which is significantly lower compared to the values for FHP6 and P6 (0.24 and 0.16, respectively). Similar tendencies were observed for the photosensitized singlet oxygen generation (see Table 2).

At this point we split our discussion into two parts due to very different photophysical properties of FP6 and both FHP6 and P6. As previously mentioned, in [5:1]fullerene hexaadducts such as FHP6, the conjugation of the fullerene π-system is broken up into a system of eight separate benzene rings causing a dramatic decrease of the electron-accepting properties compared to those of C_{60}. In P6 there are only six pyropheo chromophores covalently linked via two dendritic branches but there is no electron accepting unit in the molecule. Taking into account these facts a photoinduced electron transfer can not be the reason for the experimentally observed reduction of the fluorescence signal of FHP6 as well as P6.

The strong reduction of the fluorescence quantum yield and the appearance of additional fluorescence decay times are a prove for an interaction between the dye molecules in FHP6 and P6. It should be mentioned that the same effects were observed for pheophorbide *a* (pheo) molecules covalently linked to DAB dendrimers of different generations [50, 90, 91]. In these molecular systems a strong quenching of the fluorescence as well as of singlet oxygen generation occurs due to an efficient energy transfer from initially excited pheo molecules to energy traps (trap: two stacked pheo molecules undergo dipole-dipole excitonic interaction). Moreover, for chlorophylls and related dye molecules it was shown that the splitting of the Soret band and the red shift of the absorption Q_y-band are characteristic for face-to-face dimer formation [53, 92, 93].

To explain the experimental results obtained for the FHP6 and P6 two processes are taken into consideration: trap formation and stepwise Förster-type energy transfer along pyropheo molecules to a trap.

It is known that dipole-dipole electronic energy transfer (EET) plays a significant role in dendritic structures [15, 90, 91]. But, of course, it is well understandable that Förster EET along pyropheo molecules could not be the only reason for the strong quenching of FHP6 as well as P6 fluorescence. Moreover, the EET only could occur between pyropheo chromophores belonging to one supramolecule due to the low concentrations of FHP6 and P6 used in our experiments. Our estimations have shown that in case of FHP6, for example, for a concentration of 1 μM the distance between neighboring pyropheo molecules belonging to different fullerene moieties is approximately 100 nm and for that, the probability of EET between them is negligible. In this way, only EET to the energy traps, which have weaker fluorescence than pyropheo can reduce the fluorescence quantum yield of FHP6 and P6 compared to that of pyropheo.

In order to estimate the rate of EET, k_{EET}, between pyropheo molecules we used the well-known relation [94–96]:

$$k_{EET} = \frac{1}{\tau_0} k^2 \left(\frac{R_0}{R} \right)^6 , \tag{4}$$

where τ_0 is the lifetime of the first excited singlet state of the pyropheo in absence of EET; k is the orientational factor, which depends on the relative orientation of the dipole moments of the interacting molecules and R is the distance between them. The Förster radius for dipole-dipole EET is defined as

$$R_0^6 = \frac{9000(\ln 10)\Phi_{fl}}{128\pi^5 n^4 N_a} \int I_{fl}^n(\tilde{\nu})\varepsilon(\tilde{\nu})\frac{d\tilde{\nu}}{\tilde{\nu}^4} \tag{5}$$

where n is the refractive index (n = 1.43 for DMF); N_a is Avogadro's number; $\varepsilon(\tilde{\nu})$ is the molar absorbance of the pyropheo at wavenumber $\tilde{\nu}$ and $I_{fl}^n(\tilde{\nu})$ is the normalized fluorescence spectrum of pyropheo. Using the last two equations and experimental data the Förster radius for dipole-dipole EET between pyropheo molecules was calculated to be 52 Å.

For FHP6 compound the average distance $\overline{R}$ between two neighboring pyropheo molecules belonging to the one fullerene moiety has a value of 14 Å ($4/3\pi\overline{R}^3 = 1/C$, where $C = \frac{N}{V} = \frac{N}{4/3\pi(R_{max}/2)^3}$ is the number of pyropheo molecules in the unity volume; N = 6 is the number of pyropheo units in one supramolecule; R_{max} is the maximum distance between two pyropheo moieties, which was estimated using HyperChem program [82] as 49 Å). For P6 this distance is slightly shorter and has a value $\overline{R}$ = 13 Å [89].

According to the above data and applying k^2 = 2/3 (this average value of the orientational factor is commonly used for solutions with uniform distribution of the transition dipole moments of the interacting dye molecules) as first approximation and $\tau_0 = 5.7$ ns, the energy transfer rate constant was calculated and found to be $k_{EET} \approx 3\cdot10^{11}$ s^{-1}. This value shows that excitation remains approximately 3 ps only at the initially excited dye molecule before hopping to another pyropheo chromophore within one molecular complex. As a result, after some acts of EET the energy will be delivered to a trap. Of course, it should be taken into account, that the mutual orientation of the transition dipole moments of the interacting molecules strongly affects k_{EET} and for molecular systems with a limited number of EET-participating molecules (in case of FHP6 and P6 there are only 6) the average value of k^2 could be different from the one used.

As was discussed above, only the presence of energy traps and the efficient delivery of excitation to them can explain the strong reduction of the fluorescence quantum yield of P6 and FHP6. The possibility of dimer formation by chromophores in one dendrimer- or fullerene-dye complex was proofed using molecular modelling (HyperChem program, MM+ method at room temperature and in vacuum [82]). In Fig. 8 for FHP6 as example, an energetically optimized conformation is presented. Although this picture shows just one possible conformation, it visualizes an effect that was visible for all calculations. The covalently linked dye molecules in FHP6 or P6 have a strong tendency to stack with each other. Moreover, for P6 this stacking has a higher probability than for FHP6. It should be mentioned, that since the calculations have been carried out in vacuum, in solution the stacking effects should be reduced. Otherwise, the reduction of the fluorescence as well as the singlet oxygen quantum yields for both compounds had to be much stronger than observed experimentally. Consequently, the real conformation can not be as tight as calculated in vacuum. As a result, complexes without any trap or with non-perfect traps should also exist. Non-perfect means, that their deactivation is not that fast resulting in a certain probability of radiative deactivation or back transfer to a single chromophore in the complex.

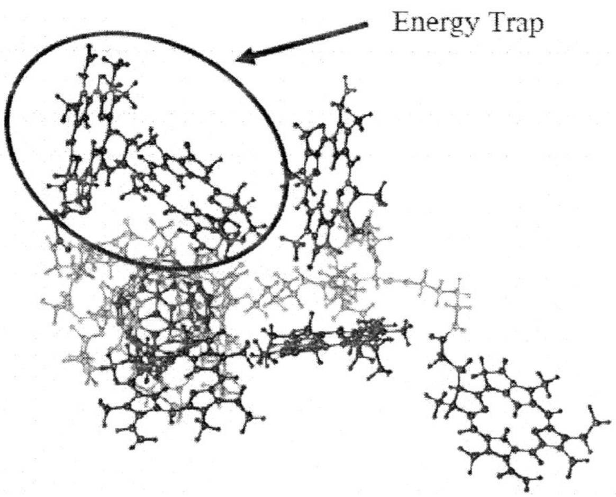

Figure 8. Possible conformation of FHP6 at room temperature; pyropheo moieties are shown in black.

Thus, two different types of energy traps were proposed to exist in the molecular systems under investigations. One of them (Trap I) is formed via face-to-face stacking of two dye molecules. The short component of the fluorescence decay of FHP6 and P6 observed in the time-resolved measurements (0.9 and 0.7 ns, Table 2) was associated with emission of this

Trap I [88–91, 97]. The second type of an energy trap (Trap II) has an oblique geometry of the interacting chromophores (the example of this type is shown in Figure 8). Due to the strength of excitonic interactions between dye molecules in Trap II, which is weaker than in Trap I, the fluorescence quantum yield of Traps II should be higher than that one of Traps I as well as the lifetime of the excited state should be longer (3.7 ns for FHP6 and 3.3 ns for P6, see Table 2). But both types of energy traps have lower fluorescence quantum yields and bathochromic shifted emission compared to the monomeric dye molecules. These conclusions are in a good agreement with results obtained for several dimers of reduced porphyrins (and among them pyropheo dimers) [53, 92, 93, 98, 99].

Due to the higher probability of the trap formation in P6 compared to FHP6 it is not surprising that the steady-state emission of P6 is more red-shifted than that one of FHP6 with respect to the pyropheo fluorescence spectrum and, moreover, the absorbance of P6 is reduced compared to FHP6 (traps have lower absorption cross-section than monomeric chromophores [93]).

In the presence of one or more traps in a complex an additional decay time arises, which represents the probability of energy transfer to the trap. Using ps-TAS technique this time was estimated to be 49 ps for FHP6 and 30 ps for P6. The faster delivery of the excitation energy to a trap in P6 is due to the fact, that the average distances between pyropheo chromophores in FHP6 are longer compared to those in P6.

The short fluorescence decay time (around 170 ps) one can attribute to the mixture of the above discussed short time observed in TAS experiments and the fluorescence decay of the energy traps where interaction between two pyropheo chromophores is very strong.

The ISC quantum yield, Φ_{ISC}, of FHP6 and P6 is only 2.1 and 2.7 times lower than that one of pyropheo, which means that the traps must be able to undergo an intersystem crossing. The probability of this process should be higher than the probability of a trap to emit light since the fluorescence quantum yields of FHP6 and P6 is 3.8 and 5.2 times lower than the one of pyropheo (Table 2). This fact correlates well with theory of excitonic interactions [100–103].

Thus, summarizing the whole discussion above we conclude that EET as well as partially excitonic interactions between chromophores covalently linked in the dendrimer-dye molecular systems significantly affect the photophysical properties of such compounds. Moreover, the strength of the interactions strongly depends on the conformation of the system and distances between pyropheo chromophores in one complex. From this point of view, the C_{60} core in FHP6 can be regarded as a spacer preventing dense folding of pyropheo molecules covalently linked to the same fullerene. In Figure 9 a schematic sketch of the transfer processes in a dendrimer-dye-complex is given.

Preliminary results of experiments with FHP6 and P6 have shown that the accumulation patterns of FHP6 and P6 in Jurkat cells are similar [104, 105]. Surprisingly, FHP6 has a much higher intracellular concentration than P6 (approximately 7 fold). In other words, FHP6 seems to have a better ability to carry pyropheo molecules into Jurkat cells. The explanation of this different uptake has to be related to the chemical structure of these two macromolecules. Both derivatives carry the same number of pyropheo chromophores and have the same dendritic structure. The only structural difference is represented by the C_{60} fullerene moiety. This leads to the conclusion that C_{60} confers to the dendron the ability to enter Jurkat cells.

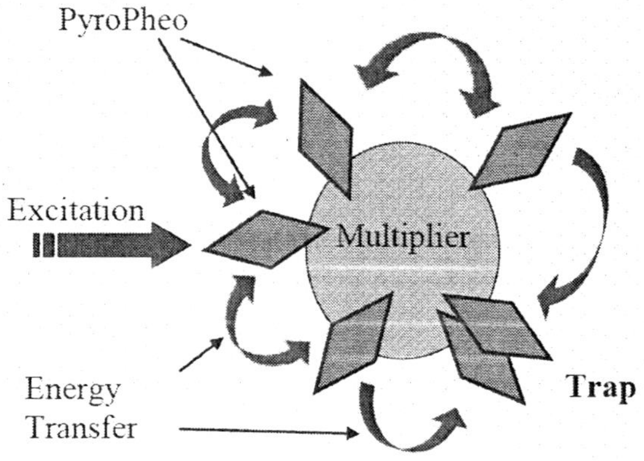

Figure 9. Schematic sketch of a dendrimer-dye-complex to illustrate the proposed energy transfer processes between dye molecules.

P6 induced at all light doses, approximately 10 % of cell death. The phototoxicity of FHP6 was significantly higher and light-dose dependent. Moreover, at complex concentrations of 5 and 1 μM, FHP6 has shown a higher phototoxicity than FHP2 [104, 105], which is due to the higher number of pyropheo molecules per one fullerene-dye molecular system. The higher phototoxicity of FHP6 with respect to P6 is due to its higher intracellular uptake as well as higher singlet oxygen generation (Φ_Δ has a value of 0.22 for FHP6 and 0.15 for P6, see Table 2). The significant improvement of both uptake and phototoxicity observed in FHP6 with respect to P6 indicates that the fullerene moiety has an important effect on the properties of these macromolecular systems. It influences the molecule three-dimensional structure by hindering the interactions between the pyropheo chromophores and therefore improving the photosensitized singlet oxygen generation. Moreover, the lipophilic fullerene sphere bound to the more hydrophilic dendron gives an amphiphilicity molecule with a strongly improved intracellular uptake.

Let us switch now to discussion of the FP6 compound and its photophysical properties which significantly differ from that of P6 and FHP6.

First of all, it is reasonable to assume that trap formation will also occur in FP6 due to its similar structure when compared with FHP6. Using molecular dynamic simulations [89] it was shown that in FP6 due to the fullerene core the pyropheo chromophores do not fold as dense as in P6. Therefore, the probability of a trap formation should be lower for FP6 than for P6. But FP6 demonstrated the lowest absorbance of all compounds carrying six pyropheo chromophores. As a larger spacing between the pyropheo chromophores mediated by the fullerene would lead to an increase of absorbance compared to P6 another effect induced by C_{60} seems to be more important. The fullerene core in FP6 with its extended π-electron system is assumed to disturb the electronic ground state properties of the attached pyropheo chromophores leading to a significant reduction of the absorbance of this compound. It should be remembered that similar effect was already observed for FP2 which has also a reduced absorbance compared to FHP2 with broken up π-system of the fullerene core (see Figure 2).

Interestingly, the existence of the energy traps alone can not explain such a strong reduction of the fluorescence, ISC and singlet oxygen quantum yields with respect to FHP6 as well as P6 (Table 2). One additional process, which needs to be taken into account is the photoinduced electron transfer between pyropheo and fullerene moieties covalently linked in

FP6. As it was shown above, for FP2 the photoinduced ET from an initially excited pyropheo moiety to the electron accepting C_{60}-fullerene is the dominant process leading to the fast deactivation of the first excited singlet state of pyropheo with a rate of 1.8×10^9 s^{-1}. The investigation of FP6 revealed a strongly reduced fluorescence quantum yield, $\Phi_{fl} = 7\times10^{-4}$, a low intersystem crossing quantum yield, Φ_{ISC}, of 0.04 and the occurrence of a fast dominant quenching process visible in the time-resolved fluorescence decay (the decay component of 110 ps with 75 % contribution in the fluorescence signal). These findings can be attributed to an additional very efficient S_1-state deactivation process, namely, ET.

Using Rehm-Weller equation (1) the forward free enthalpy for the charge separation was calculated to be -0.93 eV in DMF, showing that ET is an energetically favorable process. The rate of ET, k_{ET}, has a value of 7.2×10^9 s^{-1} as it was calculated using the approach described above (eq. (2) and (3)), which is in a good agreement with the experimentally observed quenching rate of the first excited singlet state of pyropheo moiety ($k_q\approx1/110$ ps$=8.9\times10^9$ s^{-1}).

After photoexcitation of FP6 an intramolecular competition between energy delivery to a trap and its transfer to the pyropheo chromophore located close to the fullerene core takes place. On the one hand, the energy transfer to a trap reduces the probability of an electron transfer especially in the case when Trap I is formed. On the other hand, EET facilitates delivery of the excitation from the periphery of the system to the pyropheo unit in close proximity to the fullerene. As a result, the probability of ET is increased approximately 5 times compared to that one in FP2. Due to this fact, the FP6 supramolecule can be considered as combination of a light-harvesting system consisting of several separate pyropheo molecules and a reaction center.

3. Dissipation of The Electronic Excitation in Highly Branched Fullerene-Dendrimer-Pyropheo Molecular Systems

As it was stressed above the EET and partly excitonic interactions determine the dominant pathways of the nonradiative deactivation of locally excited states of photosensitizer chromophores covalently linked in one dendrimer-like supramolecular system. In order to gain a better understanding of the possible photoprocesses and interactions in such complex systems, it seems reasonable to provide the comparative analysis of the photophysical properties of highly branched fullerene-based globular dendrimers with a different number of attached chromophores.

The synthesis of a [6:0]fullerene hexaadduct which carries twelve pyropheophorbide *a* units (FHP12) started with coupling of a malonate unit to C_{60} under template conditions to give the fullerene hexaadduct compound. After removal of protecting groups with acid a dodecakis amino-substituted precursor compound was generated and afterwards directly reacted with the *N*-succinimid ester of pyropheo which delivered the desired fullerene [6:0]-hexaadduct FHP12 (Figure 10) in excellent yield [87, 88].

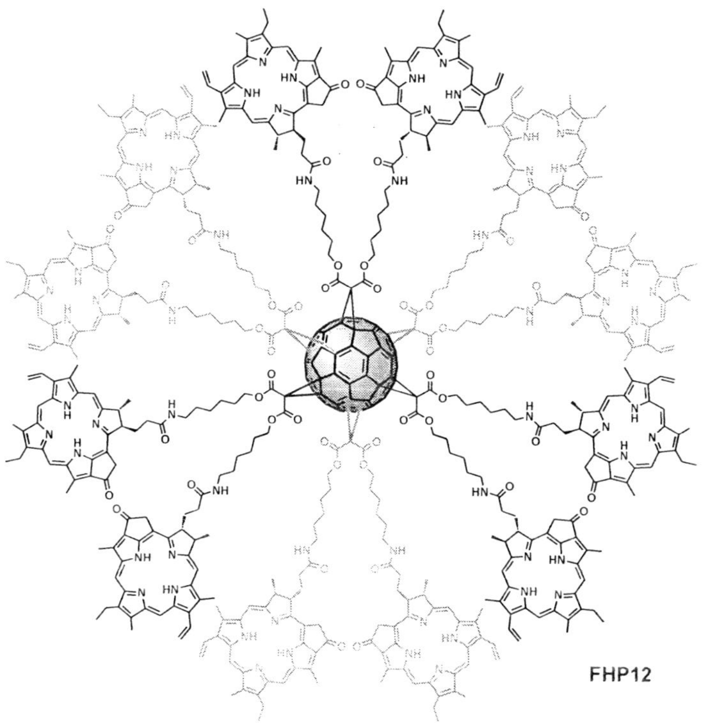

FHP12

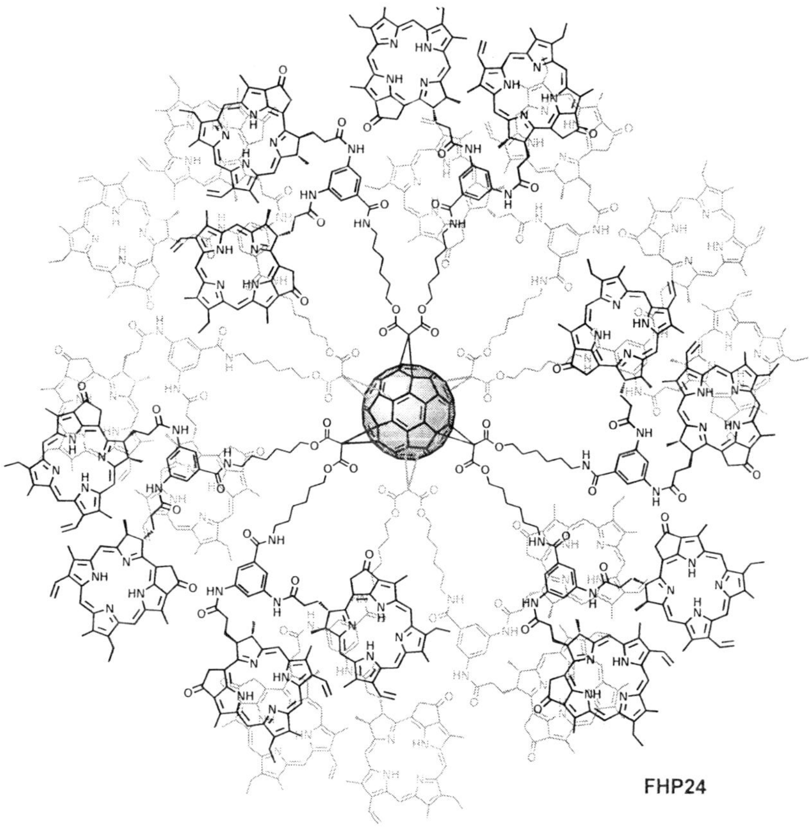

FHP24

Figure 10. Structural formula of FHP12 and FHP24.

For the preparation of FHP24, a (*bis*)pyropheophorbide system derived from 3,5-diamino benzene carboxylic acid was coupled to the same fullerene precursor which was used for FHP12 [87]. Although both FHP12 and FHP24 seem to be sterically congested, there is still plenty of space around the fullerene and enough rotational freedom for each pyropheo unit.

The steady-state absorption spectra of FHP12 and FHP24 dissolved in DMF are shown in Figure 11. The shape and spectral position of the absorption bands of FHP12 are practically the same as those of the FHP6 (see Table 3). The maximum of the Q_y-band absorption of FHP12 compared to that of FHP2 is shifted bathochromic by 1.5 nm while the Soret band of FHP12 is split and the second peak with lower absorbance has its maximum at 403 nm (see Figure 11 and Table 3).For FHP24 the strength of the interactions between the pyropheo chromophores is much stronger compared to the fullerene-based compounds carrying lower number of dye molecules (FHP2, FHP6 and FHP12). As result, strong perturbations are seen in the absorption spectrum of FHP24. The split Soret band undergoes a hypsochromic shift by 4 nm (235 cm^{-1}) and, moreover, the intensity of both peaks becomes practically identical (Fig. 11). At the same time, the Q_y-band in the absorption spectrum of FHP24 is bathochromically shifted by 12 nm (265 cm^{-1}) in respect to its position for FHP2 (Table 3). Additionally, a strong broadening of the last absorption Q-band of FHP24 occurs which follows a trend observed in the series FHP2, FHP6 and FHP12. The full width at a half maximum (FWHM) of the Q_y absorption bands is slightly increased from 21 nm for FHP2 to 23 nm for FHP12, whereas for FHP24 the bandwidth is 32 nm (Table 3).

It was observed, that neither do FHP12 reach the expected roughly 12-fold increase nor does FHP24 attain the 24-fold level of the extinction coefficient of the pyropheo molecule.

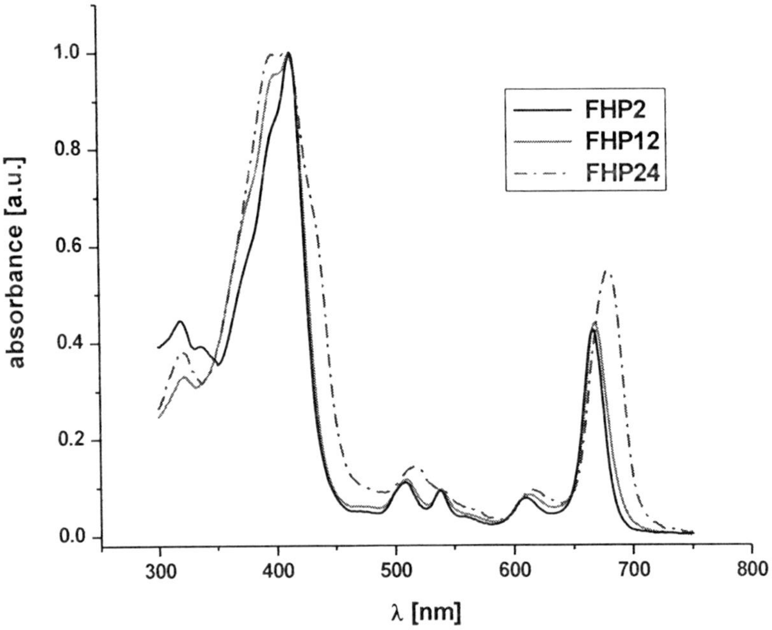

Figure 11. UV/Vis absorption spectra of FHP12 and FHP24 in DMF (the spectrum of FHP2 is also shown as a reference).

Table 3. Photophysical parameters of FHP2, FHP6, FHP12 and FHP24 in DMF

Sample	Soret [nm]	Q_y [nm]	FWHM* [nm]	λ_{fl}^{max} [nm]	Φ_{fl}^{b}	Φ_{ISC}	Φ_{Δ}
FHP2	414	668	21	675	0.23	0.49	0.43
FHP6	414 403	670	22	677	0.078	0.24	0.22
FHP12	414 403	669.5	23	679	0.023	0.14	0.13
FHP24	410 399	680	32	690	3.7×10^{-3}	0.13	0.10

*Full width at a half maximum (FWHM) of the absorption Q_y-band.

With increasing the number of pyropheo chromophores attached to the fullerene core, the maximum in the steady-state fluorescence spectrum undergoes a bathochromic shift (see Table 3). Moreover, the shape of the fluorescence spectrum of FHP12 is changed: it becomes broader and the gap between main band and vibronic shoulder is not as pronounced as was observed for FHP2 (Figure 12*a*). Surprisingly, the maximum of the emission spectrum of FHP24 has the same position as of FHP2, and the shape of the fluorescence band is not so broad as that one of FHP12. Nevertheless, it should be emphasized, that the intensity of the fluorescence of the fullerene-based supramolecular system carrying 24 pyropheo chromophores was very weak compared to that one of FHP2.

Further analysis has shown that the emission spectrum of FHP24 is a superposition of two fluorescence spectra (Figure 12*b*). One of them gives dominant contribution to the fluorescence signal and has the spectral position and shape of the band similar to the emission spectrum of parent pyropheo. The second one with much lower intensity is bathochromically shifted by 320 cm^{-1} and its maximum lies at 690 nm, correlating with abovementioned bathochromic shift of the Q(0,0) absorption band of FHP24 (Table 3).

To clarify the situation, the decay associated spectral analysis [106–108] of FHP12 and FHP24 fluorescence has been performed. For both FHP12 and FHP24 compounds four decay spectra were resolved (Figure 13). One of them has a decay time of 6.3 ns for FHP24 and 5.7 ns for FHP12. This component has the lowest amplitude, its shape and the spectral position of its maximum (675 nm) are similar to the steady-state fluorescence of non-disturbed pyropheo.

For FHP12 the decay components with lifetimes of 1.93 and 0.51 ns represent the fluorescence of Trap II and Trap I, respectively, and are located at 680 nm. It is seen, that the amplitude of the long decay component is slightly higher with respect to that of the shorter one (Fig. 13*a*). It should be mentioned that both decay times are shorter, compared to that observed in the time-resolved measurements of the FHP6 fluorescence (Table 2), indicating that the interactions between pyropheo chromophores are stronger in FHP12 compared to FHP6.

The dominant contribution in the fluorescence signal of both FHP12 and FHP24 compounds gives the short decay component belonging to the emission of the pyropheo moieties S_1 state of which is quenched by EET. For FHP24 the amplitude of this decay component is minimum five times higher compared to all other decay associated fluorescence spectra, which is not surprising if one takes into account the high local density of pyropheo molecules in the 24-mer.

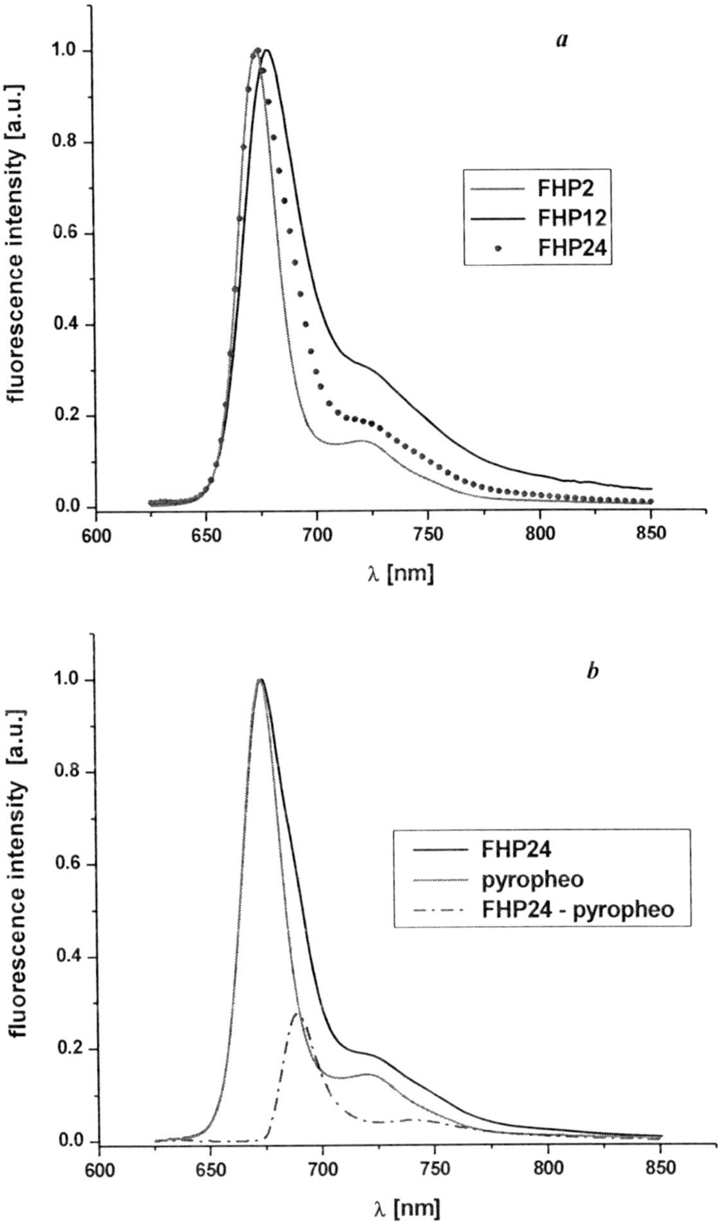

Figure 12. *a*: steady-state fluorescence spectra of FHP12 and FHP24 in DMF (the spectrum of FHP2 is also shown for comparison); *b*: deconvolution of the FHP24 and pyropheo fluorescence spectra.

In steady-state measurements of the fluorescence the signal amplitude is proportional to the fluorescence quantum yield; it is proportional to the natural fluorescence rate in the case of time resolved measurements, however. The combined analysis of the data obtained by steady-state and time-resolved measurements leads to the following interpretation: FHP24 shows the fluorescence spectra with a maximum located at 690 nm with decay times of 47 ps, 330 ps and 1.96 ns.

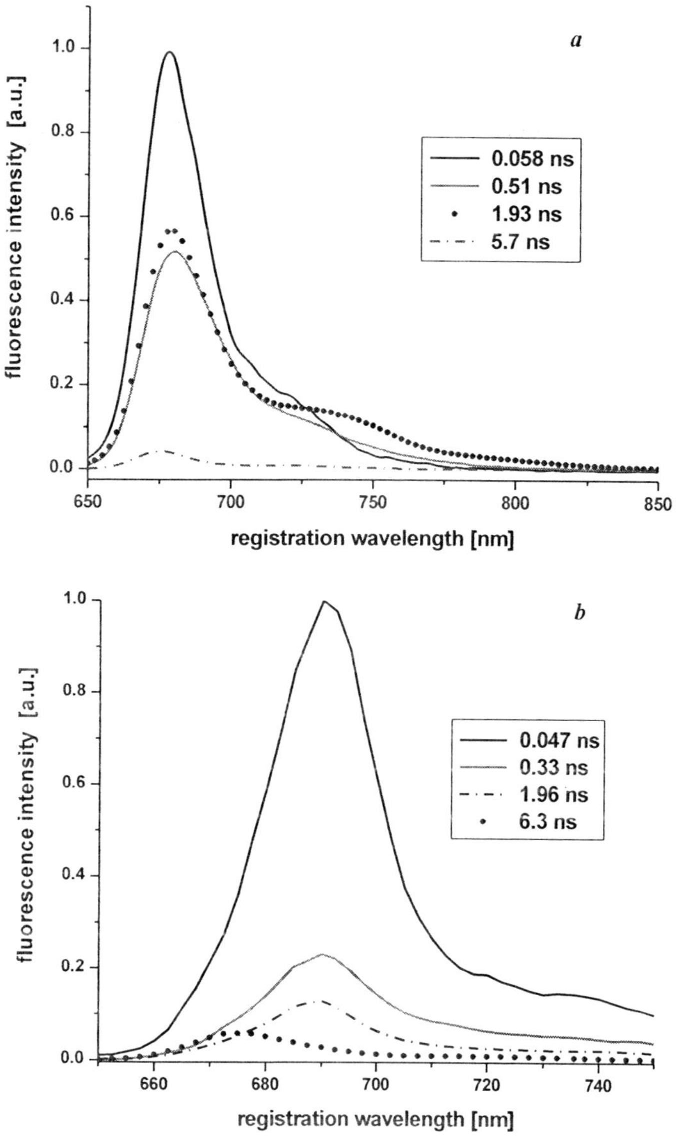

Figure 13. Decay associated fluorescence spectra of FHP12 (*a*) and FHP24 (*b*)

The long lived species in solution is assigned to a small amount of remaining educts of synthesis with comparable to pyropheo monomer high fluorescence quantum yield. Therefore, the steady-state spectrum of FHP24 is affected by these unwanted contributions and they have to be excluded from the analysis. Due to this reason, only the maximum of FHP24 fluorescence at 690 nm is compiled in Table 3. It should be emphasized, that according to the time-resolved measurements the concentration of this impurity could be estimated to be less than 5 %.

Hence, to calculate the fluorescence quantum yield of FHP24 the steady-state fluorescence spectrum was corrected by subtraction of the fluorescence from pyropheo resulting in a spectrum of pure FHP24 fluorescence.

The estimations have shown that the fluorescence quantum yield of FHP12 is about 3 and 10 times smaller compared to that of FHP6 and FHP2, respectively (Table 3). For FHP24 the quenching of the first excited singlet state is much stronger resulting in a more than 62 times reduction of the fluorescence quantum yield in respect to this value for FHP2 compound.

In general the singlet oxygen luminescence is also decreasing with rising number of dye molecules per one fullerene-based complex (Table 3). Interestingly, for all samples the quantum yield of singlet oxygen is less reduced compared to the reduction of the fluorescence quantum yield. For example, for FHP12 the fluorescence intensity is reduced by a factor of about 10 relative to that of FHP2 (for FHP24 this reduction is even much higher), the corresponding singlet oxygen quantum yield is reduced only by a factor of 3 (for FHP24 the value of Φ_Δ is about a quarter of that one for FHP2).

All results presented above are well correlated with the model described in the previous part of the paper. It was shown that stepwise intramolecular Förster energy transfer between pyropheo moieties causes a very fast and efficient delivery of the excitation to an energy trap formed by two stacked and excitonically interacting pyropheo chromophores. As result, the fluorescence as well as the singlet oxygen quantum yields of FHP6 and P6 are reduced compared to those values of the references FHP2 and P2 (see Table 2).

For FHP12, the interactions between the pyropheo chromophores should become stronger than for FHP6 due to the higher local concentration of pyropheo moieties. Indeed, based on molecular modelling studies it was shown that the average distance between two neighbouring pyropheo units belonging to one fullerene moiety ($\overline{R}$) is shorter for FHP12 than for compound FHP6 (6 Å and 14 Å, respectively [88]). For FHP24, the distances between neighbouring pyropheo chromophores are even shorter compared to FHP12, and, as result, the probability of trap formation is higher. Also, the delivery of excitation to a trap is faster. Moreover, with increasing number of pyropheo units covalently linked to one fullerene core the ratio between the concentrations of Trap I and Trap II is changing: for FHP6 this ratio is 1:4, for FHP12 it is approximately 1:1, but for FHP24 it becomes 2:1.

It should be emphasized, that similar results were observed for DAB dendrimers from 1st to 4th generation which were substituted with the maximum achievable quantity of pheophorbide *a* (pheo) molecules [90, 91]. It was shown, that with increasing number of pheo molecules per one complex the Q(0,0) absorption as well as fluorescence band undergo bathochromic shift, whereas the Soret band shifts hypsochromically. At the same time, the bandwidth of both absorption and fluorescence spectra becomes broader. The fluorescence decay has a multiexponential character for all investigated in [90, 91] dendrimer-based compounds. Moreover, both fluorescence and singlet oxygen quantum yields were strongly reduced (the reduction of the fluorescence quantum yield was much higher compared to the singlet oxygen generation) with respect to these values for monomeric pheo. It was found, that a larger number of chromophores in one dendrimer-dye-complex increases the stacking probability of pheo moieties accompanied by excitonic interactions leading to a fast non-radiative deactivation after excitation. The lack of a spacer molecule like the C_{60} in the fullerene-based systems made the energy transfer in the DAB-pheo complexes even more effective, resulting in a faster delivery of the excitation to a formed trap. Furthermore, the interactions between dye molecules within one complex are stronger in DAB-dendrimer systems compared to the fullerene-based ones carrying the same number of chromophores. For this reason the probability of traps formation is higher for DAB-pheo complexes than for fullerene-pyropheo systems. As result, for all fullerene-pyropheo compounds the reduction of the fluorescence quantum yield is not as large and changes in the absorption as well as in the fluorescence spectra are less pronounced than for DAB-pheo molecular complexes [91].

During illumination with cw-laser radiation the fluorescence quantum yield of FHP2, FHP6 and FHP12 compounds is reduced (Figure 14*a*). For all of them, this reduction has monotonic character and after one and a half hour the fluorescence quantum yields were 15%, 33% and 38% less than the initial values for FHP12, FHP6 and FHP2, respectively [88, 91]. For FHP24, these experiments were not carried out due to the very low fluorescence quantum yield of this compound, and, moreover, the strong influence of the emission of pyropheo-containing precursor moieties on the fluorescence of FHP24, as it was shown above.

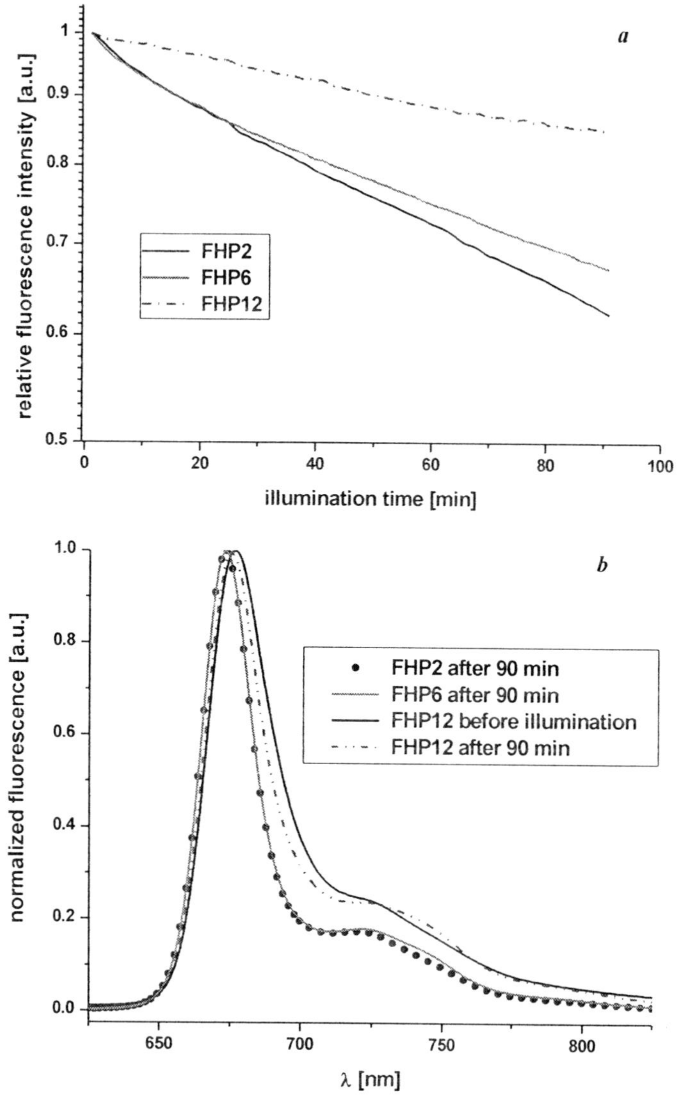

Figure 14. Fluorescence intensities during illumination (*a*) and fluorescence spectra after 1.5 hour of illumination (*b*) of FHP2, FHP6 and FHP12 in DMF [after 88]

The absorbance of the samples decreases monotonically with increasing illumination time. No changes of spectral position or shape of the absorption bands were observed. After 1.5 hour of illumination the fluorescence spectrum of FHP6 became similar to that one of FHP2 (Figure 14*b*). For FHP12 the maximum of the fluorescence spectrum is hypsochromically shifted compared to its initial position and, moreover, the bandwidth becomes narrow. It should also be mentioned that no recovery in the dark has been found. This was proofed be taking absorption as well as fluorescence spectra in 1 day after illumination. The shape, spectral position and intensity of the bands were the same as in the spectra measured after 90 minutes of illumination.

It is well-known, that the photostability of molecular systems strongly depends on their ISC as well as their singlet oxygen generation quantum yields. During illumination the excitation is efficiently delivered to the formed traps in both FHP6 and FHP12. Due to the fact that the traps have a lower ISC and singlet oxygen quantum yields compared to monomeric pyropheo, the photostability of FHP6 and FHP12 is higher than that of FHP2 (Figure 14*a*). Compared to FHP6 the interactions between pyropheo chromophores in FHP12 are stronger, leading to a higher number of formed traps. Therefore, this compound exhibits a much higher photostability than the other systems carrying only two or six dye molecules. Due to high probability of energy transfer between pyropheo chromophores in both FHP6 and FHP12 the excitation energy is very fast delivered to the traps which undergo the photodestruction in the first place. The hypsochromic shift of the fluorescence maximum and the reduction of intensity at the red edge of the fluorescence spectra, where the trap emission gives the major contribution in the fluorescence signal (see Figures 13 and 14*b*), are a proof for the above discussed model.

Conclusion

Since the last decades, following the strategy of nature, different chemical approaches were applied by numerous scientific groups to mimic primary events in photosynthesis as well as to realize the intrigues possibilities of making nanosize optoelectronic components from single supramolecules or well-organized molecular assemblies. Due to their highly branched and well-defined architecture, which consists of a high number of functional end groups, dendrimers are widely employed for supramolecular engineering. On the other hand, the unique photophysical and photochemical properties of fullerenes C_{60} have led to the synthesis of a large number of compounds in which fullerenes are covalently linked to different kinds of photoactive groups. It is reasonable to expect, that fullerene-dendrimer-dye complexes will open the way for new approaches to designing artificial light-harvesting systems as well as efficient drug delivery systems for different biomedical applications and, among them, photodynamic therapy of tumors.

Following the proposed concept of modular drug delivery system, two novel mono fullerene-(*bis*)pyropheo derivatives (FP2 and FHP2) were synthesized and there photophysical properties were investigated. It was revealed that in FP2 molecular system a strong quenching of the first excited singlet state of the pyropheo and, as result, dramatical decrease of photosensitized singlet oxygen generation occurs by efficient photoinduced electron transfer to the fullerene molecule. Due to this result, FP2 is not suitable for PDT, but it seems reasonable to assume that efficient photoconducting materials could be synthesized using FP2 motive. In contrast, FHP2 which possesses five diethyl malonate addends in the remaining octahedral positions, shows a high singlet oxygen quantum yield which is due to the reduced fullerene chromophore having lost its electron accepting capability. As result, the high phototoxic activity of FHP2 towards human leukemia T-lymphocytes was observed.

Steady-state and time-resolved experiments have shown complex energy and electron transfer dynamics in fullerene-based systems carrying six pyropheo chromophores. It was found, that due to the length and high flexibility of the hydrocarbon spacer chains neighboring pyropheo molecules stack with each other. The excitonic interactions between stacked pyropheo units lead to the formation of different energy traps, which were resolved experimentally. The dipole-dipole resonance Förster energy transfer between the dye molecules coupled to one core caused a very fast and efficient delivery of the excitation to a trap. It was found, that the probability of trap formation as well as the delivery rate of excitation to a formed trap are lower for FHP6 compared to the dendrite P6 without fullerene moiety. Due to this fact, the fullerene-hexaadduct core can be considered as a spacer unit hindering the stacking between chromophores within one molecular system.

Molecular simulations have shown that in FP6 one pyropheo chromophore can be located in closer proximity to the electron-accepting C_{60} fullerene core. This opens up the possibility of a fast electron transfer from photoexcited pyropheo to the fullerene moiety. Our measurements suggest that such a fast ET is indeed the main deactivation pathway. Moreover, the energy transfer process facilitates delivery of the excitation from the periphery of the system to the pyropheo unit in close distance to the fullerene. As result, the probability of electron transfer is increased approximately 5 times compared to that one in FP2. The FP6 supramolecule can be considered as combination of a light-harvesting system consisting of several separate pyropheo molecules and a charge-separating center.

The strength of interactions between pyropheo chromophores is higher for the complexes with higher number of dye molecules and close distance between them. Therefore, the reduction of the fluorescence as well as of ISC quantum yields of the different complexes becomes stronger with increasing number of dye molecules per complex. Nevertheless, in every case the ISC and the singlet oxygen generation was less influenced than the fluorescence quantum yield, exposing the supramolecular complex to a non-negligible amount of excited oxygen in the singlet state. Moreover, the photostability of the fullerene-dendrimer-pyropheo compounds increases with rising number of pyropheo chromophores covalently linked to one fullerene.

In the future, more experiments with these highly-branched fullerene-based molecular systems are expected to reveal still more detailed interplay between different radiative and nonradiative pathways. We hope to gather enough information to optimize the structures of supramolecular complexes and the energy conversion within them with possible applications as photoactive materials in mind.

ACKNOWLEDGEMENTS

The authors would like to thank the Deutsche Forschungsgemeinschaft (DFG) (E.A.E., B.R.: grant Nr. RO 1042/9-4, RO 1042/11-1, 2; M.H., N.J.: grant Nr. HI 468/11-1, 2) for financial support. They also would like to thank Mrs. Gisela Wöhlecke (Berlin) for technical assistance and Priv.-Doz. Dr. Andreas Humeny (Erlangen) for performing the MALDI-TOF experiments.

REFERENCE

[1] Tomalia, D. A.; Hedstrand, D. M.; Ferritto, M. S. *Macromolecules* 1991, 24, 1435-1438.

[2] Newkome, G. R.; Charles, M. N.; Vögtle, F. *Dendrimers and Dendrons*; Wiley-VCH: Weinheim, 2001.

[3] Fréchet, J. M. J. *Science* 1994, 263, 1710-1715.

[4] Fréchet, J. M. J.; Henmi, M.; Gitsov, I.; Aoshima, S.; Leduc, M. R.; Grubbs, R. B. *Science* 1995, 269, 1080-1083.

[5] Buhleier, E.; Wehner, W.; Vögtle, F. *Synthesis* 1978, 1978, 155-159.

[6] Twyman, L. J.; Beezer, A. E.; Esfand, R.; Hardy, M. J.; Mitchell, J. C. *Tetrahedron. Lett.* 1999, 40, 1743-1746.

[7] Röder, B.; Hackbarth, S.; Wöhlecke, G. US-Patent, NDN 172-0037-2622-7. Publication Nr. 00108704 WO, German Patent Nr. 199 36 997.6 02.08.1999.

[8] Jansen, J. F. G. A.; de Brabander-van den Berg, E. M. M.; Meijer, E. W. *Science* 1994, 266, 1226-1229.

[9] Li, W.-S.; Kim, K. S.; Jiang, D.-L.; Tanaka, H.; Kawai, T.; Kwon, J. H.; Kim, D.; Aida, T. *J. Am. Chem. Soc.* 2006, 128, 10527-10532.

[10] Atwood, J. L.; Davies, J. I. C.; Macnicol, D. D.; Vögtle, F.; Lehn, J.-M. *Comprehensive Supramolecular Chemistry*; Pergamon Press: Oxford, 1996.

[11] Hasobe, T.; Kashiwagi, Y.; Absalom, M. A.; Sly, J.; Hosomizu, K.; Crossley, M. J.; Imahory, H.; Kamat, P. V.; Fukuzumi, S. *Adv. Mater.* 2004, 16, 975-979.

[12] Martín-Delgado, M. A.; Rodriguez-Laguna, J.; Sierra, G. *Phys. Rev. B* 2002, 65, 155116.

[13] Andrews, D. L.; Bradshaw, D. S. *J. Chem. Phys.* 2004, 121, 2445-2454.

[14] Camps, X.; Dietel, E.; Hirsch, A.; Pyo, S.; Echegoyen L.; Hackbarth, S.; Röder, B. *Chem. Eur. J.* 1999, 5, 2362-2373.

[15] Yeow, E. K. L.; Ghiggino, K. P.; Reek, J. N. H.; Crossley, M. J.; Bosman, A. W.; Schenning, A. P. H. J.; Meijer, E. W. *J. Chem. Phys. B* 2000, 104, 2596-2606.

[16] Dolphin, D. *Can. J. Chem.* 1994, 72, 1005-1013.

[17] Röder, B. In *Encyclopedia Analytical Chemistry*; Meyers, R. A.; Ed.; Wiley: Chichester, 2000, pp 302-322.

[18] Konan, Y. N.; Gurny, R.; Allémann, E. *J. Photochem. Photobiol.* B 2002, 66, 89-106.

[19] Freeman, A. I.; Mayhew, E. *Cancer* 1986, 58, 573-583.

[20] Dressler, C.; Möller, U.; Lewald, T.; Berlien, H.-P.; Röder, B.; Risse, H.-J. *Lasers in Med. Science* 1992, 7, 174-179.

[21] Roberts, J. C.; Adams, Y. A.; Tomalia, D.; Mercer-Smith, J. A.; Lavalee, D. K. *Bioconjugate Chem.* 1990, 1, 305-308.

[22] Patri, A. K.; Thomas, T.; Baker, J. R.; Bander, N. H. *Polymer Material Science Engineering* 2002, 86, 130.

[23] Imahori, H.; Sakata, Y. *Adv. Mater.* 1997, 9, 537-546.
[24] Imahori, H.; Mori, Z.; Matano, Y. *J. Photochem. Photobiol. C: Photochem. Reviews* 2003, 4, 51-83.
[25] Rio, Y.; Nierengarten, J.-F.; *Tetrahedron Lett.* 2002, 43, 4321-4324.
[26] Tkachenko, N. V.; Rantala, L.; Tauber, A. Y.; Helaja, J.; Hynninen, P. H.; Lemmetyinen, H. *J. Am. Chem. Soc.* 1999, 121, 9378-9387.
[27] Kuciauskas, D.; Liddell, P.A.; Lin, S.; Johnson, T. E.; Weghorn, S. J.; Lindsey, J.; Moore, A. L.; Moore, T .A.; Gust, D. *J. Am. Chem. Soc.* 1999, 121, 8604-8614.
[28] Hirsch, A. *The Chemistry of the Fullerenes*; Thieme: Stuttgart, 1994.
[29] Liddell, P. A.; Kodis, G.; Andréasson, J.; de la Garza, L.; Bandyopadhyay, S.; Mitchell, R. H.; Moore, T. A.; Moore, A. L.; Gust, D. *J. Am. Chem. Soc.* 2004, 126, 4803-4811.
[30] Li, K.; Schuster, D.; Guldi, D.; Herranz, M.; Echegoyen, L.; *J. Am. Chem. Soc.* 2004, 126, 3388-3389.
[31] Sun, Y.; Drovetskaya, T.; Bolskar, R. D.; Bau, R.; Boyd, P. D. W.; Reed, Ch. A. *J. Org. Chem.* 1997, 62, 3642-3649.
[32] Suzuki, T.; Maruyama, Y.; Akasaka, T.; Ando, W.; Kobayashi, K.; Nagase, S. *J. Am. Chem. Soc.* 1994, 116, 1359-1363.
[33] Guldi, D. M.; Hungerbühler, H.; Asmus, K.-D. *J. Phys. Chem.* B 1999, 103, 1444-1453.
[34] Wolff, D. J.; Mialkovski, K.; Richardson, C. F.; Wilson, S. R.; *Biochem.* 2001, 40, 37-45.
[35] Jensen, A. W.; Wilson, S. R.; Schuster, D. I. *Bioorg. Med. Chem.* 1996, 4, 767-779.
[36] Wilson, L. J. *Interface* 1999, 8, 24-28.
[37] Da Ros, T.; Prato, M. *Chem. Commun.* 1999, 663-669.
[38] Camps, X.; Schönberger, H.; Hirsch, A. *Chem. Eur. J.* 1997, 3, 561-567.
[39] Hawker, C. J.; Fréchet, J. M. J. *J. Am. Chem. Soc.* 1990, 112, 7638-7647.
[40] Camps, X.; Dietel, E.; Hirsch, A.; Pyo, S.; Echegoyen, L.; Hackbarth, S.; Röder, B. *Chem. Eur. J.* 1999, 5, 2362-2373.
[41] Braden, B. C.; Goldbaum, F. A.; Chen, B.-X.; Kirschner, A. N.; Wilson S. R.; Erlanger, B. F. *Proc. Natl. Acad. Sci. USA* 2000, 97, 12193-12197.
[42] Ashcroft, J. M.; Tsyboulski, D. A.; Hartman, K. B.; Zakharian, T. Y.; Marks, J. W.; Weisman, R. B.; Rosenblum, M. G.; Wilson, L. *J. Chem. Commun.* 2006, 3004-3006.
[43] Jux, N.; Helmreich, M.; Hirsch, A.; Röder, B.; Ermilov, E. A. *J. Porphyrins Phthalocyanines* 2006, 10, 465.
[44] Bar-Haim, A; Klafter, J.; Kopelman, R. *J. Am. Chem. Soc.* 1997, 119, 6197-6198.
[45] Cotlet, M.; Gronheid, R.; Habuchi, S.; Stefan, A.; Barbafina, A.; Müllen, K.; Hofkens, J.; De Schryver, F. C. *J. Am. Chem. Soc.* 2003, 125, 13609-13617.
[46] Davis, B. L.; Melinger, J. S.; McMorrow, D.; Peng, Z.; Pan, Y. *J. Luminescence* 2004, 106, 301-311.
[47] Jiang, D.-L.; Aida, T. *J. Am. Chem. Soc.* 1998, 120, 10895-10901.
[48] Rana, D.; Gangopadhyay, G. *J. Chem. Phys.* 2003, 118, 434-443.
[49] Hania, P. R.; Heijs, D. J.; Bowden, T.; Pugzjlys, A.; van Esch, J.; Knoester, J.; Duppen, K. *J. Phys. Chem.* B 2004, 108, 71-81.
[50] Hackbarth, S.; Horneffer, V.; Wiehe, A.; Hillenkamp, F.; Röder, B. *Chem. Phys.* 2001, 269, 339-346.
[51] Wasielewski, M. R. *Chem. Rev.* 1992, 92, 435-461 and references therein.

[52] Kavarnos, G. J. *Fundamentals of Photoinduced Electron Transfer*; VCH: New York, 1993.

[53] Osuka, A.; Wada, Y.; Shinoda, S.; *Tetrahedron* 1996, 52, 4311-4326.

[54] Saha, S.; Johansson, E.; Flood, A. H.; Tseng, H.-R.; Zink, J. I.; Stoddart, J. F. *Chem. Eur. J.* 2005, 11, 6846-6858.

[55] Gust, D.; Moore, T. A.; Moore, A. L.; Leggett, L.; Lin, S.; DeGraziano, J. M.; Hermant, R. M.; Nicodem, D.; Craig, P.; Seely, G. R.; Nieman, R. A. *J. Phys. Chem.* 1993, 97, 7926-7931.

[56] D'Souza, F.; Ito, O. *Coord. Chem. Rev.* 2005, 249, 1410-1422.

[57] Segura, J.; Giacalone, F.; Gómez, R.; Martín, N.; Guldi, D. M.; Luo, C.; Swartz, A.; Riedel, I.; Chirvase, D.; Parisi, J.; Dyakonov, V.; Sariciftci, N. S.; Padinger, F. *Mat. Sci. Eng.* C 2005, 25, 835-842.

[58] Gonzales-Rodrigues, D; Torres, T.; Guldi, D. M.; Rivera, J.; Herranz, M. A.; Echegoyen, L. *J. Am. Chem. Soc.* 2004, 126, 6301-6313.

[59] Helmreich, M.; Hirsch, A.; Jux, N. *J. Porphyrins Phthalocyanines* 2005, 9, 130-137.

[60] Konishi, T.; Ikeda, A.; Shinkai, S. *Tetrahedron* 2005, 61, 4881-4889.

[61] *The Porphyrin Handbook*; Kadish, K. M.; Guillard, R.; Eds.; Academic Press: San Diego, 2003.

[62] Phthalocyanines – Properties and Applications; Leznoff, C. C.; Lever, A. B. P.; Eds.; VCH: New York, 1989, Vol. 1; 1993, Vols. 2 and 3; 1996, Vol. 4.

[63] McKeown, N. B. *Phthalocyanine Materials: Synthesis, Structure and Function*; Cambridge University Press: Cambridge, 1998.

[64] *The Chlorophylls*; Seely, V.; Ed.; Academic Press: New York, 1966.

[65] Pennington, F. C.; Strain, H. H.; Svec, W. A.; Katz, J. J. *J. Am. Chem. Soc.* 1964, 86, 1418-1426.

[66] Sheehan, J. C.; Hess, G. P. *J. Am. Chem. Soc.* 1955, 77, 1067-1068.

[67] Klausen, Y. S.; Bodansky, M. *Synthesis* 1972, 453-463 and literature therein.

[68] Ermilov, E. A.; Al-Omari, S.; Helmreich, M.; Jux, N.; Hirsch, A.; Röder, B. *Chem. Phys.* 2004, 301, 27-31.

[69] Al-Omari, S.; Ermilov, E. A.; Helmreich, M.; Jux, N.; Hirsch, A.; Röder, B. *Appl. Phys. B.* 2004, 79, 617-622.

[70] Dzhagarov, B. M.; Salokhiddinov, K. I.; Egorova, G. D.; Gurinovich, G. P. *Russ. J. Phys. Chem.* 1987, 61, 1281-1283.

[71] Ishii, K.; Kobayashi N. In *The Porphyrin Handbook*; Kadish, K. M.; Smith, K. M.; Guilard, R.; Eds.; Academic Press: San Diego, CA, 2003; Vol. 16, pp. 1-42.

[72] Kobayashi, N.; Higashi, Y.; Osa, T. *Chem. Lett.* 1994, 1813-1817.

[73] Lissi, E.A.; Encinas, M. V.; Lemp, E.; Rubio, M. A. *Chem. Rev.* 1993, 93, 699-723.

[74] Krasnovsky, A. A.; Bashtanov, M. E.; Drozdova, N. N.; Liddell, P. A.; Moore, A. L.; Moore, T. A.; Gust, D. J. Photochem. Photobiol. A: *Chemistry* 1997, 102, 157-161.

[75] Spiller, W.; Kliesch, H.; Wöhrle, D.; Hackbarth, S.; Röder, B.; Schnurpfeil, G. *J. Porphyrins Phthalocynines* 1998, 2, 145-158.

[76] Ermilov, E. A.; Al-Omari, S.; Helmreich, M.; Jux, N.; Hirsch, A.; Röder, B. *Opt. Commun.* 2004, 234, 245-252.

[77] Rehm, D.; Weller, A. *Ber. Bunsges. Chem.* 1969, 73, 834-839.

[78] Weller, A. *Z. Phys. Chem.* 1982, 133, 93-98.

[79] De, A. K.; Ganguly, T. *J. Luminescence* 2001, 92, 255-270.

[80] Dubois, D.; Kadish, K. M. *J. Am. Chem. Soc.* 1991, 113, 4364-4366.
[81] Imahori, H.; Hagiwara, K.; Aoki, M.; Akiyama, T.; Taniguchi, S.; Okada, T.; Shirakawa, M.; Sakata, Y. *J. Am. Chem. Soc.* 1996, 118, 11771-11782.
[82] HyperChem 6.03 evaluation release; HyperCube, Inc.: Gainesville, FL, 2000.
[83] Gould, I. R.; Ege, D.; Moser, J. E.; Farid, S. *J. Am. Chem. Soc.* 1990, 112, 4290-4301.
[84] Gould, I. R.; Young, R. H.; Moody, R. G.; Farid, S. *J. Phys. Chem.* 1991, 95, 2068-2080.
[85] Oevering, H.; Paddon-Row, M. N.; Heppener, M.; Oliver, A. M.; Cotsaris, E.; Verhoeven, J. W.; Hush, N. S. *J. Am. Chem. Soc.* 1987, 109, 3258-3269.
[86] Rancan, F.; Helmreich, M.; Mölich, A.; Jux, N.; Hirsch, A.; Röder, B.; Witt, Ch.; Böhm, F. *J. Photochem. Photobiol. B: Biology* 2005, 80, 1-7.
[87] Helmreich, M. Crown Ether-Metalloporphyrins as Ditopic Receptors and Pyropheophorbide-a Conjugates for the Photodynamic Therapy of Tumors; PhD Dissertation: Univ. of Erlangen-Nürnberg, 2005.
[88] Helmreich, M.; Ermilov, E. A.; Meyer, M.; Jux, N.; Hirsch. A.; Röder, B. *J. Am. Chem. Soc.* 2005, 127, 8376-8385.
[89] Regehly, M.; Ermilov, E. A.; Helmreich, M.; Hirsch. A.; Jux, N.; Röder, B. *J. Phys. Chem. B* 2007, DOI 10.1021/jp065626b.
[90] Hackbarth, S.; Ermilov, E.A.; Röder, B. *Opt. Commun.* 2005, 248, 295-306.
[91] Röder, B.; Ermilov, E. A.; Hackbarth, S.; Helmreich, M.; Jux, N. *SPIE* 2006, 6192, 495-507.
[92] Hoff, A. J.; Amesz, J. P. in *Chlorophylls*; Scheer, H.; Ed.; CRS Press: New York, 1991; pp. 723-738.
[93] Osuka, A.; Maruyama, K. *J. Am. Chem. Soc.* 1988, 110, 4454-4456.
[94] Förster, Th. *Ann. Physik* 1948, 2, 55-75.
[95] Förster, Th. *Z. Naturf. A.* 1949, 4a, 321-327.
[96] Förster, Th. *Discuss. Faraday. Soc.* 1959, 27, 7-17.
[97] Ermilov, E. A.; Hackbarth, S; Al-Omari, S.; Helmreich, M.; Jux, N.; Hirsch, A.; Röder, B. *Opt. Commun.* 2005, 250, 95-104.
[98] Zenkevich, E. I.; Knyukshto, V. N.; Shulga, A. M.; Kuzmitsky, V. A.; Gael, V. I.; Levinson, E. G.; Mironov, A. F. *J. Luminescence* 1997, 75, 229-244.
[99] Kozyrev, A. N.; Suresh, V.; Das, S.; Senge, M. O.;Shibata, M.; Dougherty, T. J.; Pandey, R. K. *Tetrahedron* 2000, 56, 3353-3364.
[100] McRae, E. G.; Kasha, M. *J. Chem. Phys.* 1958, 28, 721-722.
[101] McRae, E. G.; Kasha, M. in *Physical Processes in Radiation Biology*; Academic Press: New York, 1964, p. 23-42.
[102] Kasha, M. *Radiation Research* 1963, 20, 55-71.
[103] Kasha M., Rawls H. R., Ashraf El-Bayoumi M., *Pure Appl. Chem.* 1965, 11, 371-392.
[104] Röder, B.; Ermilov, E. A.; Regehly, M.; Helmreich, M.; Jux, N.; Rancan, F.; Böhm, F. *J. Porphyrins Phthalocynines* 2006, 10, 499.
[105] Rancan, F.; Böhm, F. Personal communication.
[106] Knutson, J. R.; Walbridge, D. G.; Brand, L. *Biochemistry* 1982, 21, 4671-4679.
[107] Ermilov, E. A.; Tannert, S.; Werncke, Th.; Choi, M. T. M.; Ng, D. K. P.; Röder, B. *Chem. Phys.* 2006, 328, 428-437.
[108] Lakowicz, J. R. *Principles of fluorescence spectroscopy*; Plenum Press: New York, NJ, 1999.

In: Chemical Physics Research Trends, Vol. 252
Editor: Samuel V. Arnold, pp. 247-268
ISBN: 978-1-60021-554-4

Chapter 6

CHALCONES: POSSIBLE NEW MATERIALS FOR THIRD-ORDER NONLINEAR OPTICS

***A. John Kiran*[a]*, K. Chandrasekharan*[*b]*, Balakrishna Kalluraya*[c] *and H. D. Shashikala*[a]**

[a] Department of Physics, National Institute of Technology Karnataka, Surathkal, Mangalore, India -575025

[b] Lasers and Nonlinear Optics Lab, Department of Science and Humanities, National Institute of Technology Calicut, Kerala, India - 673 601

[c] Department of Chemistry, Mangalore University, Karnataka, India -574 199

ABSTRACT

'Search for potential materials' for third-order nonlinear optics has been of continuing interest in recent years. In this context, organic molecules are increasingly being recognized as the materials of the future because their molecular nature combined with the versatility of synthetic chemistry can be used to alter and optimize molecular structure to maximize third order nonlinear response. Chalcones have received considerable interest as materials for second-order nonlinear optical applications due to their ability to crystallize in noncentrosymmetric structure and their blue light transmittance. Being charge transfer compounds, chalcones can also possess large third-order nonlinearities due to their π-conjugated structure. In this article, we discuss the structure-nonlinear response relationship among a few chalcones and the possibility of using them for third-order applications. A meager or no work, to our knowledge, has been done so far on these molecules in this regard. Z-scan and degenerate four-wave mixing techniques were employed to investigate third-order optical nonlinearities of chalcone derivatives. Some of these molecules possess large $\chi^{(3)}$ of magnitude as high as 10^{-12} esu and exhibit strong optical limiting properties. Possible mechanisms that are responsible for optical limiting property of these molecules have been discussed.

[*] Corresponding Author: - E-mail: csk@nitc.ac.in; Phone: +91-495 2287343.

INTRODUCTION

Nonlinear optics (NLO) is attracting increasing attention around the world because of its applications in telecommunications, all optical switching, sensor protections, and possibilities for optical information storage and coupling [1, 2]. To build devices for such applications, it is necessary to identify either existing materials or new materials that possess desired nonlinear optical properties. Materials for nonlinear optics have developed considerably in recent years, with novel semiconductors, organo-metallic complexes, composite glasses embedded with metal nanoparticles, and organic polymers and molecules attracting much attention [3-7]. Among these materials, organic molecular materials, because of their unique chemical structures (π bonding), exhibit the largest nonresonant (nonabsorptive) optical nonlinearities [8, 9]. The strong delocalization of electrons in these materials determines a very high molecular polarizability and thus remarkable third order nonlinearities. The most promising application of nonlinear optical activity is optical limiting, i.e. the increase of optical absorption with light intensity. Materials possessing large two-photon and reverse saturable absorptivities are being investigated for their nonlinear absorption based optical limiting applications [10].

Chalcones are 1,3-Diphenylpropenone (benzylideneacetophenone) and its derivatives formed by substitution: ArCH=CHC(=O)Ar, as shown in figure 1.

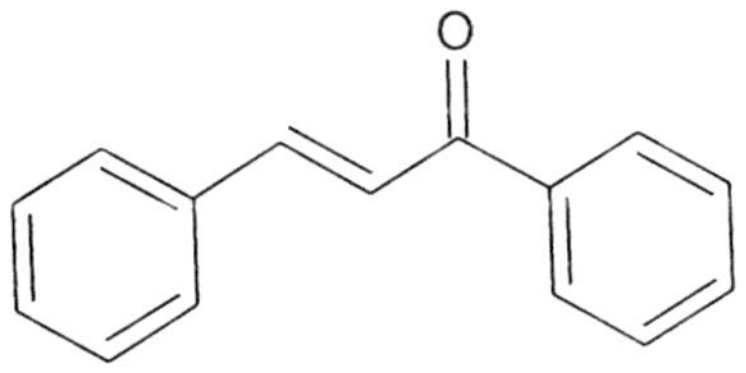

Figure 1. General structure of chalcones: ArCH=CHC(=O)Ar.

Chalcones are low molecular weight charge transfer (CT) compounds that show large nonlinear optical (NLO) properties without a longer wavelength absorption, as found in π-electron conjugated polymers. Charge transfer compounds have a large nonlinear electronic polarization that is caused by a large dipole moment change from a ground state to an excited state by optical radiation [11]. The molecular structure of intramolecular CT compounds distorts the distribution of π electrons in such a way that it results in the enhancement of optical nonlinearity. Such CT compounds are expected to possess advantages such as high solubility in organic solvents and matrix polymers and ease of sample preparation, as well as the possibility of giving high $\chi^{(3)}$ value and hence usefulness in photonic applications [11].

Till today, several CT compounds and CT complexes have been studied for their large third order nonlinearities [12-15]. By selecting an appropriate π-conjugation length and chemical structures of donor and acceptor, several new organic molecular crystals such as stilbene [13] and azobenzene derivatives [11], which exhibit $\chi^{(3)}$ larger than 10^{-12} esu, have been developed. The $\chi^{(3)}$ of 4-(N,N-diethylamino)-4′-nitrostilbene (DAENS), an intramolecular CT compound, has been reported in a solution, in a polymer matrix, and in a crystalline state [16, 17] and the maximum value obtained is $6X10^{-12}$ esu for real $\chi^{(3)}$ and $4X10^{-12}$ esu for imaginary $\chi^{(3)}$ [17]. The NLO properties of a novel CT compound, 4-(N, N-diethylamino)- 4′-nitrostyrene (DEANST), have been investigated by means of an optical Kerr shutter experiment, and a $\chi^{(3)}$ value of $7.4X10^{-12}$ esu has been measured in DEANST crystals [18, 19]. Some novel symmetrical short π-conjugated molecules, like SBA and SBAC, with large $\chi^{(3)}$ have been developed [20]. SBA and SBAC are benzylidene aniline derivatives that are symmetrically substituted by donor groups at both ends of the π-conjugations and have nitrogen atoms in their π-conjugation systems. In the off-resonant regions, the $\chi^{(3)}$ value in these molecules is measured to be as high as $4X10^{-11}$ esu [20].

Chalcones have been extensively studied for their second order nonlinear optical properties in the past few years [21]. However, being charge transfer compounds, chalcones can also possess large third-order nonlinearities due to their π-conjugated structure. Further, unlike second order nonlinearity, the advantage that the third order optical nonlinearity does not require any symmetry restrictions, has widen the range of possible structures of chalcone derivatives for consideration for third order nonlinear optical applications.

Various chalcones like, 1-(4-methylphenyl)-3-(4-methoxyphenyl)-2-propen-1-one, [22] 1-(4-chlorolphenyl)-3-(4-chlorolphenyl)-2-propen-1-one [23], 1,5-diphenylpenta-2, 4-dien-1-ones [24], 4-Br-4′-methoxy chalcone single crystals [25], arylfluranylpropenones [26], 1-(4-methylphenyl)-3-(phenyl)-2-propen-1-one [27], and 1-(4-methylphenyl)-3-(4-N, N dimethylaminophenyl)-2-propen-1-one [28, 29] have been studied for their second order NLO properties by several researchers. Chalcones have emerged as promising materials for second harmonic generation (SHG). Presently, green to blue wavelength lasers are required for high-density optical memories, color displays, and so on. Laser diodes and laser diode-pumped solid-state lasers have several advantages such as compactness, low cost, and long life. The harmonic generation of these lasers is very promising to obtain blue-green lights. As output powers of laser diodes are low, optical materials with large nonlinear susceptibilities are necessary to obtain efficient SHG. Organic crystals with conjugated π electrons are attractive candidates because of the large nonlinear optical coefficients. Among these organic crystals, the candidates must satisfy the following conditions: 1) they should be a certain size in bulk, 2) they should be transparent in the blue-green region and 3) they should be mechanically hard and chemically stable [29]. Chalcone derivatives are one of the candidates satisfying the above conditions because they have large nonlinear optical coefficients and a relatively short cutoff wavelength of transmission. Despite being recognized for second order applications, chalcone derivatives can also possess large third order nonlinearities. However, so far to our knowledge, a meager or no studies on the third order nonlinear optical (TNLO) properties of chalcones have been reported.

Apart from these optical properties, chalcones show some other interesting properties related to color reactions and biological activities. They show halochromism [30-32]. When they are wetted with concentrated H_2SO_4 they develop halochromic colors. When the intensely colored solution of chalcone in concentrated H_2SO_4 is treated with a little concentrated HNO_3, characteristic color change occurs [33]. The differing halochromic effects produced with concentrated H_2SO_4 serve to characteristic chalcones. Trans-chalcones usually show two absorption bands located at 300 nm (band I) and 330 nm (band II), of which band I results from the conjugation of the whole molecule [34]. The presence of enone function in the chalcone molecule confers antibiotic activity (bacteriostatic/ bactericidal) upon it. This property is enhanced when substitution is made at the α (nitro- and bromo-) and β- (bromo- and hydroxylic-) positions. Some substituted chalcones and their derivatives including some of their heterocyclic analogues have been reported to possess some interesting biological properties that are detrimental to the growth of microbes, tubercle bacilli, malerial parasites and intestine worms. Some of the compounds are claimed to be toxic to animals and insects and are also reported to exhibit inhibitory action on several enzymes, fungi and herbaceous plants. Compounds of the chalcone series also show a profound influence on the cardiovascular, cerebrovascular and neuromuscular systems [35-37].

We investigated some chalcones like, Dibenzylideneacetone (DBA) and derivatives, and 1-3-diaryl-propenones containing 4-methylthiophenyl moieties. We observed a good third order nonlinear response in all molecules, and interestingly a variation of the response among them that could be well associated with their molecular structure. A strong delocalization of π-electrons in these conjugated systems determines a very high molecular polarizability and hence remarkable third order nonlinearities [2]. Although great efforts have been expended on the investigation of the relationships that exist between the molecular structure of an organic material and the resulting third order nonlinearities, the complete understanding of these dependences is still incomplete. However, recent research efforts have enhanced the theoretical understanding and have thus improved processing techniques [38-40]. The study of the linear and nonlinear optical coefficients is fundamental for increasing the ability to tune nonlinear optical properties by appropriate design of organic systems at the molecular level.

Dibenzylideneacetone (DBA) has one more importance as a very good precursor for the preparation of many types of heterocyclic compounds of medical and biological importance.

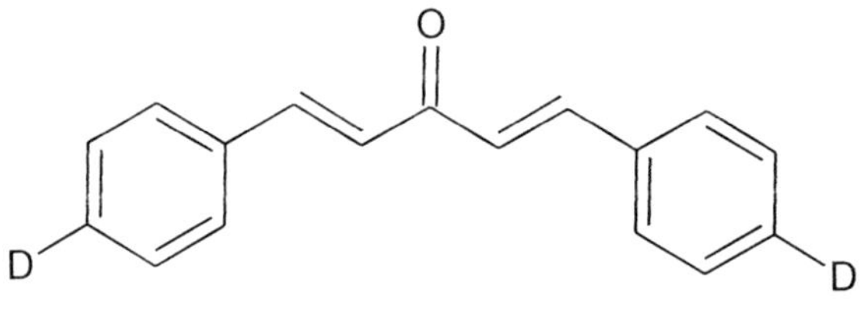

Figure 2. Structure of DBA. D is OCH_3, Cl, $N(CH_3)_2$ for p-methoxyDBA, p-chloroDBA and p-(N,N-dimethyl)DBA respectively.

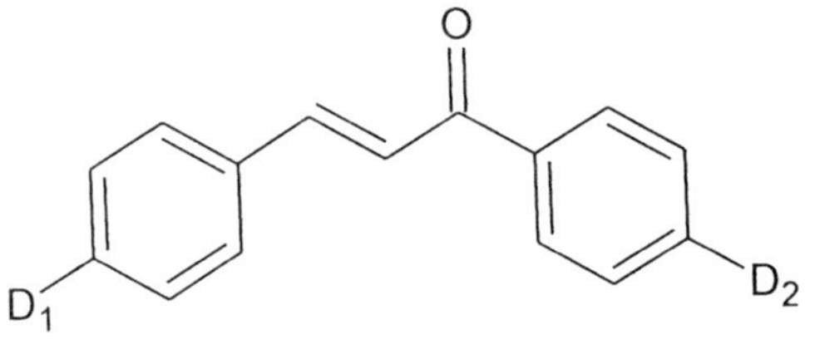

Figure 3. Structure of 1-3-diaryl-propenones containing 4-methyl thiophenyl moieties coded as mc-4, mc-6, mc-10, and mc-3. D_1 is H_3CS, common to all the four compounds. D_2 is CH_3, CH_3O, C $(CH_3)_3$, and Cl respectively for mc-4, mc-6, mc-10, and mc-3.

Practically, molecular design can play an important role in the development of third order materials by using the tremendous flexibility a molecular material offer to modify its chemical structure. Marius Alabota et al have proposed criteria to enhance the third order nonlinearity in molecular materials [41]. According to their design criteria, the third-order nonlinearity of conjugated organic compounds can be enhanced by (i) increasing the conjugation length, to increase the distance over which charge can be transferred; (ii) creating a donor-acceptor-donor motif by substitution to increase the extent of charge transfer from the ends of the molecule to the center and (iii) reversing the sense of symmetric charge transfer by substituting electron acceptors and donors, and there by creating acceptor-donor-acceptor compounds . Similar strategy was used to design these chalcones through derivatization by substituting proper electron donors and electron acceptors and thus to create donor-π-acceptor-π-donor structures. The structures of compounds are given figure 2 and 3 below. Synthesis and detailed TNLO characterization of these compounds has been reported by John Kiran et al [42, 43].

EXPERIMENT

Two experiments, a single-beam Z-scan and Degenerate Four Wave Mixing (DFWM) were employed to study TNLO properties of chalcones in their solutions in Dimethylformamide (DMF) (1×10^{-2} mol/L). DFWM was used in order to compare $\chi^{(3)}$ values of molecules with those $\chi^{(3)}$ obtained through Z-scan technique. A detailed theories and description of experiments are available in the literature [44, 45]. Z-scan is a simple and well-known technique used to measure nonlinear refractive index (n_2) and nonlinear absorption coefficient (β) of a material. The technique is particularly useful when the nonlinear refraction is accompanied by nonlinear absorption and it allows the simultaneous measurement of n_2 and β. The technique relies on the phenomenon of self-focusing or self-defocusing of an optical beam by a thin medium.

Four-wave mixing refers to the interaction of four waves in a nonlinear medium via the third order polarization. It results from the intensity dependent refractive index. When all the waves have same frequency, though their wave vectors are different, it is called as degenerate four wave mixing. It is one of the most important and versatile techniques, which provide information about the magnitude and response of the third order nonlinearity. There are several geometries used in studying this phenomenon, for example, forward geometry, backward geometry or phase conjugate geometry, nearly degenerate four wave mixing, and folded boxcar geometry [45]. However, the principle for all these geometries is the following: two beams interfere to form some type of grating and a third beam scatters off this grating, generating the fourth beam named the conjugate or signal beam. We used the standard phase conjugate geometry in our experiment.

The linear absorption spectra of DBA and its derivatives in DMF recorded with a UV-VIS Ocean Optics (Model SD 2000) spectrophotometer are depicted in figure 4. It can be seen that the excitation wavelength (532 nm), except for p-(N, N-dimethyl)DBA, is far away from the resonance wavelength, indicating that we measure only the nonresonant nonlinearity. This is true in all the samples, except for p-(N, N-dimethyl)DBA, investigated in this chapter. Linear refractive indices of samples in DMF were obtained through an Abbe refractometer.

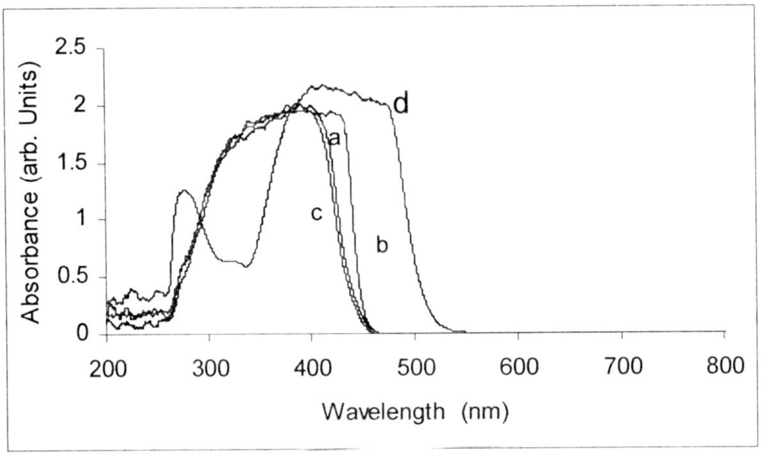

Figure 4. Linear absorption spectra of compounds dissolved in DMF ($1x10^{-2}$ mol/L). a: DBA; b: p-methoxyDBA; c: p-chloroDBA; d: p-(N, N-dimethyl)DBA.

Results

The single beam Z-scan with Nd: YAG nanosecond (7 ns) laser was performed on these chalcones dissolved in DMF. Details of experiment can be found in reference [42]. Both closed aperture and open aperture scans were performed in order to extract information on nonlinear refraction (NLR) and nonlinear absorption (NLA) of the samples in DMF solutions. Z-scan curves obtained for dibenzylideneacetone are shown in figure 5, 6 and 7 below. Solid lines in the figure are theoretical fits of experimental data. Details of equations used for curve fitting can be found in reference [44, 46]. The normalized transmittance for open aperture curve is given by

$$T(z) = \frac{\ln[1+q_0(z)]}{q_0(z)} \quad \text{for } |q_0(z)| < 1 \tag{1}$$

where $q_0(z) = \frac{\beta I_0 L_{eff}}{(1+z^2/z_0^2)}$, z is the position of sample with respect to focal plane of the lens, z_0 is the Rayleigh range of the lens and $L_{eff} = (1-e^{-\alpha L})/\alpha$ is the length of the sample. The normalized transmittance for closed aperture scan is given by

$$T(z) = 1 + \frac{4\Delta\phi_0 x}{(1+x^2)(9+x^2)} - \frac{2\Delta\varphi_0(3+x^2)}{(1+x^2)(9+x^2)} \tag{2}$$

where $\Delta\phi_0 = \frac{\Delta T_{p-v}}{0.406(1-S)^{0.25}}$ for $|\Delta\phi_0| \le \pi$ and $\Delta\varphi_0 = \frac{1}{2}\beta I_0 L_{eff}$ are phase shifts due to NLR and NLA respectively [47].

In a closed aperture Z-scan curve, both nonlinear refraction and nonlinear absorption will be present; in which case, a simple division of the closed aperture Z-scan data by the one with open aperture is followed to obtain a pure nonlinear refraction curve. In this case the normalized transmittance would become [44]

$$T(z) = 1 - \frac{4\Delta\phi_0 x}{(1+x^2)(9+x^2)} \tag{3}$$

Second Order Hyperpolarizability

The third order nonlinear susceptibility is related to the second order hyperpolarizability, which is a microscopic parameter. Second order hyperpolarizability γ_h of a molecule in an isotropic medium is related to the third order bulk susceptibility as below [48]:

$$\gamma_h = \chi^{(3)} / L^4 N. \tag{4}$$

In the above equation, N is the density of molecules in the unit of molecules per cm^3. The term L is the local field factor, which in the Lorentz approximation is given by

$$L = \left(\frac{(n^2+2)}{3}\right) \tag{5}$$

Here n is the linear refractive index of the medium.

Closed aperture curve demonstrates a strong negative refractive nonlinearity of the sample. n_2 was found to be as high as $-3.512\text{X}10^{-12}$ esu. The values of real and imaginary parts of $\chi^{(3)}$ were found to be $-0.38\text{x}10^{-13}$ esu and $0.10\text{x}10^{-13}$ esu respectively. Similar results were observed with derivatives of DBA. The values of $\chi^{(3)}$ are tabulated in table 1. It has been shown by J. Kiran et al., [42] that upon substitution of strong electron donor groups to DBA, nonlinearity can be enhanced according to the design criteria discussed by M. Alabota et al [41]. Interestingly it has been found that the third order nonlinearity in these chalcones varies according to the π electron density in the molecules.

The values of n_2, Re $\chi^{(3)}$ and Im $\chi^{(3)}$ reported in table 1 were obtained by repeating the Z-scan on each sample, and the values were found to be consistent in all trials with an error of less than 10 %. The increased nonlinear response in the case of p-methoxyDBA is mainly due to the enhanced π – electron density in the molecule [41, 49].

The results show that the increase in $\chi^{(3)}$ of p-methoxyDBA is nearly double compared to the $\chi^{(3)}$ value of DBA. Comparing to chlorine, which is an electron-withdrawing group due to high electronegativity, methoxy is an electron releasing by resonance. So, among these compounds the order of π-electron density is p-methoxyDBA > DBA > p-chloroDBA > o-chloroDBA.

Among p-chloroDBA and o-chloroDBA although both are having chlorine substituents, in the case of o-chloroDBA, chlorine is in very close vicinity of double bond, so it withdraws electrons more powerfully than chlorine present in the para -position. Consequently, o-chloroDBA showed least nonlinear response among all other compounds. It showed no detectable nonlinear response at the input energy used during our experiment. Further, to determine the contributions of the solvent to n_2 we conducted Z-scan experiment on the pure DMF, and found that neither nonlinear refraction nor nonlinear absorption was observed at the input energy used. Hence, any contribution from solvent to the nonlinearity of the samples is negligible. We did not observe any significant scattering of laser beam from sample solutions within the energy limit used during our experiments.

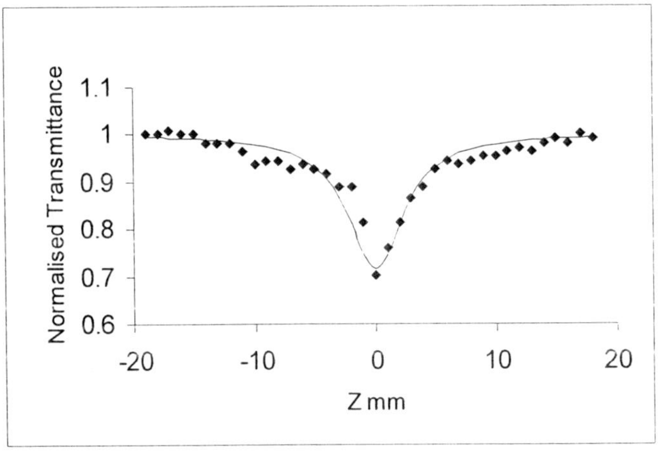

Figure 5. Open aperture curve for DBA. Solid line is a theoretical fit of experimental data with β = 0.72 cm/GW to equation (1).

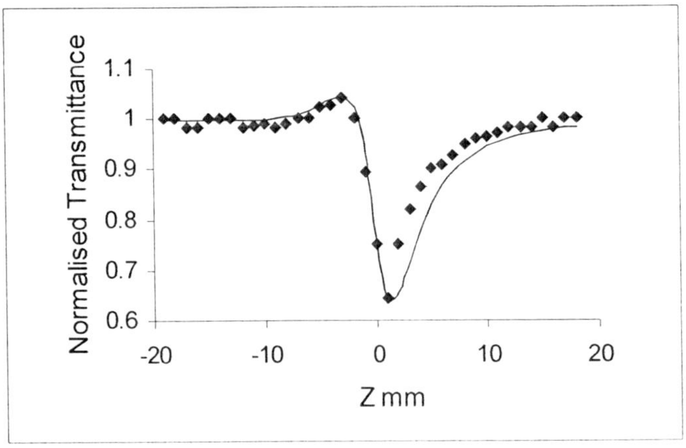

Figure 6. Closed aperture curve for DBA. Solid line is a fit of experimental data with $\Delta\Phi_0$ = 0.97 and β = 0.72 cm/GW to equation (2).

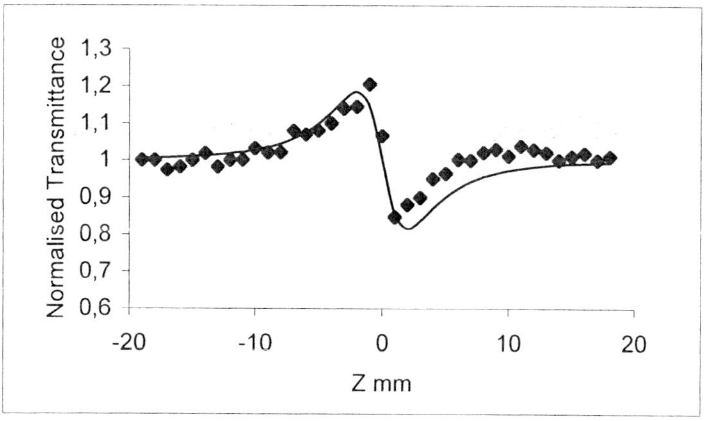

Figure 7. Pure NLR curve for DBA obtained through division method. Solid line is a fit of experimental data with $\Delta\Phi_0 = 0.97$ to equation (3).

Table 1. Values of n_2, β, γ_h, real and imaginary $\chi^{(3)}$ determined experimentally for different DBA based compounds

Name	n_0	n_2 esu	β cm/GW	$\gamma_h \times 10^{-32}$ esu	Re $\chi^{(3)}$ $\times 10^{-13}$ esu	Im $\chi^{(3)}$ $\times 10^{-13}$ esu
DBA	1.425	$-3.512\text{X}10^{-12}$	0.72	0.2	-0.38	0.10
p-chloroDBA	1.412	$-3.18\text{X}10^{-12}$	0.98	0.184	-0.33	0.11
p-methoxyDBA	1.417	$-5.745\text{X}10^{-12}$	0.77	0.325	-0.613	0.12
o-chloro DBA	1.413	No detectable response				

A similar variation of nonlinear response was observed with 1-3-diaryl-propenones containing 4-methyl thiophenyl moieties, whose structure is shown in figure 3. John Kiran et al [43] have done detailed study on these molecules. These molecules exhibit strong negative refractive nonlinearity and two-photon absorption at 532 nm. The nonlinear response was found to increase on going from mc-3 to mc-10. Among different electron donors substituted in these compounds, the electron donating ability of C $(CH_3)_3 > OCH_3 > CH_3 > Cl$. As a result, the nonlinear response is found to be the highest in the case of mc-10. In mc-10 there is an S-CH_3 group on the left hand side of the molecule and a C $(CH_3)_3$ on the right hand side. The S-CH_3 group is known to be a good electron donor and it gives away electrons easily to form a stable π-electron distributed system. This donates more electrons into the molecule and on the other side C $(CH_3)_3$ also releases electrons in to the molecule. The acceptor, C=O, at the center accepts electrons and hence there is a strong delocalization of electrons in the molecule. There is a symmetric charge transfer from the ends to the center and hence the extent of π -electron transfer is increased in the molecule. In mc-3, mc-4 and mc-6 the response was observed to vary according to the density of delocalized electrons in the molecule. Therefore, the enhanced nonlinear response of mc-10 is mainly due to the increased π-electron density in the molecule [41]. In mc-3, due to the inductive effect of chlorine, the π-

electron density and the charge transfer in the system is reduced considerably. Therefore the response is minimum compared to that of other three samples.

In order to get more information on the structure-NLO property relationships, three more similar molecules were synthesized. Their structure is shown in figure 8 below. These molecules possess one Br atom in their structure.

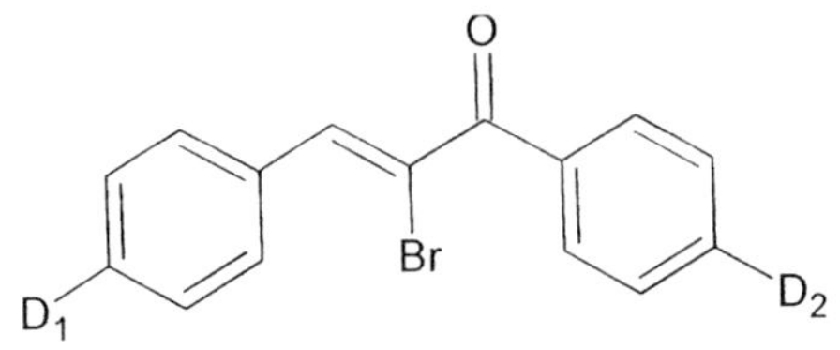

Figure 8. The structure of compounds coded mc-3m, mc-4m and mc-6m. D_1 is H_3CS common to all compounds. D_2 is Cl, CH_3 and OCH_3 respectively for mc-3m, mc-4m and mc-6m.

The nonlinear repose of molecules was found to decrease upon substitution of Br atom. The observed decrease in NLO response of mc-3m to mc-6m may be explained as follows. In compounds mc-3m, mc-4m and mc-6m there is a Bromine atom at the center of molecule. Br is a good electron acceptor due to its high electonegativity. Bromine, being a heavy atom, affects the planarity of the molecule, which in turn reduces the delocalization of electrons through out the molecule. In conjugated systems, like those of chalcones, co-planarity is required for the transfer of electrons from donor to acceptor. In α-Bromo chalcones, the bulky Bromine atom sterically interferes with adjacent cis- phenyl ring and pushes this phenyl ring out of plane. Hence there is no co-planar arrangement of two phenyl rings with the double bond and the π- conjugation is thus not effective in the transfer of electrons from one end to the other in the system [50].

Table 2. Experimentally determined values of n_0, n_2, β, γ_h, Re $\chi^{(3)}$ and Im $\chi^{(3)}$ for different Compounds

Name of the sample	n_0	n_2 10^{-11} esu	β cm/GW	γ_h $x10^{-32}$ esu	Re $\chi^{(3)}$ 10^{-13} esu	Im $\chi^{(3)}$ 10^{-13} esu
mc-3	1.423	-1.098	1.472	0.616	-1.181	0.23
mc-4	1.421	-1.21	2.046	0.681	-1.296	0.26
mc-6	1.421	-1.23	2.690	0.931	-1.556	0.42
mc-10	1.420	-2.023	3.035	1.147	-2.170	0.47
mc-3m	1.423	-0.719	1.440	0.415	-0.770	0.22
mc-4m	1.421	-0.251	0.330	0.141	-0.269	0.05
mc-6m	1.423	-0.33	0.321	0.184	-0.356	0.049

As a result, nonlinear response of these samples was found to decrease upon substitution of Br atom at position –2 of the propenones. The dependence of third-order NLO properties on donor and acceptor type of substituents clearly demonstrates the electronic origin of nonlinearity of these compounds, and hence the possibility of dominant contribution of thermal effects to the observed nonlinearity can be ruled out.

The values of determined NLO coefficients of mc-3 to mc-10 are given in table 2. These values of third order optical parameters were found to be consistent in all repeated scans with an error of less than 12 %.

Nonlinear Absorption

All compounds discussed so far are transparent at 532 nm but exhibit strong two-photon absorption (TPA). Generally, nonlinear absorption (NLA) can be caused by free carrier absorption, saturated absorption, direct multiphoton absorption, or excited state absorption. However, if the mechanism belongs to a simple two-photon absorption, β should be a constant that is independent of the on-axis irradiance I_0 [51]. Further, from basic theoretical consideration, it is seen that the TPA-induced decrease of transmissivity can be expressed as [52].

$$T(I_0) = \frac{I(L)}{I_0} = \frac{1}{1 + I_0 L\beta}, \qquad (4)$$

where I_0 is the on-axis intensity of the beam within the sample of thickness L. If the laser beam has a gausian transverse intensity distribution in the medium, then the TPA-induced transmissivity change will be

$$T(I_0) = \frac{I(L)}{I_0} = \ln(1 + I_0 L\beta) \big/ I_0 L\beta \qquad (5)$$

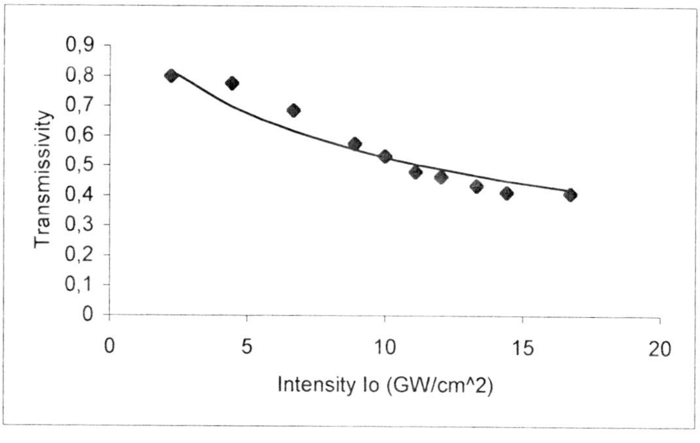

Figure 9. Transmissivity versus input intensity for p-methoxyDBA in DMF solution (1×10^{-2} mol/L). Solid line is a fit of data to equation (5) with a best-fitting parameter of β = 2.2 cm/GW.

Hence, if the transmissivity of a sample varies as in equation (5) we may assume that the dominant mechanism for nonlinear absorption is TPA. Therefore to confirm TPA we plotted transmissivity measured as a function of input intensity for p-methoxyDBA, as shown in figure 9 and a good correlation was seen between the experimental data and the theoretical fit of equation (5) [53]. Figure 10 shows a plot of β versus input intensity I_0 for p-methoxyDBA and mc-10, where it is clear that β is almost independent of I_0. A similar behavior was observed with other samples also. This confirms that the dominant mechanism of nonlinear absorption in our samples is a TPA-induced absorption.

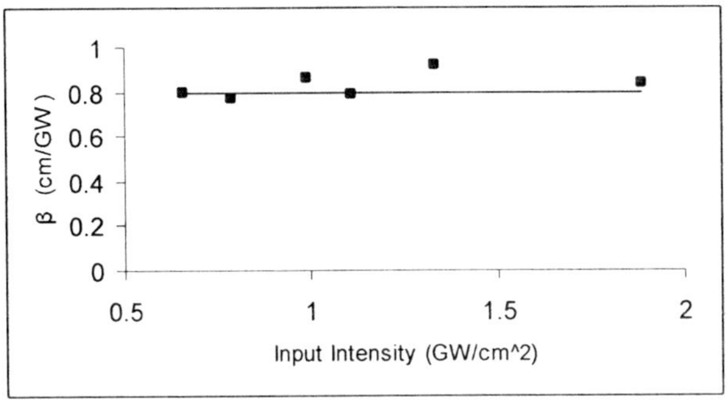

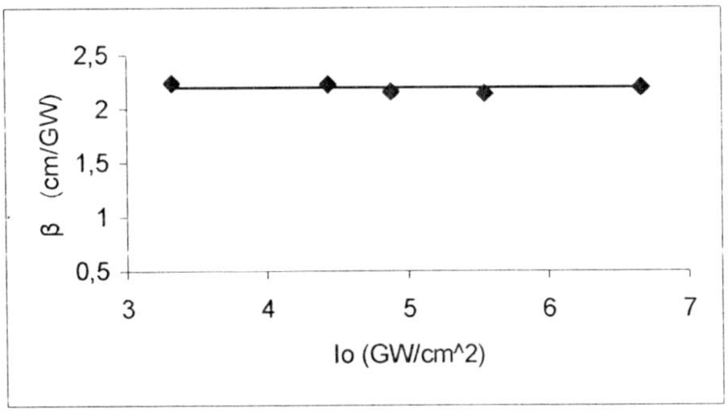

Figure 10. Plots of β versus I_0 for a) p-methoxyDBA and b) mc-10. β is nearly a constant independent of I_0.

Optical Limiting

Optical limiting is a nonlinear optical (NLO) process in which the transmittance of a material decreases with increased incident light intensity. Optical limiters have been utilized in a variety of circumstances where a decreasing transmission with increasing excitation is desirable. However, one of the most important applications is eye and sensor protection in optical systems [10]. These devices, which work due to intrinsic properties of the materials used for their fabrication, are used to control the amplitude of high intensity optical pulses. An ideal optical limiter exhibits a linear transmission below threshold, but above its threshold its output intensity is constant. For a material without significant NLO absorption (linear material) the transmitted light intensity will increase proportionally to the incoming light intensity (i.e. the relative transmission is constant). For a material with nonlinear optical absorption the transmitted intensity will increase proportionally to the incoming light intensity while the intensity is low. However, when a certain input intensity is reached the NLO absorption will start and the increase in the outgoing light will be reduced for increasing light intensities (i.e. the relative transmission decreases with increased intensity). For some materials this optical limiting effect can be so strong that the transmitted light intensity reaches a constant value independent of the incoming light intensity. This is defined as the clamping level. However, pulses with realistic temporal and spatial profile modify this response. Under realistic conditions, the limiting threshold is less well defined, and the output fluence will not be perfectly clamped at a constant value.

Based on the strong two-photon absorption, good optical limiting of nanosecond laser pulses can be expected from these chalcones. Figures 11 and 12 below show optical limiting behavior of p-methoxyDBA and some mc-series compounds respectively.

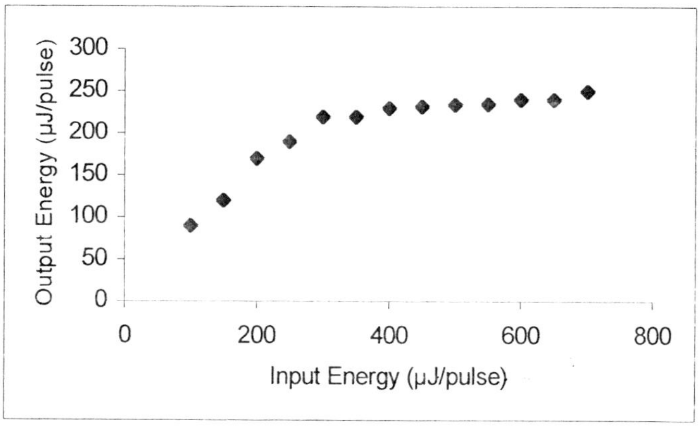

Figure 11. Optical limiting behavior of p-methoxyDBA.

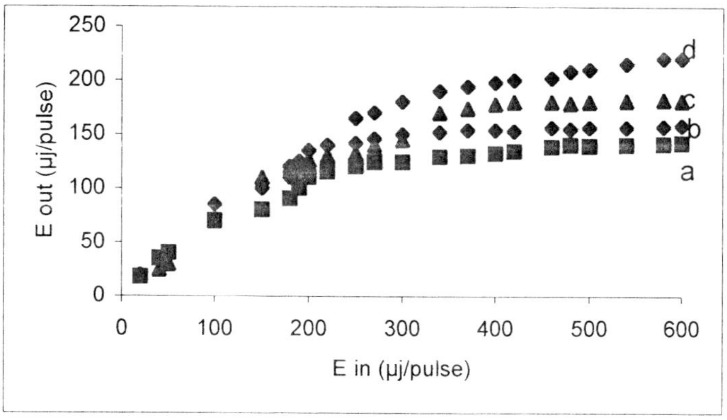

Figure 12. Optical limiting of nanosecond pulses in samples a=mc-10, b=mc-4, c=mc-6 and d=mc-3.

In the case of p-methoxyDBA, for input energies less than 300 μJ/pulse, the output energy increases linearly with incident energy. However, in excess of 300 μJ/pulse the output energy is almost clamped at 230 μJ/pulse. The nonlinear absorption was observed to increase on going from mc-3 to mc-10. In the case of mc-10, for input energies less than 220 μJ/pulse, the output energy increased linearly with the incident energy. But for energies more than 220 μJ/pulse, the output energy was almost constant assuming the value near 140 μJ/pulse. The limiting threshold increased from 220 μJ/pulse to 250 μJ/pulse in the case of mc-4 and the output was clamped at around 153 μJ/pulse. This threshold further increased to 370 μJ/pulse in the case of mc-3 and there the output became almost constant at 210 - 220 μJ/pulse. Thus, optical limiting behavior of these samples was found to vary according to the π-electron delocalization in molecules.

p-(N,N-dimethyl)Dibenzylideneacetone

It is one of the derivatives of DBA. It has been shown that the nonlinear response increased upon substitution of N, N-dimethyl group to DBA and this compound possesses the highest nonlinearity among other DBA derivatives [42]. Though electron-donating ability of N, N-dimethyl group is high compared to methoxy group and hence compound shows high NLO response, careful studies show that the compound has an absorption tail at 532 nm. This small linear absorption can also contribute to the observed large nonlinearity of the sample. Therefore, further study was done on this sample to explore possible mechanism responsible for its large nonlinearity. The linear absorption coefficient α of sample in DMF ($1\text{x}10^{-2}$ mol/L) at 532 nm was found to be 0.023 cm^{-1}. The Z-scan [with a 50% (S=0.5) aperture] was repeated on this solution at pulse energy of 16 μJ, which corresponds to a peak irradiance of 0.355 GW/cm^2. The Re $\chi^{(3)}$ value of $1.9\text{x}10^{-12}$ esu was obtained for p-(N,N-dimethyl)DBA. The $\chi^{(3)}$ value of the compound is very high and is a little more compared to $1.3\text{x}10^{-12}$ esu, obtained for tricyanovinyl azo dye and comparable with $6\text{x}10^{-12}$ esu, obtained for 4-(N,N-diethylamino)-4'-nitrostilbene (DEANS) [16]. This shows that the compound can be a promising dopant material for doping polymers for optical limiting and other photonic applications.

Since the compound absorbs at 532 nm, there could be some contribution from excited state absorption. The excited state cross section σ_{ex} can be measured from the normalized open aperture Z-scan data [54]. We assume that the molecular energy levels can be reduced to a three level system in order to calculate σ_{ex}. Molecules are optically excited from the ground state to the singlet-excited state. The molecules from this state relax either to the ground state or the triplet state, when excited state absorption can occur from the triplet to the higher triplet excited state.

The change in the intensity of the beam as it passes through the material is given by $\dfrac{dI}{dz} = -\alpha I - \sigma_{ex} N(t) I$, where I the intensity, and N is the molecules in the excited state. The excited state density of molecules appears as a result of a nonlinear absorption process whose intensity dependence can be obtained from $\dfrac{dn}{dz} = \dfrac{\sigma_{ex} I}{h\nu}$, where ν is the frequency of the laser. Combining the above two equations and solving for the fluence of the laser and over the spatial extent of the beam gives the normalized transmission for open aperture as

$$T = \ln\left(1+\frac{q_0}{1+x^2}\right) / \frac{q_0}{1+x^2} \tag{6}$$

where $q_0 = \frac{\sigma_{ex}\alpha F_0 L_{eff}}{2h\nu}$, F_0 is the fluence of the laser at the focus and $L_{eff} = \frac{(1-\exp^{-\alpha L})}{\alpha}$.

A fit of equation (6) to the open aperture data at 532 nm with q_0, yields $\sigma_{ex} = 6.24\times10^{-17}$ cm^2 for the sample solution. The ground state absorption cross-section of sample calculated from $\alpha = \sigma_g N_a C$, where N_a is Avogadro's number and C is the concentration in moles/cm^3, is found to be $\sigma_g = 3.818\times10^{-18}$ cm^2. The value of σ_{ex} is larger than the value of σ_g, which is in agreement with the condition for observing reverse saturation absorption [54, 55]. Reverse saturation absorption generally arises in a molecular system when the excited state absorption cross section is larger than the ground state cross section. If the absorption cross section of excited state is smaller than that of ground state, then the material becomes more transparent and bleaches i.e. it is a saturable absorber. If the mechanism belongs to the simple two-photon absorption, β should be a constant that is independent of the on-axis irradiance I_0. As if the mechanism is the direct three-photon absorption, β should be a linear increasing function of I_0 and the intercepts on the vertical axis should be nonzero [51]. But figure 13 shows that β is decreasing with increasing I_0. The fall-off of β with increasing I_0 is a consequence of the reverse saturable absorption [55]. Hence, the mechanism responsible for strong NLA of the compound was found to be reverse saturation absorption.

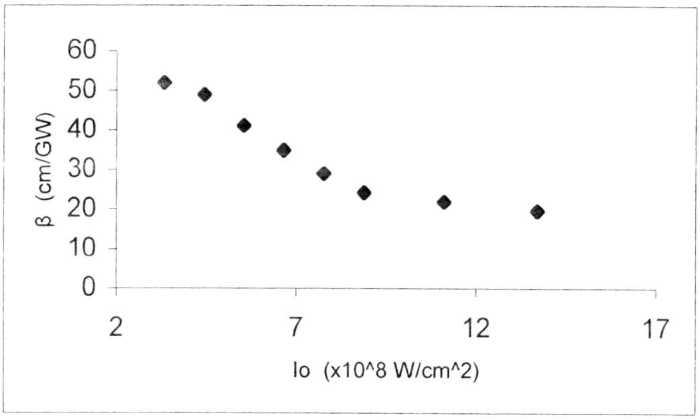

Figure 13. A fall-off of β with increase in on axis intensity within the sample I_0.

Optical limiting based on the reverse saturable absorption can be expected for the nanosecond laser pulses. Figure 14 shows strong optical limiting behavior of the compound solution. For input energies less than 15 μJ/pulse, the output energy increased linearly with the incident energy. But for energies more than 15 μJ/pulse, the measured output energy deviates from linearity, indicating the occurrence optical limiting. The linear transmittance of this sample at 532 nm was 70%.

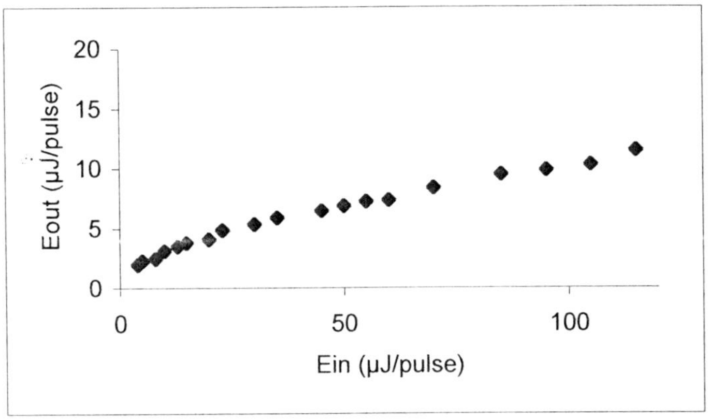

Figure 14. Optical limiting of nanosecond pulses in p-(N,N-dimethyl)DBA in DMF.

Doping Polymers

Despite these compounds possess very high third order nonlinearity, they cannot be directly used in practical devices since they can get degraded when exposed to intense laser beams. One can overcome this problem and make effective use of these materials in devices by doping them into a polymer matrix, as this can enhance the opto-chemical and opto-physical stability as well as mechanical and thermal properties, while retaining the NLO properties and linear optical transparency [56]. Here we report on doping one of these compounds, p-(N,N-dimethyl)DBA, into poly(methylmethacrylate) (PMMA). PMMA has been selected as a matrix because it is a hard, rigid but transparent polymer with glass transition temperature of 125^{0} C, and an average molecular weight of 60,000. PMMA has physical durability which is far superior to that of other thermoplastics and it is tougher than polystyrene [57].

PMMA and p-(N,N-dimethyl)DBA were taken in the powder form and dissolved in DMF. The concentration of the dopant in PMMA matrix was varied from 2.5 % to 15% [sample I: 2.5%; sample II: 5%; sample III: 10%; sample IV: 15%]. The Z-scan was obtained with a 50% (S=0.5) aperture and at pulse energy of 16 μJ, which corresponds to a peak irradiance of 0.355 GW/cm^2. In order to avoid cumulative thermal effects, data were collected in single shot mode [58]. Further, to determine the contributions of the solvent and PMMA to the observed NLO properties, we conducted Z-scan experiment on the pure DMF and PMMA dissolved in DMF. Neither nonlinear refraction nor nonlinear absorption was observed at the input energy used. Hence, any contribution from the solvent and PMMA to nonlinearity of the sample is negligible. Therefore, the nonlinear response observed was solely due to p-(N,N-dimethyl)DBA. Z-scan results obtained are given in table 3 below. It can be seen that response is increasing with concentration of the sample in PMMA. However, not much increase was observed with sample IV, where the linear transmittance became very less (~ 40 %). Figure 15 shows the optical limiting behavior of p-(N,N-dimethyl)DBA doped PMMA.

Table 3. Calculated values of β, n_2, Re $\chi^{(3)}$, and Im $\chi^{(3)}$ for doped PMMA

Sample	β cm/GW	n_2 x10^{-10} esu	Re $\chi^{(3)}$ x10^{-12} esu	Im $\chi^{(3)}$ x10^{-12} esu
Sample I	48.62	1.375	1.479	0.752
Sample II	52.90	1.552	1.602	0.801
Sample III	57.50	1.700	1.829	0.890
Sample IV	59.60	1.770	1.886	0.886

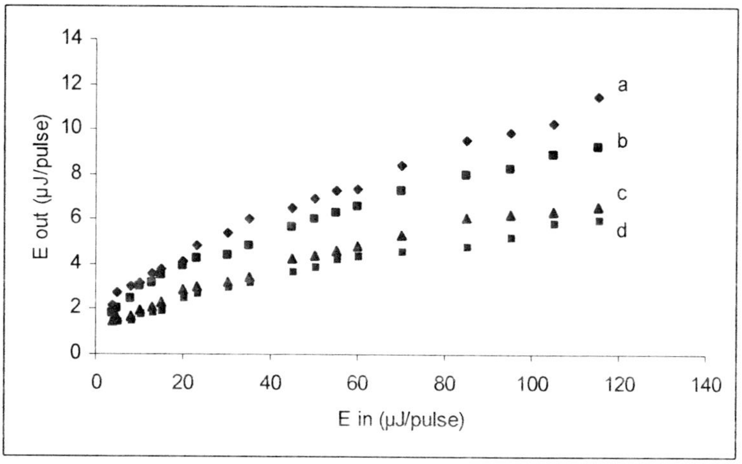

Figure 15. Optical limiting of nanosecond pulses in p-(N,N-dimethyl)DBA doped PMMA at different concentrations. a: sample I; b: sample II; c: sample III; d: sample IV.

This study shows that p-(N,N-dimethyl)DBA is an attractive optical limiting material and can be used for doping polymers for optical limiting applications.

Concentration Dependence of Nonlinearity

Concentration dependence of NLO coefficients can be analyzed to extract information on the NLO properties of the solute. The concentration of solutes in the solution was varied and Z-scan measurements were repeated on solutions at each concentration to study the variation of nonlinear response. Figure 16 shows the dependence of nonlinear absorption (β) on the concentration of DBA and p-(N,N-dimethyl)DBA in solution. The NLA absorption coefficient β varies almost linearly with sample concentration. Similar behavior was observed with all other compounds.

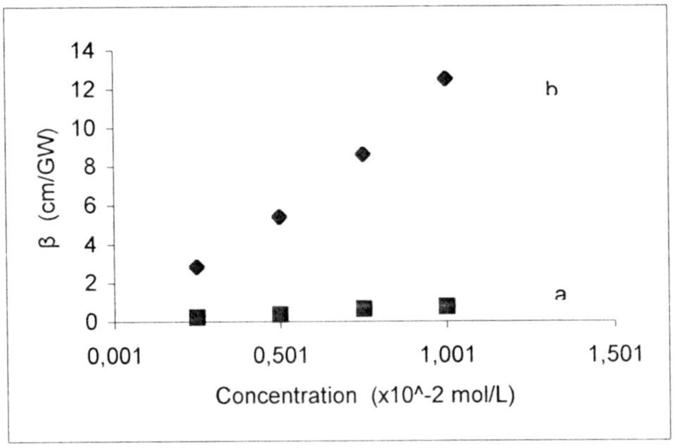

Figure 16. TPA coefficient β versus concentration of samples in DMF. a: DBA; b: p-(N,N-dimethyl)DBA.

The nonlinear absorption as well as the nonlinear refraction decreased as the concentration in solution decreased from $1x10^{-2}$ mol/L to $2.5x10^{-3}$ mol/L. It is also evident from figure 15 that optical limiting is better in PMMA doped with higher concentration of p-(N,N-dimethyl)DBA. Therefore, it is clear that the observed nonlinear response has direct relation with the concentration of samples in DMF.

Degenerate Four Wave Mixing

DFWM experiment was performed on p-(N,N-dimethyl)DBA using Nd: YAG 7 ns laser, in order to extract $\chi^{(3)}$ value and compare with that obtained through Z-scan technique. Variation of the DFWM signal as a function of the pump intensity for the compound dissolved in DMF at a concentration of $1x10^{-2}$ mol/L is shown in figure 17.

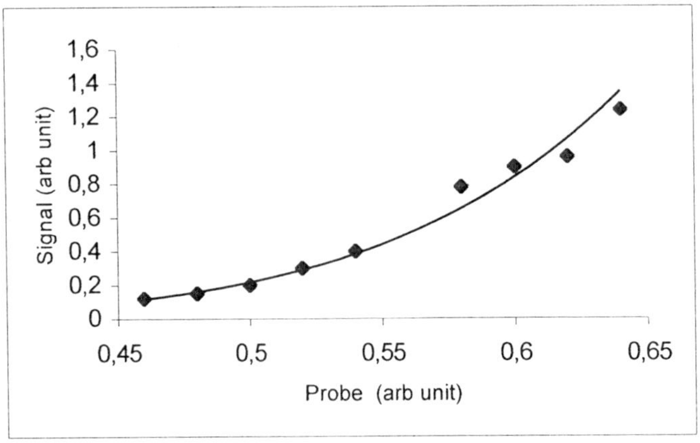

Figure 17. Phase conjugate signal versus probe for p-(N,N-dimethyl)DBA in DMF.

The signal is proportional to the cubic power of the input intensity as given by the equation

$$I(\omega)\alpha\left(\frac{\omega}{2\varepsilon_0 cn^2}\right)^2 \left|\chi^{(3)}\right|^2 l^2\, I_0^3(\omega)\ , \tag{7}$$

where $I(\omega)$ is the DFWM signal intensity, $I_0(\omega)$ is the pump intensity, l is the optical pathlength, and n is the refractive index of the medium. The solid curve in the figure is cubic fit to the experimental data. $\chi^{(3)}$ can be calculated from the equation:

$$\chi^{(3)} = \chi^{(3)}_{ref}\left[\frac{I/I_0^3}{(I/I_0^3)_{ref}}\right]^{1/2}\left[\frac{n}{n_{ref}}\right]^2 \frac{l_{ref}}{l}\left(\frac{\alpha l}{(1-e^{-\alpha l})e^{-\frac{\alpha l}{2}}}\right) \tag{8}$$

where the subscript 'ref' refers to the standard reference CS_2 under identical conditions, α is the linear absorption coefficient of the sample and $\chi^{(3)}_{ref}$ is taken to be $4.0X10^{-13}$ esu [59, 60]. The $\chi^{(3)}$ value of p-(N,N-dimethyl)DBA is found to be $1.66x10^{-12}$ esu and it matches well with the value $1.9x10^{-12}$ esu obtained for the pure compound in DMF solution by Z-scan technique.

A check on the consistency of data in the experiment can be made by assuring that the conjugate signal varies as the appropriate power of the pump power or energy. The data can be fitted by a least square technique to the appropriate power law as discussed above. The closer the fit to the theory, the more accurate will be the value of $\chi^{(3)}$ obtained. If the linear absorption becomes substantial ($\alpha L \sim 1$), then equation (8) will no longer be appropriate, because the material is then not Kerr-like, and $\chi^{(3)}$ could be then due to population redistribution. Then the data should reflect this by showing a departure from a cubic dependence on pump intensity (average power or energy).

Figure of merit F is a measure of the nonlinear response that can be achieved for a given absorption loss, and is useful in comparing nonlinear materials in regions of absorption [60]. The figure of merit F is calculated by taking α in to account. F is given by

$$F = \chi^{(3)}/\alpha \tag{9}$$

F value calculated for p-(N,N-dimethyl)DBA is $1.608\ x10^{-11}$ esu cm, which is two order of magnitude larger than that measured for some organic systems by Reji Philip et al [59].

The present study is focused on the third order NLO properties of chalcones in their solution state. However, it is worthwhile to mention here that, there is always scope for testing these results in their bulk phase, either in crystalline form or in the thin film form. Chalcones, though have low melting point, can be grown into crystals and Z-scan measurements can be repeated on them. Laser damage threshold of these crystalline materials can then be measured. The DFWM technique was employed in this study in order to compare $\chi^{(3)}$ values of samples with those obtained through Z-scan technique. However, one can also do time resolved measurements using DFWM technique.

CONCLUSION

We are introducing some chalcones as promising new third order nonlinear optical materials. Their third order nonlinear optical coefficients have been extracted through single beam Z-scan technique. An attempt has been made to explore and explain the relationships that exist between their molecular structures and the third order NLO properties. The third order nonlinear susceptibility as well as the second order hyperpolarizability of these molecules was found to vary with the extent of π-electron delocalization among them. All these molecules, except p-(N,N-dimethyl)DBA, exhibit strong two-photon absorption at 532 nm. p-(N,N-dimethyl)DBA exhibits strong reverse saturation absorption. A good optical limiting of nanosecond pulses has been demonstrated with these molecules. p-(N,N-dimethyl)DBA possesses a very high $\chi^{(3)}$ of the order of 10^{-12} esu that is comparable with those values obtained for tricyanovinyl azo dye and 4-(N, N-diethylamino)-4'-nitrostilbene (DEANS). A good comparison is observed between $\chi^{(3)}$ values of p-(N,N-dimethyl)DBA obtained through Z-scan and DFWM techniques. Present study shows that these chalcones show interesting third order response and are attractive for optical limiting and other photonic applications.

ACKNOWLEDGEMENT

Authors wish to thank Dr. Reji Philip, Scientist, Optics Group, Raman Research Institute Bangalore, for providing laboratory facility to set up Four Wave Mixing experiment and for his fruitful suggestions. They also thank Mr. Suchand Sandeep, Research Scholar, Raman Research Institute, Bangalore, for his help in the DFWM experiments. Thanks to Prof. G. Umesh, Department of Physics, NITK Surathkal, for providing excellent laboratory facility for Z-scan experiment. The generosity of Prof. Shivarama Holla, Satheesh Rai N. and Mr. Mithun Ashok, Mangalore University with interesting chalcones has been gratefully acknowledged.

REFERENCES

[1] Robert W. Boyd, *Nonlinear Optics*, Academic Press, Inc., New York, 1992.

[2] Paras N. Prasad and David J. Williams *Introduction to Nonlinear Optical Effects in Molecules and Polymers,* John Wiley and Sons, New York, 1992.

[3] J.A. Giordmaine, *Phys. Rev. Lett.* 1962, 8, 19.

[4] P.D. Maker, R.W. Terhune, M. Nisenhoff, and C.M. Savage, *Phys. Rev. Lett.* 1962, 8, 21.

[5] K. Uchida, S. Kaneko, S.Omi, C. Hata, H. Tanzi, Y. Asahara, A. J. Ikushima, T. Tokizaki, and A. Nakumara, *J.Opt.Soc. Am. B* 1994,11, 1236.

[6] S. Matsumoto, K. Kubodera, T. Kurihara, and T. Kaino, *Opt. Commun.* 1990, 76, 147.

[7] D. Ahn and S.L. Chuang, *IEEE J. Quant. Electron.* 1987, 23, 2196.

[8] Seth R. Marder, William E. Torruellas, Mirelle Blanchard-Desce, Vincent Ricci, George I. Stegeman, Sharon Gilmour, Jean-Luc Bredas, Jun Li, Greg U. Bublitz, Steve G. Boxer, *Science* 1997, 276,1233.

[9] J.W. Perry, K. Mansour, I. -Y. S. Lee, X. –L. Wu, P. V. Bedworth, C. –T. Chen, D.Ng, S.R.Marder,P. Miles, T.Wada, M. Tian, H. Sasabe, *Science* 1996, 273, 1533.

[10] Lee W. Tutt and Thomas F. Boggess *Prog. Quant. Electr.* 1993, 17, 299.

[11] L.A. Hornak, *Polymers for Light Wave and Integrated Optics Technology and Applications*, Chapter 21, Marcel Dekker Inc., New York, 1992.

[12] Q. Gong, Z. Xia, Y.H. Zou, X. Meng, L. Wei, and F.M. Li, *Appl. Phys. Lett.* 1991, 59, 381.

[13] J.E. Ehrlich, X.L. Wu, I.-Y.S. Lee, Z.-Y. Hu, H. Rockel, S.R. Marder, and J.W. Perry, *Opt. Lett.* 1997, 22, 1843.

[14] B. Sahraoui, X.N. Phu, M. Salle, and A. Gorgues, *Opt. Lett.* 1998, 23, 1811.

[15] Z. Yang, Z. Wu, J. Ma, A. Xia, Q. Li, C. Liu, and Q. Gong, *Appl. Phys. Lett.* 2005, 86, 061903.

[16] T. Kurihara, H. Kobayashi, K. Kubodera, and T. kaino, *Chem. Phys. Lett.* 1990, 165, 170.

[17] T. Kobayashi, H. Uchiki, and K. Minoshima, *Synth. Met.* 1989, 28, D699.

[18] T. Kurihara, H. Kobayashi, K. Kubodera, and T. kaino, *Opt. Commun.* 1991, 84, 149.

[19] H. Kanbara, H. Kobayashi, and K. Kubodera, *IEEE Photonics Technol. Lett.* 1989, PTL- 1, 149.

[20] T. Kurihara, Y. Mori, N. Ooba, S. Tomaru, and T. kaino, *J. Appl. Phys.* 1991, 70, 17.

[21] D.S. Chemla, J. Zyss (Eds), *Nonlinear Optical Properties of Organic Molecules and Crystals*, Vol 1, Academic Press, London, 1987.

[22] V. Crasta, V. Ravindrachari, R.F. Bhajantri, R. Gonsalves, *J. Crystal Growth*, 2004, 267, 129.

[23] V. Crasta, V. Ravindrachari, S. Lakshmi, S.N. Pramod, M.A. Shridar, and J.S. Prasad, *J. Crystal Growth* 2005, 275, e329.

[24] Di Wu, Bo Zhao, and Z. Zhou, *J. Mol. Stru. (Theochem)* 2004, 682, 83.

[25] G.J. Zhang, T. Kinoshita, K. Sasaki, Y. Goto, and M. Nakayama, *Appl. Phys. Lett.* 1990, 57, 221.

[26] B. S. Holla, B. Veerendra, J. Indira, *J. Crystal Growth 2003,* 252, 308.

[27] J. Indira, P.P. Karath, and B.K. Sarojini, *J. Crystal Growth* 2002, 242, 209.

[28] V. Ravindrachari, V. Crasta, R.F. Bhajantri, B. Poojari, *J. Crystal Growth* 2005, 275, e313.

[29] Y. Kitaoka, T. Sasaki, S. Nakai, A. Yokotani, Y. Goto, and M. Nakayama, *Appl. Phys. Lett.* 1990, 56, 2047.

[30] H. Kauffmann and F. Kieser, *Chem. Ber.* 1913, 46, 3788.

[31] W. Dilthey, L. Neuhaus, E. Reis, and W. Schommer, *J. Pract. Chem.* 1930, 124, 61.

[32] P. Pfeiffer and H. Kleu, *Chem. Ber.* 1933, 66B, 1058.

[33] G. Reddelein *Chem. Ber.* 1912, 45, 2904.

[34] V.G. Mitina, A.A. Sukhorukov, B.A. Zadorozhnyi and V. F. Lavrushin, Zn. *Obsch. Khim.* 1976, 46, 699.

[35] S. Ambelcar, S.S. Vernelcar, S.S. Acharya and S. Rajagopal, *J. Pharm. Pharmacol.* 1961, 13, 698.

[36] G. Wurm, Arch. Phar. 308 (1975) 142; R. Aries, German Patent 2,341,514 (1974), *Chem. Abstr.* 1974, 80, 1461522.

[37] N. P. Buu-Hoi, N.D. Xuong and M. Sy, *Bul. Soc. Chim. France* 1956, 1646.

[38] Benjamin, E. Z. Faraggi, Y. Anvy, D. Davidov and R. Neumann *Chem.Mater.* 1996, 8, 352.

[39] T. Cassano, R. Tomasi, M. Farrarra, F. Babudri, G. M. Farinola and F. naso *Chem. Phys.* 2001, 272, 111.

[40] R Schroeder, B Ulrich, W Graupner, and U Scherf *J. Phys.: Condens. Matter* 2001, 13, L313.

[41] Marius. Albota et. al., *Science* 1998, 281, 1653.

[42] John Kiran, K. Chandrasekharan, Satheesh Rai Nooji, H.D. Shashikala, G. Umesh and Balakrishna Kalluraya *Chem. Phys.* 2006, 324, 699.

[43] John Kiran A., Mithun Ashok, K. Chandrasekharan, B. S. Holla H. D. Shashikala and G. Umesh *Opt. Commun.* 2007, 269, 235.

[44] M. Sheik-Bahae, Ali A. Said, Tai-Huei Wei, David J. Hagan, E. W. Van Stryland, *IEEE J. Quant. Electr.* 1990, 26, 760.

[45] Richard L. Sutherland, *Handbook of Nonlinear Optics*, Marcel Dekker, Inc., New York 1996.

[46] L. Zhi-Bo, T. J. Guo, Z. W. Ping, Z.W. Yuan, Z.G. Yin, *Chin. Phys. Lett.* 2003, 20, 509.

[47] X. Liu, S. Guo, H. Wang, and L. Hou, *Opt. Commun.* 2001, 197, 431.

[48] Y.R. Shen, *The principles of Nonlinear Optics,* Wiley-Interscience, New York, 1984.

[49] T. Cassano, R. Tomasi, F. Babudri, M. Ferrarra, A. Cardone, G.M. Farinola and F. Naso, *Opt. Lett.* 2002, 27, 2176.

[50] P.S. Kalsi, *Stereochemistry-Conformation and Mechanism*, Wiley Eastern Ltd, 1990.

[51] S. –L. Guo, L. Xu, H. –T. Wang, X. –Z.You, and N. B. Ming, *Optik* 2003,114, 58.

[52] G. S. He, G. C. Xu, P. N Prasad, B. A. Reinhardt, J. C. Bhatt, and A.G. Dillard, *Opt. Lett.* 1995, 20, 435.

[53] T.F. Boggess Jr, K.M. Bohnert, K. Mansour, S.C. Moss, I.W. Boyd, and A.L. Smirl, *IEEE J. Quant. Electron.* 1986, QE-22, 360.

[54] F. Z. Henari, W. J. Blau, L. R. Milgrom, G. Yahioglu, D. Philips and J.A. Lacey, *Chem. Phys. Lett.* 1997, 267, 229.

[55] S. Couris, E. Koudoumas, A.A. Ruth, and S. Leach, *J. Phys. B: At. Mol. Opt. Phys.* 1995, 28, 4537.

[56] M. Samoc, A. Samoc, B. Luther-Davis, *J Opt Soc Am B* 1998, 15, 817.

[57] Billmeyer F.W.Jr., *Text Book of Polymer Science*, 3rd ed., John Willey and Sons, Singapore, 1994.

[58] P.Yang, Xu J., J. Ballato, R.W. Schwartz, D.L. Carroll., *Appl Phy Lett* 2002, 80, 3394.

[59] Reji Philip, M. Ravikanth, G. Ravindrakumar, *Opt. Commun.* 1999, 165, 91.

[60] J.S. Shrik, J.R. Lindle, F.J. Bartoli, M.E. Boyle, *J Phys. Chem.* 1992, 96, 5847.

In: Chemical Physics Research Trends, Vol. 252
Editor: Samuel V. Arnold, pp. 269-318
ISBN: 978-1-60021-554-4

Chapter 7

Relationships among Parameters of Molecular Potential Energy Functions: Theory and Applications

Teik–Cheng Lim
School of Science and Technology, SIM University (UniSIM),
535A Clementi Road, S599490, Republic of Singapore

Abstract

Modeling of molecular behavior can be generally categorized into two broad groups – the quantum mechanics approach and the molecular mechanics approach. Although the latter does not consider the electronic distribution, it can provide solutions that are reasonably accurate in a fraction of computation time as compared to the former. Indeed, "simple potential functions such as Morse, Rydberg, Born-Mayer, etc., give a quantitatively correct description of the potential and only modest extensions are needed to obtain functions which stand up to the most stringent experimental tests" (Murrell et al, 1984). Arising from the ease to which the molecular energy can be calculated, numerous potential energy functions have been proposed. It follows that comparative advantage exists among these potential functions. As such it is important to understand the characteristics and extent of discrepancies among normalized potential functions. An understanding of such discrepancies aid the selection of appropriate potential energy functions for computation under specific environmental conditions and for conservative computation in molecular design modeling. Although the discrepancies observation is straightforward, the methods for normalizing different potential energy functions must firstly be addressed. To normalize these potential energy functions, it is imperative that they be equated under a certain logical criteria. This is unlike bridging parameters that allow a generic potential function to be reduced into specific ones by mere substitution of numerical values into the bridging parameters [5-7]. Rather, the process of normalizing refers to the mathematical formalism that enables parameters from different potential functions to be related. In this way, a set of parameters from one potential function can be converted into parameters of another potential function. Theoretical plots of these normalized potential functions reveal their characteristics discrepancies and shed light into their relative performance in modeling under specified conditions and applications.

This chapter covers the mathematical procedure for normalizing molecular potential among 2-body, 3-body, 4-body and non-bonded potential energy functions.

1. Introduction

Though chemical physics is largely an experimental science, its scientific progress has been marked by the development of models to elucidate the measured experimental results. It has been remarked that since the middle of the 19th century, chemical physicists and chemists have used molecular structural models as a foundation for gaining insight into the physical characteristics of molecules [1]. Indeed, many physical behaviors of matter in all three states can not only be observed, but also be theoretically calculated. Although it is true that first-principles quantum mechanical approach, which considers the electronic distribution, provides better accuracy, larger computational effort is required for accurately solving the Schrödinger equation even though in a static case of 10^1 to 10^2 atoms. The empirical interatomic potentials are reputed to be able to handle the computation of equal number of atoms within a small fraction of the computation time, as well as to study the dynamic behavior of the molecular system. The correct representation of a molecular potential energy as a function of the interatomic distance is of practical importance in ensuring accuracy of the simulated results.

Towards this end, a well-balanced approach of obtaining both accuracy and computational speed is possible through providing innovative extensions to simple potential energy functions [2]. A consequence of refinement efforts lead to numerous potential energy functions. More than 80 potential functions have been reviewed by Maitland et al. [3], Varshni [4] and Erkoc [5] alone. In view of the numerous potential functions available in the literature, it follows that comparative advantage exists among them. To take advantage of the numerous available potential functions, it is required that the comparative advantage among different potential functions be understood. To pave a way for which potential functions can be compared, they must, in the first instance, be normalized amongst themselves. The normalization of the potential functions can be achieved by obtaining relationships between the compared potentials. Upon this normalization, comparison between the potential functions can be performed without bias. Broadly speaking, there exist two schools of thought in the area of potential function relationships, namely:

a) Bridging parameter approach, and
b) Parameter relation approach.

In the bridging parameter approach, a generic potential function has one or two additional parameters which, upon substituting with specific numerical values, reduce the general functions into specific ones. Hence the difference between specific potential functions that fall into the same general function is differentiated based on the bridging parameter(s). For example, the Dunham potential [6]

$$U_{DUN} = A\xi_a^{\,2}\left(1+\sum_{i=1}\alpha_i\xi_a^{\,i}\right) \quad ; \quad \xi_a = \frac{r-R}{R} \tag{1}$$

and the Simons-Parr-Finlan (SPF) potential [7]

$$U_{SPF} = B\xi_b^{\,2}\left(1+\sum_{i=1}\beta_i\xi_b^{\,i}\right) \quad ; \quad \xi_b = \frac{r-R}{r} \tag{2}$$

whereby r is the interatomic variable distance while R is the equilibrium interatomic distance, were bridged by Thakkar's relation [8]

$$\xi = sign(p)\left[1-\left(\frac{R}{r}\right)^p\right]. \tag{3}$$

The Dunham and SPF potential functions are recovered by substituting $p=-1$ and $p=+1$ respectively. Incorporation of the Ogilvie potential [9]

$$U_{OGI} = C\xi_c^{\,2}\left(1+\sum_{i=1}\gamma_i\xi_c^{\,i}\right) \quad ; \quad \xi_c = \frac{r-R}{0.5(r+R)} \tag{4}$$

was made possible through Molski's relationship [10,11]

$$\xi = \frac{r-R}{ar+(1-a)R} \tag{5}$$

whereby substituting $a=0, 1, 0.5$ leads to the Dunham, SPF and Ogilvie potential functions respectively. Here, p and a are the bridging parameters according to the Thakkar and Molski relations. It follows that the Dunham and SPF potential functions reduce to the Puppi [12] and Kratzer [13] potential functions respectively when $\alpha_i = \beta_i = 0\,;(i=1,2,...)$. Whilst bridging parameters have been considered for potential functions that give zero ground-state, recently a set of bridging parameters have been introduced for potential functions that give negative ground-state, such as the Lennard-Jones [14-16], Morse [17], Buckingham [18,19] and Linnett [20,21] potential functions. Writing a flexible potential function as [22]

$$\begin{aligned}\frac{U_{flex}}{D} &= \frac{n^x S_2^{1-x}}{m^y S_1^{1-y} - n^x S_2^{1-x}}\left(\frac{R}{r}\right)^{my} \exp\left[S_1\left(1-\frac{r}{R}\right)(1-y)\right] \\ &- \frac{m^y S_1^{1-y}}{m^y S_1^{1-y} - n^x S_2^{1-x}}\left(\frac{R}{r}\right)^{nx} \exp\left[S_2\left(1-\frac{r}{R}\right)(1-x)\right]\end{aligned} \tag{6}$$

the generalized versions of the Lennard-Jones, Morse, Buckingham and Linnett potential functions are recovered when specific values of x and y, whereby $0 \leq x \leq 1$ and $0 \leq y \leq 1$, are substituted. The scaling factors $S_1 = 16.685$ and $S_2 = 4.315$ were obtained for relating the Morse function with the Lennard-Jones function, while $S_1 = 13.772$ and $S_2 = 5.228$ were obtained for relating Buckingham and Linnett potential functions for the case where $m = 2n = 12$. For this flexible potential function, x and y are the bridging parameters. However it can be seen that these bridging parameters such as p, a, x and y only describe how the compared potential functions differ, but do not offer any criteria for normalization. The normalization of different potential functions is achieved through parameter relationships. Figure 1 depicts the different categories and sub-categories of potential function relationships, with special emphasis on parameter relationships.

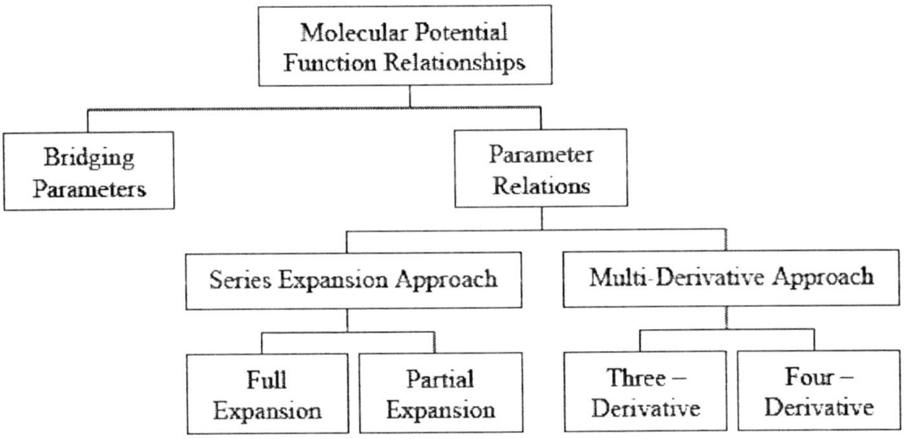

Figure 1. Categories and sub-categories of potential function relationships.

2. Review of Potential Functions

It is useful, at this juncture, to list the potential functions considered herein in order to classify them. generally potential functions can either exhibit zero ground-state

$$(U)_{r=R} = 0 \tag{7}$$

or negative ground-state

$$(U)_{r=R} = -D \tag{8}$$

whereby D is the dissociation energy. By definition of ground-state, it follows that

$$\left(\frac{\partial U}{\partial r}\right)_{r=R} = 0 \tag{9}$$

for both cases. Potential functions that exhibit zero ground-state can either converge for bond dissociation

$$(U)_{r\to\infty} = +D \tag{10}$$

or diverge

$$(U)_{r\to\infty} = \infty \tag{11}$$

while the potential functions that exhibit negative ground-state normally exhibit convergence at infinite separation

$$(U)_{r\to\infty} = 0. \tag{12}$$

For both convergence cases it follows that

$$\left(\frac{\partial U}{\partial r}\right)_{r\to\infty} = 0 \tag{13}$$

Figure 2 depicts the types of potential functions considered.

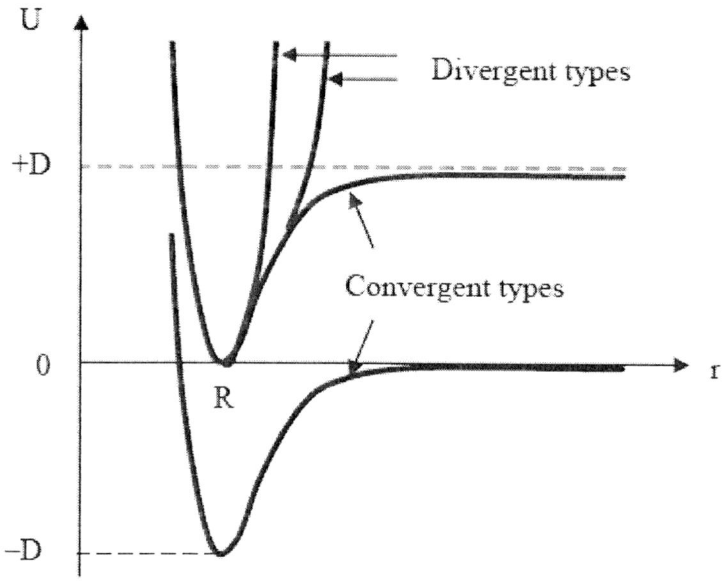

Figure 2. The general types of potential energy conceptualization.

The divergent type of potential functions include the harmonic potential

$$U_H = \frac{1}{2} k_H (r - R)^2 = \frac{1}{2} k_H (\delta r)^2 \tag{14}$$

the polynomial series

$$U_S = \frac{1}{2} \sum_{n=2}^{m} k_{Sn} (\delta r)^n \tag{15}$$

and the power series

$$U_P = \chi \xi_x^{\,2} \left(1 + \sum_{i=1} x_i \xi_x^{\,i} \right) \quad ; \quad \xi_x = \xi_x (r, R) . \tag{16}$$

The convergent types are more realistic and are more prevalent. They include the generalized Lennard-Jones function [14]

$$U_{LJ} = \frac{A_{LJ1}}{r^m} - \frac{A_{LJ2}}{r^n} \quad ; \quad (m > n) \tag{17}$$

which can be expressed in an equivalent form

$$\frac{U_{LJ}}{D} = \frac{n}{m-n} \left(\frac{R}{r} \right)^m - \frac{m}{m-n} \left(\frac{R}{r} \right)^n \quad ; \quad (m > n) \tag{18}$$

by substituting

$$\begin{Bmatrix} A_{LJ1} \\ A_{LJ2} \end{Bmatrix} = \frac{D}{m-n} \begin{Bmatrix} nR^m \\ mR^n \end{Bmatrix} . \tag{19}$$

Specific Lennard-Jones functions in molecular mechanics softwares can be obtained by substituting $m = 2n = 12$ to give the very common Lennard-Jones(12-6) function [23-26], $2m = 3n = 18$ to give the common Lennard-Jones(9-6) function [27,28], $m = 2n = 14$ to give the Lennard-Jones(14-7) function [29]. The Lennard-Jones(12-10) function, whereby $5m = 6n = 60$, is normally left as an option [25,26]. The Lennard-Jones(12-6) function is also adopted by the Pearson-Takai-Halicioglu-Tiller (PTHT) [30] for their description of 2-body interaction in silicon crystals.

In addition to the zero ground-state form

$$U_{Morse} = D[1-\exp(-\alpha\,\delta r)]^2, \tag{20}$$

the Morse function [17] can also be expressed in the form of negative ground-state

$$U_M = U_{Morse} - D = D[\exp(-2\alpha\,\delta r) - 2\exp(-\alpha\,\delta r)]. \tag{21}$$

A generalized form of the Morse potential given in Eq.(21) i.e.

$$U_{BH} = A_{BH1}\exp(-\lambda_1 r) + A_{BH2}\exp(-\lambda_2 r) \tag{22}$$

was adopted by Biswas and Hamann [31] for describing the 2-body interaction in silicon. Due to the positive sign before the attractive term, the attractive coefficient A_{BH2} has negative values in order for the function to be physically admissible. An alternate, but equivalent, form to Eq.(22) is

$$\frac{U_M}{D} = \frac{S_2}{S_1 - S_2}\exp\left[S_1\left(1-\frac{r}{R}\right)\right] - \frac{S_1}{S_1 - S_2}\exp\left[S_2\left(1-\frac{r}{R}\right)\right] \quad ; \quad (S_1 > S_2) \tag{23}$$

through the relations

$$\begin{Bmatrix} A_{BH1} \\ A_{BH2} \end{Bmatrix} = \frac{D}{S_1 - S_2}\begin{Bmatrix} +S_2\exp(S_1) \\ -S_1\exp(S_2) \end{Bmatrix} \tag{24}$$

and

$$\begin{Bmatrix} \lambda_1 \\ \lambda_2 \end{Bmatrix} = \frac{1}{R}\begin{Bmatrix} S_1 \\ S_2 \end{Bmatrix}. \tag{25}$$

The simple Morse function denoted in Eqs.(20) and (21) are commonly adopted in molecular mechanics softwares [24,27,32,33] though it is obvious that the generalized form furnished in Eqs.(22) and (23) allow better accuracy.

The Buckingham function [18] is normally adopted in its simple form known as the Exponential-6 function [33-39]

$$U_{X6} = A_{X6}\exp(-B_{X6} r) - \frac{C_{X6}}{r^6} \tag{26}$$

although a more generalized version

$$U_{Buck} = A_{Buck} \exp\left(-B_{Buck} r\right) - \frac{C_{Buck}}{r^n} \tag{27}$$

obviously allows better curve-fitting to experimental results. Equation (27) can alternatively be expressed as

$$\frac{U_{Buck}}{D} = \frac{n}{S_1 - n} \exp\left[S_1\left(1 - \frac{r}{R}\right)\right] - \frac{S_1}{S_1 - n}\left(\frac{R}{r}\right)^n \quad ; \quad (S_1 > n) \tag{28}$$

through the relations

$$\begin{Bmatrix} A_{Buck} \\ C_{Buck} \end{Bmatrix} = \frac{D}{S_1 - n} \begin{Bmatrix} n \exp(S_1) \\ S_1 R^n \end{Bmatrix} \tag{29}$$

and

$$B_{Buck} = \frac{S_1}{R}. \tag{30}$$

The Bauer-Maysenholder-Seeger potential function (BMS) [40] that was developed for FCC metals adopts Eq.(27) for describing the 2-body interaction with $\rho_{BMS} = 1/C_{Buck}$.

Although the Linnett potential function [20,21]

$$U_{Linn} = \frac{A_{Linn}}{r^m} - B_{Linn} \exp\left(-C_{Linn} r\right) \tag{31}$$

does not appear to be popular, we nonetheless propose an alternative form

$$\frac{U_{Linn}}{D} = \frac{S_2}{m - S_2}\left(\frac{R}{r}\right)^m - \frac{m}{m - S_2} \exp\left[S_2\left(1 - \frac{r}{R}\right)\right] \quad ; \quad (m > S_2) \tag{32}$$

whereby

$$\begin{Bmatrix} A_{Linn} \\ B_{Linn} \end{Bmatrix} = \frac{D}{m - S_2} \begin{Bmatrix} S_2 R^m \\ m \exp(S_2) \end{Bmatrix} \tag{33}$$

and

$$C_{Linn} = \frac{S_2}{R}. \tag{34}$$

This is so that substituting $(x, y) = (1,1)$, $(0,0)$, $(1,0)$ and $(0,1)$ into Eq.(6) reduces it into the generalized functions described by Eqs.(18), (23), (28) and (32) respectively.

A slight variation to the Eq.(22) is the 2-body interaction energy by Kaxiras and Pandey [41]

$$U_{KP} = A_{KP1}\exp\left(-\alpha_1 r^2\right) - A_{KP2}\exp\left(-\alpha_2 r^2\right) \tag{35}$$

developed for silicon. This function is similar to the others that consist of a repulsive term and an attractive term.

Apart from the type of potential functions that consists of a repulsive term deducting an attractive term to give a well-depth, another type of potential function exists through multiplication of terms, such as the Rydberg potential function [42,43]

$$U_{Ryd} = -D(1 + a\rho)\exp(-a\rho) \quad ; \quad \rho = r - R. \tag{36}$$

This potential function has been used by Murrell et al. [44-47] in the form

$$U_{MM} = -D(1 + a\rho)\exp(-a\rho) \quad ; \quad \rho = \frac{r - R}{R} \tag{37}$$

for describing the 2-body interaction in cubic and tetrahedral structures. Obviously, the shape parameter a in Eqs.(36) and (37) are in inverse Angstrom unit and dimensionless respectively due to the differing definitions of ρ. In the Extended-Rydberg version, Murrell-et al. [48,49] proposed

$$U_{ER} = -D\left[\sum_{i=0}^{3} a_i \rho^i\right]\exp(-a_1\rho) \quad ; \quad (a_0 = 1 \quad , \quad \rho = r - R), \tag{38}$$

thereby incurring the inverse, square inverse and cube inverse of Angstrom units for a_1, a_2 and a_3 respectively.

3. Series Expansion Approach

The series expansion approach adopts the use of Taylor series expansion

$$f(x)=\sum_{n=0}^{\infty} f^{(n)}(x_0)\frac{(x-x_0)^n}{n!} \tag{39}$$

for enabling potential functions to be expressed in terms of power series. In particular, the use of the Maclaurin series has been found to be highly applicable. The generalized Maclaurin series expansion is a special case of Taylor series expansion with $x_0 = 0$. As shown in Fig.1, there are two types of parameter relationships on the basis of the series expansion approach. The full expansion is discussed in the next subsection followed by the partial expansion.

3.1. Full Expansion Method

As the name suggests, the full expansion method expands all terms in the potential function, except those already in terms with integer powers of r or δr, such that the expanded function behaves as a divergent potential curve if it takes the form of power series. Alternatively, full expansion method can still give convergent potential energy curves if the related parameters are subsequently substituted into the convergent potentials.

The Morse function in Eq.(20) and the Modified Murrell (MMM) potential function

$$U_{MMM} = D + U_{MM} = D\left\{1-\left[1+a\left(\frac{r-R}{R}\right)\right]\exp\left[-a\left(\frac{r-R}{R}\right)\right]\right\} \tag{40}$$

which gives $U_{r=R} = 0$ and $U_{r\to\infty} = +D$, can be indirectly related by firstly expressing them in the form shown in Eq.(15), followed by comparison of terms. Substituting

$$\exp(\pm x)=\sum_{n=0}^{+\infty}\frac{(\pm 1)^n x^n}{n!} \tag{41}$$

into Eqs.(20) and (40) leads to [50]

$$U_{Morse} = D\left[\alpha^2(\delta r)^2 - \alpha^3(\delta r)^3 + \frac{7}{12}\alpha^4(\delta r)^4 - \frac{1}{4}\alpha^5(\delta r)^5 + \frac{31}{360}\alpha^6(\delta r)^6 - \ldots\right] \tag{42}$$

and

$$U_{MMM} = D\sum_{n=0}^{\infty}\frac{(-1)^n}{n!+(n+1)!}\left(\frac{a}{R}\delta r\right)^{n+2}. \tag{43}$$

Comparison of the first term gives

$$\alpha = \frac{1}{\sqrt{2}}\left(\frac{a}{R}\right). \tag{44}$$

For verification, we apply data from Eggen et al. [51] whereby parameters $a = 8.2$, $D = 6.298eV = 1.009 \times 10^{-18} J$ and $R = 1.507 \overset{\circ}{A}$ apply for the Murrell function. The Morse parameter is obtainable from Eq.(44) to give $\alpha = 3.848 \times 10^{10} \overset{\circ}{A}^{-1}$. A plot of non-dimensional energy versus non-dimensional interatomic distance shown in Fig.3 validates the relationship.

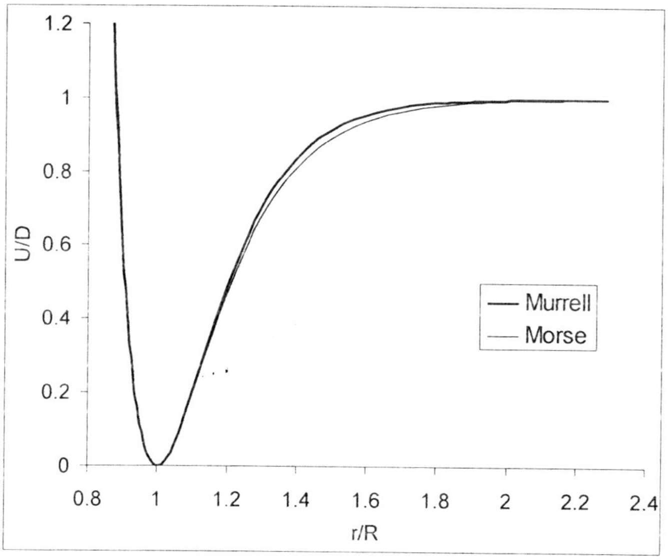

Figure 3. Relationship between Morse and Murrell potential functions based on Eq.(44) and data from Eggen et al. [51].

The Maclaurin series expansion for the exponential term and the binomial series expansion

$$\frac{R}{r} = \frac{1}{\left(1 + (\delta r / R)\right)^p} = \sum_{i=0}^{\infty} \binom{-p}{i} \left(\frac{\delta r}{R}\right)^i \tag{45}$$

can be utilized for obtaining the polynomial forms of other potential functions as

$$U = D \sum_{i=0}^{\infty} d_i \left(\frac{\delta r}{R}\right)^i \tag{46}$$

As an example, the imposition of Eqs.(41) and (45) on Eq.(18) with $m = 2n = 12$, Eq.(21), Eq.(26) and Eq.(37) gives the coefficient d_i as furnished in Table 1 [52].

Table 1. Non-dimensional coefficients for the polynomial forms of some potentials

Coefficients	LJ(12-6)	Simple Morse	Exponential-6	Murrell
d_0	-1	-1	-1	-1
d_1	0	0	0	0
d_2	+36	$+(\alpha R)^2$	$+\frac{3(BR)-21}{1}\left(\frac{BR}{BR-6}\right)$	$+\frac{1}{2}a^2$
d_3	-252	$-(\alpha R)^3$	$-\frac{(BR)^2-56}{1}\left(\frac{BR}{BR-6}\right)$	$-\frac{1}{3}a^3$
d_4	+1113	$+\frac{7}{12}(\alpha R)^4$	$+\frac{(BR)^3-504}{4}\left(\frac{BR}{BR-6}\right)$	$+\frac{1}{8}a^4$
d_5	-3864	$-\frac{1}{4}(\alpha R)^5$	$-\frac{(BR)^4-5040}{20}\left(\frac{BR}{BR-6}\right)$	$-\frac{1}{30}a^5$
d_6	+11452	$+\frac{31}{360}(\alpha R)^6$	$+\frac{(BR)^5-55440}{120}\left(\frac{BR}{BR-6}\right)$	$+\frac{1}{144}a^6$

By expanding zero ground-state potential functions and rearranging the terms into the power series form, the parameters of power series potentials can be obtained. Ogilvie [53] expressed the Dunham, Eq.(1), and Ogilvie, Eq.(4), coefficients in terms of Morse parameters as

$$A = C = \alpha^2 D \tag{47}$$

$$a_i = \left(2^{i+2} - 2\right)\frac{(-\alpha)^i}{(i+2)!} \tag{48}$$

$$c_i = \left(\frac{1}{2}\right)^i (i+1) + \sum_{j=0}^{i-1} \frac{(i+1)!\left(2^{i-2j+2} - 2^{1-j}\right)(-\alpha)^{i-j}}{j!(1+i-j)!(2+i-j)!}. \tag{49}$$

In addition, the relationship between coefficients of the Dunham, SPF and Ogilvie were given as [9,53]

$$A = B = C \tag{50}$$

$$a_j = (-1)^j (j+1) + \sum_{i=0}^{j-1} (-1)^i \binom{j+1}{i} b_{j-i} \tag{51}$$

$$b_j = (j+1) + \sum_{i=0}^{j-1} \binom{j+1}{i} a_{j-i} \tag{52}$$

$$a_j = (-1)^j \left(\frac{1}{2}\right)^j (j+1) + \sum_{i=0}^{j-1} (-1)^i \left(\frac{1}{2}\right)^i \binom{j+1}{i} c_{j-i} \tag{53}$$

$$c_j = \left(\frac{1}{2}\right)^j (j+1) + \sum_{i=0}^{j-1} \left(\frac{1}{2}\right)^i \binom{j+1}{i} a_{j-i} \tag{54}$$

$$b_j = \left(\frac{1}{2}\right)^j (j+1) + \sum_{i=0}^{j-1} \left(\frac{1}{2}\right)^i \binom{j+1}{i} c_{j-i} \tag{55}$$

$$c_j = (-1)^j \left(\frac{1}{2}\right)^j (j+1) + \sum_{i=0}^{j-1} (-1)^i \left(\frac{1}{2}\right)^i \binom{j+1}{i} b_{j-i} . \tag{56}$$

A slightly different, but nonetheless equivalent, version to Eqs.(51) and (52) are developed herein. Introducing the change in bond length as δr, then the Dunham and SPF parameters in Eqs.(1) and (2), raised to a positive integer power n, can be written as

$$(\xi_a)^n = \left(\frac{\delta r}{R}\right)^n \text{ and} \tag{57}$$

$$(\xi_b)^n = \sum_{n=1} \left[\sum_{m=1} (-1)^{m+1} \binom{m+n-2}{n-1} \left(\frac{\delta r}{R}\right)^{m+n-1} \right]. \tag{58}$$

Substituting Eqs.(57) and (58) into Eqs.(1) and (2) gives

$$\frac{U_a}{A} = \left(\frac{\delta r}{R}\right)^2 + \sum_{j=3} a_{j-2} \left(\frac{\delta r}{R}\right)^j \text{ and} \tag{59}$$

$$\frac{U_b}{B} = \left(\frac{\delta r}{R}\right)^2 + \sum_{j=3} \left[(-1)^j (j-1) + \sum_{i=1}^{j-2} (-1)^{i+j} \binom{j-1}{i+1} b_i \right] \left(\frac{\delta r}{R}\right)^j \tag{60}$$

respectively. Comparing the LHS of Eqs.(59) and (60) gives $A = B$, as partially described by Eq.(50), while comparing terms on the RHS yields the Dunham coefficients in terms of SPF coefficients gives

$$a_j = (-1)^j (j+1) + \sum_{i=1}^{j} (-1)^{i+j} \binom{j+1}{i+1} b_i \, . \tag{61}$$

Solving for SPF coefficients in Eq.(61) enables them to be expressed in terms of Dunham coefficients

$$b_j = (j+1) + \sum_{i=1}^{j} \binom{j+1}{i+1} a_i \, . \tag{62}$$

Similar method can be extended to relate these two sets of coefficients with Ogilvie coefficients. Potential functions in the power series forms are applicable only for infinitesimal bond stretching due to their divergence, hence not applicable for large bond stretching nor for van der Waals interaction.

3.2. Partial Expansion Method

The partial expansion method is an alternative to the full expansion method. Here, terms that are expanded are performed selectively before rearrangement of terms is done. It is known that while the Lennard-Jones function goes to infinity as $r \to 0$, the Buckingham potential gives an infinite negative energy. This unrealistic result is due to a finite and an infinite values to the repulsive and attractive terms respectively as the interatomic distance shrinks to zero. To give more realistic repulsion energy, the repulsive term of the Exponential-6 is incorporated with a correction factor k, raised to a power equal to the scaling factor S, so that

$$\frac{U_{X6}^{(n)}}{D} = \left(k^{(n)}\right)^S \frac{6}{S-6} \exp\left[S\left(1-\frac{r}{R}\right)\right] - \frac{S}{S-6}\left(\frac{r}{R}\right)^6 \tag{63}$$

where n refers to the order of accuracy. For $S = 12$, we have

$$\frac{U_{X6}^{(n)}}{D} = \left[k^{(n)} \exp\left(1-\frac{r}{R}\right)\right]^{12} - 2\left(\frac{R}{r}\right)^6 \tag{64}$$

such that comparison of Eq.(18) with $m = 2n = 12$ and Eq.(64) implies

$$\frac{R}{r} = k^{(n)} \exp\left(1-\frac{r}{R}\right). \tag{65}$$

Since

$$\exp\left(1-\frac{r}{R}\right)=\exp\left(-\frac{\delta r}{R}\right)\equiv\sum_{n=0}^{+\infty}\frac{1}{n!}\left(-\frac{\delta r}{R}\right)^n \text{ and} \tag{66}$$

$$\frac{R}{r}=\frac{1}{1-(-\delta r/R)}\equiv\sum_{n=0}^{+\infty}\left(-\frac{\delta r}{R}\right)^n \tag{67}$$

therefore substituting Eqs.(66) and (67) into Eq.(65) gives the correction factor of order n as [54]

$$k^{(n)}=\frac{\sum_{i=0}^{n}(-1)^i\left(\frac{\delta r}{R}\right)^i}{\sum_{i=0}^{n}\frac{(-1)^i}{i!}\left(\frac{\delta r}{R}\right)^i}. \tag{68}$$

Figure 4 shows the repulsion energy term going from a finite value at $r=0$ to amore realistic value as the order of accuracy n increases.

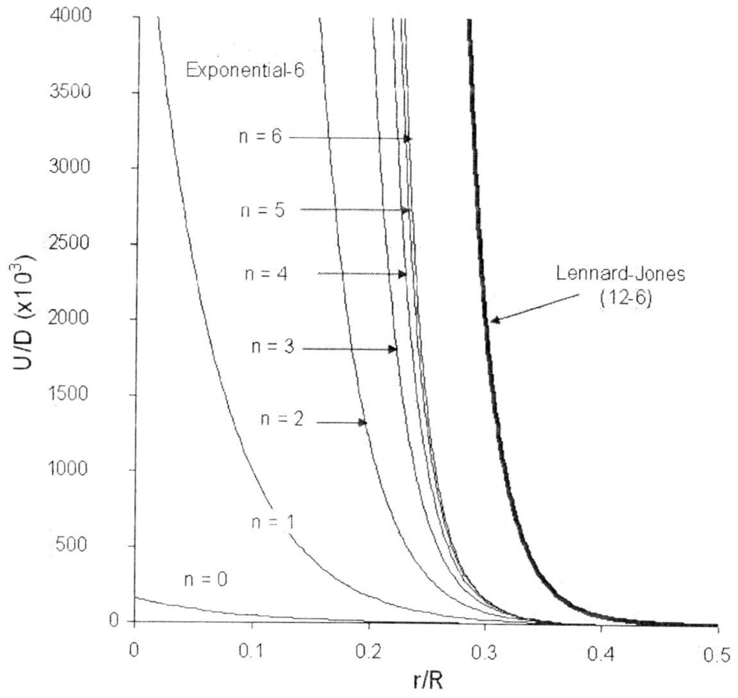

Figure 4. Comparison of the loose form of Exponential-6 (as denoted by the thin curves), at various orders of n, with the Lennard-Jones (12-6) (denoted by the thick curve).

While this example emphasizes the repulsive state, the next example focuses on the attractive state. A long range parameter relationship between the Morse and the Murrell potential functions can be obtained by firstly recasting Eqs.(20) and (40) as

$$U_M = D_M\left[\exp(-2\alpha(\delta r)) - 2\exp(-\alpha(\delta r)) + 1\right] \tag{69}$$

and

$$U_{MMM} = D_{MM}\left[(a-1)\exp\left(-\frac{a}{R}(\delta r)\right) - a\frac{r}{R}\exp\left(-\frac{a}{R}(\delta r)\right) + 1\right] \tag{70}$$

respectively. Equating the first two terms of Eqs.(66) and (67) gives

$$\frac{r}{R} = 1 + \frac{\delta r}{R} \approx \left(1 - \frac{\delta r}{R}\right)^{-1} \approx \left[\exp\left(-\frac{\delta r}{R}\right)\right]^{-1} = \exp\left(\frac{\delta r}{R}\right) \tag{71}$$

which, upon substitution into the second term on the RHS of Eq.(70) leads to

$$U_{MMM} = D\left[(a-1)\exp\left(-\frac{a}{R}(\delta r)\right) - a\exp\left(-\frac{a-1}{R}(\delta r)\right) + 1\right]. \tag{72}$$

Comparing the terms of Eqs.(69) and (72) suggest [55]

$$a = 2 \text{ and} \tag{73}$$

$$\alpha = \frac{1}{R}. \tag{74}$$

Hence substituting Eqs.(73) and (74) into Eq.(70) gives the Modified Murrell function in terms of Morse parameters,

$$U_{MMM} = D\left[1 + (1 - 2\alpha r)\exp(-2\alpha(\delta r))\right]. \tag{75}$$

Figure 5, plotted using Eqs.(20) and (75), shows that the parameter relations given by Eqs.(73) and (74) are reliable only for long range.

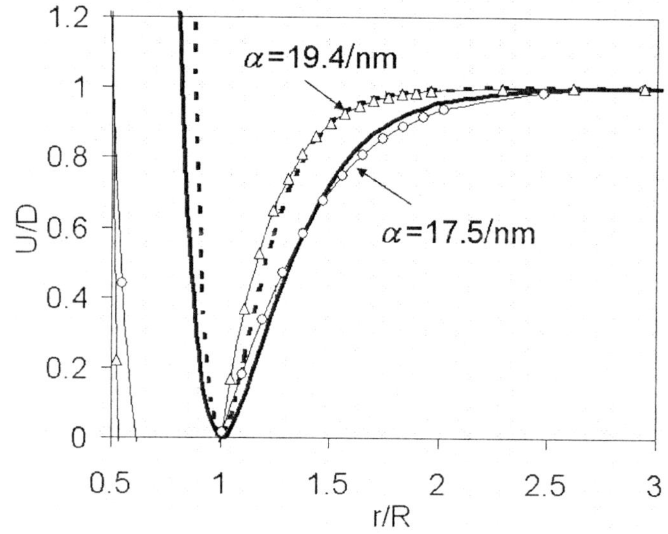

Figure 5. Murrell 2-body approximation (bold curves) to Morse potential (thin curves) using long range relationship.

Generally the accuracy of the parameter relations is highly dependent on the selected term by which expansion is performed, and the arrangement of the potential function terms in paving way for the expansion. Take, for example, the generalized Morse potential function

$$U_{GM} = A_{GM1}\exp(-\lambda_1 r) - A_{GM2}\exp(-\lambda_2 r) \quad (76)$$

where A_{GM2} is a positive value due to the negative sign before it, unlike Eq.(22). To relate this generalized Morse function with the Extended-Rydberg function described in Eq.(38), we rewrite the former as [56]

$$U_{GM} = D\left[\frac{\lambda_2}{\lambda_1 - \lambda_2}\exp(-\lambda_1 \rho) - \frac{\lambda_1}{\lambda_1 - \lambda_2}\exp(-\lambda_2 \rho)\right] \quad ; \quad (\rho = r - R) \quad (77)$$

whereby $\lambda_1 > \lambda_2 > 0$. We note that all a_1 and a_3 parameters in Eq.(38) are positive while most a_2 parameters are positive [49]. Hence parameters of the Generalized Morse and the Extended-Rydberg functions can be of two types: (i) $a_2 < 0$ and (ii) $a_2 > 0$. For the case whereby $a_2 < 0$, Eq.(77) is recasted in the form

$$U_{GM} = D\left[\frac{\lambda_2}{\lambda_1 - \lambda_2}\exp[-(\lambda_1 - \lambda_2)\rho] - \frac{\lambda_1}{\lambda_1 - \lambda_2}\right]\exp(-\lambda_2 \rho) \quad (78)$$

such that substituting

$$\exp[-(\lambda_1 - \lambda_2)\rho] \approx \sum_{n=0}^{3} \frac{[-(\lambda_1 - \lambda_2)]^n}{n!} \rho^n \tag{79}$$

into Eq.(78) gives

$$U_{GM} \approx -D\left[1 + \sum_{n=1}^{3} \lambda_2 \frac{[-(\lambda_1 - \lambda_2)]^{n-1} \rho^n}{n!}\right] \exp(-\lambda_2 \rho). \tag{80}$$

Comparing the terms in Eqs.(38) and (80) gives

$$a_n = \lambda_2 \frac{[-(\lambda_1 - \lambda_2)]^{n-1}}{n!}. \tag{81}$$

For the case whereby $a_2 > 0$, Eq.(77) is rewritten in an alternate form

$$U_{GM} = D\left[\frac{\lambda_2}{\lambda_1 - \lambda_2} - \frac{\lambda_1}{\lambda_1 - \lambda_2} \exp[(\lambda_1 - \lambda_2)\rho]\right] \exp(-\lambda_1 \rho) \tag{82}$$

so that substituting

$$\exp[(\lambda_1 - \lambda_2)\rho] \approx \sum_{n=0}^{3} \frac{(\lambda_1 - \lambda_2)^n}{n!} \rho^n \tag{83}$$

into Eq.(82) yields

$$U_{GM} \approx -D\left[1 + \sum_{n=1}^{3} \lambda_1 \frac{(\lambda_1 - \lambda_2)^{n-1} \rho^n}{n!}\right] \exp(-\lambda_1 \rho). \tag{84}$$

Upon comparison of terms in Eqs.(38) and (84), one obtains

$$a_n = \lambda_1 \frac{(\lambda_1 - \lambda_2)^{n-1}}{n!}. \tag{85}$$

Using only the first two terms $(n = 1,2)$, it would suffice to obtain

$$\lambda_1 = a_1 - 2\frac{a_2}{a_1} \tag{86}$$

and

$$\lambda_2 = a_1 \tag{87}$$

based on Eq.(81), and that

$$\lambda_1 = a_1 \tag{88}$$

and

$$\lambda_2 = a_1 - 2\frac{a_2}{a_1} \tag{89}$$

on the basis of Eq.(85). Figure 6(a) and (b) shows the validity of Eqs.(81) and (85), respectively, using the parameters furnished by Huxley and Murrell [49].

Due to the inferior accuracy for the case of $a_2 < 0$, a refined relationship is attainable by rewriting Eq.(77) as

$$U_{GM} = A_{GMi} \exp(-\lambda_i R)\left\{\exp[(a_1 - \lambda_i)\rho] - \frac{\lambda_i}{\lambda_j}\exp[(a_1 - \lambda_j)\rho]\right\}\exp(-a_1\rho) \tag{90}$$

for $i, j = 1,2$. Via the Maclaurin series expansion and rearranging of terms,

$$U_{GM} = \left(\frac{\lambda_j - \lambda_i}{\lambda_j}\right)A_i \exp(-\lambda_i R)$$
$$\left\{1 + a_1\rho + \frac{1}{2}[a_1^2 - \lambda_1\lambda_2]\rho^2 + \frac{1}{6}[a_1^3 - 3a_1\lambda_1\lambda_2 + \lambda_1\lambda_2(\lambda_1 + \lambda_2)]\rho^3\right\}\exp(-a_1\rho). \tag{91}$$

Comparing terms of Eqs.(38) and (91) leads to

$$D = \left(\frac{\lambda_j - \lambda_i}{\lambda_j}\right)A_i \exp(-\lambda_i R) \tag{92}$$

$$a_2 = \frac{1}{2}\left[a_1^2 - \lambda_1\lambda_2\right] \text{ and} \tag{93}$$

$$a_3 = \frac{1}{6}\left[a_1^{\ 3} - 3a_1\lambda_1\lambda_2 + \lambda_1\lambda_2(\lambda_1 + \lambda_2)\right]. \tag{94}$$

It is possible to solve for λ_1 and λ_2 from Eqs.(93) and (94) to give [57]

$$\begin{Bmatrix} \lambda_1 \\ \lambda_2 \end{Bmatrix} = \begin{bmatrix} 1 & 1 \\ 1 & -1 \end{bmatrix} \begin{Bmatrix} \dfrac{3}{2}a_1 + \dfrac{6a_3 - a_1^{\ 2}}{2a_1^{\ 2} - 4a_2} \\ \sqrt{\left[3a_1 + \dfrac{6a_3 - a_1^{\ 2}}{a_1^{\ 2} - 2a_2}\right]^2 - 4\left(a_1^{\ 2} - 2a_2\right)} \end{Bmatrix}. \tag{95}$$

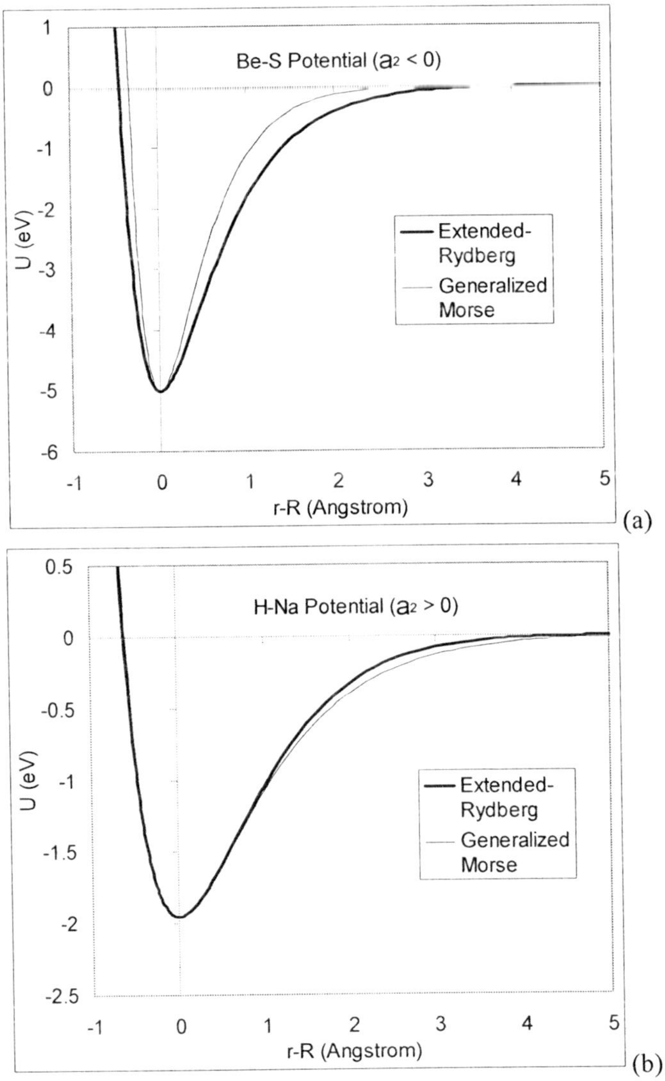

Figure 6. Generalized Morse approximation to Extended-Rydberg curve for (a) Be-S and (b) H-Na diatoms.

Figure 7(a) and (b) shows the suitability of the refined relationship for the case where $a_2 < 0$.

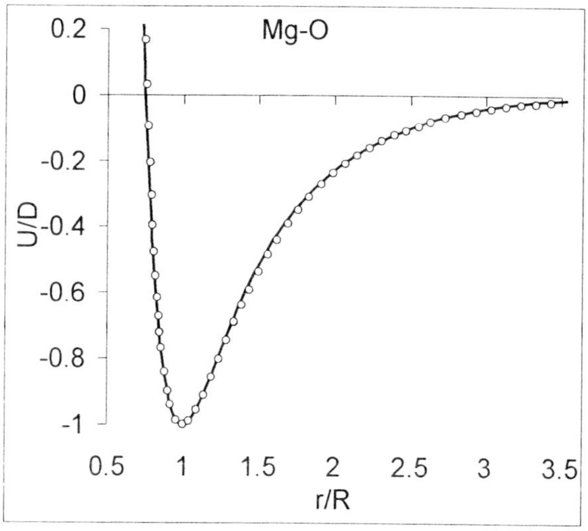

a

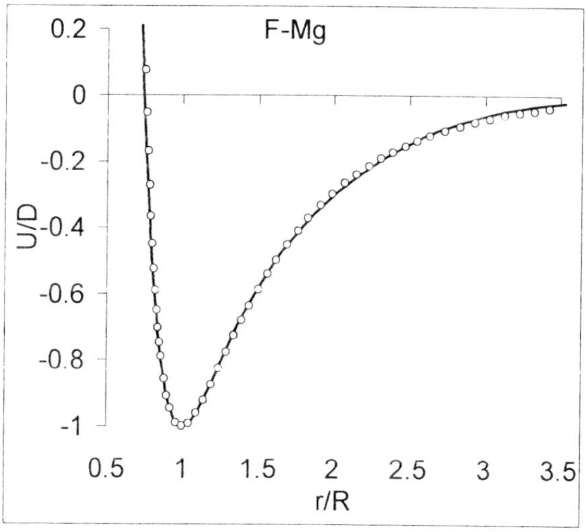

b

Figure 7. Potential curves based on Extended-Ryberg function and Huxley and Murrell's parameters (continuous curve) and the Generalized Morse function based refined relation (circles) for (a) Mg-O, and (b) F-Mg.

In addition to the Generalized Morse function, parameters of the Fayyazuddin potential [58],

$$U_{Fay} = \left[\frac{K}{x} - a - bx - cx^2\right]\exp(-tx) \quad ; \quad (x = r/R) \tag{96}$$

where $x = r / R$, was also connected with the Extended Rydberg parameters. Rewriting Eq.(96) as

$$U_{Fay} = \exp(-t)\left[-(a+b-c)-b\frac{\rho}{R}-c\left(\frac{\rho}{R}\right)^2 - 2c\frac{r}{R}+K\frac{r}{R}\right]\exp\left(-\frac{t\rho}{R}\right) \tag{97}$$

and substituting

$$\frac{r}{R} = \left[\sum_{j=0}^{3}\left(\frac{\rho}{R}\right)^j\right]^{-1} \tag{98}$$

into Eq.(97) leads to

$$U_{Fay} = -(a+b+c-K)\exp(-t)\Phi\exp\left(-\frac{t\rho}{R}\right) \tag{99}$$

whereby

$$\Phi = 1+\frac{K+b+2c}{a+b+c-K}\left(\frac{\rho}{R}\right)+\frac{c-K}{a+b+c-K}\left(\frac{\rho}{R}\right)^2+\frac{K}{a+b+c-K}\left(\frac{\rho}{R}\right)^3. \tag{100}$$

Comparing the terms in Eqs.(38) with those of Eqs.(99) and (100) expresses the Extended-Rydberg parameters in terms of Fayyazuddin parameters [59]

$$\begin{Bmatrix} a_1 \\ a_2 \\ a_3 \end{Bmatrix} = \frac{1}{a+b+c-K}\begin{bmatrix} R^{-1} & R^{-1} & 2R^{-1} \\ -R^{-2} & R^{-2} & 0 \\ R^{-3} & 0 & 0 \end{bmatrix}\begin{Bmatrix} K \\ c \\ b \end{Bmatrix} \tag{101}$$

and

$$D = (a+b+c-K)\exp(-t). \tag{102}$$

Equations (101) and (102) can then be solved to give the Fayyazuddin parameters

$$\begin{Bmatrix} a \\ b \\ c \\ K \end{Bmatrix} = D\exp(a_1 R)\begin{bmatrix} 1 & -2 & 1 & 3 \\ 0 & 1 & -2 & -3 \\ 0 & 0 & 1 & 1 \\ 0 & 0 & 0 & 1 \end{bmatrix}\begin{Bmatrix} a_0 R^0 \\ a_1 R^1 \\ a_2 R^2 \\ a_3 R^3 \end{Bmatrix}. \tag{103}$$

The validity of Eq.(103), which is based on Eqs.(101) and (102), is attested through Fig.8 based on data from Huxley and Murrell [49].

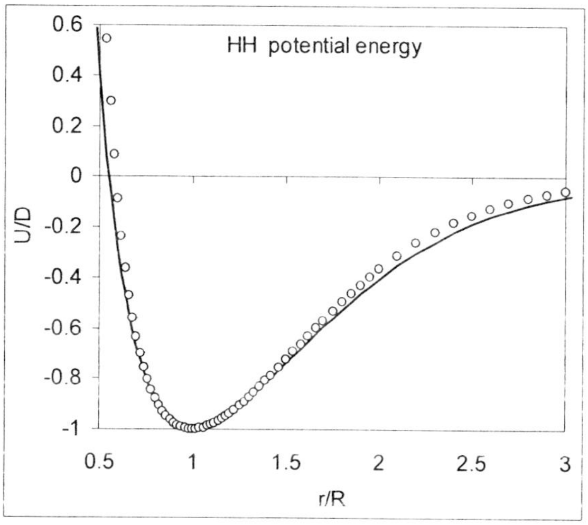

a

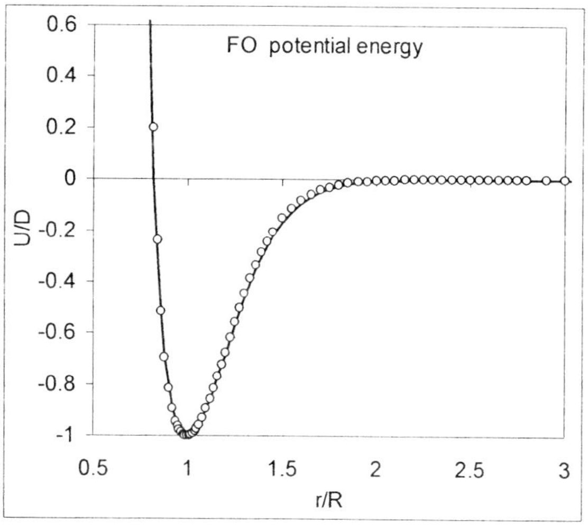

b

Figure 8. Comparison between the Extended-Rydberg (curves) and Fayyazuddin (circles) potentials based on the Huxley-Murrell parameters for (a) HH, and (b) FO.

In view of the availability of the Extended-Rydberg parameters to a number of diatomic molecules, a parameter relationship was developed to obtain those of the 2-body portion of the Kaxiras-Pandey potential. Writing the latter as [60]

$$U_{KP} = D\left\{\sum_{i=1}^{2} C_i \exp\left[-\left(\alpha_i \rho^2 + (2\alpha_i R - a_1)\rho\right)\right]\right\}\exp(-a_1\rho) \tag{104}$$

whereby the well depth magnitude is

$$D = (j-i)\frac{\alpha_i - \alpha_j}{\alpha_j} B_i \quad ; \quad (i,j = 1,2) \tag{105}$$

while the multiplier for the repulsive $i = 1$ and attractive $i = 2$ coefficients is

$$C_i = \frac{\alpha_j}{\alpha_i - \alpha_j} \tag{106}$$

allows imposition of the Maclaurin series expansion

$$\exp\left[-\left(\alpha_i \rho^2 + (2\alpha_i R - a_1)\rho\right)\right] = \sum_{k=0}^{3} b_k \rho^k \tag{107}$$

in which

$$b_0 = 1 \tag{108}$$

$$b_1 = a_1 - 2(\alpha_i R) \tag{109}$$

$$b_2 = \frac{1}{2}a_1^2 - 2a_1(\alpha_i R) + 2(\alpha_i R)^2 \text{ and} \tag{110}$$

$$b_3 = \frac{1}{6}a_1^3 - a_1\alpha_i + \left(2\alpha_i^2 - a_1^2\alpha_i\right)R + 2a_1(\alpha_i R)^2 - \frac{4}{3}(\alpha_i R)^3 . \tag{111}$$

Substituting Eq.(107) to (111) into Eq.(104) leads to

$$U_{KP} = -D\left\{\sum_{i=1}^{3} c_i \rho^i\right\}\exp(-a_1\rho) \tag{112}$$

whereby

$$c_0 = 1 \tag{113}$$

$$c_1 = a_1 \tag{114}$$

$$c_2 = \frac{1}{2}a_1^2 - 2\alpha_1\alpha_2 R^2 \text{ and} \tag{115}$$

$$c_3 = \frac{1}{6}a_1^3 - 2\alpha_1\alpha_2 R\left[(1 + a_1 R) - \frac{2}{3}(\alpha_1 + \alpha_2)R^2\right]. \tag{116}$$

Comparing the terms of Eq.(38) and Eqs.(112) to (116) gives

$$\begin{Bmatrix} a_2 \\ a_3 \end{Bmatrix} = \begin{bmatrix} 1 & 1 \\ (a_1/3) & (a_1 + R^{-1}) - 2(\alpha_1 + \alpha_2)R/3 \end{bmatrix} \begin{Bmatrix} a_1^2/2 \\ -2\alpha_1\alpha_2 R^2 \end{Bmatrix}. \tag{117}$$

Alternatively, the Kaxiras-Pandey parameters can be solved as

$$\begin{Bmatrix} \alpha_1 \\ \alpha_2 \end{Bmatrix} = \frac{1}{2}\begin{bmatrix} 1 & 1 \\ 1 & -1 \end{bmatrix} \begin{Bmatrix} f(a_1, a_2, a_3) \\ \sqrt{[f(a_1, a_2, a_3)]^2 - (a_1^2 - 2a_2)/R^2} \end{Bmatrix} \text{ whereby} \tag{118}$$

$$f(a_1, a_2, a_3) = \frac{1}{2R}\left[3\left(a_1 + \frac{1}{R}\right) - \frac{a_1^3 - 6a_3}{a_1^2 - 2a_2}\right] \text{ and} \tag{119}$$

$$\begin{Bmatrix} A_1 \\ A_2 \end{Bmatrix} = \frac{D}{\alpha_1 - \alpha_2}\begin{Bmatrix} \alpha_2 \exp(\alpha_1 R^2) \\ \alpha_1 \exp(\alpha_2 R^2) \end{Bmatrix}. \tag{120}$$

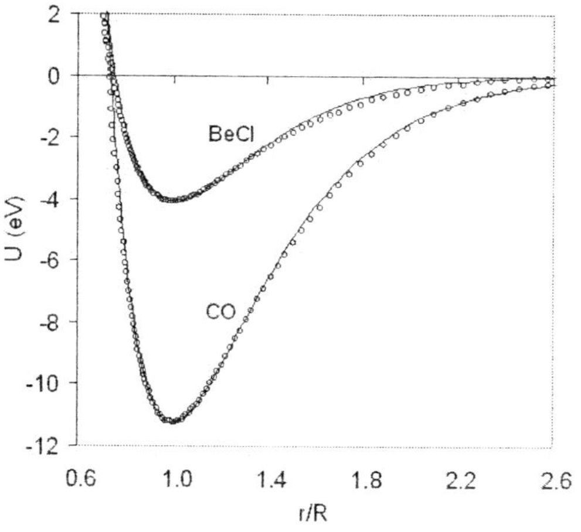

Figure 9. Plots of diatomic potential curves for BeCl and CO diatomic molecules using Extended-Rydberg function (continuous curves) and 2-body portion of Kaxiras-Pandey function (open circles) based on the former's parameters.

These relations are validated through the plots of extended Rydberg function using the Huxley-Murrell parameters [49], and the converted Kaxiras-Pandey potential function, as shown in Figure 9.

4. Multi-Derivative Approach

Other than the series expansion approach, the multi-derivative approach has been shown to be reliable for relating parameters of different potentials. An example can be seen for the case of expressing the Dunham coefficients in terms of SPF coefficients by taking derivative of both functions at $r = R$ to give [7]

$$a_j = b_j + \sum_{i=1}^{j-1} (-1)^i b_{j-i} \binom{j+1}{i} + (-1)^j (j+1). \quad (121).$$

Parameters of two potentials functions $U_{(1)}$ and $U_{(2)}$ can be related by equating

$$\left(\frac{\partial^p U_{(1)}}{\partial r^p} \right)_{r=R} = \left(\frac{\partial^p U_{(2)}}{\partial r^p} \right)_{r=R}. \quad (122)$$

4.1. Three-Derivative Method

By 3-derivative, one refers to the case whereby Eq.(122) is invoked for $p = 0,1,2$. Although the relation between the Lennard-Jones(12-6) and the Exponential-6 has been developed as

$$\frac{U_{X6}}{D} = \left(\frac{6}{S-6} \right) \exp\left[S\left(1 - \frac{r}{R} \right) \right] - \left(\frac{S}{S-6} \right) \left(\frac{R}{r} \right)^6 \quad (123)$$

with scaling factors $S = 13.772$ and $S = 12$ for short and long range respectively, a slightly more general scaling factor can be examined by considering the Lennard-Jones(m-n) described in Eq.(18). Using Eq.(122) for $p = 0,1,2$, we have

$$\begin{bmatrix} S^0 & 1 \\ S^1 & 6 \\ S^2 & 42 \end{bmatrix} \begin{Bmatrix} A_{X6} \exp(-S) \\ -C_{X6} / R^6 \end{Bmatrix} = D \begin{Bmatrix} -1 \\ 0 \\ mn \end{Bmatrix} \quad (124)$$

where

$$S = B_{X6} R . \tag{125}$$

Eliminating the terms A_{X6} and (C_{X6} / R^6) gives Eq.(123) with a more generalized expression of the scaling factor, i.e. [61]

$$S = \frac{1}{2}(m+7) + \frac{1}{2}\sqrt{m^2 - 10m + 49} \tag{126}$$

If the parameter S in Eq.(123) is a function described by Eq.(126), then Eq.(123) can be compared with Eq.(18) in which $n = 6$, i.e.

$$\frac{U_{LJ(m-6)}}{D} = \left(\frac{6}{m-6}\right)\left(\frac{R}{r}\right)^m - \left(\frac{m}{m-6}\right)\left(\frac{R}{r}\right)^6 . \tag{127}$$

While substitution of $m = 12$ into Eq.(126) gives $S = 13.772$, the novelty of that function allows one to obtain the close range relationship with the Lennard-Jones(9-6). Substituting $m = 9$ into Eq.(126) gives $S = 11.162$, which is the short range relation with the Lennard-Jones(9-6). Needless to say, the long range relationship with Lennard-Jones(9-6) is simply $S = 9$.

Even though the Lennard-Jones(12-6) and the Morse potential functions are normally applied for quantifying the van der Waals and the covalent bonds respectively, the COSMIC [62] molecular modeling software employs the Morse function for describing the van der Waals interaction. As such, a study was conducted to better understand the relations and discrepancies between these two potentials. To do so, Eq.(122) was applied on both potentials for $p = 0,1,2$ to give [63]

$$\alpha = \frac{6}{R} \tag{128}$$

although the exact relation is

$$\alpha = \frac{6}{R}\frac{\ln(r/R)}{(r/R)-1} . \tag{129}$$

Figure 10 shows the discrepancies between both potential via Eq.(128).

Applying Eq.(122) for $p = 0,1,2$ for relating the 2-body portion of Biswas-Hamann and the Murrell potentials provide three independent equations

$$\begin{bmatrix} \psi_1^{\ 0} & \psi_2^{\ 0} \\ \psi_1^{\ 1} & \psi_2^{\ 1} \\ \psi_1^{\ 2} & \psi_2^{\ 2} \end{bmatrix} \begin{Bmatrix} A_{BH1} \exp(-\psi_1) \\ A_{BH2} \exp(-\psi_2) \end{Bmatrix} = D \begin{Bmatrix} -1 \\ 0 \\ a^2 \end{Bmatrix} \quad ; \quad (\psi_i = \lambda_i R). \tag{130}$$

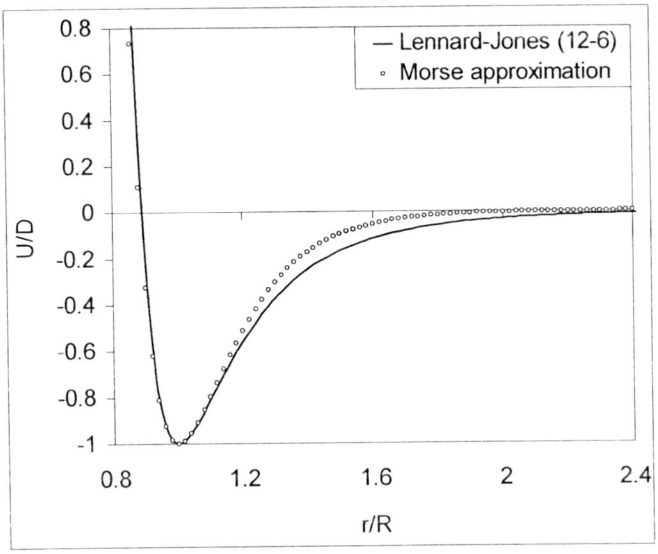

Figure 10. Comparison between Lennard-Jones (12-6) with the Morse approximation.

It can be solved to give the Biswas-Hamann parameters in terms of Murrell parameters [64]

$$\begin{Bmatrix} A_{BH1} \\ A_{BH2} \end{Bmatrix} = \frac{D}{\psi_1 - \psi_2} \begin{Bmatrix} \psi_2 \exp(\psi_1) \\ \psi_1 \exp(\psi_2) \end{Bmatrix} \text{ whereby} \tag{131}$$

$$\begin{Bmatrix} \psi_1 \\ \psi_2 \end{Bmatrix} = a \begin{Bmatrix} (\sqrt{2})^{+1} \\ (\sqrt{2})^{-1} \end{Bmatrix}. \tag{132}$$

The approximate relation is

$$\begin{Bmatrix} \psi_1 \\ \psi_2 \end{Bmatrix} = a \begin{Bmatrix} 2 \\ 1 \end{Bmatrix}. \tag{133}$$

Figure 11 compares the good relation with the approximate relation using the averaged values of silicon, i.e. $a = 6$ [44,47].

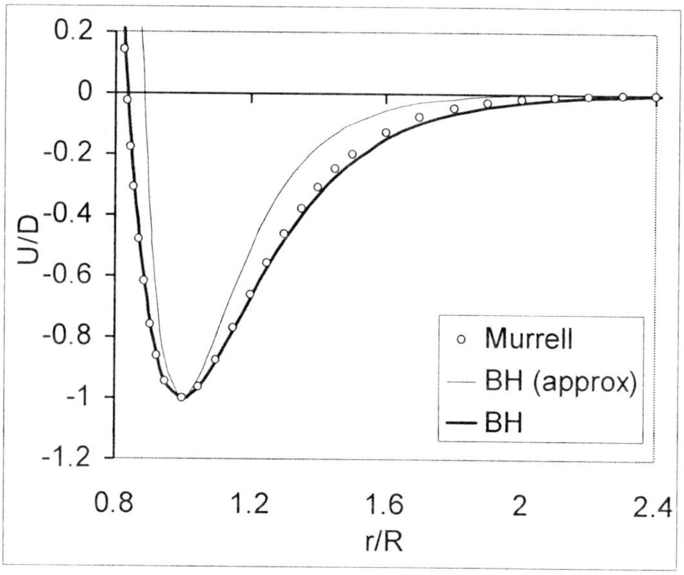

Figure 11. Plots of Biswas-Hamann curves based on conversion from Murrell parameters.

When Eq.(122) is imposed on the 2-body portion of the Kaxiras-Pandey and the 2-body portion of PTHT potential functions, which is essentially a Lennard-Jones(12-6), one obtains [65]

$$\begin{bmatrix} 1 & 1 \\ \phi_1 & \phi_2 \\ \phi_1(2\phi_1 - 1) & \phi_2(2\phi_2 - 1) \end{bmatrix} \begin{Bmatrix} +A_{KP1}\exp(-\phi_1) \\ -A_{KP2}\exp(-\phi_2) \end{Bmatrix} \equiv D \begin{Bmatrix} -1 \\ 0 \\ 36 \end{Bmatrix} \quad ; \quad (\phi_i = \alpha_i R^2). \tag{134}$$

Upon solving,

$$\begin{Bmatrix} A_{KP1} \\ A_{KP2} \end{Bmatrix} = \frac{D}{\phi_1 - \phi_2} \begin{Bmatrix} \phi_2 \exp(\phi_1) \\ \phi_1 \exp(\phi_2) \end{Bmatrix} \quad ; \quad \begin{Bmatrix} \phi_1 \\ \phi_2 \end{Bmatrix} = \begin{Bmatrix} 6 \\ 2 \end{Bmatrix}. \tag{135}$$

A schematic comparison between these two potentials are depicted in Figure 12.

Lest it be said that the use of Eq.(122) for $p = 0,1,2$ is analytically sufficient, the following examples prove otherwise. In a simultaneous development of parameter relationships amongst the generalized versions of the Lennard-Jones, Morse (2-body portion of Biswas-Hamann), Buckingham, Extended-Rydberg and the 2-body portion of the Kaxiras-Pandey functions, the multi-derivative simultaneous relations

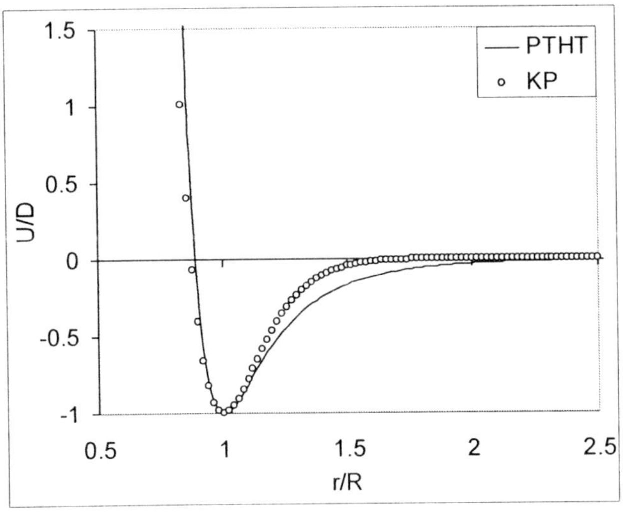

Figure 12. Comparison of 2-body interaction energy of PTHT potential, and the KP potential using PTHT parameters.

$$(-R)^p\left(\frac{\partial^p U_{Buck}}{\partial r^p}\right)_{r=R} \equiv (-R)^p\left(\frac{\partial^p U_{BH}}{\partial r^p}\right)_{r=R}$$
$$\equiv (-R)^p\left(\frac{\partial^p U_{KP}}{\partial r^p}\right)_{r=R} \equiv (-R)^p\left(\frac{\partial^p U_{LJ}}{\partial r^p}\right)_{r=R} \equiv (-R)^p\left(\frac{\partial^p U_{ER}}{\partial r^p}\right)_{r=R} \tag{136}$$

were written to test the viability of obtaining simultaneous relations across several potentials. Solving with $p = 0,1,2$ gives

$$\begin{bmatrix} S^0 & 1 \\ S^1 & n \\ S^2 & n(n+1) \end{bmatrix}\begin{Bmatrix} A_{Buck}\exp(-S) \\ -\left(\dfrac{C_{Buck}}{R^n}\right) \end{Bmatrix} \equiv \begin{bmatrix} \psi_1^{\ 0} & \psi_2^{\ 0} \\ \psi_1^{\ 1} & \psi_2^{\ 1} \\ \psi_1^{\ 2} & \psi_2^{\ 2} \end{bmatrix}\begin{Bmatrix} A_{BH1}\exp(-\psi_1) \\ A_{BH2}\exp(-\psi_2) \end{Bmatrix}$$
$$\equiv \begin{bmatrix} 1 & 1 \\ \phi_1 & \phi_2 \\ \phi_1(\phi_1-1) & \phi_2(\phi_2-1) \end{bmatrix}\begin{Bmatrix} +A_{KP1}\exp\left(-\dfrac{\phi_1}{2}\right) \\ -A_{KP2}\exp\left(-\dfrac{\phi_2}{2}\right) \end{Bmatrix} \equiv D\begin{Bmatrix} -1 \\ 0 \\ mn \end{Bmatrix} \equiv D\begin{Bmatrix} -1 \\ 0 \\ \left(a_1^{\ 2}-2a_2\right)R^2 \end{Bmatrix} \tag{137}$$

where the scaling functions are defined as

$$\begin{Bmatrix} S \\ \psi_i \\ \phi_i \end{Bmatrix} = \begin{Bmatrix} B_{Buck} R \\ \lambda_i R \\ 2\alpha_i R^2 \end{Bmatrix} \quad ; \quad (i = 1,2). \tag{138}$$

Taking advantage of the middle row of Eq.(137), which is zero, and relating all the function to the generalized Buckingham potential, then

$$\left(\frac{S-n}{n}\right) A_{Buck} \exp(-S) \equiv \left(\frac{S-n}{S}\right) \frac{C_{Buck}}{R^n} \equiv \left(\frac{\lambda_i - \lambda_j}{\lambda_j}\right) A_{BHi} \exp(-\psi_i)$$
$$\equiv (j-i)\left(\frac{\alpha_i - \alpha_j}{\alpha_j}\right) A_{KPi} \exp\left(-\frac{\phi_i}{2}\right) \equiv D \quad ; \quad (i,j = 1,2) \tag{139}$$

and

$$\frac{S}{B} \equiv \frac{\psi_i}{\lambda_i} \equiv \sqrt{\frac{\phi_i}{2\alpha_i}} \equiv R \quad ; \quad (i = 1,2). \tag{140}$$

Notwithstanding the relations to the minimum well-depth magnitude and the equilibrium interatomic distance, these two relations are not sufficient to connect all the parameters, so an additional assumption [66]

$$\frac{m}{n} = \frac{\lambda_1}{\lambda_2} = \frac{\alpha_1}{\alpha_2} \tag{141}$$

was made. As a result of this assumption, the parameter relations are limited to small deformation, as evident from Figure 13, based on Huxley and Murrell's data [49] for H-H.

In order to circumvent the disadvantage of incorporating assumptions, the analytical insufficiency can be overcome by performing numerical search. The parameters of the generalized Buckingham, Eq.(28), and the Generalized Linnett, Eq.(32), potential functions can be solved by imposing Eq.(122) for $p = 0,1,2$ to give [67]

$$S_1 = \frac{1}{2}\left(n+1+\frac{f(m,S_2)}{n}\right) \pm \frac{1}{2}\sqrt{\left(n+1+\frac{f(m,S_2)}{n}\right)^2 - 4f(m,S_2)} \tag{142}$$

and

$$S_2 = \frac{1}{2}\left(m+1+\frac{f(n,S_1)}{m}\right) \pm \frac{1}{2}\sqrt{\left(m+1+\frac{f(n,S_1)}{m}\right)^2 - 4f(n,S_1)} \tag{143}$$

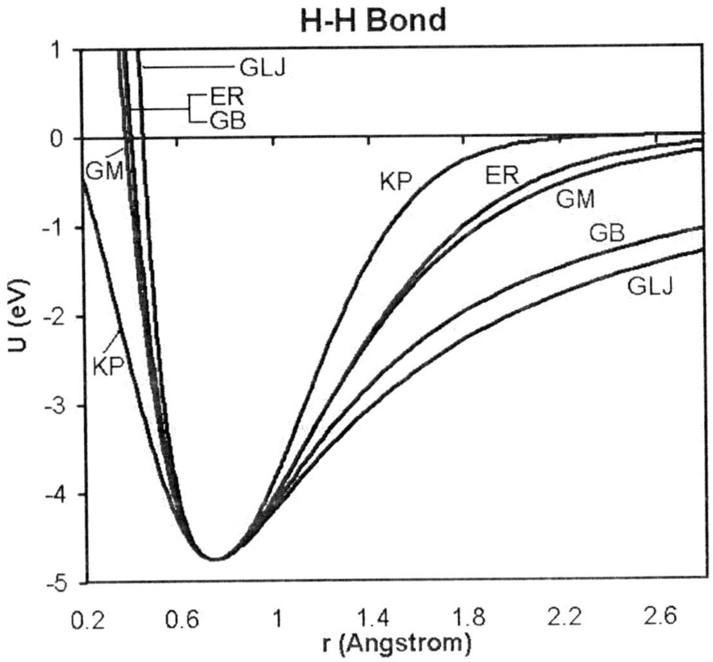

Figure 13. Interatomic potential energy curves of H–H using the five generalized potentials.

where $(S_1, S_2) = (B_{Buck}R\ , C_{Linn}R)$ and

$$f(x, y) = \frac{xy^2 - x(x+1)y}{y - x}. \tag{144}$$

Since S_2 is required to solve S_1 and vice versa, the solutions were obtained by numerical search. For the case where $m = 2n = 12$, the numerical solution is

$$\begin{Bmatrix} S_1 \\ S_2 \end{Bmatrix} = \frac{1}{2}\begin{bmatrix} 1 & 1 \\ 1 & -1 \end{bmatrix}\begin{Bmatrix} 19 \\ \sqrt{73} \end{Bmatrix}. \tag{145}$$

The validity of the solution is verified through plots of both potential energy curves at $m = 2n = 12$, as shown in Figure 14.

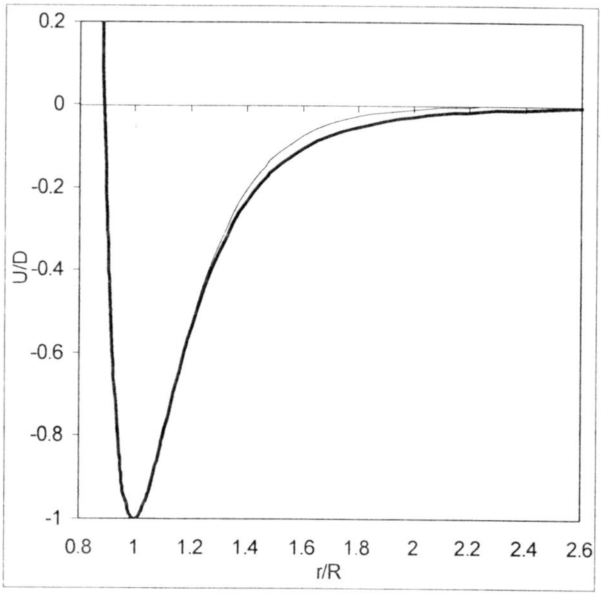

Figure 14. Comparison between Buckingham (bold curve) and Linnett (thin curve) potential functions based on parametric relationship for $m = 2n = 12$.

4.2. Four-Derivative Approach

Arising from the shortcomings associated with the 3-derivative approach in relating some potential functions, an additional order of derivative is incorporated to address the analytical insufficiency. Hence the additional derivative replaces the index assumption [66] and numerical search [67] to allow analytical sufficiency. Furthermore, this additional derivative allow a second set of parameter relationships if its corresponding 3-derivative solution is analytically sufficient.

The 2-body portion of BMS adopts Eq.(27) with the parameter $\rho_{BMS} = 1/C_{Buck}$, i.e.

$$U_{BMS} = A_{BMS}\exp\left(-\frac{r}{\rho_{BMS}}\right) - \frac{B_{BMS}}{r^n} \tag{146}$$

while the 2-body portion of the Biswas-Hamann potential is given in Eq.(22). The relation between these two potential functions were developed via Eq.(122) with $p = 0,1,2,3$ to give

$$\begin{bmatrix} S^0 & (n-1)!/(n-1)! \\ S^1 & (n)!/(n-1)! \\ S^2 & (n+1)!/(n-1)! \\ S^3 & (n+2)!/(n-1)! \end{bmatrix} \begin{Bmatrix} A_{BMS}\exp(-S) \\ -\left(\dfrac{B_{BMS}}{R^n}\right) \end{Bmatrix} = \begin{bmatrix} \psi_1{}^0 & \psi_2{}^0 \\ \psi_1{}^1 & \psi_2{}^1 \\ \psi_1{}^2 & \psi_2{}^2 \\ \psi_1{}^3 & \psi_2{}^3 \end{bmatrix} \begin{Bmatrix} A_{BH1}\exp(-\psi_1) \\ A_{BH2}\exp(-\psi_2) \end{Bmatrix} \tag{147}$$

where $S = R / \rho_{BMS}$. Imposing zero at the second row, the scaling function can be solved in the first and third rows to give

$$S = \frac{1}{2}\left(n+1+\frac{\lambda_1\lambda_2 R^2}{n}\right)+\frac{1}{2}\sqrt{\left(n+1+\frac{\lambda_1\lambda_2 R^2}{n}\right)^2 - 4\lambda_1\lambda_2 R^2} \tag{148}$$

whilst solving the third and fourth rows lead to

$$S = \frac{1}{2}(\lambda_1 R + \lambda_2 R) \pm \frac{1}{2}\sqrt{(\lambda_1 R + \lambda_2 R)^2 - 4(n+1)[\lambda_1 R + \lambda_2 R - (n+2)]} . \tag{149}$$

The BMS parameters can be expressed in terms of the Biswas-Hamann parameters as [68]

$$A_{BMS} = \left(\frac{\lambda_1 - \lambda_2}{\lambda_2}\right)\left(\frac{n}{S-n}\right) A_{BH1} \exp(S - \lambda_1 R) \tag{150}$$

$$B_{BMS} = -\left(\frac{\lambda_1 - \lambda_2}{\lambda_1}\right)\left(\frac{S}{S-n}\right) A_{BH2} R^n \exp(-\lambda_2 R) \tag{151}$$

$$\rho_{BMS} = \frac{R}{S} = \frac{1}{S}\left(\frac{1}{\lambda_1 - \lambda_2}\right)\ln\left(-\frac{\lambda_1 A_1}{\lambda_2 A_2}\right). \tag{152}$$

Adopting the values of Biswas-Hamann parameters for silicon [31], the BMS potential energy curves based on Eqs.(150) to (152) are shown in Figure 15(a) and (b), corresponding to the scaling functions described by Eqs.(148) and (149) respectively. It can be easily seen that the use of Eq.(148) gives very good agreement for small bond stretching and compression but highly unrealistic result for large compression. Some sacrifice to accuracy, as seen in Figure 15(b), enables a better agreement for large bond compression.

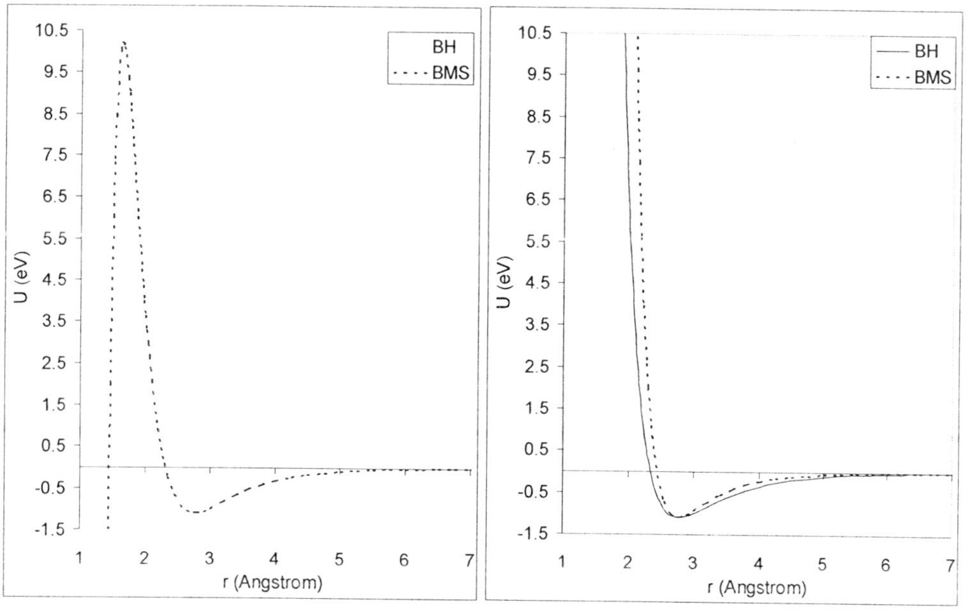

(a) (b)

Figure 15. Relationship between Biswas-Hamann and BMS function using Eqs.(150) to (152) based on (a) Eq.(148) and (b) Eq.(149).

Using Eqs.(122) for $p = 0,1,2,3$ on the 2-body portions of the Biswas-Hamann and Kaxiras-Pandey functions gives

$$\begin{bmatrix} \psi_1^{\ 0} & \psi_2^{\ 0} \\ \psi_1^{\ 1} & \psi_2^{\ 1} \\ \psi_1^{\ 2} & \psi_2^{\ 2} \\ \psi_1^{\ 3} & \psi_2^{\ 3} \end{bmatrix} \begin{Bmatrix} A_{BH1}\exp(-\psi_1) \\ A_{BH2}\exp(-\psi_2) \end{Bmatrix}$$

$$= \begin{bmatrix} 1 & 1 \\ \phi_1 & \phi_2 \\ \phi_1(\phi_1 - 1) & \phi_2(\phi_2 - 1) \\ \phi_1^{\ 2}(\phi_1 - 3) & \phi_2^{\ 2}(\phi_2 - 3) \end{bmatrix} \begin{Bmatrix} +A_{KP1}\exp\left(-\dfrac{\phi_1}{2}\right) \\ -A_{KP2}\exp\left(-\dfrac{\phi_2}{2}\right) \end{Bmatrix} \quad ; \quad \begin{Bmatrix} \psi_i \\ \phi_i \end{Bmatrix} = R\begin{Bmatrix} \lambda_i \\ 2\alpha_i R \end{Bmatrix} \tag{153}$$

for $i = 1,2$. Imposing a zero value for the second row gives

$$\psi_1\psi_2 = \phi_1\phi_2 \tag{154}$$

based on the first and third rows of Eq.(153), and

$$\psi_1 + \psi_2 + \frac{3}{2} = \phi_1 + \phi_2 - \frac{3}{2} \tag{155}$$

based on the third and fourth rows of Eq.(153) [69]. The relations can be solved to give the Biswas-Hamann parameters in terms of Kaxiras-Pandey parameters

$$A_{BH1} = +\left(\frac{\alpha_1 - \alpha_2}{\alpha_2}\right)\left(\frac{\psi_2}{\psi_1 - \psi_2}\right) A_{KP1} \exp\left(\psi_1 - \frac{\phi_1}{2}\right) \tag{156}$$

$$A_{BH2} = -\left(\frac{\alpha_1 - \alpha_2}{\alpha_1}\right)\left(\frac{\psi_1}{\psi_1 - \psi_2}\right) A_{KP2} \exp\left(\psi_2 - \frac{\phi_2}{2}\right) \tag{157}$$

$$\begin{Bmatrix} \lambda_1 \\ \lambda_2 \end{Bmatrix} = \frac{1}{R} \begin{Bmatrix} \psi_1 \\ \psi_2 \end{Bmatrix} \tag{158}$$

whereby

$$\begin{Bmatrix} \psi_1 \\ \psi_2 \end{Bmatrix} = \frac{1}{2} \begin{bmatrix} 1 & 1 \\ 1 & -1 \end{bmatrix} \begin{Bmatrix} \phi_1 + \phi_2 - 3 \\ \sqrt{\left|(\phi_1 - \phi_2)^2 - 6(\phi_1 + \phi_2) + 9\right|} \end{Bmatrix} \tag{159}$$

whilst the Kaxiras-Pandey parameters can be expressed in terms of Biswas-Hamann parameters as

$$A_{KP1} = +\left(\frac{\lambda_1 - \lambda_2}{\lambda_2}\right)\left(\frac{\phi_2}{\phi_1 - \phi_2}\right) A_{BH1} \exp\left(\frac{\phi_1}{2} - \psi_1\right) \tag{160}$$

$$A_{KP2} = -\left(\frac{\lambda_1 - \lambda_2}{\lambda_1}\right)\left(\frac{\phi_1}{\phi_1 - \phi_2}\right) A_{BH2} \exp\left(\frac{\phi_2}{2} - \psi_2\right) \tag{161}$$

$$\begin{Bmatrix} \alpha_1 \\ \alpha_2 \end{Bmatrix} = \frac{1}{2R^2} \begin{Bmatrix} \phi_1 \\ \phi_2 \end{Bmatrix} \text{ whereby} \tag{162}$$

$$\begin{Bmatrix} \phi_1 \\ \phi_2 \end{Bmatrix} = \frac{1}{2} \begin{bmatrix} 1 & 1 \\ 1 & -1 \end{bmatrix} \begin{Bmatrix} \psi_1 + \psi_2 + 3 \\ \sqrt{(\psi_1 - \psi_2)^2 + 6(\psi_1 + \psi_2) + 9} \end{Bmatrix}. \tag{163}$$

The scaling functions relations described in Eqs.(159) and (163) are obtained by simultaneously solving Eqs.(154) and (155). Figure 16 shows the Biswas-Hamann approximation to the Kaxiras-Pandey parameters [41].

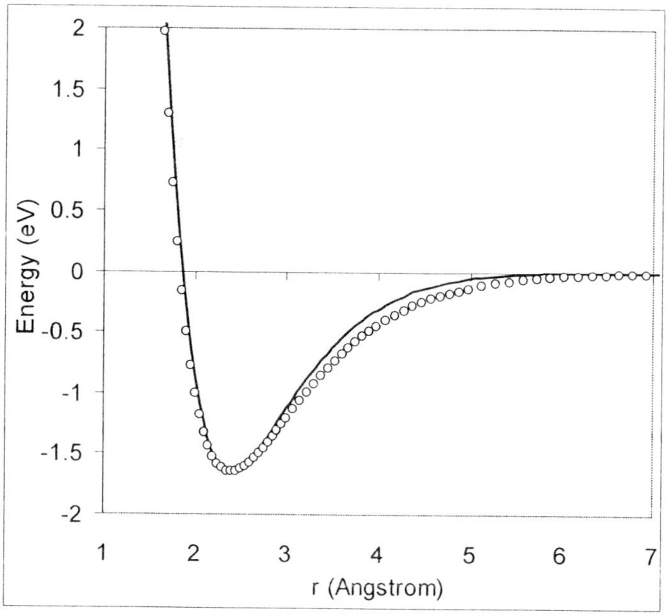

Figure 16. Two-body portion of Kaxiras-Pandey (curve) and Biswas-Hamann (circles) potentials based on the former's parameters [41]

Applying Eq.(122) with $p = 0,1,2,3$ to relate parameters of the 2-body portion of the BMS and Kaxiras-Pandey gives [70]

$$\begin{bmatrix} S^0 & (n-1)!/(n-1)! \\ S^1 & (n)!/(n-1)! \\ S^2 & (n+1)!/(n-1)! \\ S^3 & (n+2)!/(n-1)! \end{bmatrix} \begin{Bmatrix} A_{BMS}\exp(-S) \\ -\left(\dfrac{B_{BMS}}{R^n}\right) \end{Bmatrix} = \begin{bmatrix} 1 & 1 \\ \phi_1 & \phi_2 \\ \phi_1(\phi_1-1) & \phi_2(\phi_2-1) \\ \phi_1^2(\phi_1-3) & \phi_2^2(\phi_2-3) \end{bmatrix} \begin{Bmatrix} +A_{KP1}\exp\left(-\dfrac{\phi_1}{2}\right) \\ -A_{KP2}\exp\left(-\dfrac{\phi_2}{2}\right) \end{Bmatrix} \quad ; \quad \begin{Bmatrix} S \\ \phi_i \end{Bmatrix} = R\begin{Bmatrix} 1/\rho_{BMS} \\ 2\alpha_i R \end{Bmatrix} \tag{164}$$

for $i = 1,2$. Upon equating the second row to zero, consideration of the first and third rows gives the first BMS scaling function

$$S = \frac{n+1}{2} + \frac{2}{n}\alpha_1\alpha_2 R^4 + \frac{1}{2}\sqrt{\left(n+1+\frac{4}{n}\alpha_1\alpha_2 R^4\right)^2 - 16\alpha_1\alpha_2 R^4} \tag{165}$$

while considering the third and fourth rows leads to the second BMS scaling function

$$S = (\alpha_1 + \alpha_2)R^2 - \frac{3}{2} \pm \frac{1}{2}\sqrt{\left[2(\alpha_1 + \alpha_2)R^2 - 3\right]^2 - 4(n+1)\left[2(\alpha_1 + \alpha_2)R^2 - n - 5\right]}. \tag{166}$$

On the other hand, considering all four rows, with the second row being equated to zero, gives the scaling functions for Kaxiras-Pandey potential as

$$\begin{Bmatrix} \phi_1 \\ \phi_2 \end{Bmatrix} = \frac{1}{2}\begin{bmatrix} 1 & 1 \\ 1 & -1 \end{bmatrix} \begin{Bmatrix} 3 + \dfrac{(R/\rho_{BMS})^2 - (n+1)(n+2)}{(R/\rho_{BMS}) - (n+1)} \\ \sqrt{\left[3 + \dfrac{(R/\rho_{BMS})^2 - (n+1)(n+2)}{(R/\rho_{BMS}) - (n+1)}\right]^2 - \dfrac{4n(R/\rho_{BMS})\left[(R/\rho_{BMS}) - (n+1)\right]}{(R/\rho_{BMS}) - n}} \end{Bmatrix}. \tag{167}$$

The parameter relations are given as

$$A_{BMS}\left(\frac{(R/\rho_{BMS}) - n}{n}\right)\exp\left(-\frac{R}{\rho_{BMS}}\right) = A_{KP1}\left(\frac{\alpha_1 - \alpha_2}{\alpha_2}\right)\exp\left(-\alpha_1 R^2\right) \tag{168}$$

$$B_{BMS}\left(\frac{(R/\rho_{BMS}) - n}{(R/\rho_{BMS})}\right)R^{-n} = A_{KP2}\left(\frac{\alpha_1 - \alpha_2}{\alpha_1}\right)\exp\left(-\alpha_2 R^2\right) \tag{169}$$

$$S\rho_{BMS} \equiv R \equiv \sqrt{\frac{\phi_i}{2\alpha_i}} \quad ; \quad (i = 1,2) \text{ where} \tag{170}$$

$$R = \sqrt{\frac{1}{\alpha_1 - \alpha_2}\ln\left(\frac{\alpha_1 A_1}{\alpha_2 A_2}\right)}. \tag{171}$$

Based on the Kaxiras-Pandey parameters [41], the BMS approximations furnished in Figures 17(a) and (b) are based on Eqs.(165) and (166) respectively. It is clearly seen that a slight sacrifice of the bond stretching accuracy leads to a better correlation in the compressive state.

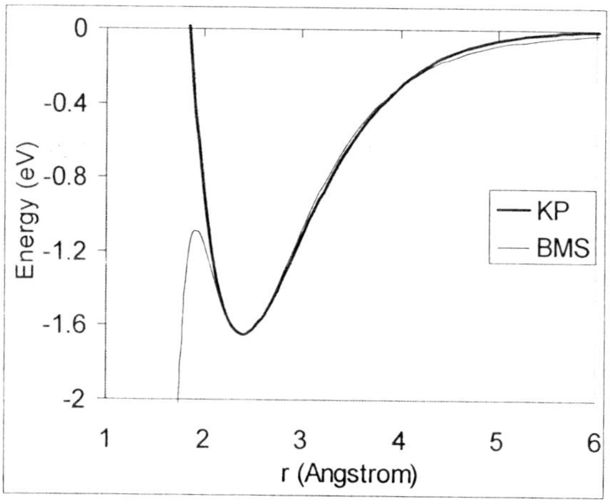

a

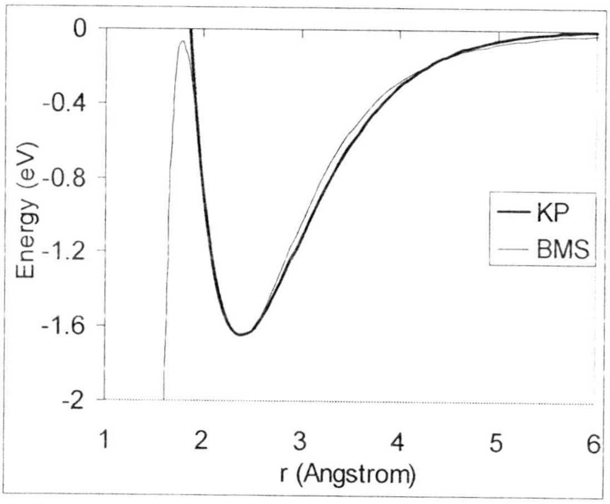

b

Figure 17. Two-body curve of BMS potential based on Kaxiras-Pandey parameters [41] and $n = 6$ considering derivatives up to the (a) second order $S = 7.805016$, and (b) third order $S = 8.094231$.

The previous three examples relate potential functions that do not explicitly express the magnitude of the minimum well-depth. In the case of parameter relations between the 2-body portions of the Kaxiras-Pandey and the Murrell potential functions, application of Eq.(122) with $p = 01,2,3$ gives

$$\begin{bmatrix} 1 & 1 \\ \phi_1 & \phi_2 \\ 2\phi_1(1-2\phi_1) & 2\phi_2(1-2\phi_2) \\ 2\phi_1^2(3-2\phi_1) & 2\phi_2^2(3-2\phi_2) \end{bmatrix} \begin{Bmatrix} +A_{KP1}\exp(-\phi_1) \\ -A_{KP2}\exp(-\phi_2) \end{Bmatrix} = D \begin{Bmatrix} 1 \\ 0 \\ a^2 \\ a^3 \end{Bmatrix} \quad ; \quad \phi_i = \alpha_i R^2 \cdot \tag{172}$$

Note the difference in the definitions of ϕ in this equation with that of Eqs.(138), (153) and (164). This system of equations can be solved to give the Kaxiras-Pandey parameters in terms of Murrell parameters [71]

$$\begin{Bmatrix} A_{KP1} \\ A_{KP2} \end{Bmatrix} = \frac{D}{\phi_1 - \phi_2} \begin{Bmatrix} \phi_2 \exp(\phi_1) \\ \phi_1 \exp(\phi_2) \end{Bmatrix} \text{ and} \tag{173}$$

$$\begin{Bmatrix} \alpha_1 \\ \alpha_2 \end{Bmatrix} = \frac{1}{R^2} \begin{Bmatrix} \phi_1 \\ \phi_2 \end{Bmatrix} \quad ; \quad (3\phi_1 = 6\phi_2 = 3 + 2a). \tag{174}$$

Figure 18 shows the plot of Murrell potential curve for $a = 6$ and its corresponding Kaxiras-Pandey function based on Eqs.(173) and (174).

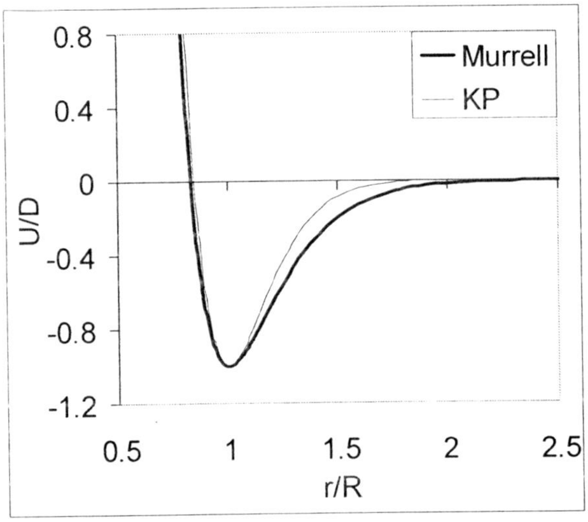

Figure 18. Comparison of Kaxiras-Pandey and Murrell 2-body interaction dimensionless energy at $a = 6$.

Although it has been shown in Fig.10 that the Lennard-Jones(12-6) function gives a more gradual increase of energy with increasing interatomic distance in comparison to the simple Morse function, the consideration of additional derivative gives agreement to their generalized potentials. Applying Eq.(122) with $p = 0,1,2,3$ on the 2-body portion of the Biswas-Hamann and the PTHT potential functions gives

$$\begin{bmatrix} \psi_1^{\,0} & \psi_2^{\,0} \\ \psi_1^{\,1} & \psi_2^{\,1} \\ \psi_1^{\,2} & \psi_2^{\,2} \\ \psi_1^{\,3} & \psi_2^{\,3} \end{bmatrix} \begin{Bmatrix} A_{BH1} \exp(-\psi_1) \\ A_{BH2} \exp(-\psi_2) \end{Bmatrix} = D \begin{Bmatrix} -1 \\ 0 \\ 72 \\ 1512 \end{Bmatrix} \quad ; \quad \psi_i = \lambda_i R \tag{175}$$

based on the conditions set forth by the RHS of Eq.(175), the Biswas-Hamann parameters can be written as [72]

$$\begin{Bmatrix} A_{BH1} \\ A_{BH2} \end{Bmatrix} = \frac{D}{\psi_1 - \psi_2} \begin{Bmatrix} +\psi_2 \exp(\psi_1) \\ -\psi_1 \exp(\psi_2) \end{Bmatrix} \text{ and} \tag{176}$$

$$\begin{Bmatrix} \lambda_1 \\ \lambda_2 \end{Bmatrix} = \frac{1}{R} \begin{Bmatrix} \psi_1 \\ \psi_2 \end{Bmatrix} \text{ whereby} \tag{177}$$

$$\begin{Bmatrix} \psi_1 \\ \psi_2 \end{Bmatrix} = \frac{1}{2} \begin{bmatrix} 1 & 1 \\ 1 & -1 \end{bmatrix} \begin{Bmatrix} 21 \\ \sqrt{153} \end{Bmatrix}. \tag{178}$$

Figure 19 shows the 2-body portion of PTHT curve and its corresponding Biswas-Hamann curve.

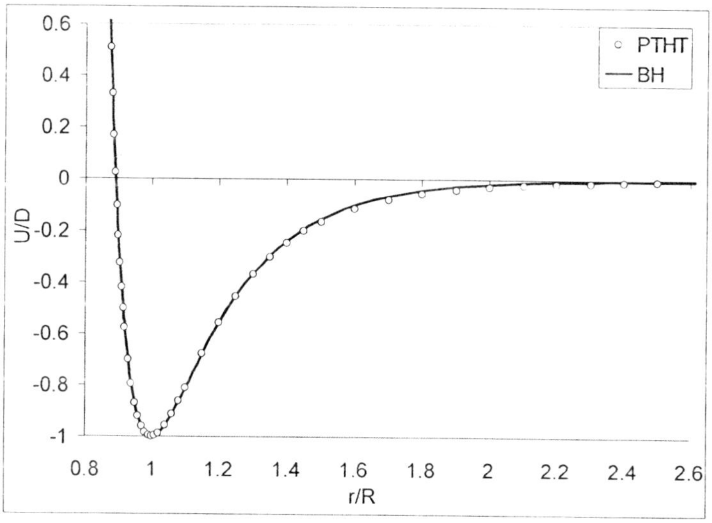

Figure 19. Comparison of 2-body energy function according to PTHT and Biswas-Hamann curves.

5. OTHER SYSTEMS

So far discussion of parameter relationships have been confined to pair potentials, which refer to the energy of interaction between two atoms due to (i) van der Waals interaction between two non-bonded atoms, (ii) bonded interaction in diatomic molecules, and (iii) 2-body interaction in multi-body systems. No comparison of potential functions is complete without considering the 3-body and 4-body interactions, which describe the change in molecular energy as a result of bond bending and twisting respectively.

Potential functions for describing the 3-body interaction normally fall into the harmonic angle function

$$U_\theta = \frac{1}{2} k_\theta (\theta - \theta_0)^2 = \frac{1}{2} k_\theta (\delta\theta)^2 \tag{179}$$

the polynomial series form

$$U_S = \frac{1}{2} \sum_{n=2}^{m} k_{\theta n} (\delta\theta)^n \tag{180}$$

the Fourier series form

$$U_F = k_F \sum_{n=0}^{m} C_n \cos(n\theta) \tag{181}$$

and the harmonic cosine angle function

$$U_C = \frac{1}{2} k_C (\cos\theta - \cos\theta_0)^2 \tag{182}$$

where θ is the angle defined from three adjacently atoms, while θ_0 refers to the angle at equilibrium. Parameters of the harmonic angle and the polynomial series form are easily related as

$$k_\theta = k_{\theta 2} \left[\sum_{n=2}^{m} \frac{k_{\theta n}}{k_{\theta 2}} (\delta\theta)^{n-2} \right] \tag{183}$$

while the polynomial series function can be related to the harmonic cosine and Fourier series potentials as [73]

$$\begin{Bmatrix} k_{\theta 2} \\ k_{\theta 3} \\ k_{\theta 4} \\ k_{\theta 5} \\ k_{\theta 6} \end{Bmatrix} = k_C \begin{Bmatrix} \sin^2\theta_0 \\ \sin\theta_0 \cos\theta_0 \\ -\frac{1}{3}\sin^2\theta_0 + \frac{1}{4}\cos^2\theta_0 \\ -\frac{1}{4}\sin\theta_0\cos\theta_0 \\ \frac{2}{45}\sin^2\theta_0 - \frac{1}{24}\cos^2\theta_0 \end{Bmatrix} = k_F \begin{Bmatrix} -(C_1\cos\theta_0 + 4C_2\cos 2\theta_0) \\ \frac{1}{3}(C_1\sin\theta_0 + 8C_2\sin 2\theta_0) \\ \frac{1}{12}(C_1\cos\theta_0 + 16C_2\cos 2\theta_0) \\ -\frac{1}{60}(C_1\sin\theta_0 + 32C_2\sin 2\theta_0) \\ -\frac{1}{360}(C_1\cos\theta_0 + 64C_2\cos 2\theta_0) \end{Bmatrix}. \tag{184}$$

The Fourier series and the harmonic cosine parameters are connected as

$$C_0 = C_2\left(2\cos^2\theta_0 + 1\right) \tag{185}$$

$$C_1 = -4C_2\cos\theta_0 \tag{186}$$

$$C_2 = \frac{1}{4\sin^2\theta_0}. \tag{187}$$

Substituting Eqs.(186) and (187) into any row in Eq.(184) leads to

$$k_F = k_C\sin^2\theta_0 \equiv k_\theta. \tag{188}$$

Figure 20 compares the bond-bending potentials upon relating their parameters based on $\theta_0 = 1.9373$ rad [74].

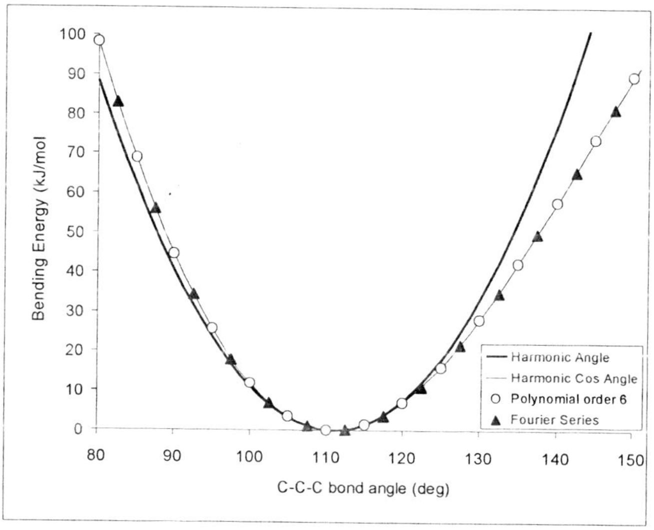

Figure 20. Results of generated harmonic cosine angle, polynomial series and Fourier series functions based on data from harmonic angle function.

Potential functions that describe 4-body interaction energy normally either take the trigonometric series form

$$U_\phi = \frac{1}{2}\sum_{n=1}^{m} k_{\phi n}(1-\cos n\phi) \tag{189}$$

or the Fourier series form

$$U_T = \sum_{n=0}^{m} k_{Tn}\cos^n\phi \tag{190}$$

where ϕ is the dihedral angle. Substituting

$$\begin{Bmatrix} \cos 1\phi \\ \cos 2\phi \\ \cos 3\phi \\ \cos 4\phi \\ \cos 5\phi \\ \cos 6\phi \\ \cos 7\phi \\ \vdots \end{Bmatrix} = \begin{bmatrix} 0 & 0 & 0 & 0 & 0 & 0 & 0 & 1 & 0 \\ 0 & 0 & 0 & 0 & 0 & 0 & 2 & 0 & -1 \\ 0 & 0 & 0 & 0 & 0 & 4 & 0 & -3 & 0 \\ 0 & 0 & 0 & 0 & 8 & 0 & -8 & 0 & 1 \\ 0 & 0 & 0 & 16 & 0 & -20 & 0 & 5 & 0 \\ 0 & 0 & 32 & 0 & -48 & 0 & 18 & 0 & -1 \\ 0 & 64 & 0 & -112 & 0 & 64 & 0 & 7 & 0 \\ \vdots & \vdots & \vdots & \vdots & \vdots & \vdots & \vdots & \vdots & \vdots \end{bmatrix} \begin{Bmatrix} \cos^0 \phi \\ \cos^1 \phi \\ \cos^2 \phi \\ \cos^3 \phi \\ \cos^4 \phi \\ \cos^5 \phi \\ \cos^6 \phi \\ \cos^7 \phi \\ \vdots \end{Bmatrix} \tag{191}$$

into Eq.(189) enables comparison of terms to be made so that k_{Tn} can be written in terms of $k_{\phi n}$

$$\{k_{Ti}\} = \left[{}^{\phi T}\mathbf{C}_{ij}\right]\{k_{\phi j}\} \tag{192}$$

and vice versa

$$\{k_{\phi i}\} = \left[{}^{T\phi}\mathbf{C}_{ij}\right]\{k_{Tj}\} \tag{193}$$

whereby the conversion matrices up to the fifth order are [75]

$$\left[{}^{\phi T}\mathbf{C}\right] = -\frac{1}{2}\begin{bmatrix} -1 & -2 & -1 & 0 & -1 \\ \mathbf{1} & 0 & -3 & 0 & 5 \\ 0 & \mathbf{2} & 0 & -8 & 0 \\ 0 & 0 & \mathbf{4} & 0 & -20 \\ 0 & 0 & 0 & \mathbf{8} & 0 \\ 0 & 0 & 0 & 0 & \mathbf{16} \end{bmatrix} \text{ and} \tag{194}$$

$$\left[{}^{T\phi}\mathbf{C}\right] = -\frac{1}{8}\begin{bmatrix} \mathbf{16} & 0 & 12 & 0 & 10 \\ 0 & \mathbf{8} & 0 & 8 & 0 \\ 0 & 0 & \mathbf{4} & 0 & 5 \\ 0 & 0 & 0 & \mathbf{2} & 0 \\ 0 & 0 & 0 & 0 & \mathbf{1} \end{bmatrix}. \tag{195}$$

Substituting $k_{\phi 1} = 3.35$ kJ/mol and $k_{\phi 3} = 13.4$ kJ/mol from Boyd et al. [76,77] into Eqs.(192) and (193) gives $k_{T0} = 8.375$ kJ/mol, $k_{T1} = 18.425$ kJ/mol and $k_{T3} = -26.8$ kJ/mol while $k_{T2} = k_{T4} = k_{T5} = 0$. This result compares well with those of Sumpter et al. [78-81] whereby $k_{T0} = 8.37$ kJ/mol, $k_{T1} = 18.4$ kJ/mol and $k_{T3} = -26.78$ kJ, while $k_{T2} = k_{T4} = k_{T5} = 0$.

6. Discrepancies Observation for Molecular Design

Having shown the formalism by which parameters of potential functions are related, the discrepancies can be observed not only in the interatomic energy-distance plane, but also in the interatomic force-distance plane. The method of comparing the interatomic force-distance plane is of importance in computational drug analysis, bottom-up nanostructure synthesis and other molecular design. The understanding of different performance of theoretical interatomic force curve in van der Waals interaction is critical for drug design for ensuring seamless ligand-receptor docking while the force curve in covalent bond interaction is essential for ensuring good strength per mass or volume ratios in nanostructural elements. The parameter relations shown in previous sections now enable more objective comparison of the characteristics of different normalized potentials. It is suffice to say that the consideration of force is taken with focus at the interatomic energy-distance plane where the inflexion point, which is the point of zero curvature, is the greatest. This is because the interatomic force is a first derivative of the interatomic energy

$$F = -\frac{\partial U}{\partial r} \tag{196}$$

whereby the negative sign depicts the direction of force being in the opposite to the direction of change in the interatomic distance. Hence the maximum interatomic force for $r > R$ occurs at the interatomic distance which gives the highest slope in the interatomic energy-distance plane. Whilst we have so far used (U/D) and (r/R) as the non-dimensionalized interatomic energy and distance respectively, it is appropriate to introduce a non-dimensionalized interatomic force, herein proposed as

$$F^* = -\frac{FR}{D} = \frac{R}{D}\left(\frac{\partial U}{\partial r}\right). \tag{197}$$

A negative sign is incorporated to produce a positive non-dimensionalized force description for $r > R$.

Figure 21 shows the interatomic force versus distance plot of the Murrell and the Kaxiras-Pandey potentials upon their normalization based on parameter relations given in Eqs.(173) and (174) with $a = 6$. The plot of F^* versus (r/R) graphically shows that the interatomic force is higher and narrower in Kaxiras-Pandey, than in Murrell curve. This clearly depicts the hard and solid nature of the Kaxiras-Pandey function, but a relatively soft or fluid-like behavior of the Murrell functional.

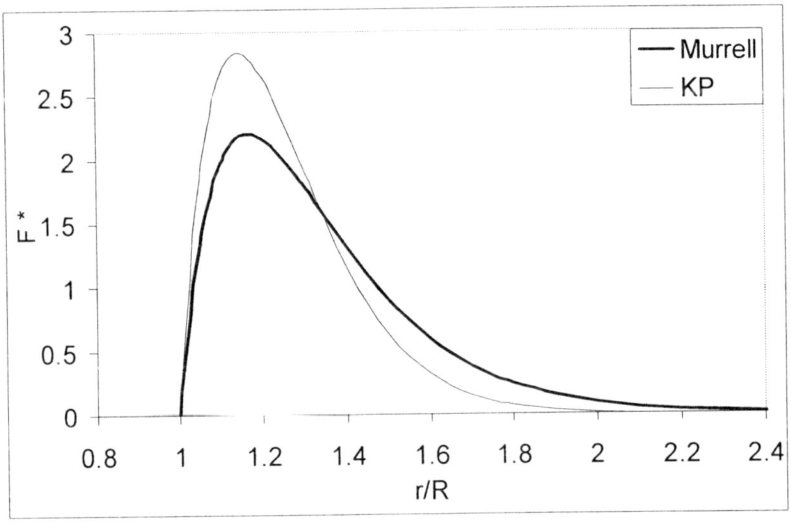

Figure 21. The non-dimensionalized interatomic force for the Kaxiras-Pandey function based on Murrell parameter $a = 6$.

The higher peak for the Kaxiras-Pandey curve in Fig.21 corresponds to a higher slope of Kaxiras-Pandey's inflexion point, as compared to that of Murrell's. Therefore we consider two approaches for design purposes, namely:

a) load is prescribed, and
b) distortion is prescribed,

followed by suggested choice of potential function on the basis of conservative design philosophy. Since computation for silicon shows that the maximum force by Kaxiras-Pandey function is higher than that of Murrell function, while the corresponding interatomic distance by Kaxiras-Pandey function is lower than that of Murrell function, then the Murrell function is more conservative when load is prescribed whilst Kaxiras-Pandey function is more conservative when distortion is prescribed. For example, when $2.207 < F^* < 2.844$ is applied to the bond, the Murrell function would predict bond dissociation while the Kaxiras-Pandey function predicts otherwise. On the other hand, when $1.145R < r < 1.167R$ is prescribed, Kaxiras-Pandey would predict the bond to be over the stability (maximum force) point but this is not so based on Murrell function.

When the discrepancies between the energy of interaction between two normalized potentials are insignificant, it follows that the maximum force and the interatomic distance where this force occurs are almost equal. Figure 22 shows the good agreement between the 2-body portion of PTHT and Biswas-Hamann force curves based on Eqs.(176) to (178), while Figure 23 shows the force distribution comparison relationship between the Buckingham and Linnett potentials based on Eq.(145) for $m = 2n = 12$.

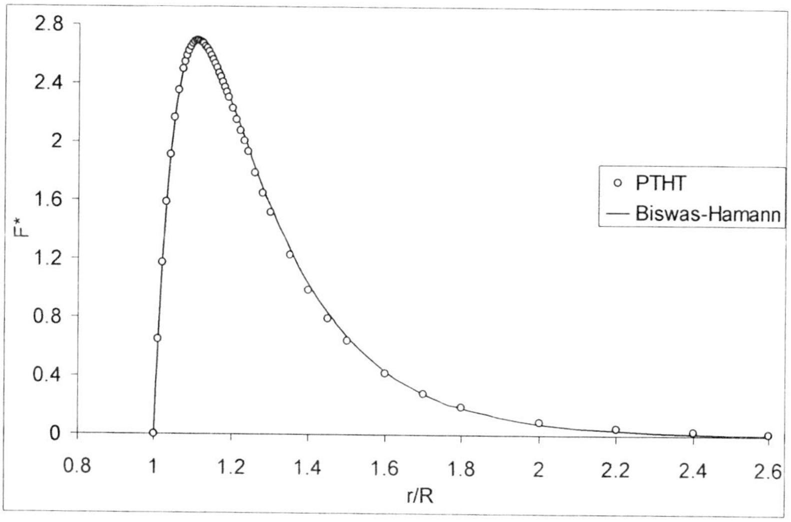

Figure 22. Comparison of 2-body interatomic force portion of PTHT and Biswas-Hamann.

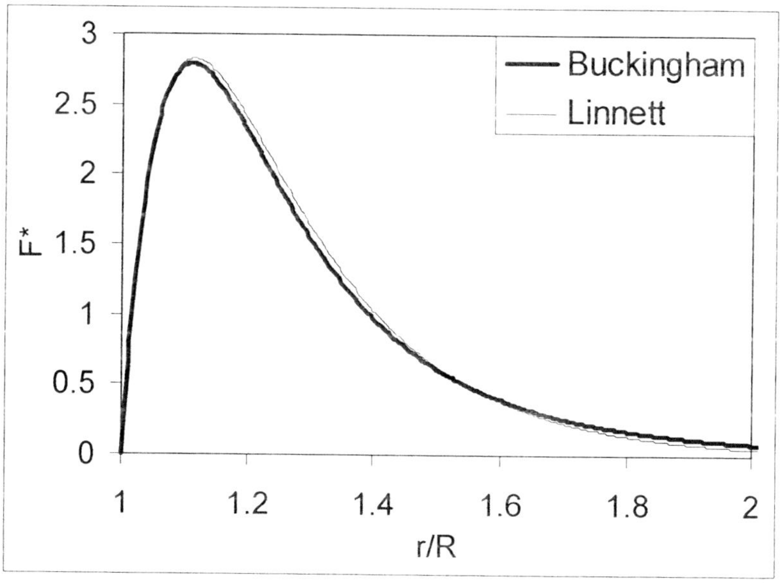

Figure 23. Comparison between Buckingham and Linnett potential functions based on their related parameter relationship for $m = 2n = 12$.

7. Conclusion

Two broad methods have been discussed, namely, the

a) Series Expansion Approach, and
b) Multi-derivative Approach

have been reported to provide reasonable parameter relationships among different interatomic potential energy functions. The series expansion is of two types: (i) the full expansion, and (ii) the partial expansion methods whereby one of the potential functions is imposed Maclaurin series expansion fully and partially respectively in order to allow comparison of terms to be made. Comparison of terms enables parameters of one set of potential to be expressed in terms of parameters of the other potential. The multi-derivative approach equates the zeroth, first and second order derivatives at the equilibrium interatomic distance. Where the three equations are analytically insufficient for obtaining the scaling functions, the scaling function can be obtained by incorporating an assumption or solving by numerical search. A physically justifiable solution is to replace the assumption with the third derivative of the two compared potentials equated at the equilibrium distance.

The potential functions that are related through their parameters can be plotted in non-dimensionalized interatomic energy versus non-dimensionalized interatomic distance plane to observe their discrepancies in an unbiased manner. Special emphasis is placed at the inflexion point whereby a higher slope and a lower slope corresponds to a harder and softer bonds regardless of the bond dissociation energy D. This inflexion point appears as the maximum point in the force-distance plane. Normally, a lower slope for the inflexion point refers to greater equilibrium interatomic distance. The gap between the two maximum points in the force-distance curves for two parameters that are related in their parameters are critical for mechanical design of molecules. The interatomic force curve that shows a lower maximum point (or greater equilibrium distance) is more conservative for load prescribed modeling whilst the interatomic force curve that shows a higher maximum point (or shorter equilibrium distance) is more conservative for distortion prescribed modeling.

REFERENCES

[1] Riddell, F.G.; Robinson, M.J.T. *Tetrahedron* 1974, *30*, 2001-2007.

[2] Murrell, J.N.; Carter, S.; Farantos, S.C.; Huxley, P.; Varandas, A.J.C. *Molecular Potential Energy Functions*; Wiley: Chichester, UK, 1984; pp 182.

[3] Maitland, G.C.; Rigby, M.; Smith, E.B.; Wakeham, W.A. *Intermolecular Forces: Their Origin and Determination*; Clarendon Press: Oxford, UK, 1981: pp 531-537.

[4] Varshni, Y.P. *Rev. Mod. Phys*. 1957, *29*, 664-682.

[5] Erkoc, S. *Phys. Rep*. 1997, *278*, 79-105.

[6] Dunham, J.L. *Phys. Rev*. 1932, *41*, 721-731.

[7] Simons, G.; Parr, R.G.; Finlan, J.M. *J. Chem. Phys*. 1973, *59*, 3229-3234.

[8] Thakkar, A.J. *J. Chem. Phys*. 1975, *62*, 1693-1701.

[9] Ogilvie, J.F. *Proc. Royal Soc. Lond. A* 1981, *378*, 287-300.

[10] Molski, M. *J. Mol. Spectrosc*. 1999, *193*, 244-248.

[11] Molski, M.; Konarski, J. *Int. J. Quant. Chem*. 2002, *90*, 183-187.

[12] Puppi, G. *Nuovo Cimento* 1946, *3*, 338.

[13] Kratzer, A. *Annalen der Physik* 1922, *67*, 127.

[14] Lennard-Jones, J.E. *Proc. Royal Soc. Lond. A* 1924, *106*, 463-477.

[15] Mecke, R. *Zeit. Phys*. 1927,*42*, 390-425.

[16] Sutherland, G.B.B.M. *Proc. Indian Acad. Sci*. 1938, *8*, 341-344.

[17] Morse, P.M. *Phys. Rev.* 1929, *34*, 57-64.
[18] Buckingham, R.A. *Proc. Royal Soc. Lond. A* 1938, *168*, 264-283.
[19] Wu, C.K.; Yang, C.T. *J. Phys. Chem.* 1944, *48*, 295-303.
[20] Linnett, J.W. *Trans. Faraday Soc.* 1940, *36*, 1123-1134.
[21] Linnett, J.W. *Trans. Faraday Soc.* 1942, *38*, 1-9.
[22] Lim, T.C. *Chem. Phys.* 2005, *320*, 54-58.
[23] Damm, W.; Frontera, A.; Tirado-Rives, J.; Jorgensen, W.L. *J. Comput. Chem.* 1997, *18*, 1955-1970.
[24] Rappe, A.K.; Casewit, C.J.; Colwell, K.S.; Goddard. W.A.; Skiff, W.M. *J. Am. Chem. Soc.* 1992, *114*, 10024-10035.
[25] Cornell, W.D.; Cieplak, P.; Baylay, C.I.; Gould, I.R.; Merz, K.M.; Ferguson, G.M.; Spellmeyer, D.C.; Fox, T.; Caldwell, J.W.; Kollman, P.A. *J. Am. Chem. Soc.* 1995, *117*, 5179-5197.
[26] Nemethy, G.; Gibson, K.D.; Palmer K.A.; Yoon, C.N.; Paterlini, G.; Zagari, A.; Rumsey, S.; Scheraga, H.A. *J. Phys. Chem.* 1992, *96*, 6472-6484.
[27] Barlow, S.; Rohl, A.A.; Shi, S.; Freeman, C.M.; O'Hare, D. *J. Am. Chem. Soc.* 1996, *118*, 7578-7592.
[28] Hwang, M.J.; Stockfisch, T.P.; Hagler, A.T. *J. Am. Chem. Soc.* 1994, *116*, 2515-2525.
[29] Halgren, T.A. *J. Comput. Chem.* 1996, *17*, 490-519.
[30] Pearson, E.; Takai, T.; Halicioglu. T.; Tiller, W.A. *J. Cryst. Growth* 1984, *70*, 33-40.
[31] Biswas, R.; Hamann, D. Phys. Rev. Lett. 1985, *55*, 2001-2004.
[32] Lifson, S.; Hagler, A.T.; Dauber, P. *J. Am. Chem. Soc.* 1979, *101*, 5111-5122.
[33] Mayo, S.L.; Olafson, B.D.; Goddard, W.A. *J. Phys. Chem.* 1990, *94*, 8897-8909.
[34] Engler, E.M.; Andose, J.D.; Schleyer, P.v.R. *J. Am. Chem. Soc.* 1993, *95*, 8005-8025.
[35] Allinger, N.L. *J. Am. Chem. Soc.* 1977, *99*, 8127-8134.
[36] Allinger, N.L.; Yuh, Y.H.; Lii, J.H. *J. Am. Chem. Soc.* 1989, *111*, 8551-8566.
[37] Dillen, J.L.M. *J. Comput. Chem.* 1995, *16*, 595-609.
[38] Comba, P.; Hambley, T.W. *Molecular Modeling of Inorganic Compounds*; Wiley: New York, NY, 1995.
[39] Allinger, N.L.; Chen, K.; Lii, J.H. *J. Comput. Chem.* 1996, *17*, 642-668.
[40] Bauer, R.; Maysenholder, W.; Seeger, A. *Phys. Lett. A* 1982, *90*, 55-58.
[41] Kaxiras, E.; Pandey, K.C. *Phys. Rev. B* 1988, *38*, 12736-12739.
[42] Rydberg, R. *Zeit. Phys.* 1931, *73*, 376-385.
[43] Rydberg, R. *Zeit. Phys.* 1933, *80*, 514-524.
[44] Murrell, J.N.; Mottram, R.E. *Mol. Phys.* 1990, *69*, 571-585.
[45] Murrell, J.N.; Rogriguez-Ruiz, J.A. *Mol. Phys.* 1990, *71*, 823-834.
[46] Al-Derzi, A.R.; Johnston, R.L.; Murrell, J.N.; Rodriguez-Ruiz, J.A. *Mol. Phys.* 1991, *73*, 265-282.
[47] Li, S.; Johnston, R.L.; Murrell, J.N. *J. Chem. Soc. Faraday Trans.* 1992, *88*, 1229-1236.
[48] MurrellJ.N.; Sorbie, K.S. *J. Chem. Soc. Faraday Trans. II* 1974, *70*, 1552-1556.
[49] Huxley, P.; Murrell, J.N. *J. Chem. Soc. Faraday Trans. II* 1983, *79*, 323-328.
[50] Lim, T.C. *J. Math. Chem.* 2003, *33*, 29-37.
[51] Eggen, B.R.; Johnston, R.L.; Li, S.L.; Murrell, J.N. *Mol. Phys.* 1992, *76*, 619-633.
[52] Lim, T.C. *J. Math. Chem.* 2005, *38*, 495-501.

[53] Ogilvie, J.F. *The Vibrational and Rotational Spectrometry of Diatomic Molecules*; Academic Press: San Diego, CL, 1998, pp 145-205.
[54] Lim, T.C. *J. Math. Chem.* 2003, *34*, 221-225.
[55] Lim, T.C. *J. Math. Chem.* 2004, *36*, 139-145.
[56] Lim, T.C. *Acta Chim. Slovenica* 2005, *52*, 149-152.
[57] Lim, T.C. *Czech. J. Phys.* 2006, *56*, 149-156.
[58] Fayyazuddin; Rafi, M.; Al-Turwirqi, R.; Bakry, A. *J. Physique II (France)* 1996, *6*, 1125-1132.
[59] Lim, T.C. *Mol. Phys.* 2006, *104*, 1827-1831.
[60] Lim, T.C. *J. Math. Chem.* (accepted)
[61] Lim, T.C. *J. Math. Chem.* 2003, *33*, 279-285.
[62] Morley, S.D.; Abraham, R.J.; Haworth, I.S.; Jackson, D.E.; Saunders, M.R.; Vinter, J.G. *J. Comput.-Aided Mol. Des.* 1991, *5*, 475-496.
[63] Lim, T.C. *Zeit. Naturforsch. A* 2003, *58*, 615-617.
[64] Lim, T.C. *Zeit. Naturforsch. A* 2004, *59*, 116-118.
[65] Lim, T.C. *Phys. Scripta* 2004, *70*, 347-348.
[66] Lim, T.C. *Chin. Phys. Lett.* 2004, *21*, 2167-2170.
[67] Lim, T.C. *Chin. J. Phys.* 2006, *44*, 19-27.
[68] Lim, T.C. *Czech. J. Phys.* 2004, *54*, 553-559.
[69] Lim, T.C. *Czech. J. Phys.* 2004, *54*, 947-963.
[70] Lim, T.C. *MATCH Commun. Math. Comput. Chem.* 2005, *54*, 29-38.
[71] Lim, T.C. *Chin. J. Phys.* 2005, *43*, 43-51.
[72] Lim, T.C. *Braz. J. Phys.* 2005, *35*, 641-644.
[73] Lim, T.C. *J. Math. Chem.* 2002, *32*, 249-256.
[74] Noid, D.W.; Tuzun, R.E.; Sumpter, B.G. *Nanotechnology* 1997, *8*, 119-125.
[75] Lim, T.C. *J. Math. Chem.* 2002, *31*, 421-428.
[76] Boyd, R.H.; Pant, P.V.K. *Macromol.* 1991, *24*, 4078-4083.
[77] Jin, Y.; Boyd, R.H. *J. Chem. Phys.* 1998, *108*, 9912-9923.
[78] Sumpter, B.G.; Noid, D.W.; Wunderlich, B. *J. Chem. Phys.* 1990, *93*, 6875-6889.
[79] Noid, D.W. Sumpter, B.G.; Wunderlich, B. *Macromol.* 1991, *24*, 4148-4151.
[80] Sumpter, B.G.; Noid, D.W.; Wunderlich, B. *Macromol.* 1992, *25*, 7247-7255.
[81] Fukui, K.; Sumpter, B.G.; Barnes, M.D.; Noid, D.W. *Comput. Theor. Polym. Sci.* 1999, *9*, 245-254.

In: Chemical Physics Research Trends, Vol. 252
Editor: Samuel V. Arnold, pp. 319-333
ISBN: 978-1-60021-554-4

Chapter 8

GAUSSIAN BASIS SETS FOR CORRELATED WAVE FUNCTIONS: APPLICATION IN HF, MP2, AND DFT CALCULATIONS OF MOLECULAR (HYPER)POLARIZABILITY

F. E. Jorge, P. A. Fantin, P. L. Barbieri and A. Canal Neto
Departamento de Física, Universidade Federal do Espírito Santo,
29060-900 Vitória, ES

ABSTRACT

Electric polarizability is a fundamental characteristic of atomic and molecular systems. The theory of electric polarizability is a key element of the rational interpretation of a wide range of phenomena, from nonlinear optics and electron scattering to phenomena induced by intermolecular interactions. Recently, segmented contracted basis sets of double, triple, and quadruple zeta valence quality plus polarization functions for the atoms from H to Ar were presented by Jorge *et al.* These sets were augmented with diffuse functions to more accurately describe the long range behavior of the wave functions. We report a systematic study of the basis sets required to obtain accurate values for the electric dipole moment and static dipole polarizability and first dipole hyperpolarizability for the five molecules H_2, CO, CH_4, NH_3, and HCP at their equilibrium geometries. Several methods were examined, including Hartree-Fock, Mϕller-Plesset second-order perturbation theory, and density functional theory. Comparison with theoretical and experimental data reported in the literature was done. One verifies that the B3LYP/AQZP mean dipole polarizabilities, anisotropies, and first dipole hyperpolarizabilities are in good agreement with the experimental values, and in very reasonable agreement with results obtained through higher theory levels.

1. INTRODUCTION

Hartree-Fock (HF) wave functions for atoms and atomic ions can be computed numerically [1]. HF-Roothaan (HFR) wave functions are algebraic approximations of the HF ones in which the radial orbitals are expanded [2,3] in a set of basis functions, such as Gaussian-type functions (GTFs). The greater convenience of the HFR wave functions has lead to a great demand for them in atomic and molecular calculations. There are several GTF basis sets reported in the literature [4-10].

An alternative to the HFR method is the improved generator coordinate HF (IGCHF) method [11], in which the radial orbitals are chosen as integral transforms. Jorge *et al.* have used with success this methodology to generate accurate Gaussian basis sets (GBSs) for atomic [11-15] and molecular [16-18] systems.

Contemporary *ab initio* methods for calculating accurate energetics and molecular properties have two major limiting factors: incomplete incorporation of electron correlation and incomplete description of the orbitals due to the use of finite basis sets [19]. Recovering substantial fractions of the correlation energy required sophisticated wave functions and large basis sets, and obtaining microhartree accuracy for absolute energies is prohibitively expensive for all but small systems. Recent research has focused on systematic hierarchies of electron correlation methods and basis sets.

The correlation consistent polarized valence basis sets (cc-pVXZ, X = D, T, Q) developed by Dunning [20] and Woon and Dunning [21] have been employed extensively as hierarchical basis sets for systematically reducing the basis set truncation error. More recently, the double (D), triple (T), and quadruple (Q) zeta plus polarization functions (XZP, X = D, T, Q) basis sets presented by Canal Neto *et al.* [22] and Barbieri *et al.* [23] have also proven to be a good choice for systematically approaching the basis set limit for correlation energy. An extensive discussion about the main differences between these two hierarchical sequences was given in Refs. [22,23]. The DZP, TZP, QZP basis sets were extended by augmenting the sets for H-Ar with diffuse functions with the objective to have a better description of electron affinities, polarizabilities, and hydrogen bonding. Recently, the augmented DZP (ADZP) set [22], when applied to study the electronic structure of some molecules, gave dipole polarizabilities [24] and optical rotations [25] in satisfactory agreement with experimental data. In addition, the augmented TZP and QZP (ATZP and AQZP, respectively) [26] basis sets were used with success in polarizability calculations of the N_2, FH, H_2O, and HCN molecules.

An accurate description of atomic and molecular polarizabilities is important in a surprisingly wide range of physical phenomena. The second-order Stark effect arises from the response of a polarizable charge distribution to an external electric field [27]. Rayleigh and Raman scattering represent interactions between photons and bound electrons [28], mediated by polarizability (Rayleigh scattering gives rise to macroscopic refraction). An accurate description of polarizabilities, as well as the permanent moments, is also critical to accurately describing intermolecular interactions and weak long-range interactions of molecules with condensed matter substrates [29]. Polarizabilities are also important in determining dielectric constants [30] via the Clausius-Mossotti relation. The experimental determination of electric polarizabilities and hyperpolarizabilities represents a difficult task [31], but the theoretical determination of these quantities is currently a remarkably active research field [32-38]. This has proved a decisive factor of progress in fields with important potential technological implications, as the simulation of fluids and also drug design. The emphasis here will largely focus upon the determination of the static dipole polarizability ($\ddot{\alpha}$) and first dipole hyperpolarizability ($\dddot{\beta}$), although electric dipole moment ($\vec{\mu}$) will also be investigated.

There have been many previous calculations of dipole and higher polarizabilities, especially for the rare gas atoms. It is well known that diffuse polarization functions are critical for calculating accurate polarizabilities. However, in spite of the wealth of previous experience, there is still a need to determine a systematic of calculation and the accuracy that can be achieved, especially for the new correlation basis sets [22,23,26].

Thus, the main objective of this chapter is to show that electric dipole moment, dipole polarizability, and first dipole hyperpolarizability can be predicted in an effective and efficient manner with the proper choice of basis set and wave function. For this purpose, the hierarchical sequence of increase size basis sets generated by Jorge *et al.* [22,23,26] is used to construct HF, Mϕller-Plesset second-order perturbation theory (MP2), and density functional theory (DFT) wave functions. Section 2 reviews the prescription for constructing correlated basis sets and describes how they have been extended for an improved description of certain atomic and molecular properties. In Sections 3 and 4, the computational details and the molecular orientations are respectively given. The results are presented in Section 5, which includes a comparison with previous theoretical and experimental results. In Section 6, conclusions are drawn.

2. Correlated Gaussian Basis Sets

The basis sets used in this work were generated to describe correlation effects in H, He, and first- and second-row atoms. Initially, the s and p exponents of these atoms were generated with the IGCHF method [11], but, as the number of exponents is small and with the objective to improve the HF total energies, they were fully-optimized. Then, a segmented contraction scheme was used. Next, p, d, f, and g polarization functions were added to the contracted basis sets and determined from correlated atomic calculations (restricted open-shell MP2, ROMP2). In the final of these optimization processes, one obtains the DZP [22] and the TZP and QZP [23] basis sets.

One way to overcome problems associated with anion calculations is to include in the basis representation one or more sets of highly diffuse functions. It has been shown that the addition of diffuse functions has dramatic effects on calculated electron affinities [21], proton affinities [39], and inversion barriers [40]. Thus, with the purpose to describe properly the long-range behavior of molecular orbitals with energies close to the ionization limit, the DZP, TZP, and QZP basis sets were supplemented with s and p functions, which were optimized for the anion ROHF energy, and with p, d, f, and g polarization functions that were optimized for the anion ROMP2 energy. Resulting in the sets nominated as ADZP [22] and ATZP and AQZP [26]. Details about the construction of the correlated basis sets has been previously presented [22,23,26].

In summary, for H, He and first- and second-row atoms, the primitive and contracted function spaces for the correlated sets are:

Species	Set	Primitive set	Contracted set
H, He	DZP	(4s1p)	[2s1p]
	TZP	(5s2p1d)	[3s2p1d]
	QZP	(6s3p2d1f)	[4s3p2d1f]

(Continued)

Species	Set	Primitive set	Contracted set
	ADZP	(5s2p)	[3s2p]
	ATZP	(6s3p2d)	[4s3p2d]
	AQZP	(7s4p3d2f)	[5s4p3d2f]
First-row atoms	DZP	(9s5p1d)	[4s2p1d]
	TZP	(10s6p2d1f)	[5s3p2d1f]
	QZP	(11s7p3d2f1g)	[6s4p3d2f1g]
	ADZP	(10s6p2d)	[5s3p2d]
	ATZP	(11s7p3d2f)	[6s4p3d2f]
	AQZP	(12s8p4d3f2g)	[7s5p4d3f2g]
Second-row atoms	DZP	(12s9p1d)	[7s4p1d]
	TZP	(13s10p2d1f)	[8s5p2d1f]
	QZP	(14s11p3d2f1g)	[9s6p3d2f1g]
	ADZP	(13s10p2d)	[8s5p2d]
	ATZP	(14s11p3d2f)	[9s6p3d2f]
	AQZP	(15s12p4d3f2g)	[10s7p4d3f2g]

3. Computational Details

The Gaussian 03 program [41] has been employed in the closed-shell restricted HF and MP2 (RHF and RMP2, respectively) and DFT molecular wave function calculations. In the MP2 calculations, we have not included the full correlation energy correction, but have limited it to the valence atomic orbitals. This approximation is known as frozen core for electron correlation. In the DFT calculations, the Becke´s three parameter hybrid method [42] using the LYP correlation functional [43] (B3LYP), and the Becke exchange functional [44] combined with the Perdew correlation functional [45] (BP86) have been employed. Throughout the calculations, we have employed spherical harmonic GTFs. At the HF level, the static dipole polarizabilities and hyperpolarizabilities were computed analytically, whereas, at the MP2 and DFT levels, these properties were evaluated analytically and numerically, respectively.

4. Molecular Orientations and Symmetry Relations for the Components of $\ddot{\alpha}$ and $\dddot{\beta}$

Calculations were carried out for the five molecules H_2 ($R_{H\text{-}H}$=0.742 Å), CO ($R_{C\text{-}O}$=1.128 Å), CH_4 ($R_{C\text{-}H}$=1.085 Å and θ_{HCH} is T_d), NH_3 ($R_{N\text{-}H}$=1.012 Å and θ_{HNH}=106.7^0), and HCP (R_{CH}=1.069 Å and $R_{C\text{-}P}$=1.540 Å) at their equilibrium geometries. Except for CO and HCP (see Refs. [46,47], respectively), the experimental equilibrium geometries of the molecules were taken from Ref. [48].

The values of the individual components of the dipole moment, polarizability, and hyperpolarizability tensors depend upon choice of coordinates. The molecular orientations were chosen that the Cartesian axes are the principal axes of $\ddot{\alpha}$, with the origin at the center of mass. More specifically, all molecules have been oriented with their permanent dipole moment on the z-axis and pointing in the positive z-direction.

Due to the high symmetry of these molecules, there are relatively few independent components of the response property tensors. One can thus verify that, with the orientations specified above, the polarizability tensor $\ddot{\alpha}$ is indeed diagonal for each of the molecules, that $\alpha_{xx} = \alpha_{yy}$ for all molecules, and that for methane all three eigenvalues of $\ddot{\alpha}$ are degenerate.

All components of the first hyperpolarizability tensor $\overleftrightarrow{\beta}$ vanish for the homonuclear diatomic (H_2), while the only nonvanishing component for the methane is β_{xyz}. For all other molecules, β_{zxx}, β_{zyy}, and β_{zzz} are nonzero. For carbon monoxide, ammonia, and HCP, $\beta_{zxx} = \beta_{zyy}$. These are the only nonzero components for the diatomic and HCP. For NH_3, there is another independent component, $\beta_{xxx} = -\beta_{xyy}$.

In addition to the Cartesian components, we calculate the mean values and anisotropies for the dipole polarizabilities $\alpha_{\alpha\beta}$ and $\beta_{\alpha\beta\gamma}$. These are defined as

$$\bar{\alpha} = (\alpha_{zz} + 2\alpha_{xx})/3$$

$$\Delta\alpha = (\alpha_{zz} - \alpha_{xx}) \qquad (1)$$

$$\bar{\beta} = \frac{3}{5}(\beta_{zzz} + \beta_{zxx} + \beta_{zyy})$$

5. Results and Discussion

In this section, the basis sets presented in Section 2 were used to calculate at the HF, MP2, and DFT levels of theory electric dipole moment, dipole polarizability, and first dipole hyperpolarizability of the H_2, CO, CH_4, NH_3, and HCP molecules.

5.1. Electric Dipole Moments

Table 1 displays the electric dipole moments computed with the hierarchical sequence of increase size basis sets constructed by Jorge *et al.* at the HF, MP2, and DFT levels, and by a numerical HF (NHF) method [33]. The correlated results are compared with the corresponding experimental values [46,49]. It is important to say that the Mulliken´s population analysis was used to calculate the atomic charges.

Inspection of Table 1 shows that except for few HCP results, the HF moments, when compared with the experimental data, are overestimate, that the addition of diffuse functions in any basis set improves the dipole moments, and that the ATZP and AQZP results are very close to the corresponding NHF values. The largest difference between the AQZP and NHF dipole moments is equal to 0.002 a.u. for CO. We recall that an error of about 0.07 a.u. is typical in such calculations, but where the dipole moment is small, the percent error can be large. An example is the carbon monoxide, for which the experimental moment is 0.043 a.u. [46] with the polarity C^-O^+, but all HF moments presented in Table 1 give wrong polarity C^+O^-. This result is not surprising, since it is known that the dipole moment of carbon monoxide has the wrong sign at the HF level, and that it is corrected only if an adequate amount of electron correlation is included.

Table 1. Electric dipole moments (in a.u.). The sign of the dipole moment has been retained for carbon monoxide where a negative sign indicates C^+O^-

Method	CO	NH_3	HCP
HF (this work)			
DZP	-0.109	0.705	0.170
TZP	-0.120	0.659	0.146
QZP	-0.109	0.649	0.145
ADZP	-0.099	0.637	0.178
ATZP	-0.105	0.636	0.150
AQZP	-0.105	0.636	0.146
NHF [33]	-0.103	0.635	-
MP2 (this work)			
DZP	0.133	0.689	0.232
TZP	0.098	0.636	0.195
QZP	0.105	0.624	0.192
ADZP	0.123	0.601	0.222
ATZP	0.111	0.600	0.195
AQZP	0.108	0.601	0.193
B3LYP (this work)			
DZP	-0.061	0.669	0.171
TZP	-0.016	0.633	0.138
QZP	-0.029	0.622	0.134
ADZP	-0.040	0.593	0.171
ATZP	0.034	0.597	0.142
AQZP	-0.034	0.597	0.135
BP86 (this work)			
DZP	0.111	0.665	0.201
TZP	0.066	0.622	0.159
QZP	-0.080	0.609	0.157
ADZP	0.092	0.577	0.201
ATZP	-0.085	0.580	0.165
AQZP	-0.085	0.581	0.159
Experimental			
	0.043[a]	0.578[a]	0.153±0.002[b]

[a]Taken from Ref. [46].
[b]Taken from Ref. [49].

From Table 1, it is clear that the MP2, B3LYP, and BP86 dipole moments can be very different from the corresponding ones obtained with the HF method. In addition, one can verify that even when the electron correlation is taken into account, the addition of diffuse functions in any basis set is crucial to improve the electric dipole moments. Except for CO, the MP2, B3LYP, and BP86/ATZP and AQZP dipole moments are in satisfactory agreement with the experimental data, being the best results obtained with the nonhybrid functional. On the other hand, for CO, the best result is obtained with the B3LYP/ATZP model, but the MP2 is the only correlated approach that gives all moments with the correct polarity. To improve the CO result, it is necessary to use higher electron correlation methods (e.g.: MP3, MP4, etc.). It is also notable the differences between the electric dipole moments calculated with the nonhybrid and hybrid functionals. Finally, the largest difference between the B3LYP/ATZP and experimental dipole moments is 0.019 a.u. for the ammonia molecule.

5.2. Dipole Polarizabilities

The components of the dipole polarizability calculated with the increase size basis sets [22,23,26] at the HF, MP2, and DFT levels are given in Table 2. The HF and correlated dipole polarizabilities were compared with the NHF [33] results available in the literature and with the experimental values taken from Ref. [33], respectively. It should be noted that while our calculations are static electronic responses properties, the experimental results [33] are for vibrating and rotating molecules and are measure at finite frequencies.

At any level of approximation, it is apparent from Table 2 that for the linear molecules the component of the polarizability parallel to the bond (z-direction) is better approximate by the non augmented basis sets than is the component of the polarizability perpendicular to the bond (x-direction). This can be explained by the phenomenon of 'basis sets sharing', in which a deficient basis on one atom is partly compensated by contributions from basis functions on neighboring atoms. Of course sharing can only occur in directions where there is something to be shared.

It is a well known shortcoming of GBSs that the orbitals fall off too quickly asymptotically (~ $e^{-\alpha r^2}$) and hence augmentation of the standard bases is important for properties which are sensitive to the large r region of the density. Now consider the augmented basis sets. A direct indication of the quality of the augmented basis sets, at the HF level, is provided by comparing results of these sets with the NHF values [33]. The augmented basis set dipole polarizabilities are fairly well converged when compared with the NHF results, and the effect of basis set sharing is not more evident. Using the AQZP basis sets, the difference from NHF is on the order of 10^{-2}, independent on the direction. At the MP2 and DFT levels, the α_{xx} of the linear molecules computed with the augmented basis sets are closer to the experimental values than the corresponding ones obtained with the non augmented basis sets.

Basis set and correlation effects require very careful consideration in the evaluation of polarizabilities and hyperpolarizabilities. In some cases correlation effects are difficult to assess because of the interplay of basis set and correlation effects. It is clear from Table 2 that with rare exceptions the electron correlation increases the components of the dipole polarizabilities and that, when compared with the corresponding experimental values, the MP2 results are underestimate and that the BP86/AXZP values are in general overestimate. Overall, the mean absolute deviations of MP2, B3LYP, BP86/AQZP from α_{xx} (α_{zz})

experimental polarizability components are 0.44 (0.23), 0.18 (0.16), and 0.30 (0.63) a.u., respectively. From these results, it is evident that the best results were obtained with the B3LYP/AQZP model.

Table 2. Components of the dipole polarizabilities ($\ddot{\alpha}$), in a.u.

Method	H_2		CH_4	CO		NH_3		HCP	
	α_{xx}	α_{zz}	α_{xx}	α_{xx}	α_{zz}	α_{xx}	α_{zz}	α_{xx}	α_{zz}
HF (this work)									
DZP	0.99	6.62	13.17	8.05	12.74	9.38	6.83	17.89	44.79
TZP	3.06	6.54	14.93	9.31	13.66	11.35	9.96	21.72	46.31
QZP	3.36	6.45	15.23	10.24	14.05	11.76	10.56	24.87	46.87
ADZP	4.42	6.53	15.86	10.85	14.39	12.47	12.77	29.49	47.98
ATZP	4.60	6.47	15.92	11.21	14.48	12.76	13.12	30.76	48.04
AQZP	4.61	6.47	15.93	11.24	14.48	12.77	13.23	30.98	48.03
NHF [33]	4.57	6.46	15.95	11.27	14.45	12.72	13.27	-	-
MP2 (this work)									
DZP	0.98	6.53	13.22	8.32	13.47	9.45	7.19	17.42	41.25
TZP	3.03	6.49	15.19	9.66	14.48	11.65	10.68	21.17	43.57
QZP	3.33	6.40	15.54	10.71	14.96	12.16	11.41	24.34	44.37
ADZP	4.37	6.54	16.40	11.55	15.62	13.28	14.90	28.76	45.96
ATZP	4.59	6.46	16.47	11.92	15.62	11.00	15.30	29.93	45.91
AQZP	4.58	6.44	16.44	11.93	15.59	13.65	15.38	30.05	45.86
B3LYP (this work)									
DZP	1.01	6.68	13.40	8.32	13.04	9.71	7.25	17.89	43.53
TZP	3.06	6.77	15.53	9.75	14.30	11.83	10.81	21.65	45.55
QZP	3.37	6.68	15.89	10.79	14.80	12.35	11.51	24.71	46.29
ADZP	4.50	6.95	16.83	11.63	15.34	13.59	15.15	29.37	47.69
ATZP	4.88	6.81	16.91	12.07	15.51	13.96	15.73	30.70	47.79
AQZP	4.89	6.80	16.93	12.10	15.50	13.99	15.87	30.93	47.84
BP86 (this work)									
DZP	1.01	6.77	13.59	8.51	13.29	9.76	7.45	18.05	43.78
TZP	3.15	6.89	15.87	9.98	14.63	12.09	11.27	21.85	45.76
QZP	3.47	6.80	16.24	11.08	15.15	12.64	12.03	25.07	46.48
ADZP	4.65	7.07	17.30	11.98	15.75	14.05	16.23	29.70	47.88
ATZP	5.01	6.94	17.34	12.44	15.94	14.40	16.91	30.94	47.92
AQZP	5.02	6.94	17.37	12.46	15.92	14.44	17.11	31.11	47.94
Experimental [33]									
	4.75	6.79	17.5	12.1	15.7	14.0	15.6	-	-

HF, MP2, B3LYP, and BP86 first dipole hyperpolarizability components calculated with the hierarchical sequence of increase size basis sets reported in Refs. [22,23,26] are displayed in Table 3 along with a selection of results obtained by other authors [34,35]. From this Table, one can observe some general trends, namely: *i)* At any level of theory, one verify that the dipole hyperpolarizabilities change significantly when diffuse functions are added in a basis set. Similar results were obtained for the dipole moments and dipole polarizabilities of these molecules. *ii)* The differences among the HF dipole hyperpolarizabilities and the corresponding ones calculated with any correlated method can be very large (e.g., see the β_{zzz} component of NH_3).

Table 3. Components of first dipole hyperpolarizabilities (in a.u.)

Method	CH_4	CO		HCP		NH_3		
	β_{xyz}	β_{zzz}	β_{zxx}	β_{zzz}	β_{zxx}	β_{zzz}	β_{zxx}	β_{xxx}
HF (this work)								
DZP	-30.44	22.33	0.29	-3.11	2.55	-9.71	-17.98	-21.96
TZP	-14.93	21.53	1.95	32.22	5.02	-4.59	-11.73	-15.37
QZP	-16.32	24.76	3.87	34.57	21.60	-2.62	-11.43	-14.47
ADZP	-13.84	31.63	5.83	43.13	17.89	-5.47	-8.78	-12.92
ATZP	-11.76	31.95	5.84	47.09	18.02	-10.54	-7.02	-10.97
AQZP	-11.29	31.03	4.97	49.05	18.66	-11.36	-7.19	-9.38
NHF [33]	-11.64	31.32	4.88	-	-	8.69	7.26	-9.02
MP2 (this work)								
DZP	-28.82	14.77	1.46	-55.20	1.68	-15.57	-19.32	-21.47
TZP	-18.74	14.86	3.06	-18.67	5.42	-11.41	-12.45	-14.72
QZP	-13.90	19.10	5.31	-13.01	20.47	-9.89	-12.41	-13.68
ADZP	-11.69	29.95	8.54	1.20	19.11	-21.43	-9.81	-13.61
ATZP	-9.12	29.76	8.40	5.92	18.73	-32.60	-7.51	-11.00
AQZP	-8.44	29.10	7.85	8.04	19.29	-35.49	-7.81	-8.26
B3LYP (this work)								
DZP	-30.74	17.94	0.88	-0.33	4.00	-15.92	-21.06	-23.76
TZP	-20.79	17.42	2.54	6.83	6.85	-11.36	-14.25	-16.75
QZP	-16.17	20.33	4.51	10.69	25.30	-9.01	-14.05	-15.94
ADZP	-14.98	32.93	7.41	16.74	20.82	-22.42	-13.25	-17.10
ATZP	-11.12	33.13	8.72	22.16	20.32	-34.09	-10.39	-14.01
AQZP	-10.29	31.42	7.73	28.06	21.94	-38.86	-10.70	-10.81
BP86 (this work)								
DZP	-32.76	17.41	1.00	-42.12	3.98	-19.21	-22.81	-24.68
TZP	-21.91	17.34	3.07	4.85	7.74	-13.74	-15.24	-17.14
QZP	-16.17	20.76	5.33	4.39	24.77	-11.01	-15.07	-16.93
ADZP	-16.78	34.30	8.48	9.08	23.81	-31.24	-17.02	-18.97
ATZP	-10.35	34.44	9.96	24.72	23.82	-48.49	-13.02	-14.86
AQZP	-9.62	32.38	8.29	25.00	23.74	-55.54	-13.16	-11.02
CCSD(T)								
	-	31.32[a]	6.60[a]	21.8[b]	12.9[b]			

[a]Maroulis, basis set [9s6p4d3f] [34]. The calculations were carried out at a bond length of 1.1283 Å.

[b]Maroulis and Pouchan, basis set [8s6p4d2f/6s4p4d2f/4s3p2d] [35]. The calculations were carried out at the bond lengths of 1.0692 Å (CH) and 1.15398 Å (CP).

At the HF level, the hyperpolarizabilities move toward the NHF values [33] whit the increase of the basis sets. The agreement between the HF/AQZP and NHF results are very good, except for the β_{zxx} and β_{zzz} components of NH_3, which give opposite signs. It is important to note that in Table 3 only the NHF approach predicts positive signs for these components.

In general, the BP86 dipole hyperpolarizabilities are in module larger than those evaluated with the MP2 and B3LYP methods. For CO and HCP, the agreement among the hyperpolarizability components obtained with any correlated method used here and the corresponding ones calculated with the single, double, and perturbative triple excitation coupled-cluster [CCSD(T)] method improve when the basis set size increase. In particular, for HCP, only the B3LYP and BP86/ATZP and AQZP hyperpolarizability components agree with the results anticipated by CCDS(T), i.e., $\beta_{zzz}>\beta_{zxx}$, but, when the β_{zzz}/β_{zxx} rate is also taken into account, the best agreement occurs between B3LYP/AQZP and CCDS(T)/[8s6p4d2f/6s4p4d2f/4s3p2d] [35].

From the growing number of calculation of molecular electrical properties, correlations with structure and bonding begin to emerge. For instance, the derivative HF calculations [50] showed that the mean dipole polarizabilities of the first row hydrides (CH_4, NH_3, H_2O, and FH) have a regular decline with the atomic number. This result agrees with our derived from the B3LYP/AQZP model (see Figure 1). The mean dipole polarizabilities for hydrides containing two first row atoms reveal an interesting pattern [50], too. The polarizability of an ABH_n molecule is well estimated (to within about 15%) by the sum of the mean polarizability of the A-hydride and the B-hydride, diminished by 6 a.u. if A and B are doubly bonded and 12 a.u. if A and B are triply bonded. There is a kind of additivity of polarizability.

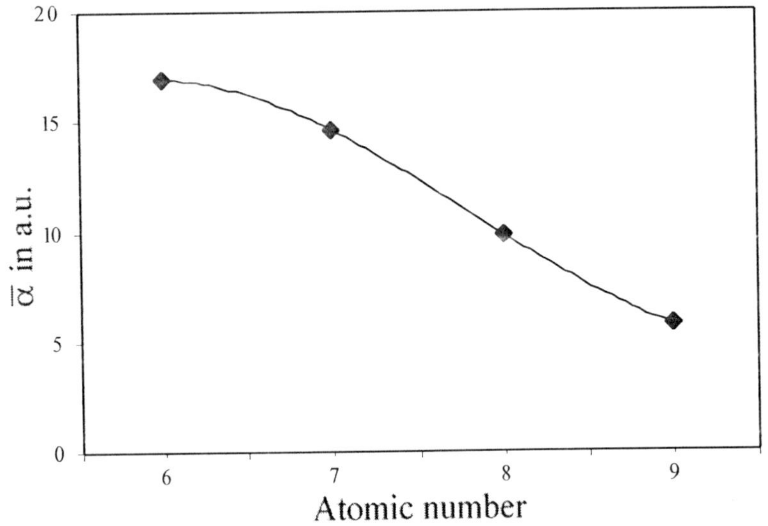

Figure 1. $\overline{\alpha}$ in a.u. of CH_4, NH_3, H_2O, and FH plotted against the atomic number of the first row element. The values were obtained from B3LYP/AQZP model. For H_2O and FH, the mean dipole polarizabilities (9.87 and 5.75 a.u., respectively) were taken from Ref. [26].

For example, the mean polarizability of HCN calculated with the B3LYP/AQZP approach 17.22 a.u. (see Ref. [26]) is well approximated as the sum of the like-calculated mean polarizability of CH_4, 16.93 a.u., and NH_3, 14.62 a.u., less 12 a.u. for triple bond, and this value is 19.55 a.u.

A selection of theoretical and experimental $\overline{\alpha}$, $\Delta\alpha$, and $\overline{\beta}$ values of the H_2, CO, and NH_3 molecules is shown in Table 4. Among the numerous theoretical works on the static dipole polarizability and first dipole hyperpolarizability of these molecules, only some theoretical efforts that include electron correlation have been shown.

For H_2, the B3LYP/AQZP mean dipole polarizability is about 5.8% higher than the CCSD (full CI for H_2) value of Sekino and Bartlett [36] obtained with the [3s2p1d] basis set, whereas for anisotropy, the two theoretical values coincide. The $\bar{\alpha}$ (5.43 a.u.) [51] and $\Delta\alpha$ (2.055 a.u.) [52] experimental data are respectively in good and satisfactory agreement with the corresponding B3LYP/AQZP results (5.53 and 1.91 a.u.).

Several studies have been reported about electrical properties of carbon monoxide. Sekino and Bartlett [36] carried out CCSD(T) calculation and used the polarizability consistent basis (POL) set recommended by Sadlej [53]. Maroulis [34] used various methods and different size basis sets in his calculations, among these, the CCSD(T)/[9s6p4d3f] model.

Table 4. Comparison of theoretical and experimental values for $\bar{\alpha}$, $\Delta\alpha$, and $\bar{\beta}$ (in a.u.) of H_2, CO, and NH_3

Method	H_2		CO			NH_3		
	$\bar{\alpha}$	$\Delta\alpha$	$\bar{\alpha}$	$\Delta\alpha$	$\bar{\beta}$	$\bar{\alpha}$	$\Delta\alpha$	$\bar{\beta}$
B3LYP[a]	5.53	1.91	13.23	3.40	28.13	14.62	1.88	-36.22
CCSD[b]	5.223	1.906						
CCSD(T)			13.04[e]	3.92[e]	23.5[e]	14.38[j]	2.01[j]	-34.3[j]
CCSD(T)[f]			13.08	3.61	26.6			
SDQ-MP4[k]						14.01	1.59	-30.47
Experiment	5.43[c]	2.055[d]	13.09[g]	3.58±0.13[h]	30.2±3.2[i]	14.56[l]	1.94[m]	-48.9±1.2[n]

[a]Present investigation, basis set AQZP.

[b]Sekino and Bartlett, basis set [3s2p1d] [36]. The calculations were carried out at a bond length of 0.7408 Å.

[c]Newell and Baird [51].

[d]Bridge and Buckingham [52].

[e]Sekino and Bartlett [36], using the POL basis set [53]. The calculations were carried out at a bond length of 1.1283 Å.

[f]Maroulis, basis set [9s6p4d3f] [34]. The calculations were carried out at a bond length of 1.1283 Å.

[g]Dipole oscillator strength distribution [54].

[h]Baas and Van den Hout [55], dynamic value at 632.8 nm.

[i]Referenced by Shelton and Rice [57], dynamic value at 694.3 nm.

[j]Sekino and Bartlett [36], using the POL basis set [53]. The calculations were carried out at a bond length (NH) of 1.012 Å and a angle (θ_{HNH}) of 106.7^0.

[k]Maroulis, basis set [6s4p4d1f/4s2p1d] [37]. The calculations were carried out at a bond length (NH) of 1.012 Å and a angle (θ_{HNH}) of 106.7^0.

[l]Zeiss and Meath [59].

[m]Bridge and Buckingham [52].

[n]ESHG at 694.3 nm, ,rescaled value, see Shelton and Rice [57].

These results were also included in Table 4. Their mean dipole polarizabilities [34,36] agree very well with our value, but the Sekino and Bartlett´s anisotropy is, as expected from the relatively small size of the POL basis, somewhat larger. Among the three theoretical values displayed in Table 4, the B3LYP/AQZP model gives the smallest $\bar{\beta}$ error. The experimental mean dipole polarizability is 13.09 a.u. [54]. To compare our value for $\bar{\alpha}$ to the experimental one, we add to B3LYP/AQZP result the zero-point vibrational correction

(ZPVC) estimated in Ref. [34]. We obtain B3LYP/AQZP+ZPVC value of $\overline{\alpha}$ =13.28. Our value is a mere 1.4% higher. No reliable experimental value of the static dipole polarizability anisotropy exists. One shows in Table 4 an experimental result [55] measure at 632.8 nm. A reasonable agreement between theory and experiment is readily observed, but an accurate determination of the dispertion of $\Delta\alpha$ and the corresponding static limit would help the formulation of a more quantitative appreciation. It is worth mentioning that theoretical study [56] of the dispertion shows that between 632.8 nm and the static limits, the value of $\Delta\alpha$ changes by small 1-2%. The available experimental first dipole hyperpolarizability [57] has been determined by electric field induced second harmonic generation (ESHG) at 694.3 nm. The value is 30.2±3.2 a.u. Adding to our B3LYP/AQZP value the dispersion for this property estimated by Sekino and Bartlett [36] from the SCF percentage dispersion, we obtain 32.32 a.u., which is within the experimental error bars.

Accurate static dipole polarizability and first dipole hyperpolarizability for NH_3 have been reported previously. Sekino and Bartlett [36] obtained CCSD(T) values of $\overline{\alpha}$ =14.38, $\Delta\alpha$=2.01, and $\overline{\beta}$ =-34.3 a.u. with the POL basis set [53]. For ammonia, Maroulis [37] reported HF and MP4 (Møller-Plesset fourth-order perturbation theory) calculations. His SDQ-MP4 values of $\overline{\alpha}$ =14.01, $\Delta\alpha$=1.59, and $\overline{\beta}$ =-30.47 a.u. were calculated with a [6s4p4d1f/4s2p1d] basis set. In a somewhat smaller basis the triples (T4) contribution to the fourth-order correction for the above properties was estimated to be 0.29, 0.25, and 4.42 a.u. [37], respectively. Adding these quantities to the Maroulis′ model, one obtains the static SDQ-MP4/[6s4p4d1f/4s2p1d]+T4 values of 14.30, 1.84, and 34.89 a.u., in very reasonable agreement with the values obtained by Sekino and Bartlett and in this work. We recall that our results pertain to the equilibrium geometry. Then, comparison with experiment would necessitate averaging over the ground vibrational state. We rely on Werner and Meyer [58] for estimate of the vibrational correction for the dipole polarizability. Our static mean dipole polarizability is 14.62 a.u. and the vibartional correction 0.26 [58]. This results in an estimate of 14.88 a.u. for the ground vibrational state, which is about 2% higher than the semiempirical estimate of the static value reported Zeiss and Meath [59], 14.56 a.u.. The only experimental value of the anisotropy is a dynamic value of 1.94 a.u. [52]. Our static B3LYP/AQZP estimate is 1.88 a.u.. Relying on measurements of the dispersion of the anisotropy for similar molecules [60] we would expect the static limit to be 5%-10% lower than the dynamic value of Bridge and Buckingham [52]. There is no reliable estimate of the vibrational correction to $\Delta\alpha$. It would probably lower our value. The available experimental first dipole hyperpolarizability [57] is dynamic ESHG value pertaining at 694.3 nm, $\overline{\beta}$ =-48.9 ±1.2 a.u. An estimate of the dispersion correction for this molecule was also done by Sekino and Bartlett [36]. Considering the dispersion, our estimate is -51.80 a.u., in good agreement with the experimental value.

6. Conclusions

The hierarchical sequence of increase size basis sets for systematically reducing the basis set truncation error DZP, TZP, QZP, ADZP, ATZP, and AQZP [22,23,26] was used to estimate electric dipole moment and static dipole (hyper)polarizability of H_2, CO, CH_4, NH_3, and HCP. It is shown that this series of correlation basis sets when used in HF, MP2, and DFT calculations of the electrical properties cited above, they present a monotonic convergence toward the recommended or observed value, and that the dipole (hyper)polarizability improves significantly when diffuse functions are added in any basis set.

At the HF level, the ATZP and AQZP electric dipole moments are very similar and are in excellent agreement with the NHF values. The largest difference among the HF/AQZP and NHF dipole moments is 0.002 a.u. for CO. When correlation is taken into account, the MP2 is the only method that gives the correct polarity for the CO molecule with any basis set. On the other hand, for NH_3 and HCP, the best agreement with the experimental data was obtained with the BP86/AQZP model.

Using the AQZP basis set, we obtained near NHF values for the dipole polarizability components of all molecules studied here. Considering the electron correlation, the best mean absolute deviations between the theoretical and experimental α_{xx} (α_{zz}) values were obtained with B3LYP/AQZP. We also verified that the B3LYP/AQZP mean dipole polarizability and anisotropy are close to the results computed with higher levels of theory and in good agreement with the experimental values.

Differences among HF first dipole hyperpolarizability components and the corresponding ones calculated with any correlated method can be very larger. In module, the BP86 hyperpolarizability components when compared with the corresponding MP2 and B3LYP results are larger. For CO and NH_3, the $\overline{\beta}$ values obtained with the B3LYP/AQZP model are larger than those computed with the CCSD(T) method. Adding the dispersion to the B3LYP/AQZP hyperpolarizability values, one obtains estimates of $\overline{\beta}$ in good agreement with the experimental ESHG values.

The complete set of the s, p, d, f, and g parameters of the DZP, TZP, QZP, ADZP, ATZP, and AQZP basis sets for H and Li-Ar are available through the internet at http://www.cce.ufes.br/qcgv/pub/.

REFERENCES

[1] Froese-Fischer, C. *The Hartree- Fock method for atoms*; Wiley: New York, 1977.

[2] Roothaan, C. C. J. *Rev. Mod. Phys.* 1951, *23*, 69.

[3] Roothaan, C. C. J. *Rev. Mod. Phys.* 1960, *32*, 179.

[4] Mohallem, J.R.; Dreizler, R.M.; Trsic, M. *Int. J. Quantum Chem. Symp.* 1986, *20*, 45.

[5] De Castro, E. V. R.; Jorge, F. E. *J. Chem. Phys.* 1998, *108*, 5225.

[6] Huzinaga S. *Approximate atomic functions II*; Department of Chem. Report., Univ. of Alberta: Edmonton, 1971.

[7] Kobus, J.; Moncrieff, D.; Wilson, S. *J. Phys. B* 1994, *27*, 5139.

[8] Koga, T.; Takkar, A. J. *Theor. Chim. Acta* 1993, *85*, 391.

[9] Jensen, F. *J. Chem. Phys.* 2001, *115*, 9113.

[10] Weigend, F.; Furche, F.; Ahlrichs, R. *J. Chem. Phys.* 2003, *119*, 12753.

[11] Jorge, F. E.; de Castro, E. V. R. *Chem. Phys. Lett.* 1999, *302*, 454.

[12] De Castro, E. V. R.; Jorge, F.E.; Pinheiro, J. C. *Chem. Phys.* 1999, *243*, 1.

[13] Canal Neto, A.; Jorge, F. E.; de Castro M. *Int. J. Quantum Chem.* 2002, *88*, 252.

[14] Centoducatte, R.; Jorge, F. E.; de Castro, E. V. R. *Int. J. Quantum Chem.* 2001, *82*, 126.

[15] Librelon, P. R.; Jorge, F. E. *Int. J. Quantum Chem.* 2003, *95*, 190.
[16] Pinheiro, J. C.; Jorge, F.E.; de Castro, E. V. R. *J. Mol. Structure (Theochem)* 1999, *491*, 81.
[17] Pinheiro, J. C.; Jorge, F.E.; de Castro, E. V. R. *Int. J. Quantum Chem.* 2000, *78*, 15.
[18] Pires, J. M.; Jorge, F. E. *Int. J. Quantum Chem.* 2003, *95*, 144.
[19] Hehre, W. H.; Radom, L.; Schleyer, P. v. R.; Pople J. A. *Ab initio molecular orbital theory*; Wiley: New York, 1986.
[20] Dunning Jr., T. H. *J. Chem. Phys.* 1989, *90*, 1007.
[21] Woon, D. E.; Dunning Jr., T. H. *J. Chem. Phys.* 1993, *98*, 1358.
[22] Canal Neto, A.; Muniz, E. P.; Centoducatte, R.; Jorge, F. E. *J. Mol. Structure (Theochem)* 2005, *718*, 219.
[23] Barbieri, P. L.; Fantin, P. A.; Jorge, F. E. *Mol. Phys.* 2006, *104*, 2945.
[24] Muniz, E. P.; Jorge, F. E. *Int. J. Quantum Chem.* 2005, *106*, 943.
[25] Canal Neto, A.; Jorge F. E. *Chirality* 2007, *19*, 67.
[26] Fantin, P. A.; Barbieri, P. L.; Canal Neto, A.; Jorge, F. E. 2007, submitted for publication.
[27] Schiff, L. I. *Quantum Mechanics*; McGraw-Hill: New York, 1968.
[28] Svanberg, S. *Atomic and Molecular Spectroscopy: Basic Aspects and Practical Applications*; Springer-Verlag: Berlin, 1992.
[29] Buckingham, A. D. *Adv. Chem. Phys.* 1967,*12*, 107.
[30] Aschcroft, N. W.; Mermin, N. D. *Solid State Physics*; Saunders College: Philadelphia, 1976.
[31] Shelton, D. P. *Phys. Rev. A* 1980, *42*, 2578.
[32] Guthmuller, J.; Simon, D. *J. Phys. Chem. A* 2006, *110*, 9967.
[33] Guan, J.; Duffy, P.; Carter, J. T.; Chong, D. P.; Casida, K. C.; Casida, M. E.; Wrinn, M. *J. Chem. Phys.* 1993, *9*, 4753.
[34] Maroulis, G. *J. Phys. Chem.* 1996, *100*, 13466.
[35] Maroulis, G.; Pouchan, C. *Phys. Rev. A* 1998, *57*, 2440.
[36] Sekino, H.; Bartlett, R. J. *J. Chem. Phys.* 1993, *98*, 3022.
[37] Maroulis, G. *Chem. Phys. Lett.* 1992, *195*, 85.
[38] Zuevand, M. B.; Nefediev, S. E. *J. Comput. Meth. Sci. Engrg.* 2004, *4*, 481.
[39] Kollmar, H. *J. Am. Chem. Soc.* 1978, *100*, 2665.
[40] Duke, A. J. *Chem. Phys. Lett.* 1973, *21*, 275.
[41] Frisch, M. J.; Trucks, G. W.; Schlegel, H. B.; Scuseria, G. E.; Robb, M. A.; Cheeseman, J. R.; Montgomery Jr., J. A.; Vreven, T.; Kudin, K. N.; Burant, J. C.; Millam, J. M.; Iyengar, S. S.; Tomasi, J.; Barone, V.; Mennucci, B.; Cossi, M.; Scalmani, G.; Rega, N.; Petersson, G. A.; Nakatsuji, H.; Hada, M.; Ehara, M.; Toyota, K.; Fukuda, R.; Hasegawa, J.; Ishida, M.; Nakajima, T.; Honda, Y.; Kitao, O.; Nakai, H.; Klene, M.; Li, X.; Knox, J. E.; Hratchian, H. P.; Cross, J. B.; Adamo, C.; Jaramillo, J.; Gomperts, R.; Stratmann, R. E.; Yazyev, O.; Austin, A. J.; Cammi, R.; Pomelli, C.; Ochterski, J. W.; Ayala, P. Y.; Morokuma, K.; Voth, G. A.; Salvador, P.; Dannenberg, J. J.; Zakrzewski, V. G.; Dapprich, S.; Daniels, A. D.; Strain, M. C.; Farkas, O.; Malick, D. K.; Rabuck, A. D.; Raghavachari, K.; Foresman, J. B.; Ortiz, J. V.; Cui, Q.; Baboul, A. G.; Clifford, S.; Cioslowski, J.; Stefanov, B. B.; Liu, G.; Liashenko, A.; Piskorz, P.; Komaromi, I.; Martin, R. L.; Fox, D. J.; Keith, T.; Al-Laham, M. A.; Peng, C. Y.; Nanayakkara, A.;

Challacombe, M.; Gill, P. M. W.; Johnson, B.; Chen, W.; Wong, M. W.; Gonzalez, C.; Pople, J. A. *Gaussian 03, Revision A.1*; Gaussian, Inc.: Pittsburgh, 2003.
[42] Becke, A. D. *J. Chem. Phys.* 1993, *98*, 5648.
[43] Lee, C.; Yang, W.; Parr, R. G. *Phys. Rev. B* 1988, *37*, 785.
[44] Becke, A. D. *Phys. Rev. A* 1988, *38*, 3098.
[45] Perdew, J. P. *Phys. Rev. B* 1986, *33*, 8822.
[46] Editor-in-chief Lide D. R. *CRC Handbook of Chemistry and Physics*; CRC Press: London, 1994.
[47] Strey, G.; Mills, I. M. *Mol. Phys.* 1973, *26*, 129.
[48] Newton, M. D.; Lathan, W. A.; Hehre, W. J.; Pople J. A. *J. Chem. Phys.* 1970, *52*,4064.
[49] Tyler, J. K. *J. Chem. Phys.* 1964, *40*,1170.
[50] Liu, S. Y. Dykstra, C. E. *J. Phys. Chem.* 1987, *91*, 1749.
[51] Newell, A. C.; Baird, R. C. *J. Appl. Phys.* 1965, *36*, 3751.
[52] Bridge, N. J.; Buckingham, A. D. *Proc. R. Soc. London Ser. A* 1966, *295*, 334.
[53] Sadlej, A. J. *Collect. Czech. Chem. Commun.* 1988, *53*, 1995.
[54] Parker, G. A.; Pack, R.T. *J. Chem. Phys.* 1976, *64*, 2010.
[55] Bass, F.; Van den Hout, K. D. *Physica* 1979, *95A*, 597.
[56] Oddershede, J.; Svendsen, E. N. *Chem. Phys.* 1982, *64*, 359.
[57] Shelton, D. P.; Rice, J. E. *Chem. Rev.* 1994, *94*, 3.
[58] Werner, H. J.; Meyer, W. *Mol. Phys.* 1976, *31*, 855.
[59] Zeiss, G. D.; Meath, W. J. *Mol. Phys.* 1977, *33*, 1155.
[60] Bogaard, M. P.; Buckingham, A. D.; Pierens, R. K.; White, A. H. *J. Chem. Soc. Faraday Trans.* 1978, *74*, 3008.

In: Chemical Physics Research Trends, Vol. 252
Editor: Samuel V. Arnold, pp. 335-341

ISBN: 978-1-60021-554-4

Chapter 9

CALCULATION OF PHASE TRANSITIONS IN ICE WHICH DO NOT OBEY THE CLAPEYRON EQUATION

I. A. Stepanov[1]
School of Chemistry, University of Nottingham,
University Park, Nottingham, NG7 2RD, UK

ABSTRACT

The Clapeyron equation cannot describe a number of phase transitions in ice: it gives huge mistakes. Even the sign of dT/dP is often wrong. It is shown that the modified Clapeyron equation earlier obtained by the author gives much better results. The following transitions are considered: Low-Density Amorph (LDA) → I_c, LDA → High-Density Amorph (HDA) at 0.1 and 600 MPa, ice III → ice V, ice V→ ice VI, ice VI → ice VII, ice VI → ice VIII.

Keywords: Ice; Amorphous materials; Phase diagrams; Phase transitions; Clapeyron equation; Thermal expansion; The first law of thermodynamics.

1. INTRODUCTION

The Clapeyron equation

$$dT/dP = T\Delta V/\Delta H = \Delta V/\Delta S, \tag{1}$$

where ΔH is the change in the enthalpy, ΔV is the change in the volume and ΔS is the change in the entropy, fails to describe a number of phase transitions in ice. This equation is derived from the first and second laws of thermodynamics without simplification and must give correct results. However, for substances with negative thermal expansion coefficient α, it

[1] Permanent address: Juglas 1a-20, Riga, LV-1024, Latvia. igstepanov@yahoo.com

often gives wrong results deflecting greatly from the experimental values. Often happens that even the sign of dT/dP is wrong [1]. Earlier the author derived the modified Clapeyron equation [1]:

$$dT/dP = T(\mathrm{sign}\alpha_2 V_2 - \mathrm{sign}\alpha_1 V_1)/\Delta H. \qquad (2)$$

Here α_1 and α_2 are thermal expansion coefficients of the two phases. Preliminary tests gave very promising results. For example, the tests always predicted the sign of dT/dP correctly. In this paper this equation is applied to analysis of some phase transitions in ice which do not obey Eq. (1).

2. Theory

In [1] a paradox was considered:

$$(\partial S/\partial V)_T = \alpha/\beta \qquad (3)$$

[2] where β is the isothermal compressibility $-(1/V)(\partial V/\partial P)_T$. (One can easy derive Eq. (3): $\alpha = (1/V)(\partial V/\partial T)_P = -(1/V)(\partial V/\partial P)_T(\partial P/\partial T)_V = \beta(\partial S/\partial V)_T$). Therefore, if $\alpha < 0$ then $(\partial S/\partial V)_T < 0$. On the other hand,

$$\delta Q = TdS = dU + PdV \qquad (4)$$

and

$$(\partial S/\partial V)_U = P/T > 0 \qquad (5)$$

for all α.

However, $(\partial U/\partial T)_X \neq 0$ and $(\partial U/\partial X)_T \neq 0$ where X = P, V, S, H, F or G, H is the enthalpy, F is the Helmholtz energy and G is the Gibbs energy [2]. It means that in almost all processes, $dU/dT = (\partial U/\partial T)_X + (\partial U/\partial X)_T dX/dT$ is not equal to zero. Therefore, almost always, if U is constant then T is constant. If U is constant and T is constant, then the right sides of Eqs. (3) and (5) must have the same sign. So, there is a contradiction in thermodynamic equations. (In heat exchange, one introduces the quantity of heat δQ in substance, and U and T always vary. In compression of substance without introducing heat, of course, T can vary when U = const).

For example,

$$dU = (\partial U/\partial P)_V dP + (\partial U/\partial V)_P dV = \beta/\alpha \cdot C_V dP + (C_P - \alpha PV)/(\alpha V)dV \qquad (6)$$

[2]. From this, if dU/dT = 0 then

$$dP/dV = (\alpha PV - C_P)/(\beta C_V V). \qquad (7)$$

For every process which does not satisfy Eq. (7), $dU/dT \neq 0$ and if U is constant then T is constant.

For water at 273 K, $\alpha = -68.05 \cdot 10^{-6}$ K^{-1}, $C_P = 4217.6$ J/(kg·K), $C_P - C_V \approx 2.5$ J/(kg·K) and $\beta = 5.09 \cdot 10^{-10}$ Pa^{-1}. Substituting these values in Eq. (7) one gets $dP/dV = -1/(\beta V) \sim -10^{12}$ Pa·kg/m^3 with the highest accuracy. Only for processes with $dP/dV = -10^{12}$, $dU/dT = 0$ and T is not constant when U is constant. For all other processes, when U is constant then T is constant, thus the paradox exists. One can explain this paradox assuming that for substances with negative thermal expansion

$$\delta Q = TdS = dU - PdV. \tag{8}$$

If we derive the Clapeyron equation by using Eq. (8) then it will look like Eq. (2).

There is a well-known differential formula widely used in thermodynamics [2]:

$$(\partial x/\partial y)_z = (\partial x/\partial y)_w + (\partial x/\partial w)_y(\partial w/\partial y)_z. \tag{9}$$

From it follows that

$$(\partial S/\partial V)_U = (\partial S/\partial V)_T - (\partial U/\partial V)_T/T \tag{10}$$

[2]. If in the left part of Eq. (10) U is constant then in the right part of Eq. (10) U is also constant. If in calorimeter U is constant and one wants to calculate $(\partial S/\partial V)_U$ according to Eq. (10), one has to assume that U is constant in both parts of Eq. (10). Thus $dU = (\partial U/\partial V)_T + (\partial U/\partial T)_V = 0$. If U is constant and T is not constant then two cases are possible. The first one: $(\partial U/\partial T)_V = 0$ and $(\partial U/\partial V)_T = 0$. The second one: $(\partial U/\partial T)_V \neq 0$ and $(\partial U/\partial V)_T \neq 0$. One can show that the second case cannot happen.

According to [2], $(\partial U/\partial V)_T = (\alpha T - \beta P)/\beta$. If to introduce this into Eq. (10) then one obtains

$$(\partial S/\partial V)_U = (\partial S/\partial V)_T - (\alpha/\beta - P/T). \tag{11}$$

This equation has only one solution: $\alpha/\beta = P/T$. It is clear if to introduce it into the second term on the right hand side of this equation and compare with Eqs. (3) and (5): $(\partial S/\partial V)_U = P/T$ and $(\partial S/\partial V)_T = \alpha/\beta$. If to suppose that $\alpha/\beta = P/T + \Delta$ and to introduce it into Eq. (11), then it turns to identity and is not informative at all. Pay attention that $(\partial S/\partial V)_U$ equals $(\partial S/\partial V)_T$ for all processes, even for those with constant U and varying T.

3. Results and Discussion

Water has two different amorphous phases called the low-density amorph (LDA) and the high-density amorph (HDA). The pressure - induced LDA → HDA transition at ~600 MPa and 77 K, and the temperature-induced HDA → LDA transition at 0.1 MPA and ~120 K are reported [3]. They happened with a similar volume change and heat evolution. It is important to stress: evolution of heat is detected by differential thermal analysis both in the LDA → HDA transitions at ~600 MPa and HDA → LDA transitions at 0.1 MPA [3].

In [3] a transition LDA → I_c with small volume change was also studied. For it $\Delta V \approx -0.04 \cdot 10^{-3}$ m^3/kg [3], T ≈140 K and $\Delta H \approx -79167$ J/kg [4]. According to experiment, dT/dP ≈ -0.09 K/MPa. The Clapeyron equation gives dT/dP ≈ 0.08 K/MPa. One can suppose that here α_1 and α_2 are less than 0 thus from Eq. (2) dT/dP ≈ -0.08 K/MPa.

For the temperature-induced HDA → LDA transition, dT/dP ≈ 0.18 K/MPa [3]. In this case the change in the volume ΔV for the sharp transition is $0.11 \cdot 10^{-3}$ m^3/kg [5], T ≈ 120 K and $\Delta H \approx -30222$ J/kg [4]. From Eq. (1), dT/dP ≈ -0.43K/MPa. One can assume that α_1 is less than 0, from the previous result α_2 is less than 0, thus from Eq. (2), dT/dP ≈ 0.43 K/MPa.

For the pressure - induced LDA → HDA transition, dT/dP ≈ -0.26 K/MPa [3]. Here T = 77 K, $\Delta V \approx -0.11 \cdot 10^{-3}$ m^3/kg and $\Delta H \approx -30222$ J/kg (similarly to the previous transition), and from Eq. (1) dT/dP ≈ 0.28 K/MPa. It is very possible that α_1 and α_2 are less than 0, and from Eq. (2), dT/dP ≈ -0.28 K/MPa.

In Table 1 the results of this paper are compiled. A not very good agreement of Eq. (2) with the experiment for the HDA → LDA transition can be due to the lack of experimental points.

Phase transition ice III → ice V. From [6], dT/dP = 3.6 K/MPa. The volumes are $V_1 = 0.87 \cdot 10^{-3}$ m^3/kg and $V_2 = 0.81 \cdot 10^{-3}$ m^3/kg [7]. The experimental enthalpies are ΔH(III → I_c) ≈ -16760 J/kg at T ≈ 150 K [8] and ΔH(I_c → V) ≈ 50833 J/kg at T ≈ 145 K [9] thus ΔH(III → V) ≈ 34073 J/kg. From Eq. (1), dT/dP ≈ -0.26 K/MPa. If to suppose that $\alpha_1 < 0$, then from Eq. (2), dT/dP ≈ 7.1 K/MPa. A not very good agreement with experiment is due only to a very small number of experimental points.

Phase transition ice V → ice VI. From the phase diagram, dT/dP = -8.7 K/MPa [10]. (Here and throughout the paper I made enlarged copies of the figures and measured dT/dP). The measured enthalpies are ΔH(V → I_h) ≈ -53667 J/kg and ΔH(I_h → VI) ≈ 80778 J/kg at T ≈ 140 K [9]. Thus ΔH(V → VI) ≈ 27111 J/kg. The volumes are $V_1 = 0.81 \cdot 10^{-3}$ m^3/kg and $V_2 = 0.76 \cdot 10^{-3}$ m^3/kg [7]. From Eq. (1), dT/dP ≈ -0.26 K/MPa. If to assume that $\alpha_2 < 0$, then from Eq. (2), dT/dP ≈ -8.1 K/MPa.

Phase transition ice VI → ice VII. From the phase diagram, dT/dP = 3.7 K/MPa [10]. Here $V_1 = 0.76 \cdot 10^{-3}$ m^3/kg, $V_2 = 0.64 \cdot 10^{-3}$ m^3/kg [7], ΔH(VI → I_c) ≈ -78000 J/kg at T ≈140 K [9] ΔH(I_c → VII) ≈ 129890 J/kg at T ≈125 K [8] and ΔH(VI → VII) ≈ 51890 J/kg. From Eq. (1), dT/dP ≈ -0.30 K/MPa. According to the previous result, $\alpha_1 < 0$, then from Eq. (2), dT/dP ≈ 3.5 K/MPa.

Table 1. Some phase transitions in ice measured experimentally and described by the Clapeyron equation (1) and by the modified Clapeyron equation (2)

Transition	Pressure, MPa	Temperature, K	dT/dP, experiment, K/MPa	dT/dP, Clapeyron equation (1), K/MPa	dT/dP, Equation (2), K/MPa
LDA $\rightarrow$ I_c	0.1	~140	-0.09	0.08	-0.08
HDA $\rightarrow$ LDA	0.1	~120	0.18	-0.43	0.43
LDA $\rightarrow$ HDA	600	~77	-0.26	0.28	-0.28
III $\rightarrow$ V	344	~250	3.6	-0.26	7.1
V $\rightarrow$ VI	600	~260	-8.7	-0.26	-8.1
VI $\rightarrow$ VII	2000	~320	3.7	-0.30	3.5
VI $\rightarrow$ VIII	1900	~240	0.21	-0.23	0.23

Phase transition ice VI $\rightarrow$ ice VIII. From the phase diagram, dT/dP = 0.21 K/MPa [10]. The volume difference ΔV is $-0.12 \cdot 10^{-3}$ m^3/kg [7]. The enthalpy change $\Delta H(VI \rightarrow I_h) \approx -80778$ J/kg at T $\approx$ 140 K and $\Delta H(I_h \rightarrow VIII) \approx 150111$ J/kg at T $\approx$ 120 K [9]. Therefore, $\Delta H(VI \rightarrow VIII) \approx 69333$ J/kg. From Eq. (1), dT/dP $\approx$ -0.23 K/MPa. Earlier it was obtained that α_1 is less than 0 and one can assume that α_2 is less than 0. Then from Eq. (2), dT/dP $\approx$ 0.23 K/MPa.

It is interesting to investigate melting of ice I_h. Water has negative expansion till P $\approx$ 14 MPa [12]. At this pressure ice melts at T $\approx$ 272 K. In [7] a few dT/dP values are given in the interval 272 - 273 K, however, they are calculated by an equation and not found experimentally. The data in [7] are based on experimental values [11, 13]. In [13] there are no experimental points in this interval but in [11] there is only one: 4.9 MPa and T = 272.8 K. It obeys Eq. (1): dT/dP = -0.074 K/MPa. Equation (2) gives dT/dP = -1.71 K/MPa. Nevertheless, one may not still say that Eq. (2) is not applicable to melting of ice. It is necessary to check, how was it obtained that water has negative expansion at 272.8 K: by direct measurement or by an indirect one, and to make many measurements of dT/dP for pressures less than 4.9 MPa.

4. CONCLUSIONS

The modified Clapeyron equation (2) again proved to be successful in describing phase transitions in ice where the Clapeyron equation is helpless. It is necessary to obtain many experimental points of dT/dP for melting of ice I_h in the region 272 – 273.16 K.

An important result is obtained in [14]. There it is found that the difference between the heats of chemical reactions measured by the Van't-Hoff equation, ΔH^0, and obtained by calorimetry, ΔQ^0, is $P\Delta V^0$. The Van't-Hoff equation must give the same results as calorimetry because it is derived from the first and the second laws of thermodynamics without simplifications. But the difference between both methods is far beyond the error limits. Nobody succeeded to explain this phenomenon. It is supposed that for endothermic chemical reactions with increasing volume and for exothermic chemical reactions with decreasing volume the law of conservation of energy is

$$\Delta U = \Delta Q + P\Delta V + \sum \mu_i \Delta N_i, \tag{12}$$

(notice the change of sign in front of P and that $(\partial S/\partial V)_{U,Ni} = -P/T < 0$), here μ_i are chemical potentials and ΔN_i are changes in the number of moles of the substance i. If to derive the Van't-Hoff equation using Eq. (12) then it will be

$$d/dT \ln K = \Delta H^0/RT^2 = (\Delta Q^0 + P\Delta V^0)/RT^2. \tag{13}$$

The traditional Van't-Hoff equation is

$$d/dT \ln K = \Delta H^0/RT^2 = (\Delta U^0 + P\Delta V^0)/RT^2. \tag{14}$$

Other contradictions to equilibrium thermodynamics were found by the author in [15-18].

REFERENCES

[1] I. A. Stepanov, *Physica* B 349 (2004) 251 - 253.

[2] M. Kh. Karapetyants, Himicheskaja Termodinamika, Vysshaja Shkola, Moscow, 1974, pp. 108-110. [English translation: M. Kh. Karapetyants, Chemical Thermodynamics, (translated by G. Leib), *Mir*, Moscow, 1978].

[3] O. Mishima, *J. Chem. Phys.* 100 (1994) 5910 - 5912.

[4] Y. P. Handa, O. Mishima, E. Whalley, *J. Chem. Phys.* 84 (1986) 2766 - 2770.

[5] O. Mishima, Y. Suzuki, *Nature* 419 (2002) 599 - 603.

[6] D. Eizenberg, V. Kautzman, Structure and Properties of Water, *Gidrometeoizdat*, Leningrad, 1975. Russ. translation.

[7] CRC *Handbook of Chemistry and Physics*, 80th Ed. CRC Press, Inc., Boca Raton, Florida, 1999 - 2000.

[8] J. E. Bertie, L. D. Calvert, E. Whalley, *Can. J. Chem.* 42 (1964) 1373 – 1378.

[9] Y. P. Handa, D. D. Klug, E. Whalley, *Can. J. Chem.* 66 (1988) 919 – 924.

[10] D. D. Klug, Y. P. Handa, J. S. Tse, E. Whalley, *J. Chem. Phys.* 90 (1989) 2390 – 2392.

[11] S. J. Henderson, R. J. Speedy, *J. Phys. Chem.* 91 (1987) 3069 – 3072.

[12] http://webbook.nist.gov.

[13] P. W. Bridgman, *Proc. Am. Acad. Arts Sci.* 47 (1912) 441 - 558.

[14] I. A. Stepanov, The Difference between the Heats of Chemical Reactions Obtained by the Van't- Hoff Equation and by Calorimetry. In: *Trends in Chemical Physics Research.* Ed. A. N. Linke. Nova Science Publishers, Inc. Hauppauge, NY. 2006, pp. 117 - 127.

[15] I. A. Stepanov, Negative Thermal Expansion of ZrW_2O_8 and HfW_2O_8: Contradiction to a Thermodynamic Relation. In: A. N. Linke (Ed), *Progress in Chemical Physics Research.* Nova Science Publishers, Inc., Hauppauge, NY, 2005, pp. 107 - 114.

[16] . A. Stepanov, Paradoxes of Thermal Expansion. In: A. N. Linke (Ed), *Progress in Chemical Physics Research.* Nova Science Publishers, Inc, Hauppauge, NY, 2005, pp. 115 – 120.

[17] I. A. Stepanov, $(\partial S/\partial V)_U$ is negative for substances with negative thermal expansion. In: A. N. Linke (Ed), *Trends in Chemical Physics Research.* Nova Science Publishers, Inc., Hauppauge, NY, 2006, pp. 129 - 134.

[18] I. A. Stepanov, Supercooled Water: Contradiction to a Thermodynamic Relation. - In: A. N. Linke (Ed), *Chemical Physics*: New Research. Nova Science Publishers, Inc., Hauppauge, NY., 2006, pp. 83 - 87.

INDEX

A

B

C

D

E

F

G

H

I

J

K

L

M

N

O

P

Q

R

S

T

U

V

W

Y